Beltsville Symposia in Agricultural Research

A Series of Annual Symposia Held at
THE BELTSVILLE AGRICULTURAL RESEARCH CENTER
Agricultural Research Service
United States Department of Agriculture

[13] Biotic Diversity and Germplasm Preservation, Global Imperatives

SYMPOSIA IN THIS SERIES

[1] Virology in Agriculture
May 10–12, 1976

[2] Biosystematics in Agriculture
May 8–11, 1977

[3] Animal Reproduction
May 14–17, 1978

[4] Human Nutrition Research
May 8–9, 1979

[5] Biological Control in Crop Production
May 18–21, 1980

[6] Strategies of Plant Reproduction
May 17–20, 1981

[7] Genetic Engineering: Applications to Agriculture
May 16–19, 1982

[8] Agricultural Chemicals of the Future
May 16–19, 1983

[9] Frontiers of Membrane Research
May 20–24, 1984

[10] Biotechnology for Solving Agricultural Problems
May 5–9, 1985

[11] Instrumentation in Agricultural Research
May 4–8, 1986

[12] Biomechanisms Regulating Growth and Development
May 3–7, 1987

[13] Biotic Diversity and Germplasm Preservation
May 9–11, 1988

FORTHCOMING SYMPOSIUM

[14] The Rhizosphere and Plant Growth
May 8–11, 1989

Beltsville Symposia in Agricultural Research

[13] Biotic Diversity and Germplasm Preservation, Global Imperatives

Lloyd Knutson and Allan K. Stoner,
Editors

Invited papers presented at a symposium held
May 9–11, 1988, at the Beltsville Agricultural
Research Center Beltsville, Maryland

Sponsored by
THE BELTSVILLE AGRICULTURAL RESEARCH CENTER
Agricultural Research Service
United States Department of Agriculture
with the cooperation of
Friends of Agricultural Research
Beltsville, Inc.
And the assistance of
Committee on Managing Global Genetic Resources,
Board on Agriculture,
National Research Council

KLUWER ACADEMIC PUBLISHERS
DORDRECHT / BOSTON / LONDON

Library of Congress Cataloging-in-Publication Data

CIP

Biotic diversity and germplasm preservation, global imperatives /
edited by Lloyd Knutson and Allan K. Stoner.
p. cm. -- (Beltsville symposia in agricultural research ; 13)
"Beltsville Symposium XIII was sponsored by the Beltsville Area, Agricultural Research Service, U.S. Department of Agriculture, with the cooperation of Friends of Agricultural Research-Beltsville"--Pref.
Includes index.
ISBN 0-7923-0178-1
1. Biological diversity conservation--Congresses. 2. Germplasm resources--Congresses. 3. Nature conservation--Congresses.
I. Knutson, Lloyd V., 1934- . II. Stoner, Allan K.
III. Beltsville Agricultural Research Center. IV. Friends of Agricultural Research Beltsville, Inc. V. Series.
QH75.A1B57 1989
333.95'16--dc19 89-2671

Published by Kluwer Academic Publishers
P.O. Box 17, 3300 AA Dordrecht, The Netherlands

Kluwer Academic Publishers incorporates the publishing programmes of D. Reidel, Martinus Nijhoff, Dr W. Junk and MTP Press.

Sold and distributed in the U.S.A. and Canada
by Kluwer Academic Publishers,
101 Philip Drive, Norwell, MA 02061, U.S.A.

In all other countries, sold and distributed
by Kluwer Academic Publishers Group,
P.O. Box 322, 3300 AH Dordrecht, The Netherlands

printed on acid free paper

Printed in The Netherlands.

CONTENTS

Section III: Frontiers in Germplasm Utilization

Section IV: Germplasm Collection and Data Management

1988 DISTINGUISHED SCIENTIST AWARDS

Beginning in 1987, The Friends of Agricultural Research-Beltsville (FAR-B) have presented a Distinguished Scientist Award to a Beltsville Scientists. The 1987 award recipient was Dr. Mary Shaw Shorb. For 1988, in keeping with the subject of Beltsville Symposium XIII, FAR-B presented Distinguished Scientist Awards in the areas of biotic diversity and germplasm preservation to Dr. Edward W. Baker and Dr. John G. Moseman, respectively.

Edward W. Baker

Dr. Edward W. Baker retired in 1987 after 47 years with the Systematic Entomology Laboratory. He continues to pursue an active research program at Beltsville.

His research has been at the forefront of basic and applied studies on the biotic diversity and systematics of mites. Through his efforts, the first comprehensive classification system of mites was developed. Due to his intensive work on plant-feeding spider mites, these important pest species are now well known. He has published more than 140 research papers and books and has described over 700 species of mites previously unknown to science. His research serves as the basis of nearly all programs dealing with the control, quarantine, and study of behavior and ecology of mites in United States crop systems.

Dr. Baker has been a leader in international congresses of acarology, an organizer and teacher in the long-term Institute of Acarology, and has been invited by international organizations to establish acarology programs in five foreign countries. His several text books serve as the basis of acarology curricula throughout the world.

John G. Moseman

Dr. John G. Moseman spent his 36-year career with the U.S. Department of Agriculture conducting basic research on host-parasite interactions and applying the findings to barley cultivar and barley germplasm improvement. His research resulted in the identification of genes in barley germplasm for resistance to powdery mildew, stripe rust, leaf rust, stem rust and loose smut. The genes identified have been or are being used by scientists around the world to develop disease-resistant barley cultivars. Another of Dr. Moseman's important contributions was the discovery that genes identified or resistance in barley germplasm relate directly to corresponding genes for virulence in the fungus causing powdery mildew.

During his career, Dr. Moseman presented his research findings at international scientific meetings; coordinated uniform barley and wheat disease nurseries at international, national and regional levels; served on national and international committees; and served as sponsoring scientist on several binational and PL 480 research projects.

Dr. Moseman served as Barley Investigations Leader, Agricultural Research Service, and Chairman, Plant Genetics and Germplasm Institute, Beltsville Agricultural Research Center. In the latter capacity he was responsible over a period of eight years for coordinating the research of approximately 70 scientists involved in research in most plant science disciplines to improve the quality and productivity of agronomic and horticultural crops.

THEODORE C. BYERLY, GEORGE W. IRVING
Co-Chairmen, Symposium Committee, Friends of Agricultural Research-Beltsville

FOREWORD

Topics for the Beltsville Symposia are selected to highlight specific areas of research and science policy that are of concern to scientists in the Beltsville Area as well as to the general scientific community. Each symposium in the series is structured to provide a realistic appraisal of current findings, research progress, and relevant policy issues within the constraints established by the organizing committee. Thus, the presentations and discussions that have marked these symposia have had a strong appeal to the broad community of scientists.

Knowledge of the diversity of living organisms is still quite limited. Since the time of Linnaeus, about 1.7 million species have been described. The actual number has been estimated between 5–50 million. Many species, land races, and strains are vanishing. Clearly, the world's scientific institutions are inadequately equipped to attain sufficient knowledge of a significant fraction of the diverse living forms. Also, efforts in the collection and preservation of germplasm of plants and animals urgently need to be strengthened. These mattes are critically important to future generations.

This symposium addresses vital concerns of biotic diversity and germplasm preservation from diverse perspectives. Many of the parts provide concrete recommendations for action, and they call attention to areas of research that must be pursued with intensity.

WALDEMAR KLASSEN, Director
Beltsville Area, Agricultural Research Service

PREFACE

Symposia have been held annually over the past thirteen years at the Beltsville Agricultural Research Center to address major areas of research and issues in agriculture. In recognition of the accelerating threats to biotic diversity, Beltsville Symposium XIII, May 9–11, 1988, focused on the needs and opportunities for preservation and utilization of our natural heritage. We are pleased to present here the papers given at the symposium. Unfortunately, we are not able to present the lively and informative discussions, computer demonstrations, poster presentations, and other aspects that also contributed importantly to the symposium.

The primary purposes of the symposium were to discuss recent research results and to identify future research needs for biotic diversity and germplasm preservation, in relation to major national and international agricultural issues. Biotic diversity aspects of the symposium included all organisms important to agriculture – plants, animals, and microorganisms. Forging new linkages between agriculture and conservation biology, by including speakers addressing societal issues and representing theoretical and applied aspects, was one of the underlying objectives of the symposium.

Discussion included new technologies, those available and others needed; traditional research areas, such as systematics, that require renewed emphasis and support; and information processing and distribution. Relationship among basic and applied interests, research areas, organizational entities, and national and international concerns were emphasized. General as well as specific recommendations for emerging research opportunities, approaches, and applications also were objectives of the symposium.

Beltsville Symposium XIII was sponsored by the Beltsville Area, Agricultural Research Service, U.S. Department of Agriculture, with the cooperation of Friends of Agricultural Research-Beltsville and the assistance of the Committee on Managing Global Genetic Resources, Board on Agriculture, National Research Council. We are grateful to the speakers, audience, sponsors, organizing committees, and support staff that helped to take this step in considering an issue that stands at the crossroads of the future well-being of mankind.

LLOYD KNUTSON, ALLAN K. STONER, Co-Chairmen, Beltsville Symposium XIII
Beltsville Area, Agricultural Research Service

PRESENTERS AND THEIR AFFILIATIONS

John A. Acree
Senior Staff Veterinarian
Veterinary Services
USDA, APHIS
Room 846AAA, Federal Building
Hyattsville, MD 20782
U.S.A.

Rob Bertram
Agricultural Research Specialist
Office of Agriculture
Agency for International Development
Washington, DC 20523
U.S.A.

James A. Briggs
Manager/Plant Materials Specialist
National Plant Materials Center
USDA-Soil Conservation Service
Building 509, BARC-W
Beltsville, MD 20705
U.S.A.

Howard J. Brooks
National Program Leader for Horticulture
USDA, ARS
Room 238, Building 005, BARC-W
Beltsville, MD 20705
U.S.A.

William L. Brown
Board on Agriculture
National Academy of Sciences
2101 Constitutional Avenue
Washington, DC 20418
U.S.A.

Jeffrey R. Burkhardt
Assistant Professor of Range Management
University of Florida
Gainesville, FL 32611
U.S.A.

Lawrence M. Busch
Professor of Sociology
University of Kentucky
223 Scovell Hall
Lexington, KY 40506
U.S.A.

Te-Tzu Chang
Principal Scientist and Head
International Rice Germplasm Center
International Rice Research Institute
P.O. Box 933
Manila, Philippines

Joel I. Cohen
Biotechnology Specialist
Office of Agriculture
Agency for International Development
SA 18, Room 413E
Washington, DC 20523
U.S.A.

Peter R. Day
Director
Center for Agricultural Molecular Biology
Cook College, P.O. Box 231
Rutgers, The State University of New Jersey
New Brunswick, NJ 08903
U.S.A.

Gordon E. Dickerson
Research Geneticist
USDA, ARS
U.S. Meat Animal Research Center
A-218 Marvel Baker Hall
Lincoln, NE 68583
U.S.A.

Sherl M. Dietz
Regional Coordinator, W-6
USDA, ARS
Johnson Hall, Room 59
Washington State University
Pullman, WA 99164–6402
U.S.A.

Donald N. Duvick
Senior Vice President - Research
Pioneer Hi-Bred International, Inc.
700 Capital Square
400 Locust Street
Des Moines, IA 50309
U.S.A.

Steve Eberhart
Director, National Seed Storage Laboratory
USDA, ARS
Colorado State University
Fort Collins, CO 80523
U.S.A.

James H. Elgin, Jr.
Deputy Area Director
Plant Sciences Institute
USDA, ARS
Room 201, Building 003, BARC-W
Beltsville, MD 20705
U.S.A.

Paul J. Fitzgerald
Agricultural Science Advisor
USDA, ARS
Plant Germplasm, Agronomy Department
Room 202, Poultry Science Building
Purdue University
W. Lafayette, IN 47907
U.S.A.

Rodrigo Gámez
Professor
Cellular and Molecular Biology Research Center
University of Costa Rica
Ciudad Universitaria 2060
San Jose, Costa Rica

Keith D. Gregory
Research Geneticist
Roman L. Hruska U.S. Meat Animal Research Center
USDA, ARS
P.O. Box 166 State Spur 18D
Clay Center, NE 68933
U.S.A.

Kevin J. Hackett
Research Entomologist
Insect Pathology Laboratory
USDA, ARS
Room 9, Building 465, BARC-E
Beltsville, MD 20705
U.S.A.

Lawrence A. Johnson
Research Physiologist
Reproduction Laboratory
USDA, ARS
Room 22, Building 200, BARC-W
Beltsville, MD 20705
U.S.A.

Shung-Chang Jong
Curator and Head
Mycology and Botany Department
American Type Culture Collection
12301 Parklawn Drive
Rockville, MD 20852
U.S.A.

Joseph H. Kirkbride, Jr.
Botanist
Systematic Botany, Mycology, and Nematology Laboratory
USDA, ARS
Building 265, BARC-E
Beltsville, MD 20705
U.S.A.

Lloyd V. Knutson
Research Entomologist
Systematic Entomology Laboratory
USDA, ARS,
Building 003
Beltsvikke, MD 20705
U.S.A.

William B. Lacy
Professor of Sociology
University of Kentucky
223 Scovell Hall
Lexington, KY 40546
U.S.A.

Roger A. Leopold
Research Entomologist
Metabolism and Radiation Research Laboratory
USDA, ARS
P.O. Box 5674
Fargo, ND 58105
U.S.A.

Benjamin F. Matthews
Plant Biochemist
Plant Molecular Biology Laboratory
USDA, ARS
Room 109, Building 006, BARC-W
Beltsville, MD 20705
U.S.A.

Douglass R. Miller
Research Leader
Systematic Entomology Laboratory
USDA, ARS
Room 4, Building 003
Beltsville, MD 20705
U.S.A.

Phillip A. Miller
National Program Leader for Fiber, Oil, and Tobacco
USDA, ARS,
Room 207, Building 005
Beltsville, MD 20705
U.S.A.

Jimmie D. Mowder
Database Manager
Germplasm Resources Information Network
Germplasm Services Laboratory
USDA, ARS
Room 130, Building 001, BARC-W
Beltsville, MD 20705
U.S.A.

Charles F. Murphy
National Program Leader for Grain Crops
USDA, ARS
Room 239, Building 005, BARC-W
Beltsville, MD 20705
U.S.A.

Gene Namkoong
Research Geneticist
USDA, FS
Department of Genetics, Box 7614
North Carolina State University
Raleigh, NC 27695–7614
U.S.A.

H. Duane Norman
Research Leader
Animal Improvement Programs Laboratory
USDA, ARS
Room 1A, Building 263, BARC-E
Beltsville, MD 20705
U.S.A.

John A. Pino
Project Director
Committee on Managing Global Genetic Resources
Board on Agriculture
National Academy of Sciences
2101 Constitution Avenue, N.W.
Washington, DC 20418
U.S.A.

Rex L. Powell
Research Geneticist
Animal Improvement Laboratory
USDA, ARS
Room 11d, Building 263, BARC-E
Beltsville, MD 20705
U.S.A.

Vernon G. Pursel
Research Physiologist
Reproduction Laboratory
USDA, ARS
Room 2, Building 200, BARC-E
Beltsville, MD 20705
U.S.A.

Caird E. Rexroad, Jr.
Research Physiologist
Reproduction Laboratory
USDA, ARS
Room 13, Building 200, BARC-E
Beltsville, MD 20705
U.S.A.

Robert R. Rojas
Research Biologist
Matabolism and Radiation Research Laboratory
USDA, ARS
P.O. Box 5674
Fargo, ND 8105
U.S.A.

Eric E. Roos
Research Leader
National Seed Storage Laboratory
USDA, ARS
Fort Collins, CO 80523
U.S.A.

Amy Y. Rossman
Research Leader
Systematic Botany, Mycology, and Nematology Laboratory
USDA, ARS
Room 304, Building 011A, BARC-W
Beltsville, MD 20705
U.S.A.

Setijati Sastrapradja
Director and Botanist
National Centre for Research in Biotechnology
Jalan Juanda No. 18
P.O. Box 323
Bogor, Indonesia

James A. Saunders
Research Biochemist
Germplasm Quality and Enhancement Laboratory
USDA, ARS
Room 5C, Building 009, BARC-W
Beltsville, MD 20705
U.S.A.

Henry L. Shands
National Program Leader for Germplasm
USDA, ARS
Room 140, Building 005, BARC-W
Beltsville, MD 20704
U.S.A.

Allan K. Stoner
Research Leader
Germplasm Services Laboratory
USDA, ARS
Room 127, Building 001, BARC-W
Beltsville, MD 20705
U.S.A.

Leigh E. Towill
Plant Physiologist
National Seed Storage Laboratory
USDA, ARS
National Seed Storage Laboratory
Ft. Collins, CO 80523
U.S.A.

Robert J. Wall
Research Physiologist
Reproduction Laboratory
USDA, ARS
Room 6, Building 200, BARC-W
Beltsville, MD 20705
U.S.A.

Melvin N. Westwood
Professor Emeritus
National Clonal Repository
Department of Horticulture
Oregon State University
Corvallis, OR 97331
U.S.A.

Robert F. Whitcomb
Research Entomologist
Insect Pathology Laboratory
USDA, ARS
Room 10, Building 465, BARC-E
Beltsville, MD 20705
U.S.A.

George A. White
Plant Introduction Officer
Germplasm Services Laboratory
USDA, ARS
Room 322, Building 001, BARC-W
Beltsville, MD 20705
U.S.A.

Garrison Wilkes
Professor
Department of Biology
University of Massachusetts
Boston Harbor Campus
Boston, MA 02125
U.S.A.

J. Trevor Williams
Director, International Board for Plant Genetic Resources
Food and Agriculture Organization of the United Nations
Via delle Terme di Caracalla
00100 Rome, Italy

Zhang Yihua
Director
Institute of Crop Germplasm Resources
Chinese Academy of Agricultural Sciences
30 Bai Shi Qiao Road
Beijing, People's Republic of China

Dong Yuchen
Professor, Plant Genetics and Breeding
Institute of Crop Germplasm Resources
Chinese Academy of Agricultural Sciences
30 BA; Shi Qiao Road
Beijing, People's Republic of China

BELTSVILLE SYMPOSIUM XIII COMMITTEES

PROGRAM
L. Knutson – Systematic Entomology Laboratory, and
A. K. Stoner – Germplasm Services Laboratory, Co-Chairmen
L. W. Briggle – Germplasm Services Laboratory
R. J. Gerritts – National Program Staff
D. R. Miller – Systematic Entomology Laboratory
J. D. Mowder – Germplasm Resources Information Network
H. L. Shands – National Program Staff
G. A. White – Germplasm Services Laboratory

POSTER PRESENTATIONS
A. Y. Rossman – Systematic Botany, Mycology, and Nematology Laboratory

COMPUTER DEMONSTRATIONS
J. H. Kirkbride – Systematic Botany, Mycology, and Nematology Laboratory

PUBLICITY
E. W. Herbert – Beneficial Insects Laboratory, Chairman
S. M. Berberich – Information Office, Agricultural Research Service
T. C. Byerly – Friends of Agricultural Research – Beltsville, Inc.
T. R. Dudley – National Arboretum
J. R. Lichtenfels – Biosystematic Parasitology Laboratory
J. Wall – Friends of Agricultural Research – Beltsville, Inc.

LOCAL ARRANGEMENTS
T. H. Barksdale – Vegetable Laboratory, Chairman
A. A. Atchley – Germplasm Services Laboratory
G. T. Carpenter – Visitor's Unit
R. W. Spjut – Systematic Botany, Mycology, and Nematology Laboratory

FINANCE
F. J. Longen – Friends of Agricultural Research – Beltsville, Inc., Chairman
R. R. Rhodes – Beltsville Area Office
F. C. Thompson – Systematic Entomology Laboratory

REGISTRATION
J. R. Weirman – Secretary, Beltsville Symposium XIII

WITH THE COOPERATION OF FRIENDS OF AGRICULTURAL RESEARCH BELTSVILLE, INC.

SECTION ONE

Introduction and delineation of issues

Systematics, diversity, and germplasm

DOUGLASS R. MILLER[1], AMY Y. ROSSMAN[2] and JOSEPH KIRKBRIDE[3]

[1] *Research Leader, Systematic Entomology Laboratory, Plant Sciences Institute (PSI), Agricultural Research Service (ARS), U.S. Dept. Agric., Rm. 1, Bldg.003, BARC-W, Beltsville, Maryland 20705, U.S.A.*

[2] *Research Leader, Systematic Botany and Mycology, Laboratory (SBML), PSI, ARS, U.S. Dept. Agric., Rm. 312, Bldg. 011-A, BARC-W, Beltsville, Maryland 20705, U.S.A.*

[3] *Research Botanist, SBML, PSI, ARS, U.S. Dept. Agric., Bldg. 265, BARC-E, Beltsville, Maryland 20705, U.S.A.*

Abstract Systematics is the foundation of nearly all biological research, providing the framework for repeatability in natural science. Systematic classifications also provide a marvelously efficient system for communication and they enable various levels of prediction. Unfortunately, systematics and biological diversity are generally misunderstood because the general populace holds a simplistic view of the complexity and size of the world's biota. Decimation of the diversity of life is occurring at an alarming rate. From an agricultural point of view, the impact of this destruction is a constant decrease in the germplasm available for manipulation and utilization to improve the welfare of humankind. Every organism that becomes extinct removes from the global genetic pool, thousands or even millions of genes of potential value to the agricultural community of the future. With new technology in gene transfer, this loss becomes even more significant. Equally alarming is the fact that only one out of every 10-30 organisms is known to science. Therefore, the first order of business in biological diversity research must be to make a complete inventory of the world's biota. This task is essential to nearly all conservation and germplasm plans for the future. A major complication in undertaking this task is overcoming problems of collection space, collection management, and efficient retrieval of specimen label information. Unfortunately, support for systematics has eroded over the past several decades. Therefore, increased funding of systematics must begin immediately to meet the major need for taxonomic information in the future.

Key words: systematics, collections, biological diversity, germplasm, classifications, predictions

The planet earth supports a vast diversity of organisms. These organisms, which live in such diverse habitats as the sparse, lichen-dominated areas of the arctic tundra to the lush lianas of the tropical rain forests constitute the global pool of germplasm that is not only esthetically pleasing and 'cute and cuddly' but is of critical importance to the future of humankind.

L. Knutson and A.K. Stoner (eds.). Biotic diversity and germplasm preservation, global imperatives.
 ISBN 0-7923-0178-1

The fundamental relationship of diversity and germplasm is evident: The diversity of organisms is the source of all germplasm. Systematics is the discipline that provides the means of communicating about the basic characteristics and interrelationships of organisms. This information forms the framework for most comparative biological studies.

The following thoughts of leading scientists are especially pertinent to the subject of systematics, diversity, and germplasm:

> E. O. Wilson 'Systematics, the study of biological diversity, is sometimes portrayed as the mere classification of organisms, but in fact its range and challenges are among the greatest in biology.' [1].
>
> L. R. Brown 'Each morning I wake up to a five minute National Public Radio newscast. Invariably I am disappointed...because it often deals with news events that do not matter... What about the four plant species that disappeared yesterday? Their passing went unmentioned. How do I know there were that many? I don't. It may have been 40, since estimates of extinction rates vary that widely with long-term consequences that no one can forsee.' [2].
>
> E. O. Wilson 'And how many species of organisms are there on earth? We do not know, not even to the nearest order of magnitude.' [3].
>
> C. Darwin 'There is grandeur in this view of life..., and that, whilst this planet has gone cycling on according to the fixed law of gravity, from so simple a beginning endless forms most beautiful and most wonderful have been, and are being evolved.' [5].

Facts and figures on the size of the world's biota are presented throughout this volume; please note the extensive differences in these statistics. The point of noting these disparites is that reasonable knowledge of the extent of biotic diversity is lacking, as well as essential information that will aid in its preservation and use in improvment of human existence. Estimates of the number of extant species range from 1 or 2 million [6], to as many as 30 or even 50 million [7].

Biotic diversity and systematics

But why should this symposium be concerned with systematics? What does this discipline have to do with biological diversity and germplasm? To answer these questions we first assert that everyone that reads this paper is a systematist. Thus, if you observe a specimen of a mushroom, poison ivy, or a crocodile, you as a systematist assess certain critical characters, make a taxonomic decision about the identity of the organism, and make predictions that will influence your interactions with the organism. In the case of a snake, you observe a narrow, legless body and your taxonomic knowledge allows you to decide that you are seeing a snake. You observe a rattle at the caudal apex and an enlarged head and you decide that you are seeing a rattlesnake. Based on previous communication and accumulated systematic knowledge, you know that if bitten you will receive a dose of poison that is delivered by fangs that unfold from the front of the mouth. Because you have identified the organism as a rattlesnake, you associate other characters with the creature, e.g., it is a good predator, overwinters in dens, is cold blooded, etc.

Communication

Consider the communication aspects of this observation. By your use of the word 'rattlesnake' or its scientific name, *Crotalus viridis* (Rafinesque), you have conveyed an enormous amount of information that goes unstated in most conversations.

Unfortunately, the lay person, and many scientists as well, usually are ignorant systematists. The general public has an incredibly simplistic view of biological diversity and systematics. We characterize this phenomenon as the 'dog and cat' syndrome. The lay person sees dogs and cats, maples and oaks, cows and pigs, and easily recognizes them as distinct. Unfortunately, in its daily routine the public does not see, and does not need to recognize, the hundreds of species of leafhoppers or cutworm moths that all look alike. Nor does the public understand that the many color forms of certain species of beetles, butterflies, and other organisms actually represent variants of a single species. It is unaware that certain organisms that look alike actually are different species and that other organisms that look different are the same species. Such lack of awareness stems from the fact that people do not deal with these kinds of complex systematic problems in their daily lives. Further, it is beyond their imagination that the actual number of species of insects could be as many as 50 million. If you query an individual on the street about how many different kinds of insect species there are, the answer generally will be 100 to 1,000. Because of this distorted perception of biological diversity, the discipline of systematics is underrated in its complexity and necessity.

Thus, systematics provides an essential organization of names that is a marvelously efficient system of information communication–an important aspect of diversity research and germplasm exploitation.

Predictability

Let us return to the rattlesnake example and consider the predictions that could be made if we were to discover a species unknown to science. Assume that an organism is observed that has the same narrow, legless body as the rattlesnake mentioned earlier, and has a rattle on the tail and a broad head. We identify it is a rattlesnake. From this we predict that its bite would cause the injection of a poison through fangs, even though we did not observe this phenomenon. We also would hypothesize that it is predaceous on small mammals, that it is primarily nocturnal, that it overwinters in dens, etc. We would make these predictions on the basis of our knowledge of previous systematic research findings.

Our example may be a bit simple, but it does make the point that the classification is a predictive one. In germplasm research, such predictability may be the only source of information available to begin the development of a new agricultural product or to solve an agricultural problem. In searching for biological control agents of a major pest, the Russian wheat aphid, introduced into the United States in 1986, systematists predict that the best place to look for natural enemies will be in the southern USSR and surrounding areas. Other predictions are that the most likely arthropod parasites and predators will be certain genera of wasps and beetles, and that the natural enemies of two apparently closely related aphid species may be effective against the Russian wheat aphid [8].

Consider what might happen without the predictive components of a classification system. A recent publication in Mycolaxon, a mycological journal modelled after the more serious Mycotaxon, contained an article by a fictitious noteworthy mycologist, Dr. Sigmund J. Lumper. Dr. Lumper proposed an extremely pragmatic approach to the identification of fungi, namely recognition of only one taxon: *Fungus*. Using this practical system, the problem of providing the 'correct' determination was eliminated. However, within Lumper's concept of *Fungus*, all information content and predictability was lost. Within this taxon the range of diversity includes: Strains of *Penicillium* that produce the antibiotic penicillin; another fungus *Penicillium roquefortii* in cheese; and *Aspergillus flavus* a fungus that produces aflatoxins. Even the fungus that is used to make beer would be equivalent to the deadly *Amanita* mushroom. Obviously there are some problems with the Lumper system.

Identification

Having discussed some general characteristics of systematics, we now will provide examples of more specific contributions made by this discipline. The most obvious of these is taxonomic services. The importance of an accurate identification cannot be overstated. In science, determinations are basic to the repeatability of experimentation. If the name of the species used in an experiment is not given, the possibility of repeating the results is greatly diminished. If an experiment is conducted with different species, there is a significant chance that discrepancies from previous experiments are the result of species differences. Without repeatability, research is not science but is more in the realm of art.

In this regard, correct identification is the key to accessing all of the literature and past knowledge about an organism. Without a name for a taxon, the known information about that taxon is virtually unavailable.

The body of information that allows the systematist to provide accurate identifications is contained in 1. reference organisms stored in collections and 2. publications.

Collections

Collections of both living and non-living organisms are the reference resources for biology, particularly systematic biology. Like a library, collections serve as standards. They form the basis for the knowledge of systematics, libraries of information that are the permanent reference resources for interpreting the living world.

For example, if a breeder hypothesizes that one germplasm accession should interbreed with another but finds that it will not do so, the voucher specimens for those accessions can be consulted to determine if either one was misidentified or if cross-pollination has occurred in increasing one of the accessions. Collections house 'evidence' for drawing conclusions in biological 'cases' involving unexpected or unexplainable results. Specimens in collections are the 'truth' behind the conclusions of systematic accounts published as keys for identification and descriptions of taxa.

Some of the collections in the Beltsville Area are: The U.S. National Fungus Collection, which contains about one million specimens; the U.S. National Seed Herbarium, the largest collection of seeds in the world; the USDA Nematode Collection, which contains about 20,000 slides and vials; the U.S. National Parasite Collections, the largest collection of parasites of animals and humans; the U.S. National Arboretum Herbarium, which contains more than 400,000 specimens of cultivated plants; and the National Collection of Insects, which contains more than 30 million specimens; this collection is the property of the Smithsonian Institution, but USDA employees in collaboration with Smithsonian entomologists curate major parts of the collection.

The role of collections in biological diversity research is of critical significance, but only under rare circumstances are the major, collection-oriented problems considered in the planning for an inventory of the world's biota. Imagine the added museum space that will be required to deal with the 10 to 30 times more species than currently are stored in the world's collections. Effective storage, management, and accessability of such increased loads are problems that must be given serious consideration in the near future.

Publication

Along with collections, published research serves as the primary resource for identification. Obviously, determinations can only be as good as the research on which they are based. For flowering plants and other conspicuous groups, published keys and descriptions often are sufficient to make accurate determinations. However, for the less conspicuous organisms the situation is quite different. In these groups, even the species that are known to science often are recorded only once and are based on a brief examination by the expert.

Our knowledge of the most diverse groups of organisms is wholly inadequate. For example, only 1 out of every 5 [9] or perhaps 1 out of every 30 [10] species of insects is known to science; for nematodes the relative numbers of species unknown to science may be even larger. This situation is complicated by the fact that most species of organisms have several different life stages that all must be studied so that they can be recognized and identified accurately.

The lack of sufficient systematic information is frustrating to those who wish to conduct research with poorly known groups. Dan Janzen, an insect ecologist working in Costa Rica, often has been unable to obtain the taxonomic information that is critical to his research. To circumvent this problem, he has hired systematists to study the insect groups that are of major significance to his studies. In other cases, scientists in disciplines other than systematics have purposely avoided studying groups that have major taxonomic problems since an inadequate classification of the group could seriously confound or invalidate their research conclusions.

Systematic research is not restricted to hypothesizing classifications and publishing revisions and monographs. Systematics has a long history of providing synthetic discoveries that have given a new appreciation to the natural world. The research on island biogeography by MacArthur and Wilson [11] is an example of research that directly uses, and which in some cases is limited by a lack of systematic knowledge.

In recent years, studies of habitat diversity have developed because of strong systematic underpinnings. It is the research of such imaginative programs as the U.S. National Museum of Natural History's BIOLAT that will ultimately give good estimates of how many organisms exist on Earth and their interrelationships in the complex, tropical ecosystems.

Systematics and germplasm

We have discussed the relationship of systematics and biological diversity. Now we will address the interaction of systematics and germplasm.

Some startling statistics again merit mention. In the next 25 years, if 'development' continues at the current pace, most tropical lowland forests will be destroyed [12]. With these forests will perish more than one-quarter of the total existing diversity on Earth!!! An even larger number of species will become extinct in the first part of the next century. The rate of extinction will escalate to considerably more than 100 species per day. Very likely, most of these species will never be known to science. Never since the Cretaceous epoch, 65 million years ago, has there been such traumatic decimation [4]. AND WE ARE THE CAUSE.

Consider the implications of this extraordinary loss to the potential well-being of humankind. To quote Peter Raven [4]

> 'The thousands of antibiotics that we obtain from bacteria, the drugs that permit open heart surgery or prevent cancer, the food that we eat, the clothes that we wear, the wood and pulp that we use, even our beer and wine – all of these come directly or indirectly from plants, animals, or microorganisms. We use a few hundred of the 250,000 kinds of plants for food; yet it is estimated that tens of thousands more may be suitable. Even in terms of our newfound ability to move genes..., the loss of each organism represents not the loss of a single entity, but the loss of thousands or tens of thousands of genes, each of which might be valuable to us when placed in another kind of organism...'

With each hundred species gone we lose a wealth of germplasm that someday might have been extremely valuable.

Earth's biotic inventory

How does systematics fit in this drama? The mandate presented by Wilson [3] to make 'a complete catalog of life on Earth' is an obvious first step in addressing the biodiversity crisis. Clearly this is a function for the systematist. Wilson gives statistics supporting the hypothesis that such a staggering project is achievable even with traditional techniques and standard forms of publication. An important consideration that has not been taken into account is the needs of the world's museums. The travesty is that with an impending significant demand for increased systematic expertise, the systematist is an endangered resource. Curricula in systematics continue to diminish or disappear. Recent technological developments in computerized analysis, data storage and management, and automated identification systems should be beneficial in expediting the completion of Earth's life catalog (perhaps database would be better). New methods in character analysis such as phylogenetics and molecular technology also will greatly facilitate this endeavor.

With a synthesis of the Earth's biotic inventory, we will have a clearer understanding of where to maintain 'natural germplasm banks' and to understand the optimum size, structure, and diversity of these banks so that we can maintain some semblence of our biotic resources into perpetuity.

With a classification system of the entire world's biota, we will be able to predict sources of anticancer drugs without dredging through the thousands of plants studied by the National Institute for Health for this purpose. Substances such as the recently-discovered antibiotic secretion of the African clawed frog [13] will remain for science of the future as will the additional, undiscovered, naturally-occurring proteins like arcelin that was recently found in wild beans and that has been shown to have a toxic effect on beetle pests [14].

Conclusion

Systematics provides the foundation upon which is based a functioning system of storing and utilizing germplasm, both natural and artificial. Through precise characterization of organisms, the effective use of germplasm is possible. Only if the diverse organisms that exist today are rescued from extinction can they be studied and further utilized as the germplasm of tomorrow.

References

1. Wilson, E. O. 1985. Time to revive systematics. Science 230: 1227.
2. Brown, L. P. 1988. Editor's page. World Watch 1: 2.
3. Wilson, E. O. 1985. The biological diversity crisis. BioScience 35: 700–706.
4. Raven, P. H. 1987. Forests, people, and global sustainability.Keynote address, National Audubon Society, Biennial Convention, Bellingham, Washington, August 24, 1987, 15 pp.

5. Darwin, C. 1895. On the Origin of Species by Means of Natural Selection, or the Preservation of Favoured Races in the Struggle for Life. John Murray, London, 502 pp.
6. Parker, S. P. 1982. Synopsis and Classification of Living Organisms. McGraw-Hill, New York, 496 pp.
7. Erwin, T. L. 1988. The tropical forest canopy. Pp. 123–129 in E. O. Wilson and F. M. Peter (eds.), Biodiversity. National Academy Press, Washington, DC. 521 pp.
8. Durr, H. J. R. 1983. Diuraphis noxia (Mordvilko), a recent addition to the aphid fauna of South Africa. Phytophylactica 15: 81–83.
9. Wilson, E. O. 1988. The current state of biological diversity. Pp. 3–18 in E. O. Wilson and F. M. Peter (eds.), Biodiversity. National Academy Press, Washington, DC. 521 pp.
10. Erwin, T. L. 1983. Beetles and other insects of tropical forest canopies at Manus, Brazil, sampled by insecticidal fogging. Pp. 59–75. in S. L. Sutton T. C. Whitmore and A. C. Chadwick (eds.), Tropical Rain Forest: Ecology and Management. Blackwell, Edinburgh, 432 pp.
11. MacArthur, R. H. and Wilson, E. O. 1967. The Theory of IslandBiogeography. Princeton University Press, Princeton, New Jersey, 203 pp.
12. Raven, P. H. 1988. Our diminishing tropical forests. Pp. 119–122 in E. O. Wilson and F. M. Peter (eds.), Biodiversity. National Academy Press, Washington, DC. 521 pp.
13. Zasloff, M. 1987. Magainins, a class of antimicrobial peptides from Xenopus skin: Isolation, characterization of two active forms, and partial cDNA sequence of a precursor. Proc. Natl. Acad. Sci. 84: 5449–5453.
14. Osborn, T. C., D. C. Alexander S. S. M. Sun C. Cardona and F. A. Bliss 1988. Insecticidal activity and lectine homology of arcelin seed protein. Science 240: 207–210.

Germplasm preservation: objectives and needs

GARRISON WILKES
Biology Department, UMASS/BOSTON, Massachusetts 02125, U.S.A.

Abstract Germplasm is the hereditary message which dictates the development of plants and animals. This biological information is passed from generation to generation in an unbroken chain of living cells. Unlike other resourses, once cell life is lost, the germplasm is lost, therefore for all practical purposes it is stored only in living systems. For convenience germplasm resources can be grouped into several levels of usefulness and management:

1. Varieties and breeds in current use,
2. Obsolete cultivars and minor breeds,
3. Primitive cultivar/landraces and folk breeds of indigenous agriculture,
4. Wild and weedy/feral taxa and near relatives of useful animals and crops,
5. Special genetic stocks which are the tools of breeders,
6. Special coadapted genetic stocks where two or more taxa have a symbiotic relationship (pollinator, nodule forming bacteria etc.) and
7. Lastly organisms of no apparent immediate usefulness but which are part of the web of life.

Germplasm of the last category is best held in natural systems or entire biomes. The remaining more immediately useful plants and animals once held in the now disappearing peasant agriculture are assembled *in situ* or more extensively *ex situ* in genebanks. The problem with the easier to manage *ex situ* banking is that it withdraws seed and semen/ova from observation and evaluation. In my view genebanks are only banks if the holdings are used, and depositories if their mission is preservation. Unfortunately, germplasm is being lost in some holding institutions and they have become, sadly, germplasm morgues. If the expense and efficacy of storage is to be proven the breeders are going to have to do a significantly better job of evaluating the material held and making this information/material freely and universally available to all, both nationally and internationally, for crop and animal improvement.

Key words: germplasm, plant breeding enhancement, genetic evaluation of plant genetic resources

The future is uncertain. Nature deals with uncertainty through genetic diversity. Agriculture and food security deals with uncertainty through germplasm preservation. Plant breeding is a future–looking activity. True future–looking plant breeding is germplasm enhancement.

Germplasm enhancement is as dependent on creative and insightful plant breeding as it is on genetic diversity found in the crop plant greater

L. Knutson and A.K. Stoner (eds.). Biotic diversity and germplasm preservation, global imperatives.
 ISBN 0-7923-0178-1

> gene pool of exotic materials and wild relatives. Deep stirring of the gene pool is the art of plant breeding; production of finished varieties is the craft of plant breeding.

Germplasm is the hereditary message which dictates the development of plants and animals. This biological information is passed from generation to generation in an unbroken chain of living cells. Unlike other resources, once cell life is lost, the germplasm is lost, therefore for all practical purposes it is stored only in living systems. Biological and genetic diversity maintenance have been one of the unrecognized public service functions of natural systems. The use of germplasm to promote food security is an obvious challenge between now and the end of this century as more of the world becomes human–managed, agro-urban-ecosystems and not natural systems. Suddenly we have come to realize that our most productive cultivars are stretching around the world and the elastic of cultivated genetic diversity is very thin. For us the central question is how much of the world can be made over for food production and still keep enough biological diversity in the form of genebanks to promote stability.

The kinds of genetic diversity that can be stored in genebanks, are not uniform and exist in varying levels of usefulness and management:

1. Varieties and breeds in current use,
2. Obsolete cultivars and minor breeds; primitive cultivars/landraces and folk breeds of indigenous agriculture,
3. Wild and weedy/feral taxa and near relatives of useful animals and crops,
4. Special genetic stocks which are the tools of breeders,
5. Special coadapted genetic stocks where two or more taxa have a symbiotic relationship (pollinator, nodule forming bacteria) and
6. Organisms of no apparent immediate usefulness but which are part of the web of life.

Germplasm of the last category is best held in natural systems. The remaining more immediately useful plants and animals, once held in the now disappearing peasant agriculture, have been assembled *in situ* and more extensively *ex situ* in genebanks [1–4].

The germplasm imperative

The imperative for genetic conservation of germplasm develops from three processes in our current world:

1. An increased human population which leads to further alteration of natural ecosystems and the expansion of food–producing agriculture,

2. The widespread adoption of elite germplasm and common agricultural technology which promotes genetic uniformity, sometimes worldwide, and
3. The centers of genetic variability are moving from natural systems and primitive agriculture to genebanks and breeders working collections with the liabilities that a concentration of resources implies.

The germplasm of crop plants is in a much more advanced state of genebanking than either forestry or micro-organisms although many of the principles are the same. For animals the situation is even less advanced with very limited knowledge of the variation that exists worldwide. The problems with animals are unique because of the size and expense of maintaining herds and flocks, not to mention diseases and quarantine. Because of the more advanced world-wide effort and planning for plant germplasm preservation the rest of the paper will focus on crop plants.

Currently there is widespread support for the building of crop germplasm preservation institutions by The International Board for Plant Genetic Resources (IBPGR) but the system operates on goodwill which is subject to sociopolitical uncertainty. The system works well now because it doesn't make excessive budgetary demands, but costs are going to rise over time especially when the much needed seed rejuvenation is undertaken. In 1975 there were only eight institutions in the world with refrigerated storage suitable for maintaining seed at $-18\,^{\circ}C$ necessary for long term (base) storage as against $0\,^{\circ}C$ suitable for medium-term (active) storage of working collections. Now that the number has more than tripled, will nations discover they are too expensive to maintain and wish to cut back their commitments?

An even more powerful question is: Will seed be available freely from all the collections? Will developing nations cut off the germplasm supply claiming they are germplasm rich, yet it is being laid claim to by developed nations without proper economic considerations, especially by the multinational seed companies. This is an interesting point, because the claim is made that the great number of accessions in genebanks are primitive cultivars and landraces from peasant agriculture. Therefore the peasants of the world should be paid for this germplasm. There is an element of truth here, but a deeper look will reveal that most of the primitive cultivars and landraces are unproven germplasm. The proven 'elite' germplasm of genebanks are the obsolete cultivars – quite possibly half of the bread wheat planted in the world has some genetic relationship to the obsolete variety 'Turkey.' The claim that multi-nationals are profiting from peasant germplasm is without warrant but developing nations have contributed their primitive cultivars and multi-nationals have not contributed their obsolete elite germplasm (note this is not current elite lines but proven lines from the past) and for this they can be faulted. The free and open exchange of germplasm will only work if the contributions are made from both sides. The U.S. seed industry could take leadership here and set a precedent for other developed nations by contributing our obsolete 'elite' germplasm.

The current major issue in germplasm preservation is no longer how to get the collections made, store the seed or computerize the bank passport data [5–6]. It is instead, to establish the efficacy of the collections by growing the material, extensive genetic evaluation and selective hybridization to get these genes back into the dynamic process of crop evolution. Very few gene banks are doing this while most are little better than static gene depositories.

By the year 2000, the human population of 6 billion will produce more food in one generation (20 years) than we did in the previous 10,000 years since the beginnings of agriculture. One in every ten humans that has ever existed will be sitting down to dinner dependent on this supply. To meet this demand, much of it in the developing countries, we are going to have to be better at food production with the basic crops than we are now [7–9].

Since new arable land in both the developed and developing nations will become steadily more scarce, higher yields from land already in production will be the only way to support the population increase. Higher yields means more fertilizer, plowing and water lifting energy, and improved plant materials. Breeding of better crop plants will be the focal point around which all strategies to increase crop yield will develop. It is the positive response of these seeds and root stocks to soil, water, fertilizer, pest, pathogen, weather, farmer skill, and social institutions that will determine the success of our future agricultural advancement.

Biological diversity

Biological diversity is the key element in most natural systems to minimize risk, yet what we do most often in our crop plants is to promote biological and genetic uniformity, which is the opposite [10, 11]. The elements of biological uniformity/diversity in agriculture are too complex for simple pin-pointing of problems and quick fix solutions. I will in this presentation be primarily examining the problems surrounding the evaluation of germplasm to broaden the genetic base of agriculture. Overall the productivity of American agriculture has been nothing short of fantastic, so the main thrust of this paper is not that 'the sky is falling in', but that for marginal addition expenditure, the system could be a quantum leap ahead.

Human made agricultural systems that mimic natural ecosystems appear to be most sustainable. Primitive systems in peasant agriculture had a diverse number of genotypes in the cropping fields. Often, in addition, there was hybridization with wild relatives and weedy forms along the margin of the field increasing the genetic variance. Contrast this diversity with advanced agricultural systems which plant a highly uniform genotype across the entire field, town, county, even state. Usually this genotype is not cropped very many years in a row before it is replaced by an advanced line thus resulting in genetic diversity over time. In these advanced agricultural systems the plant breeder is the keystone for this agricultural arch into the future, because plant breeders alone are the ones stirring the genetic diversity 'pot'. We have come so far down the path of domestication and restructuring our environment that natural systems no longer supply the genetic diversity for our crops.

Much of the current discussion which focuses on biological diversity mentions the three negative processes that affect the crops we now cultivate. These three processes – GENETIC EROSION, GENETIC VULNERABILITY, and GENETIC WIPEOUT [12–17] are not mutually exclusive, but are in fact, interlocked by the demands of an increasing human population and rising expectations.

Genetic erosion

The technological bind of improving varieties is that they eliminate the resource upon which they are based. Over the past 10,000 years crop plants have proliferated an innumerable number of locally adapted genotypes. These landraces and folk varieties of indigenous and peasant agriculture have been the genetic reservoir for the plant breeder in crop improvement. Suddenly this genetic diversity is being replaced with a relatively small number of varieties bred for high yields and other adaptions necessary for high input agriculture. In addition, the scarcity of land is forcing changes in land use and agricultural practices resulting in the disappearance of the habitat to maintain wild progenitors and weedy forms of the basic food plants. As a result of these two trends, there is urgent need to collect and conserve the diverse genetic materials that remain. In a world where per capita resources are decreasing as the human population grows, the concept of a sustainable future is becoming increasingly more important. Biological diversity in the form of crop plant genetic resources is one of the components of any sustainable future that includes humans.

Genetic vulnerability

Genetic vulnerability is a risk for any organism that finds itself incapable of fighting off a pest/pathogen or environmental condition. What makes the condition so serious for agriculture is the large number of genetically identical individuals and therefore the greatly magnified risk fact. The loss of a wheat plant or even a wheat field would not affect the wheat market in the U.S. The failure of 10%–15% of the crop would profoundly effect the market as witness the 1970 corn blight. Genetic vulnerability is the risk of high imput agriculture with common commercial seed sources typical of developed nations. Genetic erosion, the gradual persistent loss of plant genetic resources is most typically, but not exclusively, a phenomena of land races in developing nations. Genetic vulnerability in developed nations is the thin ice of a narrow genetic base which is fine as long as there are no breaks or holes in the ice. Never before have there been such widespread monocultures (dense uniform stands of billions of plants) covering thousands of acres all genetically similar. The narrowness of the genetic base means that the superior traits will be expressed by all the plants but this also carries a magnified risk of crop failure, as occurred in the wheat stem rust of 1954 or the southern corn blight of 1970 in the U.S. The Irish potato famine in the 1840's is a classic example of genetic vulnerability.

Genetic wipeout

The third threat to crop plant germplasm is the rapid and wholesale destruction of genetic resources. Social disruptions such as political instability or crop failure and famine can eliminate genetic resources. Quite literally, the genetic heritage of a millennium in a particular place can disappear in a simple bowl of porridge if the seeds are cooked and eaten instead of saved as seed stock. Equally dramatic is the discarding of a genetic collection because a curator retires or the collection is no longer of use to the institution. A classic case of the above is the USDA melon breeding program. The crop was threatened by a mildew problem and plant explorers assembled a world collection of melons. The genes for resistance were located and bred into the crop and seed of susceptible melon types discarded. No sooner had the collection been thrown away than attacks of a virus threatened the crop and plant explorers went out for a second collection. At the present time there is no institutional arrangements by which the perpetuation of genetic collections can be coordinated. The US has no policy, no clearing house for privately and/or publicly held research and working collections of genetic stocks. The even bigger genetic wipeout will start taking place in the next decade if about two million genebank accessions are not successfully regenerated with rejuvenated seed. About a quarter million of these accessions are in the U.S. We know the seed needs to be rejuvenated, because cold storage held seed, although it ages more slowly than at room temperature, it does age to the point where it doesn't possess sufficient energy reservoirs to germinate. We do know of and recognize this problem but still we have no policy or masterplan to handle this major undertaking of a genebank grow out – our half hearted commitment to genetic resource maintenance may result in one of the greatest genetic wipeouts of historic times. Another related problem is the loss of collections when breeders and University faculty retire and collections are discarded because their worth is not realized. There needs to be a central clearing house that Deans and other administrators can turn to for guidance on what to do with these genetic resources. Ideally there should be a budget fund in the USDA to prepare these collections with appropriate documentation for transfer to the National Germplasm System. This fund could cover up to a half year's salary post retirement to have these collections made available and useful to others by the retiree.

These three processes are negative, in the sense that they tend to decrease biological diversity.

The fourth process in biological diversity of cultivated plants, HYBRIDIZATION AND GENETIC RECOMBINATION, is positive, and results in an increased diversity. Because of allelic interaction this diversity is often greater than the sum of the two parents. An often overlooked fact is that we probably have more crop plant genetic diversity now, because of plant breeders than at any time since agriculture began. Plant breeders do a lot more than just stir the genetic pot through hybridization. There is both the skill of selecting parents and rouging out the unwanted progeny, but more importantly the art of the 'mind's-eye', of seeing potential in unlikely parents and progeny.

It is often said, agriculture is dependent on the *SUN* (renewable energy), the *SOIL* (the nutrients) associated with water and air, the *SEED* (plant genetic resources), and the knowledge of how to put them all together in a planted field. The agricultural establishment makes frequent mention of the sun, soil and seed as the basis of our agricultural productivity but *SKILL* and knowledge of how to manage the seed in the sun and soil is seldom credited. To adjust to chance and anticipate the future is one of our crowning achievements as humans and we should be doing more of it in agriculture.

The technological bind of improved varieties is that they eliminate the resource upon which they are based. Over the past 10,000 years crop plants have proliferated an innumerable number of locally adapted genotypes and an equal number of products of hybridization with wild relatives and more primitive cultivars, that have fueled the selection process. These land races, folk varieties and weedy forms/wild relatives of indigenous and peasant agriculture have been the genetic reservoir for plant breeders in crop improvement. Suddenly this genetic diversity is being replaced with a relatively small number of varieties bred for high yields and other adaptations necessary for high input agriculture. A prime example are the dwarf stature rice varieties from IRRI and wheat from CIMMYT.

In addition, the scarcity of land is forcing changes in land use and agricultural practices resulting in the habitat disappearance of wild progenitors and weedy forms of our basic food plants. As a result of these two trends, an urgent need to collect and conserve the genetic diversity of crop plants developed in the 1960's & 70's.

Germplasm in genebanks

There currently are reasonably good genebanks for most of the major crops and a good number of the lesser crops (See Tables 1 and 2) [3, 4]. For some crops there is more diversity in the genebanks than in peasant agriculture and now the fear is that if these genebanks are not better managed, they will be the next wave of genetic erosion as the stored seed loses viability and must be rejuvenated in regeneration growouts.

Only about 5000 plant species have fed the human population since the origin of agriculture. (Less than 1% of the vascular plant flora of the world.) As the human population has grown, we have depended increasingly upon a shorter list containing the most productive plants. Today only about 150 plants species with about 250,000 local races and/or named cultivars are important globally in meeting the caloric and dietary (vitamin and mineral) needs of humans.

The process of plant breeding is a dynamic one of genetic selection in response to changing diseases, parasites, agricultural techniques and human use. The earliest domesticated crops were probably not much more productive than their wild progenitors but the act of cultivation was a radical break with the past. As plants were moved from their old habitats to new sites, yields often were greater because they were isolated from their co-adapted pest and diseases. Sometimes there were new gene combinations that were favored by new soil types, better water supply, different daylengths and weather patterns. Genetically what was significant was the change in gene frequencies that took place in small isolated interbreeding populations. In-breeding increased the fixation of recessive alleles and the act of harvesting and sowing meant only a part of the population was represented in the subsequent generation. The next step was probably the most significant for the evolution under domestication and that was the bringing together of different seed from small isolated and quite possibly habitat distinct locations to a common field where they hybridized and farmer selection was made from the subsequent introgressive hybridizatoin and/or explosive evolution of polyploidy. For me this explosive evolution created the highly productive plant within a plant that we see in the tetraploid and hexploid wheats and in heterosis of hybrid corn. For me the origin of a crop wasn't

Table 1. International Board Plant Genetic Resources (IBPGR) network of designated base collections of seed crops.[a]

Crop	Species covered	Scope of collection	Institute
Cereals			
Barley		Global	PGR, Ottawa, Canada
		European	NGB, Lund, Sweden
		African	PGRC/E, Addis Ababa, Ethiopia
		Asian	NIAR, Tsukuba, Japan
Maize		New World	NPGS[b], USA
		Asian	NIAR, Tsukuba, Japan
		Asian	TISTR, Bangkok, Thailand
		European	VIR, Leningrad, USSR
		South European	Portuguese Genebank, Braga
Millets	Pennisetum	Global	NPGS, USA
	Pennisetum	Global	PGR, Ottawa, Canada
	Pennisetum	Global	ICRISAT
	Eleusine	Global	PGRC/E, Addis Ababa, Ethiopia
	Eleusine	Global	ICRISAT
	Minor Indian millets	Regional	NBPGR, New Delhi, India
	Eragrostis	Global	PGRC/E, Addis Ababa, Ethiopia
	Panicum miliaceaum	Global	ICRISAT
	Setaria italica	Global	ICRISAT
Oats		Global	PGR, Ottawa, Canada NGB, Lund, Sweden
Rice	Oryza sativa-indica	Global	IRRI
	javanica	Global	IRRI
	japonica	Global	NIAR, Tsukuba, Japan
		African	ITA, Ibadan, Nigeria
		Mediterranean, temperate and intermediate forms from the USA	NPGS, USA
	Wild species		IRRI
Rye	Global	Polish Genebank,	
	Global	Radizkow	

Table 1. (Continued).

Crop	Species covered	Scope of collection	Institute
Sorghum		Global	NPGS, USA
		Global	ICRISAT
Wheat	Cultivated species		VIR, Leningrad, USSR
		Global	CNR, Bari, Italy
		Global	NPGS, USA
	Wild species (Triticum and Aegilops)	Global	NPGS, USA Plant Germplasm Institute, University of Kyoto, Japan
Food legumes			
Chickpea		Global	ICRISAT
Faba bean		Global	CNR, Bari, Italy
Groundnut		Global	ICRISAT
		South American	INTA, Pergamino, Argentina
	Wild perennial species		CENARGEN, Brazil
Lupin		Global	ZIGuK, Gatersleben. German Democratic Republic
		European	INIA, Madrid, Spain
Pea		Global	NGB, Lund, Sweden
		Mediterranean	CNR, Bari, Italy
		Central and East European	Polish Genebank, Radzikow
Phaselous	Wild species	Global	Faculté des Sciences Agronomiques de l'Etat, Gembloux, Beligum
	Cultivated species	Global	CIAT
	Cultivated species	Global	NPGS, USA
		European	FAL, Braunschweig, Federal Republic of Germany
Pigeonpea		Global	ICRISAT

Table 1. (Continued).

Crop	Species covered	Scope of collection	Institute
Soybean		Global	NIAR, Tsukuba, Japan
		Global	NPGS, USA
	Wild perennial	Global	CSIRO, Canberra, Australia
Vigna	Wild species	Global	Faculté des Sciences Agronomiques de l'Etat, Gembloux, Belgium
	V. radiata	Global	IPB, Los Banos, Philippines, AVRDC, Taiwan, China
	V. unguiculata	Global	IITA
		Global	NPGS, USA
Winged bean		Global	IPB, Los Banos, Philippines
		Global	TISTR, Bangkok, Thailand
Root crops			
Cassava (seed)		Global	CIAT
Potato (seed)		Global	CIP
Sweet potato (seed)		Global	NPGS, USA
		Asian	AVRDC, Taiwan, China
		Global	NIAR, Tsukuba, Japan
Vegetables			
Allium		Global	NVRS, Wellesbourne, UK
		Global	NPGS, USA
		South and East European	RCA, Tapiozele, Hungary
		Asian	NIAR, Tsukuba, Japan
Amaranthus		Global	NPGS, USA
		Asian	NBPGR, New Delhi India
Capsium		Global	CATIE, Turrialba, Costa Rica
		Global	IVT, Wageningen, Netherlands

Table 1. (Continued).

Crop	Species covered	Scope of collection	Institute
Cruciferae	Brassica carinata	Global	FAL, Braunschweig, Federal Republic of Germany
	B. carinata	Global	PGRC/E, Addis Ababa, Ethiopia
	B. oleracea	Global	NVRS, Wellesbourne, UK
	B. oleracea	Global	IVT, Wageningen, Netherlands
	Raphanus	Global	NVRS, Wellesbourne, UK
	Wild species	Global	Universidad Politecnia, Madrid,
		Global	Spain
		Global	Tohoku University, Sendai, Japan
	Oilseeds and green manures:		
	B. campestris, B. juncea	Global	PGR, Ottawa, Canada
	B. apus, Sinapis alba	Global	FAL, Braunschweig, Federal Republic of Germany
	Vegetables and fodders: B. campestris, B. juncea, B. napus	Global	NVRS, Wellsbourne, UK
	B. napus	Global	FAL, Braunschweig, Fedral Republic of Germany
	All Cruciferae crops	East Asian	NIAR, Tsukuba, Japan
Curcurbita	Benincasa, Luffa, Momordica, Trichosanthes	Global	IPB, Los Banos, Philippines
	Cucumis, Citrullus, Cucurbita	Global	NPGS, USA
	Cucumis, Citrullus	Global	INIA, Madrid Spain
Eggplant		Global	IVT, Wageningen, Netherlands
		Global	NPGS, USA
Okra		Global	NPGS, USA

Table 1. (Continued).

Crop	Species covered	Scope of collection	Institute
Tomato		Global	CATIE, Turrialba, Costa Rica
		Global	ZIGuK, Gatersleben, German Democratic Republic
		Global	NPGS, USA
		Asian	IPB, Los Banos, Philippines
Southeast Asian Vegetables		Regional	IPB, Los Banos, Philippines
Industrial crops			
Beet		Global	FAL, Braunschweig, Federal Republic of Germany
		Global	NGB, Lund, Sweden
		Mediterranean	Greek Gene Bank, Thessaloniki
Cotton		Mediterranean	Greek Gene Bank, Thessaloniki
Sugarcane (seed)		Global	NIAR, sukuba,
		Global	Japan
			NPGS, USA
Tobacco		Mediterranean	Greek Gene Bank, Thessaloniki
Forages			
Legumes:			
Centrosema		Global	CIAT
		Global	CENARGEN, Brazil
Desmodium		Global	CIAT
Stylosanthes		Global	CIAT
		Global	CSIRO, Brisbane,
Leucaena			Australia
Lotononis		Global	NPGS, USA
		Global	ILCA
Macroptilium		Global	Seed Bank, RBG,
Neonotonia			Kew, UK
		Global	CENARGEN, Brazil
Zornia		African	ILCA
			Seed Bank, RBG,
Trifolium			Kew, UK
		Global	NPGS, USA
			CIAT, Colombia
		African	ILCA
		Global	Seed Bank, RBG, Kew, UK

Table 1. (Continued).

Crop	Species covered	Scope of collection	Institute
Grasses:			
Cynodon		Global	NPGS, USA
Cenchrus		Global	Seed Bank, RBG,
			Kew, UK
			ILCA
			CSIRO, Brisbane,
Digitaria			Australia
		Global	ILCA
			CSIRO, Brisbane,
			Australia
Pennisetum			Seed Bank, RBG,
Paspalum			Kew, UK
		Global	NPGS, USA
		Global	NPGS, USA
		Global	
Others			
Tree species		(Fuel and environmental stabilization in arid areas)	Seed Bank, RBG, Kew, UK

[a] Other base collections are being designated, particularly in developing countries, but are not featured in the table due to protracted negotiations.
[b] The USA has a large coordinated National Plant Germplasm System (NPGS). Seeds for base storage are normally stored in NSSL, Fort Collins, Colorado.

the first day the seed from the wild was planted but the time frame in which the critical accumulated mass of small variants that had been established in genetic isolation gave rise to explosive evolution when hybridized. This process has continued right up to the present but is essentially stopped now that the crop genetic diversity is held in genebanks and not farmer fields.

The world was quick to rally round the concept of genebanks when the concern was genetic erosion but has been very slow to realize that effective preservation of germplasm is only half the job [18–21]. The importance and effective use of plant genetic diversity in a productive and sustainable agriculture is one of the least appreciated and poorly managed aspects of the U.S. food production system. Public sector research and development planners lack a management plan and the general public is unaware of our lack of direction, which translates into 'no wind in our sails of crop improvement' fifteen years hence.

We have become so enmeshed in our urban lifestyle that we tend to

Table 2. IBPGR network of field genebanks (active collections for vegetative material).

Crop	Geographical representation	Institute
Roots and tubers		
Cassava	Global	CIAT
	Latin American	CENARGEN, Bazil[a]
	Central American	INIA, Mexico
	African	IITA
Sweet potato	Asian and Pacific	AVRDC, Taiwan, China
	Global	IITA
Fruits		
Banana	Global	Banana Board, Jamaica
	Southeast Asian	PCARRD, Philippines
	African	Délégation Générale de la Recherche Scientific et Technique, Cameroon
Citrus	East Asian	Fruit Tree Research Station, Tsukuba, Japan
	Mediterranean	INIA, Valencia, Spain
	Mediterranean and African	IRFA, Corisca, France[a]
	North American	USDA[b], USA
	Latin American	CENARGEN, Brazil[a]
	South Asian	IIHR, India[a]
Industrial crops		
Cocoa	Global	University of the West Indies, Trinidad and Tobago
	Global	CATIE, Costa Rica
Sugarcane	Global	Sugarcane Breeding Institute Coimbatore, India[a]
	Global	USDA, Florida, USA
Perennial species		
Allium	Short-day species	Hebrew University, of Jerusalem, Israel[a]
	Long-day species	Research Institute for Vegetable Growing and Breeding, Olomouc, Czechoslovakia

Table 2. (Continued).

Crop	Geographical representation	Institute
A. cepa,		
A. ampeloprasum	Global	IVT, Wageningen, Netherlands
Groundnut		
	Wild perennials	CENARGEN, Brazil
Soyabean		IVT, Wageningen, Netherlands
	Wild perennials	CSIRO, Austalia

[a] Under discussion or awaiting formal agreement.
[b] Location being agreed.

over-look the primacy of plants as primary producers in our food supply. As a nation we spend less than 25% of our income for food. Contrast that with a figure of 75% to 90% of our labor for food through much of human history. Ten thousand years ago and pre-agriculture we were hunter gatherers living at a density of less than 1 person/sq. km. Today with a highly productive assured food production from planted crops we live at a density that exceeds 25 people/sq km and in the Boston to Washington, D.C. corridor our density is about 600 people/sq km. Our population increase has made us absolutely dependent on our plowed fields and the crops planted in them. We can ill afford to have them fail because we can't go back to hunting and gathering. We are absolutely dependent on crop plants and yet the management of risk from crop failure is poor at best.

Our restructuring the food supply has set in motion numerous interlocking forces, many not at the time consciously intended, that have directed the evolution of the crops and the societies that attend and are dependent upon them. As the human population increased, the growing of these crops expanded into many different environments and an enormous wealth of genetic variation was created and survived over the centuries in locally adapted populations. Only a small fraction of this variation has been sampled and included in the present leading crop varieties. A great deal of this variation is locked up for the major crops in the approximately two million genebank accessions (about half are estimated to be duplicates) world wide. Trevor Williams has presented evidence at this conference that two-thirds of these approximately 2 million bank accessions are in fact duplicates. Yet, to be blunt these 300,000 accessions are useless until we grow out and use this variation for crop improvement. Genes in the bank are like candles under a basket.

Preservation in perpetuity

The way crop plant genetic resources are preserved will influence both what is held in the collection and how it is evaluated. There are essentially three modes of increasing human management to conserve in perpetuity these resources:

Entire biomes

The preservation of entire vast tracts with *in situ* conservation of the animal and plant ecosystems. This level of preservation will be extremely important in slowing extinct rates especially in the tropical forests but will have little impact on genetic resources of useful plants with the exception of timber trees. Some wild relatives of food/fiber crops will be affected but most of these can be accounted in the next level management.

In situ

Preservation of agro-ecosystems of village farming where both fields and forest are skillfully managed, land races and wild relatives in ecographic pockets of genetic diversity, and of natural vegetation where near relatives and weedy forms exist are all prime candidates for *in situ* conservation. Mexico has already taken the necessary steps to set aside the Sierra de Manantlan for the perennial maize relative z. diplioperennis. The United States has not explored this mode either domestically with say cranberries or internationally with such programs as those of AID within the State Department. Probably the best case for *in situ* agro-ecosystems rest in the developing countries where tremendous potential exist to combine *in situ* preservation with tourism and demonstrating the best of local cultural heritage. A second type of *in situ* preservation are evolutionary gardens where crops and their wild relatives hybridize and the resulting variation is in dynamic evolution with evolving pest and pathogens. A third is land set aside for the wild relatives and weedy forms which are losing habitat thru more intense land use patterns. The interesting aspect of *in situ* preservation is that it does not freeze the seed in a genebank generation but is grown-out under constant genetic evaluation to both natural selection and artificial selection. For preservation *in situ* what is being preserved is a living population, if it is in place there are no questions of sampling strategy of the population; if it is big enough there is no problem with holding allelic frequencies of 1 in a 1000, and no question about selection shift at the time of seed regeneration. *In situ* has a lot of practical considerations with management, economics, communication and local public support but it does conserve genes at the ideal population level. This mode deserves much more consideration than it has enjoyed in past discussions.

Ex situ

Preservation as seed or *in vitro* cell lines/clones in gene banks under appropriate conditions for long-term storage is often seen as the easiest and most preferred. The problems are that the genebank is only as complete as the starting sample and seldom can alleles at a frequency of less than 1 in 100 be either originally collected or maintained in rejuvenated seed (a more true statement for outcrossers than inbreeders at regeneration). The more significant problem is that hidden variation in genebanks is useless. This mode for preservation draws genes out of circulation and therefore to be useful requires documentation and evaluation so that users will have enough information to know what to request. Information management will be as important as the physical arrangements of the genebank. Of the 2 million genebank accessions world wide, about half lack basic passport information and only 100,000 have this passport on machine driven management systems. Genebanks slow down crop plant evolution and so the hybridization and breeding process becomes a necessary part in making *ex situ* preservation useful. *Ex situ* preservation has three aspects:

1. Exploration, collection and banking,
2. Evaluation and documentation, and
3. Breeding for enhancement.

Exploration, collection and banking

The best sample in a genebank is no better than the starting collection from the field. Many of the accessions in genebanks are grabbag roadside stops and not carefully planned sampling of the geographic distribution with special attention to unique microhabitats or cultural traditions within the distribution range. Compounding the lack of adequate data for bank accessions is the fact that the samples are small and rare alleles probably were left in the field and are not present in the sample. There is a literature now on how to optimize the collecting process to catch the rare allele at the 0.01 frequency depending on the outcrossing-inbreeding nature of the taxa but for many areas of the world, 'grab bag' collected are the only accessions the banks have. The next most important bottle neck in losing genes is the first bank regeneration of seed. Again a literature is building but the guideline appears to be as many plants as possible (effectiveness decreases after 500) and either reciprocal pair crossing in an outcrosser or selfing in inbreeders, with equal seed from every plant in the bank accession storage sample generation packet. Later regenerations of seed can probably be at the 250 plant population level or smaller depending on the breeding system and initial collection heterozygosity [22].

Passport data is the first line of evaluation. Genebank managers would be well advised to not accept seed without basic passport information but alas they seldom have a say or choice in this matter. First, the passport is a management tool to make a guess on the best conditions under which to regenerate the seed and secondly, it identifies ecogeographically possible genetic attributes the accession might possess. This later function has always been clear to plantsmen who knew the crop variation world wide (Harry Harlan in Barley or Glen Anderson in Wheat) but this knowledge is fading from the plant breeding ranks. Hopefully genebanks will be able to ecogeographically characterize the major growing areas for the individual crops they hold. Certainly this characterization is very important for a bank to designate a core collection from its holdings for intensive genetic evaluation.

Banking conditions will depend on the crop and the ultimate goals of the collection. Some collections are maintained in *ex situ* clonal depositories with the plants grown in the ground (*in vivo*) or in meristem culture (*in vitro*). The really large collections are seed stored at very low seed moisture. These collections can be maintained above freezing for short-term storage (approximately 20 years) or below freezing (– 18 °C) for long-term storage (50 years to 500 years and beyond in liquid nitrogen). This method removes seed from use and evaluation and is fine for seed depositories but for genebanking there must be the ability to regenerate the seed and evaluate it for use in crop improvement. Using my definitions, crop germplasm is held worldwide in about 50–60 gene depositories and about 10 to 15 gene banks. The crop specific gene banks in the International Agricultural Research Centers (IARC's) are, as a group, probably in the lead as true gene banks. Some national program gene depositories containing seed collected with international funds are so poorly financed/managed that the seed is no longer viable and will never be regenerated. Major Goodman has called them gene morgues.

Evaluation and documentation

The passportdata of where collected, elevation and cultural conditions such as length of season and rainfed or irrigated are the first line in guessing the potential use of an accession. Very few banks have published catalogues of their holding because most honestly do not know what they have. Even fewer have machine driven data bases that can be queried such as the publically available system for the 29 character catalogue of the CIMMYT maize bank. This data base on CD/ROM is available to anyone with a PC and a compact disc reader and covers all the accessions (10,500). Communication between banks is a very weak link and in need of considerable improvement.

Basic passports contain much more hidden information for primitive cultivars/landraces of indigenous agriculture and wild and weedy near relatives of crop plants than for varieties and cultivars in current use or previously elite lines and obsolete cultivars. But for this last class the pedigree charts being developed by the CIMMYT wheat bank which show the origins of the accession in the past and contributions to later breeding progress works like a telescope into the past and a crystal ball into later uses of the accession. These pedigree charts are as powerful in evaluating germplasm as identifying the microhabitat of a landrace. This is the first line of evaluation and not to be confused with gene bank characterization, which is usually a basic list of seed characteristics that help the bank staff to verify that the seed in the can is in fact the accessions the books say it is i.e. white dent with white cap with a seed weight of 330 grams per 1000 seed. Characterization data is a management tool of the gene bank, while evaluation data points the way for breeding in the crop improvement process. Germplasm depositories characterize, whereas true germplasm banks evaluate accessions so as to make them more available to the larger community and breeding process.

Gene banks and in fact many breeders' collections are too big to intensively investigate and record their traits in the screening process and recently the core collection concept has become popular [23]. In the core a selected sample of appropriately 10% of the holdings (not to exceed approximately 500 entries) is carefully chosen to represent the agro-eco-geographic variation of the crop world wide. When a specific sample in the core shows promise in evaluation for a desired attribute/trait then the geographic cluster in the bank from which that core came is focused in upon for further evaluation. Many single valuable genes are often present in populations at very low levels (below 0.005) and remain undetected except for a single accession or region where the frequency may be very high (as high as 80–90%). An example is stem borer and fall army worm resistance from Antiqua (108-sq.-mile island in the Leeward group in the West Indies). Once discovered this attribute can be searched for in other materials in trials designed to uncover very rare alleles. The core idea is as good as the completeness of the core in capturing the crop variation worldwide but for a starting point in evaluation it looks promising. The recent work of John Peeters indicates the world collection core size for a crop such as barley could be in the 300 accession range [24]. My personal guess is that the Latin American maize core is also in the 300–500 range.

The first priority in the documentation of bank accessions should be a published catalogue, then cyclic updates of core evaluations. Data exchange between banks is extremely poor and many banks are reluctant to go public with a catalogue for fear seed will be requested and they would not be able to find it, or worse, it would not be viable. To me this is positive proof that the number of true gene banks worldwide is very limited.

Breeding for enhancement

The 21st century is going to see many new industrial uses for plant products and many high technology substitutes for conventional plant products (lumber and fabrics for instance) but the human dietary requirements are going to remain unchanged and met by basically the same short list of crop plants that feed us now: wheat, corn, rice, soybeans, potatoes, beans, sugar cane and sugar beets, tomatoes, onions, carrots, cabbages, apples and oranges. And these are the crops that demand new germplasm for improvement and to which we commit our plant breeding.

The yield increase of most cultivated crops over the wild ancestor is approximately 100-fold, but this figure is based on an oversimplification. An examination of the changes involved in the jump from wild plants in the native vegetation to domesticated plants in cultivated fields will highlight the role of genetic evaluation in the plant breeding process. Under the most ideal conditions selected individual wild plants of many of our crop plants approach the productivity of their modern counterparts. What has increased one hundred fold is the yield per acre or all the plants on average and this has been achieved by selection of plants that grow well in crowded pure stands and are reasonably dependable in a wide variety of climates and soils year after year. The generally heard statement is 'bred for high yields,' while in fact [25] what has been bred is dependability, stability, and wide adaptability. Supporting of this statement is that after almost a century of genetic analysis with crop plants, we have yet to identify a 'yield gene.'

Some of the most significant advances in the past have been made by plant breeders who had a life time of experience with the crop and also had a first hand experience with genetic variation that existed in peasant agriculture and even the wild relatives of the crop. Most of these old-school generalists are now gone or nearing retirement – Charles Rick in tomatoes, Virgil Johnson in wheat, Raymond Baker and Bill Brown in maize. Most breeders stay with the familiar, proven germplasm where increments of improvement can be expected but not radical leaps forward. The rest of the crop's worldwide genepool is called exotic because it remains unknown and untried. There are some real blocks to the use of exotic germplasm – it is usually poorly adapted to sites where the breeding program is being carrried out and requires a great deal of selection, meaning many breeding cycles before it is ready for farmers fields. Sometimes the problems with daylength control or other growth requirements are just too great for an average plant breeder to challenge when considering all the other demands of the program. For example selection of germplasm from peasant fields in the mountains along the Tropic of Cancer (23 1/2°N) can raise considerable obstacles to its incorporation in a temperature zone breeding program. The analogy to the breaking of a wild horse is not inappropriate.

The first step in using exotic germplasm is illustrated by the daylength problem of a short day plant not flowering in our long days of summer. Soybeans were in the United States a hundred years before they became a successful crop for exactly this reason. To break these genetically controlled traits of daylength, seed dormancy, lodging under even marginally enriched soils, or extreme susceptibility to diseases long ago eliminated in the U.S. crop, there needs to be pre-breeding to produce usable material which is day neutral, germinates readily or has better standability and disease resistance. We have very few plant breeders who have broad experience with pre-breeding, and yet this is one of the most pressing needs. The power of deep stirring through hybridization of differing parents is tremendous. In plant breeding the skill to make varietal hybrids is not where broad experience is needed but rather in choosing which varieties from the thousands maybe tens of thousands around the world to use as the initial parents.

The second step after the germplasm has been 'broken-in' (pre- breeding finished) is to evaluate the material for useful or potentially useful traits. Everyone now is saying that we have all the germplasm for the major crops in gene banks, but we have to evaluate it and know what is there. A more truthful statement could not be made, but no botanist/plant person to my knowledge is paid exclusively to evaluate germplasm for a gene bank collection. Rewards in plant breeding are for released varieties, so who can afford to set up computer print out for someone else to get the plum. True there are a few genebank curators like those for soybeans at the University of Illinois, for potatoes at Wisconsin or in the IARC's the CIMMYT Maize Gene Bank and the IRRI Rice Gene Bank, that have regenerated, characterized [identified traits that help with the identification and verification (seed color, weight and endosperm type as examples) that the accession number does in fact belong to this packet of seed] and undertaken preliminary evaluation [plant height, length of growing cycle, aspects of root system growth, yield] but the testing for combining ability of gene bank accessions is beyond most curators. The story of what happens upon hybridization and deep stirring of the genetic pot is where we want to be and we are making little national or international effort to get there.

This is an instant on/off culture and there are those that think that germplasm in gene banks could be used like a parts catalog. All that has to happen is evaluate all the traits for the accessions and enter this mass of information in a computer. Then all you do is 'dial a trait' – plug into the database to find the desired trait – then plug it in the plant. This plug-in idea is rather popular with the media accounts for the promise of biotechnology also. The real truth is that genetic factors that make for local adaptation and yield, the genetic margin of difference for the farmer of one variety or hybrid over another, is practically never due to a single gene. About the only 'magic bullet' single gene traits have been those for disease resistance and most of these become obsolete before a decade has passed.

Need for more broadly trained plants people

I have yet to find a replacement for the insights of a plants person who knows the crop well in its distribution around the world. When the number of working plant breeders decreases, so will the number of insights. Unfortunately, the general population and demand for food is not decreasing and most plant breeding faculty positions in land grant universities are being replaced with more narrowly defined molecular/biotechnology candidates. Given the fifteen year lag time between the initial hybridization and farmers' fields, our present situation is serious and growing worse.

Another good reason to be concerned about the decreasing number of germplasm evaluating plants people is the training function. One of the services at which the American land-grant universities has excelled, is the training of plant breeders which started soon after the rediscovery of Mendels papers in 1900. Most of our current crop varieties trace back to the product of breeders trained in the period 1920-1960. Not only does this apply to the United States but to a remarkable degree to the developing world. The Green Revolution wheats in India have been improved by Punjab Agricultural University plant breeders who are graduates of UC/Davis. In other parts of the world it has been Cornell, Purdue, and the Universities of Minnesota, Wisconsin, Nebraska, Illinois and from North Carolina State, Iowa State and Kansas State Universities. These wells of excellence are drying up in the academic shift to biotechnology. This administrative shift is often the best use of limited funds for both the academic program and the interest of the state, but not in the best interest of the nation or world community.

My view isn't that good plants people/breeders are made by more courses in Plant Breeding I and II. Training in classical systematics, development morphology and the ability to see variation in the crop is what is needed to fill the ranks of evaluation geneticist. The trained eye can spot an early onset 'major gene' that controls the growth pattern of an accession, the untrained eye will only see that there is genetic segregation in the flower color. Now that a major shift has occurred to move crop variation from peasant agriculture to gene banks, we need well trained eyes to see the same potential for change that indigenous folk cultivators, who developed the crop up to this point, have seen and exploited in the past. To achieve this goal a college degree is not the issue, but an ability to see and mentally process variation is absolutely essential. I will use the term 'plant breeder' in the rest of the paper but what I really mean is 'student of crop plant variation' who is willing to work hard to make it available to the world's farmers.

The work of U.S. trained plant breeders around the world is one of the major reasons that the predicted famine of the 1970's never happened. These plant breeders are not being replaced by a second wave as well trained. Funds for plant breeding training at the university level are hard to come by and developing nations can't afford the cost of U.S. tuitions. Even an advanced country, like Mexico, with a very well developed agricultural research matrix, can not afford to send nationals for three years of Ph.D. training. Mexico has had to go to a program of a six month post-doc for their Mexican trained key personnel. Ninety- five percent of the developing world can not pay for U.S. university tuition cost. That they are being dealt out is tragic enough, but this relationship cuts two ways. Many of our leading plant breeders have former students around the world who have kept in touch and the two way flow of germplasm has been significant. Even when the political situation has been cold, personal contact and mutual respect has resulted in a free and open exchange of seed materials. This infra-structure of goodwill has been one of the most powerful forces in plant breeding in the 60's and 70's. Plant breeding is a human activity and without communication and exchange of seed the process comes to a dead stop. That such a small cadre of plant breeders have been so effective in doubling the world's food supply is a tribute to genetics and the land grant universities.

Before I leave the topic of plant breeding there is a serious problem that needs mention. Most of our state and USDA plant breeders and major germplasm collections are in major market commodities. This is because the system is user driven. A good idea, but the problem is that a large number of minor crops do not count in a user driven system. Many of these are kitchen vegetables which are extremely important for nutrition and/or the cooking of ethnic dishes, but the total dollar value isn't sufficient to merit a plant breeder. In many cases these are exactly the crops that should have a breeder in the public sector. In theory the state experimental stations are expected to pick up these minor/regional crops. When the crop for the nation comes from only one or a few states (cranberries/Massachusetts for existence) the state does pick up the program, but when the crop is supplied by many states (Asian Brassicas for example) it becomes an orphan. Also many of the USDA breeders work and breed varieties for the zones of highest production but seldom for marginal habitats at the edge of the growing zone. Contrast this with the first USDA plant introduction of a hundred years ago searching for cold hardy plants. The first USDA plant introduction was a cabbage from Russia for North Dakota.

By now, I think it is clear that the U.S. plant breeding establishment needs to be rejuvenated. The training of young broadly based plants people not only for our own country but for the world community needs to go hand in hand with established long standing crop improvement programs. More than just the market/industrial value of the crop needs to be factored into the crop priority system. My suggestion is for nutritional importance and social value of a crop to be a weighted factor. In addition we need to expand our new crops and plant exploration section to better serve the changing American diet which now is incorporating Asian vegetables, new cooking oils, and new crops from Latin America into our kitchens.

New plants people in pre-breeding and genetic evaluation

There needs to be two new kinds of plant breeders. A prebreeding/plant-evolutionist/plant collector who can work with wide ranging variation patterns and break or adapt exotic germplasm so it grows reasonably well under U.S. conditions (Walt Galinat's day neutral teosinte is an example – teosinte is the closest wild relative of maize and is a short day plant which doesn't normally flower in the U.S. until the time of the killing frost in September/October). Once this exotic germplasm has been reasonably tamed it can be made available to all users both public and private. This is also cost effective, because once the job has been done it will be good for as long as seed of the stock is saved. This system also decreases duplication of effort when several breeding programs might have to pre-breed the same exotic material if it wasn't available from a lead source.

The other new type of plant breeder would be an evaluation breeder who doesn't just try to assemble a genetic-trait parts catalog for the genebank accessions, but also works with geographic patterns for the crop, and looks for adaptive complexes in the genetic diversity. Certainly one of the aspects about which we could use a lot more data is the combining ability of differing landraces. In maize for instance, most of the heterotic vigor that has been exploited is that of New England Flints and Southern Dents (related to Tuxpeno) and yet there are known potentials from other landraces. For maize there would be the heterotic interaction of Cuban Flints, Harinoso de Ocho of western Mexico, Celaya of central Mexico, and Comiteco of southern Mexico. Unfortunately these new sources of variation take too long to develop for plant breeders whose productivity is measured by released varieties. Evaluation breeders would aim primarily, if not exclusively, for intermediate products and not finished varieties. This is a team approach to the art of plant breeding, but it answers many of the demands of the times.

My proposal to define the roles of pre-breeding and evaluation breeding in the conventional USDA system would expand the number of person years devoted to working with crop plant variation by 50%. There are approximately 500 scientific person years in the USDA and USDA/state cooperative breeding programs (Table 3) and this number would expand to approximately 750. Since U.S. agriculture feeds an excess of 750 million (our population plus exports), the effectiveness ratio for public sector breeding is 10^6:1, an efficiency ratio for research and development few institutions or industries could match. Taking into consideration that U.S. plant breeding training affects the seed feeding about one and half billion people; then my level of 750 plant breeders has an effectiveness ration of $2x10^6$:1.

Gene banks around the world now store 85 to 95% of the genetic variation found in landrace agriculture for the major crops [1], but it remains for plant breeders to define what is useful in these collections. Knowing what is in the world collection has been the major stumbling block to their use. In part this is because most of the major genebanks are depositories with no or very limited ability to grow out their holdings (as witness the National Seed Storage Laboratory). If their function is storage the material will never be evaluated and its potential will not be discovered. To evaluate seed means to look at the plant in the field and not count the number of seed and their viability which is what most depositories do now. Because of this flaw, most germplasm worldwide is in storage and not seen by the plant breeding process.

I would like to see germplasm actively being worked and to accomplish this aim the 80 major genebanks around the world need to add a minimum of a thousand to two thousand evaluation plant breeders. Half of these banks are in developing nations and no single factor could increase food self-sufficiency more than an increase in regional plant breeding staff in the tropics and an infrastructure to support their proven elite seed. The IARC's already have well developed training courses but none in plant germplasm or genetic evaluation. Lead institutions such as CIMMYT and IRRI would be logical starting points to strengthen the international gene banking and the evaluation plant breeder matrix. Only a few developing nations have an adequate cadre of plant breeders and interestingly these nations are food sufficient. The impact of a gene depends on it being discovered and on it being valued. The plant breeder is involved in the discovery and the farmer/consumer establishes the value. If we don't expand the effort in evaluation plant breeding then the tremendous potential of seed now in storage will go unrealized. Plant breeders clearly are going to have to establish the efficacy of genebanks. To do this there has to be more plant breeders and they must be associated with major genebanks.

Table 3. Scientific man years (SMY) asssigned to plant breeding research (1969–1970, 1979–1980), and most recently (1986–1987).

Commodity agronomic	1969–1970 Total (SMY)	1979–1980 Total (SMY)	1986–1987 Total (SMY)	Commodity horticultural	1969–1970 Total (SMY)	1979–1980 Total (SMY)	1986–1987 Total (SMY)
Corn	65.1	68.0	54.5	Potato	20.0	15.6	17.2
Grain sorghum	16.4	17.5	15.8	Carrot	1.5	2.6	2.8
Rice	7.5	10.5	10.8	Tomato	20.8	13.1	14.4
Wheat	51.2	58.5	62.8	Bean-pea	20.1	19.6	23.5
Barley	20.0	13.9	13.8	Sweet Corn	3.3	4.8	3.7
Oat	15.2	13.7	10.2	Cucurbit	10.5	16.4	10.8
Small grains	23.3[a]	11.1[a]	7.3[a]	Sweet potato	3.7	3.9	4.1
Soybean	35.9	42.4	45.0	Crucifer	1.2	2.8	1.6
Cotton	45.8	42.0	34.8	Onion	1.0	1.4	2.5
Tobacco	32.3	16.5	14.1	Vegetable crops	35.1[b]	26.6	24.5
Alfalfa (and other legumes)	24.3	42.7	31.6				
Grasses (and other forages)	33.0	35.9	29.8				
Total	370.0	372.7	330.5	Total	117.4	105.8	87.9

[a] Scientific man years assigned to small grains without specification crop. This value is in addition to the assigned values for wheat, barley and oat.
[b] Includes lettuce and other crops itemized separately; in addition, this value would include commitment to the listed vegetables without specifically designation of the programs by crop.
From analysis of 1969 CRIS reports by H.J. Hodson, Coop. State Res. Serv., USDA; analysis of 1979 CRIS report by W.C. Schaefer, USDA-ARS, and analysis of 1986 CRIS report by John Myers USDA-ARS.

Conclusion

Most of the elements in the current world network – the genebanks, the uniform common language of descriptors, the free exchange of germplasm – have been molded by the plant breeding tradition and are on a voluntary basis. Will this good will to share our genetic heritage continue or will, as resources become scarce, the heritage be laid claim to by private or nationalistic groups. Clearly, the future of germplasm, plant breeding and therefore, agriculture and world's food supply, rest on the plant genetic resources of our most productive crop plants. The responsibilities goes two ways; both those that hold the genetic diversity and the breeders that enhance the genetic architecture have an obligation to insure that these resources are used to improve the human condition for all people and that germplasm remains freely available to all.

Investment in plant breeding, which results in better seed, means greater yields with no more investment in land, fertilizers or labor. There is nothing biologically unsound about breeding for high yields. It was Thomas Jefferson who said 'The greatest service a citizen can do for his country is to add a new crop for his Countrymen.' I think we can modernize this to read: 'The greatest service plant breeders can do for their fellows is to improve the seed upon which we all depend.' The world's farmers demand the best seed, but are we doing the best job of getting it to them?

References

1. Lyman, J.M., 1984. Progress and Planning for Germplasm Conservation of Major Food Crops. Plant Genetic Resources Newsletter 60: 3–21.
2. Wilkes, G. 1983. Current status of crop plant germplasm. CRC Critical Reviews in Plant Science Vol. 1: 133–181.
3. Wilkes, G. 1987. Plant genetic resources: Why privatize a public good? BioScience 37: 215–217.
4. IBPGR Annual Report. 1986. International Board for Plant Genetic Resources, Rome. 91 pp.
5. Hawks, J.G. 1983. The Diversity of Crop Plants. Harvard Univ. Press, Cambridge, Massachusetts 184 pp.
6. Plucknett, D.L., N.J.H. Smith, J.T. Williams and N.M. Anishetty. 1987. Gene Banks and the World's Food. Princeton University Press, Princeton, New Jersey 247 pp.
7. IUCN. 1980. World Conservation Strategy. International Union for the conservation and Natural Resources. Gland, Switzerland. 43 pp.
8. NAS. 1978. Conservation of Germplasm Resources: An Imperative. National Academy of Science. Washington D.C. 118 pp.
9. FAO. 1981. Agriculture Toward 2000. Food and Agriculture Organization. Rome 134 pp., and statistical annex XXV.
10. Wilson, E.O. (ed). 1988. Bio Diversity. National Academy Press. Washington D.C. 521 pp.

11. Frankel, O.H. and Soule, M. 1981. Conservation and Evolution. Cambridge University Press, Cambridge. 327 pp.
12. Frankel, O.H. 1967. Guarding the plant breeders' treasury. New Scientist 35: 538–540.
13. Frankel, O.H. 1974. Genetic conservation; our evolutionary responsibility. Genetics 78: 53–65.
14. Harlan, J.R. 1971. Agricultural origins: centers and non-centers. Science 174: 468–474.
15. Harlan, J.R. 1975. Our vanishing genetic resources. Science 188: 618–622.
16. Wilkes, G. 1977. The world's crop plant germplasm – an endangered resource. Bulletin of Atom. Sci. 33: 8–16.
17. Wilkes, G. and Wilkes, S.K. 1972. The green revolution. Environment 14: 32–39.
18. Whyte, R.D. and Julen, G. 1963. Proceedings of a technical meeting on plant exploration and introduction. FAO. Genetica Agraria 17: 1–573.
19. Frankel, O.H. and Bennet, E. (eds). 1970. Genetic Resources in Plants: their exploration and conservation. Blackwell, Oxford and Edinborough. IBP Handbook No. 11. 538 pp.
20. Frankel, O.H. and Hawkes, J.G. (eds). 1975. Genetic Resources: For Today and Tomorrow. Cambridge University Press, Cambridge. 492 pp.
21. Holden, J.H.W. and J.T. Williams (eds). 1984. Crop Genetic Resources: Conservation and Evaluation. Geo Allen, London. 296 pp.
22. Crossa, J. 1988. Dterminnation of sample size for genetic conservation of out breeding crops. CIMMYT Working Paper. El Baton, Mexico. 48 pp.
23. Brown, A.H.D. 1989. 'The case for core collectors' in The Use of Plant Genetic Resources. AHD Brown, O.H. Frankel, D.R. Marshall and J.T. Williams (eds). Cambridge University Press. 382 pp.
24. Peeters, J.P. 1988. The emergence of new centers of diversity: Evidence from barley. Theoretical and Applied Genetics (in press).
25. Duvick D.N. 1986. Plant breeding past achievements and expectation for the future. Economic Botany 40: 289–297.

Culture and care: ethical and policy dimensions of germplasm conservation

LAWRENCE BUSCH[1], WILLIAM B. LACY[2], and JEFFREY BURKHARDT[3]

[1] *Professor of Sociology, Committee for Agricultural Research Policy, 223 Scovell Hall, University of Kentucky, Lexington, Kentucky 40546-0064, U.S.A.*

[2] *Professor of Sociology, Committee for Agricultural Research Policy, 223 Scovell Hall, University of Kentucky, Lexington, Kentucky 40546-0064, U.S.A.*

[3] *Associate Professor and Director, Ethics and Policy studies Program, 1001 McCarty Hall, IFAS, University of Florida, Gainesville, Florida, 32611, U.S.A.*

Abstract The very moment when the first human being began to cultivate the soil or tend animals marks the beginning of a great responsibility that we collectively as a species took upon ourselves. Until recently, however, the maintenance of biotic diversity required little active care. The simple activities of farmers and herders have sufficed to insure that nature and all its diversity would remain safe from harm.

It was not until the last several decades that the full import of this restructuring of nature began to pose serious problems. The vast improvements in science, communications and transport that have recently occurred have made possible the restructuring of nature. In the near future the advances of molecular biology threaten to dwarf even these achievements by remolding the biosphere to our needs. Ironically, these very improvements are themselves the source of our contemporary problems, for by them we have begun to create a second nature in which the maintenance of biotic diversity becomes a part of our responsibility for the world. Moreover, this second nature is no longer sacred.

The dilemma that we face today is that the tools we have employed to accomplish these feats have all but blinded us to the potential problems that may arise as a result of their use. Similarly, while it is easy to create public interest over a declining whale population, it is difficult to arouse comparable concern over the loss of an unusual seed variety. Programs to conserve biotic diversity must recognize this as a central issue in their success. Only by working to raise the problem of species conservation to the level of the sacred can success be achieved.

Key words: ethics, policy, culture, care, germplasm, conservation

> 'The technological bind of improved varieties is that they eliminate the resource upon which they are based.' – Wilkes [1:134]

Introduction: Agriculture and culture

Though cultivated fields appear to us as manifestations of the great outdoors, of nature's bounty, they are in fact manifestations of the diligence

L. Knutson and A.K. Stoner (eds.). Biotic diversity and germplasm preservation, global imperatives.
 ISBN 0-7923-0178-1

and hard work of human beings who have transformed nature to serve our ends. The cultivated field is a place where nature is cultured. In very real terms, agriculture is a form of culture and an expression of culture.

The intimate connection between agriculture and culture is poignantly illustrated in the various rites and rituals associated with food–harvest festivals, great feasts, and celebrations. Such rituals affirm our reliance on agriculture, as well as celebrate our achievements in culture. However, in Western nations, and especially in the U.S., we have tended to conceal this relation between agriculture and culture over the last century. We have reduced our farm population to less than 2% of the total. We have separated the production of food from its processing and preservation. To the general public, food appears simply as products that are presented in endless quantities in neatly stocked rows on our supermarket shelves. Food also appears as the endless supply that streams from our fast-food restaurants. Even the necessity for preparation of food products at home has been reduced. We eat more meals away from home and spend less time preparing foods in the home. At the same time, the celebration of life associated with food and agriculture has all but vanished. No longer do many restaurants restore; no longer do we pare or prepare in the home.

Changes in how society views nature and agriculture, we shall argue, have a profound and continuing effect upon our concern for the conservation and preservation of genetic resources. In demystifying nature, in subjecting it to the regularized routines of industrial production, we have both gained and lost. In particular, what we have gained in terms of efficiency and convenience, we may have lost in terms of culture.

Notably, we may have lost a sense of care – for ourselves, our offspring, and the 'garden' into which we have been placed. In fact, in this paper we begin at the beginning, with the problem of the Garden of Eden, for it is there that our ancestors posed the problem that we have inherited. Then we examine the contrast between sacred and secular: the secularization of food and agriculture becomes the philosophical problem we ultimately must address. We will argue that three factors are responsible for the secularization of food and agriculture: science, technology, and capitalism. Understanding this can give us pause for reflecting on our moral and political imperatives with respect to germplasm conservation.

The Garden of Eden

The very moment when the first human being began to cultivate the soil or tend animals marks the beginning of a great responsibility that we collectively and as a species took upon ourselves. From that day on, the perpetuation of the bounty of nature became not something that happened by itself but something that required our constant care and vigilance. Gradually, and almost unconsciously, we transformed nature such that many plants and animals became literally domesticated, part of our household. They came to be wholly dependent upon our intervention to insure their existence, not just as individuals but as species. Until recently, however, the maintenance of biotic diversity required little active care. Though virtually all societies have held nature in reverence, the simple activities of farmers and herders have usually sufficed to insure that nature and all its diversity would remain safe from harm.

In Jewish, Christian, and Islamic traditions, the story of the Garden of Eden is well-known. One way to understand the story of the Garden is as the mythologized recounting of the origins of sedentary agriculture. The Garden represents nature, a benevolent nature that is provided by God and that is sufficient to fulfill human needs without any need for tilling the soil or caring for animals. When Adam and Eve ate from the tree of knowledge, they lost the innocence of the Garden and learned something of what heretofore had been known only to God. As a result they were banished and made to till the soil, to eek a living from it, as an eternal punishment for their curiosity.[1]

The story of the Garden of Eden is a story of the birth of agriculture told from the perspective of the hunters and gatherers and/or pastoralists.[2] Agriculture is depicted as the defilement of the sacred earth, the result of a quest for power over nature, of the failure to respect nature as something sacred, even divine [2]. Today, we can neither return to hunting and gathering nor reenter the garden. Nevertheless, the image of the Garden of Eden can provide considerable insight in rethinking about what nature–and our environment–ought to be.

Importantly, the story of the Garden illustrates the concern of our ancestors for diversity. The Garden is not one in which only a few plants and animals are found but one in which all the plants and animals are to be found, protected, and cherished. The same message is displayed even more clearly in the story of Noah. Noah does not select which animals and plants to take with him on the ark. He takes all species and always in pairs, for the importance of pairs to reproduction was already well understood.

[1] Graves and Patai [26:79–80] noted some years ago that 'The *Genesis* story, in which agricultural work is represented as a curse laid upon man because of Eve's inquisitiveness and disobedient mischief, mythically expresses the age-old Mediterranean point of view which regards physical labor (symbolized by tillage of the soil) as an unmitigated and unavoidable hardship'.

[2] As Schneidau [27] notes, this issue is taken up again in the story of Cain and Abel.

In addition, the dispute between the hunters and pastoralists and agriculturalists represents a turning point in human history and the beginning of the secularization of agriculture and food. Nor is the dispute entirely over. Surely, the remaining hunters and gatherers and even pastoralists are few, weak, and despised by nation-states as they tend to ignore secular authority. Yet, it may be that these few remaining hunters and gatherers and pastoralists still have an important message for us to which now we must finally listen.

The Sacred and the Secular

The 'victory' of the agriculturalists over the hunter/gatherers in contemporary Western societies is most apparent in the deconsecration of food. Early agriculture, and agriculture even through the 17th and 18th centuries, retained a notion of 'nature's bounty' as evidenced by the American Thanksgiving celebration and similar ones throughout the world. Today we have increasingly lost that understanding. Though our food is perhaps more beautiful and more attractive than that of previous generations, we have achieved this only at the expense of making food into something mundane, ordinary and most importantly so abundant as to be simply expected. Such abundance was not always the norm. The great feasts depicted in the art and literature of past epochs, and in which people often over-indulged themselves, were celebrated in the hope of abundance; they were contrasted with the scarcity that marked everyday life. They were the dream of plenty realized only momentarily.

How did our food and agriculture become secularized? What factors account for it? At least three major factors appear to have contributed, however unwittingly, to the secularization of food and agriculture: science, technology,[3] and capitalism. Let us briefly examine each in turn.

Science

The founders of modern science were particularly concerned to contrast it with knowledge acquired in previous epochs. Galileo, Bacon, and Descartes all shared the desire to demythologize the natural world, to clear away what they saw as the veil of ignorance and to show the world as it really was. No longer would magic and superstition prevail; no longer would the traditions of the past simply be accepted as handed down to us. Instead, everything

[3] We separate here science and technology in order to examine them analytically. In current practice, much of modern science is dependent upon technology and much of modern technology is dependent upon science.

would be opened to questioning. Nature would not be merely accepted but examined anew. No longer would nature be seen as a purposive force in the world but as merely an inert mass to be molded to human desires [3].

In their attempt to avoid the questions of metaphysics that had plagued their colleagues in the past, the founders of modern science also attempted to isolate, to control, and to objectify nature by removing it from context and studying it in pieces rather than as a whole. As a consequence of this, scientists tended to focus, and still do focus, on things that gain their status as objects only as a result of the special procedures that science itself uses. Biologist Erwin Chargaff [4:56] explains the dilemma that this poses:

> 'The wonderful, inconceivably intricate tapestry is being taken apart strand by strand; each thread is being pulled out, torn up, and analyzed; and at the end even the memory of the design is lost and can no longer be recalled.'

Once nature had been divided into an immensely complex jigsaw puzzle, it appeared to have lost its mystery. No purpose, no teleology, no final cause could be found within the natural world. Thus the contrast between humankind and nature was magnified; for both scientists and the general public, nature was the raw materials out of which we would fashion and refashion a world of our making.

In so doing, however, science hastened the secularization of nature. It removed the sense of the sacred that accompanied nature in the past and replaced it (or at least attempted to replace it) with carefully constructed theories that claimed to explain all natural phenomena. Once this sense of the sacred was removed, however, there appeared to be no longer any reason to respect or revere nature. Nature was merely an elaborate array of physical elements that could be arranged in any way that we might deem appropriate to *our* needs. Indeed, Liebig, Pasteur, and many of the other early agricultural and food scientists attempted to arrange nature–in the field–to duplicate their laboratory conditions [5, 6].

With respect to the diversity of plants, animals, and other living organisms on the earth, science merely told us that they were all the result of a process of evolution. Evolution had no purpose, no plan, and hence no meaning outside that which we, as human beings, might wish to assign to it. However, science alone was insufficent to create our present dilemma; technology was the obverse side of the coin.

Technology

If modern science provided the intellectual justification for the reconstruction of nature along the lines desired by human beings, modern technology provided the means. Technology allowed us to treat the world of nature in a new way, as a collection of tools and materials to be called on when necessary to further our human ends. It allowed us to treat nature as what Heidegger [7:17] called standing-reserve:

> 'Everywhere everything is ordered to stand by, to be immediately at hand, indeed to stand there just so that it may be on call for a further ordering. Whatever is ordered about in this way has its own standing. We call it the standing-reserve'

Thus, as Heidegger notes, with the addition of a dam, the Rhine River becomes merely a source of energy. Similarly, the soil becomes merely a convenient place in which we can plant crops. Plants and animals become merely convenient things that we may also use and even redesign to meet our tastes. They lose their sacred standing as the works of a being beyond our comprehension and become merely 'resources' that we use to satisfy our needs, our desires, and even our whims. The biblical conception of stewardship or husbandry gives way to a newer one of control, of domination.

It is important to understand that the problem we face with respect to germplasm conservation is fundamentally the result of the successes of science and technology. Seeds, animals, and microorganisms are reduced merely to the materials and tools of research. Their sudden scarcity is the result of our extraordinary success in convincing farmers that these scientific tools and materials could also become the tools and materials for an improved agriculture. As these products of science became widely adopted, they replaced their predecessors–the landraces of plants for centuries adapted to human use by peasants, the farm animals domesticated by those same peasants, the wild relatives of cultivated plants located on the edges of fields, and the wild plants displaced by the gradual (and not so gradual) encroachment of our industrialized civilization upon virgin forests.

Similarly, the mechanization of food production and processing removed the mystery of food and agriculture and replaced it with scientific explanations and technical processes. The food production process was standardized such that food products have become indistinguishable from other products sold in our consumer society. The creation of bread was once a varied household or village activity requiring that wheat be planted, harvested, milled into flour, kneaded, made into dough, and baked. All of these varied activities might take place within a single family or village. Today, few of us even see the various processes by which the bread is produced. It is simply one more item to be purchased at the local supermarket.

The secularization of food through technology is perhaps nowhere more apparent than in the meat industry. Where once the process of killing animals took place in a butcher shop or on the farm, today it occurs only in great meat factories, far from the clean, almost sterile environments found in the local supermarket. Animals are now killed in large stockyards where they are butchered with machine-like regularity. Hence, the sacredness of life, once painfully apparent to the butcher, is neutralized and demystified. Giedion's [8] comparison of the stockyards of Paris and Chicago at the turn of the century is instructive. In Paris, the yards were well planned, and each animal had its own stall; its individuality, the sanctity of its life, was respected. In Chicago, the yards were built haphazardly but with an eye to slaughtering animals as rapidly as possible. The mystery of life was removed. Today, however, the process of butchering is remystified through science and technology. The neatly wrapped packages of meat that grace our supermarket shelves provide virtually no clue as to their origin. We must explain to our children that beef comes from cattle, pork from hogs; the organic link between the two has somehow been severed as the process of butchering has been mechanized.

The creation of the fast food industry has also eliminated much of the sacred element found in food. The very uniformity, the rapidity of production, the sterile environment for dining, the speed at which the food is consumed, all convert food into merely an instrument, a tool, a resource for our survival, little different from the fuel used to run an automobile.

Science and technology have also broken the link between seeds and fertility. Whereas seeds remain related to fertility, the ancient linkage between the seed and the earth is no longer understood or appreciated by the vast majority of the population.

Capitalism

The third and final factor in the secularizing of nature is capitalism. Science and technology together *lead toward* the profanizing of nature. Capitalism, by its very nature, exacerbates this trend. On one point the political economists of the nineteenth century were agreed. Both Adam Smith and Karl Marx celebrated the enormous salutory effects that competitive markets had on the development of science and technology. Under competitive market conditions, capitalists would constantly seek ways to lower the costs of production, cut prices below those of their competitors, and increase their market shares. The introduction of new technology was a key feature of this process. Thus, while science and technology provided the wherewithal, it was capitalism that provided the societal incentive, indeed, imperative.

Today, with the spread of Western ideas and ideologies, even systems in apparent opposition to capitalism have adopted its fascination with the products of science and technology. Perhaps Lenin [9:333] was right (and therefore wrong!) when he said that, 'The possibility of building socialism will be determined precisely by our success in combining the Soviet government and the Soviet organization of administration with the modern achievements of capitalism.' Without a doubt, the Soviet model poses the same problems for genetic diversity as does ours. Yet since the Soviet model was derived as a response to capitalism, it lacks the dynamism that impels scientific and technical change under capitalism.

In sum, then, it is the combined effects of science, technology, and capitalism that have led us to secularize food and agriculture and to create, unwittingly, the problem of germplasm conservation. Therefore, Wilkes [1] is only partially correct when he notes that 'The forces that promote uniformity, and, therefore, vulnerability, are market place forces and consumer preferences' [1:138]. In fact, without scientific and technical advance, itself promoted by the marketplace, we would have little to discuss today.

Yet Durkheim [10] and other sociologists have argued that the existence of both sacred and secular[4] worlds is essential to all human societies. Indeed, the sacred is defined by its separation from the secular world of everyday life. In previous epochs the sense of the sacred was maintained by a host of rites and rituals surrounding food and agriculture. Great festivals marked the spring sowing and the fall harvest. Animals were slaughtered according to certain carefully regulated practices. Bread was broken only after certain prayers were performed. In short, the reliance on and limitations imposed by nature were understood and respected. By eliminating the sacred from food and agriculture and replacing it with the secular, we risk not only a significant component of our culture but indeed may threaten a vast reduction in the diversity of the biological world in which we live.

Germplasm conservation: A problem created by science itself

The development of plant and animal breeding based on Mendelian genetics at the turn of the century gradually began to create uniformity on a scale previously unimaginable. 'Scientific' agricultural management and technology development early in this century provided further impetus to a reshaping of agriculture and the transformation of nature. However, it was not until the last several decades that the full import of this restructuring of nature began to pose serious problems. The vast improvements in science, communications, and transport that have recently occurred have made possible:

1. The sowing of much of the world's farmland in high yielding varieties,
2. An increase in the amount of marginal land under cultivation,
3. An increased intensity of cultivation,

[4] Durkheim [10] actually used the term 'profane' instead of secular. However, as that term has a pejorative cast that Durkheim did not intend, we use the term 'secular' as an appropriate substitute.

4. The development and spread worldwide of farm animals that can gain weight very rapidly, and
5. The rapid harvesting of timber from huge forests to provide furniture and housing in the West and fuelwood in the Third World.

In the near future the advances of molecular biology may well dwarf even these achievements by remolding the biosphere to our needs. Ironically, these very improvements are themselves the source of our contemporary problems, for by them we have begun to create a second nature in which the maintenance of biotic diversity becomes a part of our responsibility for the world. Macer [11:355] explained the problem more than a decade ago:

> 'The few successful varieties usually have outstanding attributes and, therefore, tend to predominate in large areas of production and to become a potential hazard in themselves. Also, because of their very success these varieties are used repeatedly by breeders as parents in new breeding programmes, thus further narrowing the genetic base of future varieties.'

Moreover, the widespread adoption of high yielding varieties of plants and animals restructures the world of microorganisms. The complex ecosystems that existed before our very arrival on the planet are replaced by greatly simplified agroecosystems. Within these simplified systems,

> 'It must now be accepted that the activities of plant breeders determine the distribution of virulence types of many major pathogens–particularly the air-borne, foliar pathogens such as the cereal rusts–and provide the 'substrates' which favour the selection and multiplication of pathogenic forms with new combinations of virulence.'

In short, the pests and diseases as well as the crop plants and farm animals become our responsibility as we replace traditional agriculture with the products of modern science and technology.

Seedbanks vs. nature preserves: The problems of the hunters and the agriculturalists revisited

A major point of disagreement today among those who would preserve germplasm concerns the overall approach that should be taken. On one side are those who argue that only some form of storage can be successful. Without question, most agricultural scientists take this position because of the greater access that such storage provides. On the other side are those who argue for the creation of nature preserves in which exotic species can be left alone, away from human contact. Ironically, both positions harken back to the age-old dispute the between agriculturalists and the hunters and gatherers. Those advocating storage are the latter day agriculturalists, insisting that nature be reformed according to human desires. Those advocating nature preserves are the latter-day hunters insisting that nature be left in its pristine condition, maintaining its sacredness.

Yet, paradoxically, each of these two positions today contains the other. Like the Yin and Yang of Chinese philosophy, the two are today inseparable solutions to a complex problem. On the one hand, the seedbanks are the contemporary temples in which we carefully preserve our heritage. On the other hand, the gene parks are not nature 'out there'; they are not a wilderness in the usual sense of that term, for they are carefully bounded and actively protected. A true wilderness is beyond the pale of civilization. It needs no protection to exist; it is simply there. Gene parks and nature preserves are within civilization; they are of our making.[5] Thus, though the two positions still reflect the millenia-old clash between the agriculturalists and the pastoralists over the proper role to be accorded to nature, the two positions today no longer can be distinguished easily. They are two ends of a continuum or two complementary strategies for achieving the same goal.

Each position has its strengths: Germplasm banks allow one to know what it is that is stored. They permit the careful classification of natural materials, their long-term storage under controlled conditions, and their duplication and preservation in locations far removed from their places of origin. This permits conservation even under conditions of political turmoil or environmental degradation. Since many landraces could not survive without human intervention, germplasm banks are essential to maintaining certain cultivars. Yet, as Marshall and Brown [12:26] state, 'there are definite limits to the number of accessions which can be handled effectively in programs for the conservation and utilization of genetic resources. These limits arise from the restricted [financial] resources available....' In contrast, nature preserves permit evolution to go on unhindered. They permit the continuation of the complex interactions that occur both within and between species. Perhaps most importantly, they permit the preservation of the enormous, perhaps infinite, diversity that is found only in nature. At the same time, the very infinity of the material available limits the utility of nature preserves.

Though different in practice, the underlying intent of seedbanks and nature preserves suggests that the two positions must now become complementary. It may be too late to return to the sacred world of our ancestors.

[5] The recent report of the OTA [28] implicitly takes this point of view by placing management systems on a continuum from those that emphasize natural processes to those that emphasize human intervention. Yet, the report fails to note the way in which each approach incorporates aspects of the others. In contrast, see Turner's [29] examination of the apparent paradox of the restoration of the prairie.

Once profaned it cannot be reconsecrated. However, we cannot maintain the hubris that we can control and reorganize nature entirely to our liking. Even with the construction of huge germplasm banks and the enormous power that the new biotechnologies surely will release, we must concede that the world is infinitely more complex than we can even imagine. Thus, we must begin to protect nature from our apparently incessant desire for power and control. Means must be found to reunite 'agriculturalists' and 'hunters and gatherers' so as to preserve biological diversity, the ultimate source of both culture and agriculture.

The ethics of conservation: For what shall we care?

The secularization of nature that accompanied the rise of modern science and technology and capitalism finds its clearest philosophical expression in the principle of utility, the basis for 'utilitarianism.' This philosophy espouses that things have value only in so far as they 'work': both human artifacts and natural things–including perhaps human beings themselves–are to be judged worthy to the extent that they can be used for the production of 'maximum utility,' a sum of benefits over costs. What counts, of course, as a benefit or a cost has long been debated; as this philosophy now stands, however, benefits are defined as 'satisfied preferences' of individual human beings. Actions, institutions, practices, even whole cultures are judged, therefore, on the basis of whether or not 'the greatest good of the greatest number' is achieved through them. Whatever ethical responsibilities we might have are dictated by whether we can and do bring about this greatest good. This philosophy has formed the philosophical or ethical foundation for 'modern agriculture' and its attendant scientific and technological handmaidens. Agriculture and modern agricultural research is indeed thought justified by the tremendous gains achieved in improving the material well-being of great numbers of people. Germplasm preservation/conservation is conceived as a means toward continued agricultural productivity and is therefore also justified on utilitarian grounds. Our only responsibility in this regard is to be 'efficient' in the provision of usable seed for continued agricultural production.

As pervasive as this viewpoint might be, however, there is a growing unease both with the 'trade-off' or 'cost-benefit' analysis of our institutions and practices and with the underlying notion that things have value only in so far as they produce utility. Indeed, nature preserves or unspoiled wilderness, for example, may have no such utility. Or, if they do, their utility is not immediately clear. Some have suggested that the utility might lie in the future as a result of some as yet undiscovered benefit. However, this is ultimately an unsatisfactory position. As Dahlberg [13:368–69] explains:

> 'The process of 'discovering' what are seen to be 'new biological resources is really a process of 'commodification,' that is, the transferring and incorporating into formal and utilitarian systems of species that previously were outside human awareness or were drawn upon by indigenous peoples in ways that acknowledged a mutual symbiosis, rather than a one-way exploitation'.

Such a procedure is unsatisfactory because it retains the anthropocentric view of the world that is central to the utilitarian position.

In a world in which scarcity in an absolute sense is pervasive, it arguably makes sense to think largely in terms of costs versus benefits. A value system structured in this way focuses attention on the problem of scarcity. However, in a world of abundance–created in part by science, technology, and capitalism–the dilemma we face is the erosion of the very culture (and agriculture) that we value most highly. Continuing to act and think only in a narrow utilitarian perspective only exacerbates the problem.

In short, other values, or as Sagoff [14] suggests, other metaphors, beyond utility must also be considered. Perhaps it is a primitive sense of the sacred, or the Garden, that we carry within us which leads us to appreciate, even cherish, these 'nonutilitarian goods.' Regardless, these intuitions may provide us with a different model or way of viewing our cultural and agricultural practices and nature itself. These intuitions may force us to reassess the kinds of responsibilities we have beyond the production of utilitarian goods. No longer may we view our actions and institutions solely in terms of their 'productivity.' Instead they can be understood in terms of deeper responsibilities we have to present and future generations and perhaps to nature itself.

Philosopher Hans Jonas [15] argues that the two major approaches to ethics in our time fail to account adequately for the growing dimensions of our responsibilities to the future. As we stand neck-deep in the modern scientific and technological pool, to talk of satisfying preferences or even of respecting the rights of people in the future is not enough. Indeed, as critics have long noted, cost-benefit analysis *requires* that future effects of our present actions be 'discounted.' Philosophers have shown the conceptual hollowness as well as political inefficacy of attributing 'rights' to nonexistent future persons. To have a right is to be able to make a claim against others. Presently, those in the future can neither express their preferences nor their indignation.

Jonas proposes a 'new imperative' for the ethical guidance of modern culture: the imperative of responsibility. This imperative is best understood not in terms of specific obligations we might have to act in particular ways but instead in terms of an attitude or outlook we must adopt. This attitude is care.

'Care', according to standard dictionary definitions, means a variety of things: on the one hand, grief, distress, anxiety, concern; on the other hand attention, heed, caution, watchful oversight, circumspection. An ethic of care includes all these elements. To be sure, attention and circumspection with respect to present and future human beings may not conflict with maximizing their utility nor with respecting their rights. Grief and concern may lead to a felt responsibility not to violate rights or undermine the environmental quality requisite for the satisfaction of preferences for material comforts. Nevertheless, a redirected concern for the full range of consequences that our scientific and technological practices will impose on present and future generations will force us to reassess the relevance of economic welfare or rights in our ethical scheme. A resurrected care for others' lives may lead us to rediscover at least a small part of the sacredness of nature.

With respect to germplasm, care requires both conservation and preservation, both management and hands-off appreciation. Philosopher and environmentalist Bryan Norton [16] has recently argued–and we have suggested above–that the distinction between conservation and preservation is more a distinction in strategy than one in philosophy. In reality, both are based on a recognition of our responsibility for and to the ecosystem of which we are a part. We are responsible, in John Locke's [17] words, to use only what we truly need, and leave 'as much and as good' for posterity. Some wild, some compiled, we might say. We, primary among the animals, have the capacities for annihilating, debilitating, or enriching whole orders of living things, including ourselves. With this power comes the imperative of care.

That we have a responsibility to care for ourselves, the future, and nature itself is certainly not a given. Sophisticated arguments and considerable reflection are required before the intuitively appealing dimensions of this notion are given more rational, scientific foundation. One such argument, a thought experiment really, runs thus: let us set aside our particular preferences and biases and ask ourselves what sort of environment and cultural practices, including science and technology, we would think justifiable if we did not know who we were, where, or in what time period we lived. Would it not be reasonable to wish for a culture wherein there were many opportunities for people to experience the richness of life? Would it not be reasonable to wish for an environment populated with an ever-growing diversity of plants and animals and people? To walk, so to speak, in everyman's shoes is to understand the foundations of an ethic of care.

Although we cannot even begin to speculate on how may species have become extinct since the beginnings of life, human actions have directly or indirectly contributed to the extinction or threat of extinction of somewhere in excess of 200 species [18]). This is morally unacceptable. We as a species are clearly responsible not to reduce arbitrarily or uncaringly diversity.

Thoughtful utilitarians and rights-oriented ethicists agree with those who hold to the imperative of responsibility that diversity is generally a good thing. Yet maintaining and promoting diversity can have its drawbacks. For example, philosopher Kathryn George [19] has recently argued that in some situations biodiversity is not an unqualified good; in some cases it is precisely the controlled reduction in diversity in the field that brings about the best result for humans. Agriculture is a case in point. Some have suggested the eradication of the boll weevil (*Anthonomus grandis* Boheman) through a massive biological control program. Clearly, it is to our advantage that our cotton crop not be ravaged by this plant pest. This can be avoided by employing measures that vastly reduce the population of insects in the field. On the other hand, the extinction of the species would violate our responsibilities and is unnecessary as well for successful cotton production. However, apart from such clear cases of the need for careful reduction in diversity, the imperative suggests conservation and preservation, circumspective use, and caution.

Diversity, of course, can be maintained at many levels: individuals, phenotypes, genotypes, individual genes. Among plants, we have cultivars, landraces, breeders' stocks, and wild species about which to be concerned. Among animals, we have domesticated and wild species to address. What, and which, diversity should we conserve? In some cases, our conservation/preservation resources may meet competing demands: landraces or wild relatives? individual mutants or clearly useful cultivars? We cannot hope to resolve these issues here. Let us instead reaffirm the principle that the careful conservation of diversity requires a longer-term vision than the present-focused utilitarian philosophy will admit. Perhaps, in the end, we can even avoid the question of 'how much and how good' we can leave the future. There will simply be enough, and as needed.

Moreover, the question of what we should conserve is inextricably bound to the problem of how to conserve. If care is to be the central concern in germplasm conservation, and if our obligation is to species rather than to individuals, then our responsibility must extend to everything in nature. No longer can priorities be established solely upon the current value of an organism to us. On the other hand, we can establish short and long term priorities. Short term priorities can be based upon the means we have available to us immediately and the urgency of the situation (i.e., the likelihood of extinction if we do not act). This further suggests that no one means (i.e., germplasm banks, natural preserves, etc.) will be the sole solution for all organisms in danger of extinction. Proponents of germplasm banks must begin to recognize that much material can be preserved only in its natural habitat. Conversely, proponents of nature preserves must recognize that the evolution of certain species is so closely linked to human culture that once we cease cultivating them, we can only preserve those materials in germplasm banks. In the longer term, however, neither of the two 'solutions' is really likely to solve the problem. As long as our culture is pitted against nature rather than a part of it, no amount of financial resources and scientific expertise is likely to reverse the trend toward reduced diversity. The long term will require a thorough rethinking of the goals and direction of human civilization, an admittedly enormous task. How can our collective, institutional care and concern for the world of nature be revitalized? How can we take on the imperative of responsibility?

Several proposals for increasing care and concern

It is curious if not shocking that a nation that could consider spending $4 billion for a particle accelerator would find itself short of funds to insure that germplasm is adequately conserved. Yet, since until recently it appeared that germplasm conservation had little or no utility–and for reasons noted above, utility was the central measure of worth–there was little incentive to finance it. This situation must be reversed in the U.S. and other nations. In the past, agricultural research institutions have been extraordinarily effective in building the clienteles necessary to support important research endeavors. For example, the American 'Land Grant' universities actively organized farmers during the second and third decade of this century to support agricultural research and education [20]. Even today the importance of client groups to the effective functioning of agricultural research systems throughout the world cannot be underestimated. Meier [21] estimated that there are 117 client groups for the U.S. Department of Agriculture alone. Though some have rightly criticized the narrowness of the constituencies that these groups often represent [22], they are not a luxury to be discarded; they are instead an essential part of the effectiveness of agricultural scientific institutions.

The proposals we offer are modest ones that might be articulated before legislative and administrative bodies so as to insure that we do not destroy diversity and that we live up to our responsibilities to the future and to nature itself.

Food security

Though the nations of the world spend billions of dollars each year on weapons that will supposedly protect us against each other, we spend little on germplasm conservation/preservation. Plucknett *et al.* [23] estimated that worldwide germplasm expenditures annually total only $55 million. A single submarine, by way of comparison, now costs considerably more. Given that the continuous production of our crops and livestock depend upon the ability to respond rapidly to diseases and pests, and that this in turn depends upon our ability to maintain a diverse stock of germplasm, it is obvious that germplasm conservation must receive higher priority on grounds of national defense. Trading one submarine for better germplasm maintenance would be a bargain indeed.

Building new constituencies for germplasm maintenance

Six major groups have a common interest in germplasm conservation:

1. Nation-states,
2. Seed and livestock breeding companies and their parent corporations,
3. Voluntary associations of seed savers,
4. Environmental groups,
5. Farmers, and
6. Plant and animal breeders.

Nations have an interest and have done much to establish germplasm banks and nature preserves over the last several decades. Though there are clear areas of dispute among nations, there are far more points of agreement. However, few nations have drawn on the potential support that exists or may be created within the other five groups. For example, seed and livestock companies have a vested interest in ensuring that adequate stocks of germplasm are available for use in future breeding projects. They can be strong supporters of government financing of germplasm conservation, especially for germplasm banks.

In many nations there also exist voluntary associations of seed savers. These groups range in size from a handful of persons to several thousand members and engage in a variety of activities designed to keep local landraces in use, save unusual cultivars, or protect wild species of plants and animals. Currently, their activities are largely divorced from those of governments. They receive no financial support, and only rarely do they receive technical support. Yet, given that they often work with minor crops and wild species–materials often ignored by governments–they are a natural ally in conservation. Simple pamphlets on obtaining and growing unusual plant varieties, distributed with government funds through extension or other offices, would serve both to increase public awareness and to encourage active public participation in conservation programs. For relatively small sums, such private voluntary associations might become distribution centers for rare varieties. The key ingredients in such a program are both public education and conservation. Such a program would encourage duplication and redundancy. It would not involve storage but rather the active planting and growing of unusual varieties. It would serve to build public support while appealing to the desire on the part of many to grow something exotic.

Some scientists might argue that such groups are inconsequential and unlikely to understand the need for special care in maintaining exotic species. Others would argue that their informal status makes them poor keepers of our genetic heritage. However, the facts appear to belie this position. Many of the members of these groups are already actively consulted in regard to the conservation of some species. In addition, illiterate peasants not only conserved germplasm for centuries but actually made major improvements. To think that our much better educated population of today is incapable of doing that is to be victim to an erroneous scientific elitism. Moreover, while we like to think of our governmental institutions as permanent and enduring, we should remember that the record of history shows that governments and their institutions are quite impermanent themselves.

Also, environmental groups should be drawn into the debates over germplasm conservation. Some of these groups are already concerned about large animals (*e.g.*, the World Wildlife Fund). Ironically, whales receive enormous attention while seeds are virtually ignored. The large size of whales and their apparent suffering force themselves upon our attention while the last seeds of a rare grass disappear unnoticed under the plow. More education is needed that shows how size can blind us to the many other species that face extinction. Unlike the smaller seed saver groups, these large environmental organizations have already shown that they can mobilize large numbers of persons and significantly influence government policies. Their support for germplasm conservation is vital.

Both seed savers and environmental groups could benefit from stronger linkages with botanical and zoological gardens. With proper funding gardens would serve a much greater educational and client-building role than is currently the case. In particular, gardens in Third World nations could educate wealthy tourists (in return for a contribution) as to the gravity of the problem of species loss.

Farmers represent yet another group with a potentially strong interest in germplasm conservation. Indeed, it is farmers who have much to lose immediately if germplasm is not conserved. Yet, the importance of germplasm conservation will have to be explained to farmers as most are unaware of its necessity even when they actively participate in it through the cultivation of landraces.

Finally, plant and animal breeders are a logical constituency for germplasm conservation. However, most are only likely to become ardent supporters to the extent that germplasm is not merely conserved but evaluated as well. This suggests that new ways need to be developed to engage breeders actively in the evaluation process. And, appeals to breeders should clearly emphasize the inadequate support for evaluation as well as pointing to the successes with a few crops that have been well-studied (e.g., tomatoes).

Taxing seeds and animals

A third possibility is a tax, the proceeds from which would be specifically devoted to germplasm conservation. Such a tax could be levied on the sales of bulk seed, semen, and breeding stock. Given the relatively low cost of germplasm conservation as a proportion of seed and animal sales, such a tax would be virtually unnoticeable to the purchasers or the general public. Any country could begin immediately to impose such a tax with the goal of having a worldwide tax by some target date (for example, ten years). Proceeds from the tax would be used to support collection, creation, and maintenance of genebanks; and creation of nature preserves. Some system also might be devised by which more Third World nations would be helped to establish their own institutions for the conservation of genetic resources.

Such a system of taxation would also have the advantage that the revenues would grow in proportion to the volume of improved seeds and animals produced. Thus, the sums available for genetic conservation would increase in proportion to the replacement of landraces by higher yielding organisms.

Revaluing traditional knowledge

A misconception often found in the West is that traditional knowledge is mostly superstition, wrong, and irrelevant to the needs of the modern world. Unfortunately, even the leaders of many non-Western nations share this view, having themselves been trained in the West and taught to share this belief [24]. One tragic result of what Gadamer [25] in another context has called the modern 'prejudice against prejudice,' is the destruction of the stock of knowledge of indigenous plants, animals, and soils in many parts of the world. Yet this kind of ethnobotanical and ethnozoological knowledge can provide us with many important insights into the natural world developed over centuries, even millenia, by trial and error and observation. Admittedly, this knowledge does not fit easily into modern taxonomic schemas. It is often very specific to particular ecological niches. Nevertheless, it is far more developed than any modern taxonomy for the realms of nature that it maps. Moreover, it often incorporates an understanding of nature as a whole, of which human beings are but a part. By codifying and synthesizing indigenous knowledge into our Western worldview, we perhaps can come to understand the natural world better, educate ourselves about the infinite diversity that surrounds us, and even learn more about how diversity may be conserved. In short, the conservation of natural diversity requires the conservation of cultural diversity.

New models of development

Beyond these specific remedies, we must begin to question the model of development that we in the West have embraced these past 300 years and have foisted upon many of the peoples of the Second and Third Worlds. We refer here not to a specifically economic model, *e.g.*, capitalism or socialism, but to the underlying philosophical assumptions upon which Western development is based. It should be obvious to all that the problems we face today that threaten genetic diversity, *e.g.*, industrial pollution, deforestation, fertilizer runoff into our streams and rivers, are the result of the success of a particular way of looking at the world. The Soviets are no better than we are despite their adherence to a different ideological position. Thus, what we are talking about here is far deeper than any mere ideology. It is a set of root beliefs, of myths, about the abilities that we possess to restructure the entire planet.

No matter how well we may run our germplasm banks, no matter how many protected wilderness areas we may create, no matter how many species we protect, we shall never be able to take on the active responsibility for them all. That is simply beyond the capabilities of the human species. Therefore, we shall have to abandon our war with nature, we shall have to reconsider the limits of human ingenuity, and we shall have to revalue –if it is not already too late– the indigenous knowledge that the non-Western peoples of the world have. This is a challenge of perhaps greater magnitude than any mere system for conserving genetic diversity. Yet it is essential that we respond to it if we are to take our responsibilities to other species and to our own future generations seriously. It is essential if our culture is to care.

Acknowledgements

The authors wish to thank Fred Buttel and Mark Sagoff for their helpful comments on a previous draft of this paper. This paper was written with support from the National Science Foundation's Ethics and Values Studies Program (EVS) and the Cooperative State Research Service (USDA) under grant No. BSR-8608719. Opinions, findings, and conclusions are solely those of the authors, and do not necessarily reflect the views of the sponsors.

References

1. Wilkes, G. 1984. Germplasm Conservation Toward the Year 2000, Pp. 131–164 in C. W. Yeatman, D. Kafton, and G. Wilkes, ed. Plant Genetic Resources: A Conservation Imperative. Westview Press, Boulder.
2. Hamblin, D. J. 1987. Has the Garden of Eden been Located at Last? Smith. 18 (May): 127–135.
3. Leiss, W. 1972. The Domination of Nature. Braziller, New York.
4. Chargaff, E. 1978. Heraclitean Fire. Rockefeller University Press, New York.
5. Krohn, W. and W. Schafer. 1983. Agricultural Chemistry: The Origin and Structure of a Finalized Science, Pp. 17–52 in W. Schafer, ed. Finalization in Science: The Social Orientation of Scientific Progress. D. Reidel Publishing Co., Dordrecht.
6. Latour, B. 1984. Les Microbes: Guerre et Paix suivi de Irreductions. Editions A. M. Metaillie, Paris.
7. Heidegger, M. 1977. The Question Concerning Technology and Other Essays, Harper and Row, New York.
8. Giedion, S. 1975. Mechanization Takes Command. W. W. Norton, New York.
9. Lenin, V. I. 1937. Selected Works, P. 333. Lawrence and Wishart, London. Volume VII.
10. Durkheim, E. 1975. Durkheim on Religion. Routledge and Kegan Paul, London.
11. Macer, R. C. F. 1975. Plant Pathology in a Changing World, Trans. Brit. Mycol. Soc. 65 (Dec.): 351–67.
12. Marshall, D. R. and A.H.D. Brown. 1981. Wheat Genetic Resources, Pp. 21–40 in L. T. Evans and W. J. Peacock, ed. Wheat Science–Today and Tomorrow. Cambridge University Press, Cambridge.
13. Dahlberg, K. A. 1987. Redefining Development Priorities: Genetic Diversity and Agroecodevelopment, Bull. Sci., Tech., and Soc. 7(3): 367–382.
14. Sagoff, M. 1988. The Economy of the Earth. Cambridge University Press, Cambridge.
15. Jonas, H. 1985. The Imperative of Responsibility. University of Chicago Press, Chicago.
16. Norton, B., ed. 1986. Preservation of Species. Princeton University Press, Princeton.
17. Locke, J. 1967. Two Treatises of Government, P. 329. Edited by Peter Laslett. Cambridge University Press, New York.
18. Gunn, A. 1984. Preserving rare species, Pp. 289–335. in T. Regan, ed. Earthbound. Random House, New York.
19. George, K. P. in press. Biotechnology and biodiversity, J. Ag. Eth. 1(3).
20. Busch, L. and W. B. Lacy. 1983. Science, Agriculture, and the Politics of Research. Westview Press, Boulder; Colorado.
21. Meier, K. J. 1978. Client Representation in USDA Bureaus: Causes and Consequences, Pol. Stu. J. 6(4): 484–87.
22. Hightower, J. 1973. Hard Tomatoes, Hard Times. Schenckman, Cambridge, Massachusetts.
23. Plucknett, D.; N. J. H. Smith; J. T. Williams; and N. M. Anishetty. 1987. Gene Banks and the World's Food. Princeton University Press, Princeton, New Jersey.
24. Goonalilake, S. 1982. Crippled Minds: An Exploration into Colonial Culture. Vikas Publishing House, New Delhi.
25. Gadamer, H. 1975. Truth and Method. Seabury, New York.
26. Graves, R. and R. Patai. 1964. Hebrew Myths: The Book of Genesis. Doubleday, Garden City.
27. Schneidau, H. N. 1976. Sacred Discontent: The Bible and Western Tradition. University of California Press, Berkeley.
28. Office of Technology Assessment. 1987. Technologies to Maintain Biological Diversity. U.S. Gov. Print. Off., Washington.
29. Turner, F. 1988. A Field Guide to the Synthetic Landscape, Har. Mag. (April): 49–55.

Ecology of agroecosystems in relation to natural ecosystems

SETIJATI SASTRAPRADJA
Center for Research in Biotechnology, Indonesian Institute of Sciences, Bogor, Indonesia

Abstract In the last two decades food production has increased tremendously, even in the developing countries where problems of poverty and hunger are not uncommon. Despite successes, advancement of agriculture is still required so that the level of food production can keep pace with population growth. In the process of advancing agriculture, we utilize and modify natural ecosystems more and more to suit our needs, using the available technologies and inventing new ones. Quite often, these modifications have gone too far and have resulted in degradation of environmental quality. The objective of eco-development is to optimize the productive utilization of natural resources while considering impacts on the environment or agroecosystems. Due to the uneven distribution of natural resources, and following the availability of water, types of soils, and kinds of biological resources, various natural ecosystems are in existence. The transformation of these ecosystems into agroecosystems largely depends on cultural background, level of technology, and political decisions. In consequence, a wide variation of agroecosystem development, ranging from the simplest to the most advanced, exists today. Quite often, such variation can be found within a developing country. To illustrate various factors influencing the development of agroecosystems, Indonesia is selected as an example. The reason behind this selection is the fact that Indonesia not only has many agricultural problems, but it has also succeeded in overcoming some of them.

Key words: eco-development, natural resources, food production, developing countries

Introduction

Global conservation of biological resources has long been considered. In its early stages of development, conservationists concerned mainly with protecting specific plants or animals were rare. Scientific judgment was used as a tool to select which species were worth protecting. The World Wildlife Fund and The International Union for the Conservation of Nature and Natural Resources are among the best known international agencies dealing with conservation.

With the rapid growth of human population, needs for food and housing have increased proportionately. These needs in turn have transformed natural ecosystems into manmade ecosystems and have had a negative impact on the environment. As a consequence, the habitats in which much of our biological resources live were sacrificed for human habitation. Eventually the conservation effort attained a new dimension, concerned not only with

L. Knutson and A.K. Stoner (eds.). Biotic diversity and germplasm preservation, global imperatives.
 ISBN 0-7923-0178-1

biological resources but with the ecosystem as a whole. The purpose of conserving an area is no longer just for scientific study, but also to guard against exploitation and to preserve a place for human enjoyment. The United Nations Educational, Scientific and Cultural Organization, through its programs on Man and the Biosphere, offers alternatives to conserving ecosystems to enable a harmonious interaction between man and nature.

In the field of agriculture, the production sector requires more and more advanced technology if yield is to increase to cope with rapid population growth. In developing countries, where most people are still linked closely with nature, the application of new technology for food production does not always yield the expected results. Along with the newly introduced high yielding varieties, simultaneous application should be made of chemical fertilizers, pesticides, and herbicides. By failing to apply these chemicals with the seeds, farmers meet with disappointment; their harvests are far less than what was expected. Such disappointment is a common phenomenon when chemicals cannot be supplied on a timely basis. On the contrary, when chemicals are easily accessible, the tendency is to apply an excess. Should this happen, the impact on the environment might be negative, such as water pollution which in turn affects people who use the polluted water in their daily lives. The effect of pesticides on wildlife has been discussed at length [1].

High yielding crop varieties grow well on the best available land, which is mostly already under cultivation. In developing countries, such land is subjected to other uses in the progress of national development so that the amount of best agricultural land area decreases with time [2]. To increase food production for the future, it is imperative to intensify the production systems per unit of land and/or to expand such systems to less favorable areas. In the latter case, the newly opened forestland quite often is developed.

In the past, agricultural development seemed to neglect ecological principles and confrontation with conservationists could not be avoided [3]. Today, agricultural scientists are as aware as the conservationists of the environmental consequences of agricultural development, so much so that one of the goals in agricultural development is to optimize the use of natural resources (soil, water, forests, etc.), while protecting environmental quality [4]. The application of ecological principles to agricultural systems is described by the term, 'agroecosystem'. With these systems, ecological considerations are applied to modify natural ecosystems for agricultural production. Ecological processes shared by both systems are discussed herein and the relationship between agroecosystems and genetic resources are examined in the context of their use for food production.

Organization of the systems

Like any other system, an ecosystem is composed of various components that interact with each other to function as a unit. These components are broadly classified into abiotic and biotic components. Soil, water, inorganic and organic substances, temperature, etc., form the abiotic components, while plants, animals, and other organisms form the biotic components, the biological resources. In discussing natural ecosystems, I will exclude man from the system.

In any ecosystem, the biological components are arranged in a hierarchical order, from genes to cells, tissues, organs, organisms, populations, and communities. Each level of arrangement interacts with its abiotic components to form a functional system, characteristic to that level. In this way, genetic systems, cell systems, etc., are recognized. An ecosystem, then, deals with the functional systems at the community level [3].

The biological components in manmade agroecosystems can be arranged in an equivalent hierarchy [5]. Since crops are the main concern in these systems, an assemblage of individuals of a single species is comparable to a population of a wild species in a natural system. Likewise, a cropping system that involves two or more crop species is comparable to a community in a natural system. In a given area, more than one cropping system may be practiced to form an agroecosystem. Conway [5] illustrates the hierarchies of the two systems in Fig. 1.

In reality there are no sharp boundaries anywhere in the hierarchy if interrelationships, survival, and interdependence are considered. Therefore, Odum [3] presents such an organization in a horizontal rather than a vertical arrangement. In this way, a biological spectrum is illustrated with genes at one end and the community at the other. In this spectrum, ecology deals mainly with the portion of the spectrum that begins with organisms and ends with communities. However, the functional integration of all levels of organization and their characteristics should be understood to comprehend the functioning of an ecosystem as a unit.

Crops as biological components in manmade ecosystems follow the same hierarchical arrangement as do species in ecology. Therefore, to understand well the functioning of an agroecosystem, an agricultural ecologist should not be content working only with one level of organization. In fact, agroecosystems offer much more complex problems than do natural ecosystems because cultural values form an element in agroecosystems.

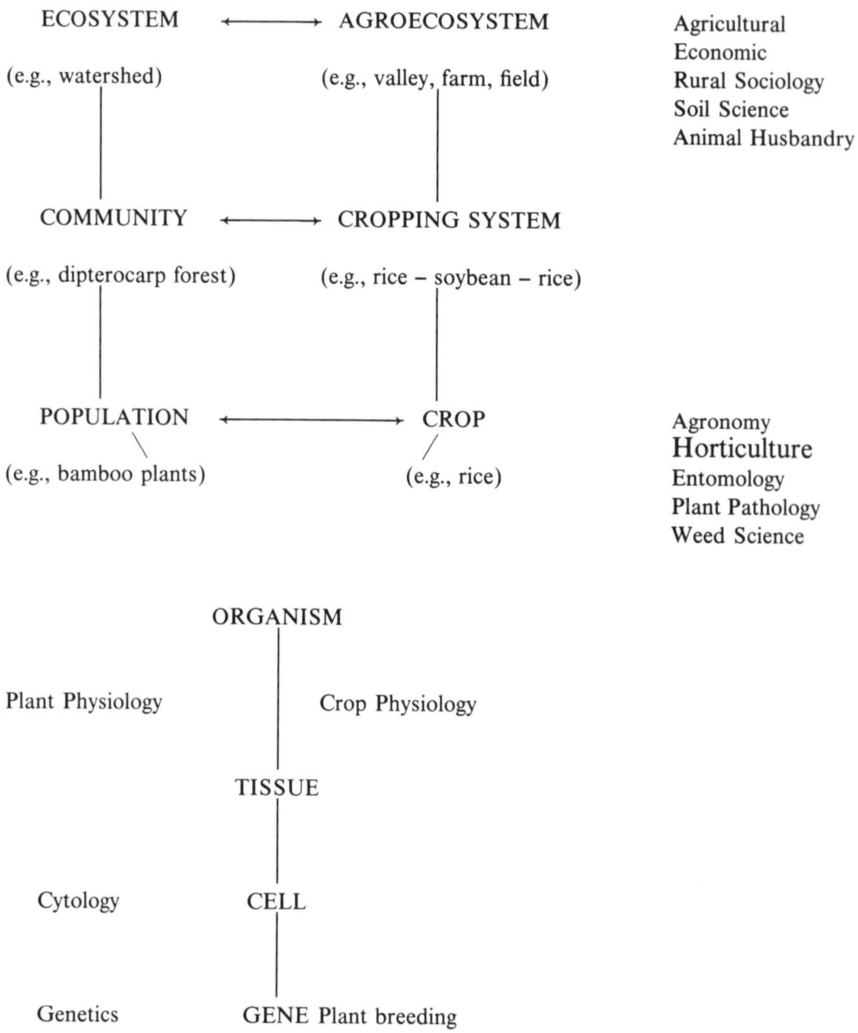

Figure 1. Hierarchies of biology and agriculture and their related disciplines (Conway, G. R. 1984. Applying ecological concepts to the study of the intensification of the use of Indonesia's agroecosystems. In: KEPAS, 1983. The Sustainability of Agricultural Intensification in Indonesia.)

Productivity

One of the goals of agricultural development is to increase food production, which in turn increases the purchasing power of farmers. Increased food production is especially necessary in the densely populated regions of the world. Most of these regions are located in the humid tropics and are characterized by temperature regimes that allow year-round crop production [6]. However, the average yield of crop species in these regions generally is lower than that of the same species in temperate regions.

That science and technology have played a major role in enhancing agricultural productivity is shown in the increase in U.S. productivity during the past 200 years [7]. The increase was tremendous after World War II when chemicals became available for agricultural use as fertilizers and pesticides. Coupled with these chemicals was the development of high yielding seeds and the use of mechanical power; however, the area of cropland in cultivation remained almost the same as before. The effect of science and technology on agriculture has been felt in the developing world as well as in the U.S. [8].

Crop productivity, as an element of agricultural productivity, is defined by Wittwer [9] as the ratio of the value of all the outputs divided by the total of all the inputs. Included in the inputs are the energy required for irrigation as well as factors such as fertilizers, pest control, etc. The supply of many of these inputs are dependent on human factors, while the supply of those provided by nature (land, water, solar energy, air, and, to some extent, genetic resources) are independent of man. The latter inputs are freely available, whereas the former are subsidized by man.

In comparing the productivity of manmade systems with those of the natural ecosystems, Odum [3] showed that the gross productivity of the former does not exceed that of the latter. In other words, high productivity in manmade systems is achieved through high energy subsidies. Failure to supply energy subsidies in the manmade systems results in low crop yields. Odum further pointed out that agricultural productivity in the U.S. is about three times per hectare as in Asia or Africa. But in terms of the cost of energy subsidies, the U.S. spent ten times more on energy to produce a three-fold amount. Many developing countries cannot afford to spend such a sum. An alternative is offered for tropical and low resources regions in the use of complex intercrop systems [10].

Species diversity

The history of agriculture began approximately 10,000 years ago when people discovered that it was possible to grow plant species they needed from the many species they employed from the wild population [11]. Only a relatively small number of plants were chosen for domestication and later on they became cultivated plants. In fact, out of ± 3,000 useful species, only 100 are really important crops today, and of those only 7 species are considered to be major food crops. McNeely [11] pointed out that these species were selected during the first few thousand years after their cultivation. Further artifical selection and manipulation of genetic variations resulted in their domestication. Through domestication, plants become better suited to human environments and gradually lose their ability to compete in natural environments.

In traditional agriculture, the process of both cultivation and domestication may be observed. Such agriculture is still practiced today in many parts of the humid tropics and is recognized as a subsistence level of agriculture. In this system many species are grown together on the same piece of land. As such, natural selection operates, and in times of environmental stress, some species continue to produce. Productivity per unit of land and time, of course, is low.

Traditional agriculture, with its rich diversity, provides man with ways to adapt himself to his environment. In areas where monsoons occur, clearing of land and planting of crops follow the cycle of rain. Traditional cropping systems allow man to plant crops year round, with various species adapted to seasonal changes. The Tegalan system, which is practiced in Java and Madura, offers intensive cultivation of land without the burden of chemical input. Another type of traditional agriculture is the home garden practiced in many places in Indonesia, where fruit trees, tuber crops, and various plants that serve as vegetables or medicines are planted together.

By observing the crop component in the traditional agricultural system, one can deduce that species diversity is the heart of the system. This diversity reflects the richness of the natural species in the environments in which the system operates. In the tropics, the flora is rich in species diversity, but the diversity, however, varies with the type of natural ecosystems.

One type of natural ecosystem that is the subject of recurrent discussion is the lowland humid forest. This kind of forest is very rich in biological resources both in number of species and individuals. Such forests are considered to be fragile ecosystems because of their susceptibility to environmental change, hence the attention given them. Without interference from man, the mature forest is in a state of more or less static equilibrium.

It is difficult to say how many species of plants are present in the world today. A rough approximation is between 400,000 and 700,000 species [12]. That number is only an estimation because the inventory is incomplete. In the tropics, the number of species present is much less certain than in temperate regions. Studies have shown that in the Malaysian region there are 25,000 to 30,000 species, in West Africa 7,000 species, in the Cameroon 8,000 species, in the Canal Zone 2,000 species, etc. Species diversity is very complex in the tropical forests as shown by the following statistics [12]. In 60 acres in the Malaysian Peninsula are found 183 species of trees from a total of 379 individual trees occurring in the area. On one hectare of forestland in Sumatra, 60 species were found from a total of 320 individuals. On one acre in Cameroon, 230 species of plants are found from a total of 1,300 individual plants. The number of species comprising a community is referred to as its intrinsic or α diversity. In such a community in the tropics, a few species are represented by many individuals, whereas most species have small numbers of individuals; hence the status of dominant species in the community.

β diversity is a measure of the variation within a habitat. It is also termed extrinsic diversity. On a micro scale, diversity can be observed from changes in the microflora; certain species growing in the shade differ more significantly from those growing in areas of increasing sunlight. Such phenomena also may be observed on a large scale, involving major climatic zones, regions, or continents. The number of dominant species may change in a short distance with a change in soil type. This kind of diversity contributes to the richness of tropical ecosystems.

Man has used plant species for many purposes, yet many useful species have never been cultivated, especially those exploited for timber or traditional medicine. If agroecosystems are going to be developed in various places in the tropics, certainly the knowledge of both α and β diversities will be valuable in providing a basis for their development, although species diversity will certainly be greatly reduced by such development.

Interaction

Species of plants occupying a common area interact in either positive or negative ways. Sometimes, however, the relationship seems neutral, meaning that neither species is affected by the presence of the other. Odum [3] listed nine different types of interactions on the basis of their general nature. Of these various types, competition, predation or parasitism, and mutualism are important in agroecosystems.

Competition occurs when two organisms overlap in their use of limited resources (water, nutrients, light, etc.). The process of competition itself is difficult to observe, but the ultimate results are well understood as one of the mechanisms of natural selection. Only those individuals or species winning the competition will be successful in adapting themselves to their environment.

Both intraspecific and interspecific competition are possible. In the first instance, individuals within a species compete with each other in using limited resources, while in the latter, competition is between individuals of different species. Natural populations may give evidence that both types of competition operate simultaneously. Both types are important in manmade environments.

Intraspecific competition is best observed when too many individuals are planted together in a limited space. Later in their development it becomes obvious that not every individual reacts the same with regard to available resources. Some establish themselves well, but others develop slowly. Therefore, spacing between individuals is an important factor in cultivation. Intraspecific competition in nature may be seen more clearly in animal species than in plant species.

When two species are in competition, the ultimate result is not always that one species is successful in displacing another. Only when competition is severe can such a condition be achieved. Sometimes, however, two species are able to coexist in a state of equilibrium after a period of adjustment.

Natural processes such as competition also operate in manmade systems. Therefore, when cultivating plant species, either in a single-crop or multicrop system, consideration should be given to the possible effects of competition between and within crops. Moreover, many species of weeds may invade the system at any time during the growth period. If weeds are unattended, crop yield is reduced.

Traditional agroecosystems generally have been established in temperate areas. In these systems, more than one crop is cultivated, involving not only annuals but perennial crops as well. Sometimes trees are included in the system, especially if it is a home garden. From the viewpoint of genetic resources, most crops grown in these systems belong to primitive cultivars, although some taxa may be undergoing a process of cultivation. Therefore, most still retain many of their wild characters, among which is the ability to compete. The type of competition most commonly operating in these systems is interspecific competition, especially for water and light. If several growth forms are involved, the crops can stratify as do tropical forests. A combination of legumes and corn or of legumes and cassava are often planted in agroecosystems involving only annual crops. Ecologically, legumes do not compete with non-legumes because of their different nutritional requirements. Some species of tuber crops tolerate shade and hence are grown among food trees. Their yield is lower than those grown under full sun, but through this means farmers may derive an extra harvest from their land.

One problem in modern agriculture is how to control pests and diseases. To overcome this problem, chemicals are applied to the systems. What and when to apply, how much and how to apply, and the effect on crops and the environment have been the subjects of many studies. An area with a direct application to agroecosystems is the relationship between pests (insects, fungi, viruses) and crops. This is best illustrated in plant-insect interactions.

Insects are by far the most important agricultural pests. They cause damage by acting as herbivores, and also act as vectors for other pests and pathogens. Herbivorous insects may consume almost all parts of plants. Many eat the young parts, but some insects attack the mature portions as well. Insects may damage the plant at each level of its development. Defoliation sometimes is caused by larvae while mature insects may chew stems or leaves. They also may secrete chemicals that are toxic to the plant.

Insects as pests and vectors of diseases in forests have been discussed at length in the report 'Tropical Forest Ecosystems' [12]. As herbivores, tropical insects are much more host specific than are temperate region insects. Moreover, in tropical forests the distance between individual trees of a species is relatively greater than it is in the temperate zones. These two conditions may explain the lower frequency of insect outbreaks in the tropical forests. Host specificity explains why an introduced insect may grow better in a region into which it is introduced than in its place of origin.

Insects play important and beneficial roles in forest ecosystems. Many outcrossing species are dependent on insects for pollination. As in herbivorous insect-plant relationships, some insects are host specific in pollination. Other insects visit flowers without regard to species or pollination. Carpenter bees are among these generalists. Regarding lowland tropics, Heinrich and Raven [13] noted that because of the great distances between trees of a species, their flowers should offer energy rewards to the most important pollinators necessary for cross-pollination. Only with excessive rewards are insects attracted. The classic example of plant-insect dependency is illustrated in fig species. Without figs, the associated insects cannot live, and without the insects, the figs cannot set seeds.

In discussing the roles of insects in an ecosystem, it is useful to consider insects as one of the elements in the food chain. Like many organisms, in such a system insects are subject to predators. Eggs, larvae, imagoes, and mature insects are much sought after by birds, geckos, spiders, etc. This kind of knowledge is very useful for biological control. The question is whether or not in practice such a control method is sound both economically and technically.

As already indicated, interactions between two species of organisms may benefit both species as in the classic example between legumes and *Rhizobium*. The result of this association (nitrogen fixed in the form of HNO_3) is shared by the nonlegume species that are planted together with the legumes. In traditional agroecosystems minor legume species such as lima beans, winged beans, and lablab beans commonly are planted between other crops. More advanced agroecosystems use *Calliandra*, *Sesbania*, and *Callopogonium* to enrich soil nutrients for other crops. In this way the energy subsidy comes from plants rather than from fossil energy. Biological nitrogen fixation species have been used for rice cultivation for the past several years in the International Rice Research Institute.

Symbiosis between fungi and plant roots recently has gained much attention. As in the case of the association between *Rhizobium* and legume species, fungal mycelia penetrate the root, then interact with the root to form organs capable of extracting minerals from the soil. This symbiosis is referred to as mycorrhizal symbiosis. These mycelia appear in three forms:

1. Ectotrophic mycorrhizae – envelop the roots and only partially enter the root tissue.
2. Endotrophic mycorrhizae – penetrate the root cells.
3. Extramatrical mycorrhizae – form loose mantles around the root.

Extramatrical mycorrhizae are believed to be important in transforming insoluble minerals to forms that can be absorbed by plants [3].

The importance of mycorrhizal associations for forest trees is well known. An example is the growth of pine trees in soil too poor to support other crops. Inoculation with fungal symbionts enables the tree to grow vigorously. Recently it has been learned that tropical lowland timber (dipterocarps), which previously had been grown only with difficulty can be grown easily if inoculated with the proper fungus. Attempts are now underway to use mycorrhyzal symbiosis in rice. If successful, rice-growing areas can be expanded to include areas with problem soils.

In summary, competition, predation or parasitism, and mutualism operate in agroecosystems as in natural ecosystems. In the former, man can control these interactions according to his knowledge of insecticides, but the growth of symbionts is encouraged by use of fungal inoculants.

Limiting factors

Man can manipulate natural systems to suit what he thinks is best for him. His desire to improve his situation changes with time, as does his progress in science and technology. Is there a limit to what he can think or do? If not, then what is now impossible as regards growing a crop because of limiting factors may be possible tomorrow.

The success of a species in adapting itself to a certain environment depends on many factors that may act independently or collectively to form complex conditions.

For its growth and reproduction, an organism requires at least a minimum amount of essential resources. Long ago it became apparent that crop yield often was inhibited by the absence of substances needed in trace amounts and that the same plants were inhibited by the presence of an excess of the same substances. The range between the two amounts represents the limits of tolerance for the organism. Only within these limits can it grow and reproduce. Each species has its own limits of tolerance.

Limits of tolerance help explain why one species is widely distributed while another species has a limited distribution. Those plants with a wide range of tolerance for many ecological factors tend to spread widely. Quite often such widely distributed species develop ecotypes that are well adapted to local conditions. From a genetic point-of-view, each ecotype may be different in certain characters. Ecotypes related to cultivated species may form sources for locally adapted lines.

In crops, studies on physiological stress (drought, temperature, trace element requirements, etc.) have provided valuable information as to the nature of tolerance. With this information at hand the limit of tolerance can be extended by means of genetic and environmental manipulation. Through genetic manipulation it may be possible to create more tolerant crops, including changes in plant types. Environmental conditions also can be manipulated to suit the growth of the new variants [14].

The progress made in biotechnology during the past few years has provided new opportunities to explore further physiological mechanisms for increasing plant tolerance. Before DNA-recombinant techniques can be applied, however, the genes responsible for tolerance must be identified. Identification and isolation of such genes will help plant physiologists and geneticists breed stress-tolerant crops.

Agroecosystems and food security

Despite the fact that science and technology have succeeded in increasing food production worldwide during the past thirty years, poverty and hunger still prevail in some regions. Ecological constraints are one reason for failure of harvests. However, all nations wish to one day achieve self sufficiency in their food supply.

Although the global food supply may be sufficient to feed today's population, the distribution of the supply is uneven worldwide. Therefore, the phrase 'charity begins at home' applies to food production, meaning that the solution to food insufficiency problems should begin at the national level [8]. Swaminathan states that among the essential components to be considered in establishing national food security is ecological security.

As for natural resources, food production requires water, land, sunlight, air, and genetic resources. For a long time, the availability of these resources has been taken for granted. The recent rapid increase in the human population forces us to think, however, that difficulties are liable to arise. All that remains free today for agricultural development in many areas is air and sunlight.

The extent of arable land is not expanding. To increase production in most countries, resources should be manipulated on existing agricultural land. Meanwhile, cultivation of food crops on problem soils in the tropics remains a challenge. Wherever population is dense, extension of agriculture to marginal land is unavoidable. Moreover, in some countries cultivation has crept into areas formerly considered unsuitable for agriculture because of its extreme slope.

Water is the most critical natural resource for future agricultural development [14]. In China, India, the U.S.S.R., the U.S. and Indonesia, water already is a limiting resource. In Indonesia, for example, many regions depend on rainfall alone for crop cultivation. Crop loss is certain in those regions if the pattern of rainfall continues to be as uncertain as it is today. Quite often water is excessive during the rainy season but hardly available at all in the dry season, when streams cease to flow. It is generally understood that river flow is regulated by forest watersheds, yet forests are not left undisturbed. Dams have been built in many developing countries to manage water for agriculture and domestic uses but they often become quickly filled with silt, resulting in a shorter life span.

One of the most important factors affecting plant life, but one mostly taken for granted by man, is sunlight. Only green plants can trap energy from the sun and convert it directly into products of value to both man and animals through photosynthesis. Studies on photosynthesis clearly showed that 90–95% of the dry matter of crop plants is derived from photosynthesis. Crop plants, however, are not so efficient in converting sunlight into products as are wild plants [15]. In the laboratory, plants can convert as much as 10% of the incoming light energy to sugars. This percentage has never been achieved in agricultural practice. The highest percentage that plants can achieve in the field is 1–2%. It is possible, however, to improve the energy conversion rate in plants if they are grown under intensive cultivation. In this way, the efficiency of harvesting solar energy through the plant is increased.

Carbon dioxide, which is used in the process of photosynthesis, is absorbed by green plants from the atmosphere. Hence its availability in the air affects the rate of photosynthesis, which in turn influences the crop production. Air quality is becoming a matter of concern to agriculturists as well as to conservationists because of the increasing degree of air pollution.

Other assets of agriculture that attract the attention of botanists and agriculturists alike are crop genetic resources. The importance of genetic resources for crop improvement cannot be overemphasized. But there are limitations preventing their full use in developing countries, in which these resources are present in abundance and in which food production should be enhanced.

If all the essential assets of agriculture are available in a country, and if science and technology have identified their potentials as well as their limitations, how then can ecological security for food security systems be achieved? The production system was one of the topics discussed in an international conference on Crop Productivity – Research Imperatives [14]. The participants felt that the greatest need in agricultural research is to know the relationship between all the biological parts that compose a production system. However, the participants also were fully aware that applied research is site specific, meaning that generalizations are difficult to draw. Moreover, what makes the systems work are the human resources. Human resources, as the most important asset of any country, should be addressed in developing the concept of ecological security if a national food security system is expected to be developed.

Like soil type, rainfall and genetic resources, which change significantly over even short distances in the tropics, so do human resources. The variability is observed not only in physical features, but also in language and culture. Cultural differences affect food habits, which in turn are reflected in agricultural systems. Therefore, interactions between genetic resources, abiotic components (soil type, rainfall, altitude), and human resources have resulted in the present agroecosystems. The crop species cultivated fit not only the environmental conditions but also the regional culture. From a productivity point of view, the results of a system may be low, but for purposes of ecological security it offers a good deal of food alternatives and stability. So the challenge is how to increase the productivity of the system without changing the crop components. Research is underway to provide alternatives for improving the present agroecosystems in Indonesia. A similar effort is being made for forestry. An agroforestry system can be developed by use of both tree species and annuals. Quite often the annuals planted are food crops. To the farmers without suitable agricultural land this system is as good as the traditional agroecosystems.

Agroecosystems and genetic resources

One way to increase food production is to convert barren or abandoned lands to cultivation. In the past, such undertakings have been only duplications of systems of cultivation necessary to produce the most required staple food, without consideration whether or not these lands are suitable for cultivation. Acceptance is increasing, however, that the concept of eco-development should be applied if agriculture is to be successful. Therefore, agroecosystems are encouraged to offer alternatives for land use. At the same time, local minor crops can be intercropped with the major staple food species.

In the midst of enthusiasm to conserve plant genetic resources, an idea arose to conserve traditional varieties of both major or minor crops on the farmers' land. While it is true that agroecosystems may utilize local genetic resources because of the cultural background of the farmers, and these crops have been adapted to local conditions, no one can guarantee the continuation of their plantings. Most farmers using traditional agroecosystems are not as well educated as are those who live in cities, but they are sensitive to progress and would like to improve their quality of life if they have an opportunity to do so. Scientists often are compelled to show them the way to achieve a higher qualtity of life through the efforts of extension workers. Many scientists believe that a revolution in biological sciences will soon be here, that in turn will alter agriculture. Results of biotechnology research so far are very convincing. The gap between traditional farmers and modern farmers will widen if the above revolution takes place. What, then, will happen to the offering hands?

References

1. Moore, N.W. (ed). 1966. Pesticides in the environment and their effect on wildlife. J. Appl. Ecol. 3 [Suppl]: 312 pp.
2. Syarifuddin, K.A., T.G. Ismail, and P.H. Soetjipto, 1988. Investigation of development of a system of farm operation with insight into the ecosystem. (Unpublished paper)
3. Odum, E.P. 1971. Fundamentals of Ecology, W.B. Saunders Company. 574 pp.
4. Manwan, I. 1984. Present effort to intensify agriculture in Indonesia. *In*: KEPAS, 1983. The Sustainability of Agricultural Intensification in Indonesia.
5. Conway, G.R. 1984. Applying ecological concepts to the study of the intensification of the use of Indonesia's agroecosystems. *In*: KEPAS, 1983. The Sustainability of Agricultural Intensification in Indonesia.
6. Gomez, A.A. and K.A. Gomez. 1983. Multiple Cropping in the Humid Tropics of Asia. Ottawa, Ontario I.D.R.C. 284 pp.
7. Hess, C.E. 1985. Past, Present and Future of Agricultural Research. *In*: M. Gibbs and C. Carlson (eds.), 1985. Crop Productivity – Research Imperatives Revisited, an international conference held at Boyne Highlands Inn, October 13-18, 1985, and Airlie House, December 11-13, 1985.
8. Swaminathan, M.S. 1981. Building A National Food Security System. Indian Env. Society. 138 pp.
9 Wittwer, S.H. 1985. *In*: Gibbs, M. and C. Carlson (eds.). Crop Productivity-Research Imperatives Revisited, an international conference held at Boyne Highlands Inn, October 13-18, 1985, and Airlie House, December 11-13, 1985.
10 Harwood, R.R. 1979. Small Farm Development. Understanding improving farming systems in humid tropics. Westview Press, Boulder, Colorado. 160 pp.
11 McNeely, J.A. 1984. Biosphere Reserves in Human Ecosystems. *In*: Conservation, Science and Security. UNESCO-UNEP, pp. 492–508.

12 UNESCO/UNEP/FAO. 1978. Tropical Forest Ecosystems. *In*: Natural Resources of Humid Tropical Asia, 456 pp.
13 Heinrich, B. and P.H. Raven. 1972. Energetics and Pollination Ecology. Science 176 (4035): 597–602.
14 Gibbs, M. and C. Carlson (eds.) 1985. Crops Productivity-Research Imperatives Revisited, an international conference held at Boyne Highlands Inn, October 13-15, 1985, and Airlie House, December 11–13, 1985.
15 Williams, C.N. and K.T. Joseph. 1976. Climate, Soil and Crop Production in the Humid Tropics. Revised Edition. Oxford University Press, London, 177 pp.

SECTION TWO

Basic research and Germplasm status

Plant germplasm preservation: a global perspective

J. TREVOR WILLIAMS
International Board for Plant Genetic Resources, c/o FAO of the United Nations, Via delle Terme di Caracalla, 00100 Rome, Italy

Abstract An enormous amount of crop genetic resources has been collected in the past two decades. Emphasis now is shifting to the post-collecting phases and to problems of storing material not easily conserved as seed. Coordination of work among genebanks is badly needed to carry out conservation and evaluation work efficiently. However, the budgets to support necessary research and development will be large, but are not yet in evidence. There is a growing awareness of the importance of germplasm conservation, but translation into financial support must come from various national, international, and industry sources. This paper relates action taken by the International Board for Plant Genetic Resources since its creation in 1974; draws out the lessons from this experience; and outlines future needs in research and organization.

Key words: genetic resources, conservation, global network, national programs, evaluation, funding, *in vitro* technology

Introduction

The past fifteen years have witnessed a tremendous development of new programs around the world dealing with various aspects of plant genetic resources conservation. This has stemmed from the realization, during the late 1960s and 1970s, that there is an urgent need to collect and preserve primitive forms of cultivated plants before they are lost to rapid agricultural development, particularly the introduction of modern crop varieties.

Much of this work has followed directly and indirectly the establishment of the International Board for Plant Genetic Resources (IBPGR) in 1974. The IBPGR was created by the Consultative Group on International Agricultural Research (CGIAR) to salvage threatened germplasm and to ensure that it is described, documented, and conserved. This activity was not to be undertaken solely by the IBPGR, which was and continues to function as a secretariat organization, but through the purposeful build-up of institutions throughout the world.

L. Knutson and A.K. Stoner (eds.). Biotic diversity and germplasm preservation, global imperatives.
 ISBN 0-7923-0178-1

When IBPGR was created, the existing programs were largely in the developed countries of the world and in some of the other centers of the CGIAR. Many had grown out of plant introduction programs or were centered around historic collections of germplasm that had been assembled for breeding and scientific research. Few conservation facilities existed and materials were not, on the whole, well preserved nor, in many cases, documented [1]. Yet the conservation of plant genetic resources had come to be recognized as a global necessity requiring international initiatives to safeguard a heritage of mankind, and to make it available to all who could use it.

In the early years the most urgent action necessary was to collect the landraces of major cereals. However, by the late 1970s genetic erosion had extended to minor cereals, food legumes, vegetables, and fruits; and by the early 1980s to many parts of forage genepools as well. In consequence, over a 15-year period, the IBPGR has organized some 400 missions in over 100 countries to collect germplasm for conservation.

Strategy for the development of a global program

The IBPGR has a responsibility to assist in establishing new genetic resources programs, integrate these with ongoing ones, and to fill gaps as expediently as possible. Thus, while a rescue operation was underway to collect endangered germplasm, effort was also placed on building up adequate conservation facilities and on providing training to increase the number of personnel to staff them. This approach was adopted because the IBPGR is not a technical assistance agency as such but rather is charged with acting as a catalyst. It provides pump-priming funding and scientific expertise to set programs in motion, but does not support them indefinitely.

In carrying this out, IBPGR is unusual in that it has been able to work with nearly all countries of the world, whether developing or developed and regardless of political affiliation. However, funding is largely channeled to developing countries or to research, that will result in benefits to the Third World. Currently 96 countries have developed some form of genetic resources program, and a further 21 carry out related activities. IBPGR works with all, apolitically and on an equal-partnership basis. Not all of these programs yet have the right strengths, nor do they meet all the desirable scientific standards, but the groundwork has been laid for further development and for other organizations to assist.

Concurrently, IBPGR has helped to mobilize the scientific community to serve on committees and advisory groups. Objectives for collecting and priorities for follow-up work were established for specific crops. For other aspects, such as seed conservation, it was necessary to assess the current state of the relevant sciences, arrange for transfer of technology or initiate new research as needed.

As a scientific center, therefore, IBPGR developed a central coordination role and also became a repository for information providing know-how on collecting, conservation, characterization, and documentation. It has also developed and reorganized its own structure from time to time to better serve the world's genetic resources community. For example, IBPGR locates Field Staff to liaise with scientists and governments in the developing world where national programs need to be stimulated and assisted. Even so a greater awareness of the value of genetic resources work still needs to be created. More recently, growth in the IBPGR has led to the differentiation of Field and Research Programs.

It has become clear that several prerequisites are necessary for the success of a national program:

1. A strong scientific community of users, particularly breeders,
2. An adequate infrastructure of genebanks, documentation and other facilities,
3. Sufficient financing, and
4. Trained and conscientious personnel committed to doing the work to recognized standards.

Over the past decade, it has become apparent that an effective global network comprises individual groupings of scientists and breeders working on particular crops or problems. The national genetic resources units (genebanks, information systems, etc.) are administrative, logistic, and service components of this, not generally themselves the prime movers. Opinion has generally moved away from regional frameworks that work in particular areas only. However, regional frameworks have proven effective for specific tasks such as the European Cooperative Programme for the Conservation and Exchange of Crop Genetic Resources (ECP/GR) and the Latin American Maize Project (LAMP).

Complementary methods of conservation

Plant genetic diversity can be conserved in a number of ways. For any crop or genepool, more than one method may be applicable.

Most of the staple food crops, vegetables, and forages can be conserved effectively as 'orthodox' seed, by drying to low moisture and storage at low temperatures. Seed is always preferred to vegetative material where possible; because the storage technology is tried and tested, seed is known to be genetically stable, represents the most robust method of distribution, and is more free of disease problems. Even for many vegetatively propagated crops, genetic conservation (of the genes but not the genotype) is adequately cared for by storage of seed. Hence, it is worth researching some crops such as yam to determine how to cause plants to produce seed. Estimates of materials currently conserved as orthodox seed (approaching 2,000,000 samples) are shown in Table 1.

There are, however, a number of important species whose seed is 'recalcitrant', intolerant of drying or chilling. These include cocoa, rubber, many tropical fruits and many forest species. Some others, such as banana clones, are completely sterile. Additionally, clones of some (fertile) vegetatively propagated crops are required as starting points in breeding. Presently, all these must be conserved in field gene banks (clonal repositories or plantations), and for the wild populations, *in situ* conservation may well be the best method. However, a great effort is being put into developing conservation *in vitro*. Because of the greater complexity and expense of these approaches when compared with seed storage, the numbers of clones of any particular crop to be conserved will be limited.

Conservation may be long-term in base collections for security, or medium-term and for use in active collections (on the whole, breeders' working collections are excluded from the formal conservation framework). Currently there are adequate long-term conservation methods for most orthodox seeded crops, but not for *in vitro* cultures, where much more research on cryopreservation is needed. For *in situ* genetic reserves, little effort has yet been put into their selection and design to conserve maximum diversity, and even less has been expended on monitoring and management.

The various methods of conservation are complementary, not exclusive alternatives, and the requirements for any one crop may be met by a mixture of approaches. Thus, while particular potato clones may be conserved in field genebanks or as plantlets grown *in vitro*, their genetic variation can also be conserved in medium and long-term storage through the orthodox seed they produce.

With the large amount of collecting carried out over the past decade, much of which must be considered irreplaceable, it has been important to arrange a formal security element in conservation. Accordingly, IBPGR has reached agreement with a number of institutions to hold, on behalf of the international community, designated base collections of seed crops. Thirty-eight centers in all parts of the world have, to date, accepted such designation. These collections cover all the major food crops, vegetables, and forages. Some improvements are still necessary, particularly more systematic duplication for safety and, in some cases, a need to upgrade standards. These changes are under way since IBPGR inaugurated a registration and monitoring process of the network.

Table 1. Estimated number of germplasm accessions of major crops world-wide updated from [3].

Crop	Number of	
	Accessions	Distinct samples
Cereals		
Wheat	410,000	120,000
Barley	280,000	40,000
Rice	215,000	90,000
Maize	100,000	25,000
Sorghum	95,000	20,000
Oats	37,000	10,000
Pearl millet	31,500	15,000
Finger millet	9,000	3,000
Other millets	16,500	5,000
Rye	18,000	8,000
Pulses		
Phaseolus spp.	105,000	40,000
Soybean	100,000	18,000
Groundnut	34,000	11,000
Chickpea	25,000	10,000
Pigeonpea	22,000	10,000
Pea	20,000	6,500
Cowpea	20,000	10,000
Mungbean	16,000	5,000
Lentil	13,500	5,000
Faba bean	10,000	5,000
Lupin	3,500	2,000
Root crops		
Potato	42,000	20,000
Cassava	14,000	5,000
Yams	10,000	5,000
Sweet potato	8,000	5,000
Vegetables		
Tomato	32,000	10,000
Cucurbits	30,000	15,000
Cruciferae	30,000	15,000
Capsicum spp.	23,000	8,000
Allium spp.	10,000	5,000
Amaranths	5,000	3,000
Okra	3.600	2,000
Eggplant	3,500	2,000

Materials held in active collections, typically medium-term storage, are those that are grown out for description, documentation, and distribution to users. Currently IBPGR is developing a strategy for the identification and coordination of those active collections that meet scientific standards and which hold significant parts of crop genepool diversity. This would be done on a crop by crop basis.

The most up-to-date information on holdings for most major crops is shown in Appendix tables 1-8.

The current global network

It is estimated that, irrespective of genebank facilities, there are about 400 significant collections of crop germplasm. To improve the world network, a number of points need urgent attention:

1. There is excessive and redundant duplication at all levels. Within the same genebank the amount of duplication varies between 10 to 50%, depending upon the crop. Excessive duplication of parts of crop collections between genebanks is a common phenomenon. Conversely, adequate systematic duplication for safety is an essential requirement, and is not always being met.

2. Managers of individual collections rarely correspond with their counterparts in other countries to attempt to divide responsibilities and lead to more efficient and more useful work. It is urgent, therefore, to establish central information databases for each crop as this will lead to more efficient and effective collaboration among the individual programs. Such cooperative networks have already been started for some crops in certain regions, e.g., the ECP/GR and LAMP.

3. Insufficient attention has been paid to passport data. Since evolutionary theory implies an ecogeographical basis for diversity, data on the origin of samples is critical. Only a few centers holding germplasm collections have made major efforts to gather such data. Further, many centers have not verified the information they do have. Thus, in some cases the origin of the sample refers not to the location of collection, but only to the donor institute or country that provided the material. Since 1980, IBPGR has been encouraging its collaborating centers to acquire or update passport data on their collections as matter of urgency.
   Similarly, species identifications and verifications rarely seem to enter the databases. This will become a major handicap as more and more atten-

APPENDIX tables updated from [3].

Appendix *table 1.* Wheat (*Triticum* spp.) in genebanks.

No. of accessions	Type[a] of storage	Institution	Location
74,500	M, L	VIR	Leningrad, USSR
39,000	M	USDA	Aberdeen, ID, USA
37,477	L	NSSL	Fort Collins, CO., USA
31,144	M, L	CIMMYT	El Batan, Mexico
31,000	M	ARO	Bet Dagan, Israel
26,000	M, L	IG	Bari, Italy
22,100	L	NSWDA	Tamworth, Australia
20,000	M, L	CGI	Beijing, P.R. China
16,596	M	ICARDA	Aleppo, Syria
16,000	M, L	IARI	New Delhi, India
13,600	M, L	IPIGR	Plovdiv, Bulgaria
10,875	M, L	FAL	Braunschweig, F.R. Germany
10,000	M, L	ZGK	Gatersleben, German D.R.
8,000	M, L	IHAR	Radzikov, Poland
7,201	M, L	PGRC	Addis Ababa, Ethiopia
7,000	S	CNPT	Passo Fundo, Brazil
6,774	M	PGI	Kyoto, Japan
6,000	S	RICTP	Fundulea, Romania
6,000	M	PARC	Islamabad, Pakistan
5,000	M, L	IGPB	Prague, Czechoslovakia
4,852	M, L	INTA	Pergamino, Argentina
4,506	M, L	PBI	Cambridge, UK
4,200	M, L	NIAR	Tsukuba, Japan
4,000	M	SPA	Wukung, P.R. China
4,000	M, L	ARARI	Menemen, Turkey
4,000	M, L	CGN	Wageningen, Netherlands
4,000	M, L	NIAVT	Tapioszele, Hungary
2,500	S	INRA	Versailles, France
2,000	S	UC	Riverside, CA, USA
1,726	M	DARS	Kabul, Afghanistan
1,200	M	UNA	Lima, Peru

[a] S = short-term; M = medium-term; L = long-term.

tion is given to the wild species of the wider genepools. It is already a major constraint for organizations promoting *in situ* conservation.

4. Even though the tables included here list numbers of accessions, far too much attention is paid to shear numbers rather than to estimates of what patterns of diversity are encompassed by the collections. Since determining these patterns involves systematic characterization and evaluation of samples, which is demanding on resources (financial and manpower),

Appendix *table 2*. Barley (*Hordeum* spp.) in gene banks.

No. of accessions	Type of storage	Institution	Location
25,284	L	NSSL	Fort Collins, CO, USA
26,441	M	USDA	Aberdeen, ID, USA
22,316	L	PGRO	Ottawa, Canada
19,500	S	CNPT	Passo Fundo, Brazil
23,500	M, L	VIR	Leningrad, USSR
15,195	M	ICARDA	Aleppo, Syria
13,900	L	NGB	Lund, Sweden
10,266	M, L	ZGK	Gatersleben, German D.R.
5,569	M, L	CIMMYT	El Batan, Mexico
5,314	M, L	NIAR	Tsukuba, Japan
8,274	M, L	FAL	Braunschweig, F.R. Germany
6,421	M, L	PGRC	Addis Ababa, Ethiopia
9,458	M, L	UC	Cambridge, UK
2,321	M, L	IHAR	Radzikow, Poland
3,500	S	RICTP	Fundulea, Romania
3,200	M, L	IPIGR	Plovdiv, Bulgaria
3,100	S	SCRI	Edinburgh, UK
3,000	M, L	IARI	New Delhi, India
2,912	M, L	CGN	Wageningen, Netherlands
2,600	M	PARC	Islamabad, Pakistan
2,000	M, L	CSIRO	Canberra, Australia
1,500	M	UNA	Lima, Peru
1,278	M, L	RIPP	Prague, Czechoslovakia
1,333	M, L	GGB	Thessaloniki, Greece
1,275	S	CNIA	Buenos Aires, Argentina
1,240	S, M	IPB	Jokioinen, Finland
1,000	S, M	ARO	Bet Dagan, Israel
1,242	M, L	IG	Bari, Italy

many of the centers have been unable to give much attention to such exercises nor to the subsequent rationalization of collections they may allow. This work needs to be encouraged and, if necessary, supported.

5. Methods of regeneration of samples leave much to be desired. More thought needs to be given to handling wild species and landraces, especially when such germplasm is outbreeding or mixed inbreeders/outbreeders, in order to preserve the diversity captured. New standards and guidelines are needed.

6. Funding is, and will continue to be, a major problem. A conservative estimate of immediate needs for evaluation of major crop germplasm is US$ 30 million over a 5-year period. It is assumed that about 60% of

Appendix *table 3*. Rice (*Oryza* spp.) in genebanks.

No. of accessions	Type of storage	Institution	Location
82,000	M, L	IRRI	Los Banos, Philippines
18,063	L	NSSL	Fort Collins, CO, USA
18,000	M, L	NIAR	Tsukuba, Japan
13,511	M	CRIFC	Bogor, Indonesia
13,050	S	CRRI	Cuttack, India
11,230	S, M	USDA	Aberdeen, ID, USA
10,556	M, L	IITA	Ibadan, Nigeria
8,226	M	WARDA	Monrovia, Liberia
6,675	M, L	CENARGEN	Brasilia, Brazil
6,650	M	IRAT	Montpellier, France
6,000	S, M, L	RRI	Bangkok, Thailand
5,100	M	BRRI	Dhaka, Bangladesh
3,592	S	IRA	Tananarive, Madagascar
3,500	M, L	VIR	Krasnodar, USSR
3,200	S	INTA	Cordoba, Argentina
3,130	M, L	KSB	Penang, Malaysia
2,745	S	CARI	Peradeniya, Sri Lanka
2,500	M, L	CGI	Beijing, P.R. China
2,080	S	ARI	Yezin, Burma
1,500	S	NIAS	Hanoi, Vietnam
1,419	M	PARC	Islamabad, Pakistan
1,400	S	CARS	Lilongwe, Malawi
1,027	M, L	AES	Suweon, Rep. of Korea
1,000	M	ICA	Bogota, Colombia
1,000	M	ORSTOM	Paris, France

the costs could be mobilized as an international input. Such figures should be compared to the total annual expenditures worldwide on crop genetic resources, which is a little in excess of US$ 55 million. The international provisions are at present 25% of this total, and they deal largely with organizational development and national research needs. From where will the inducement come to channel such funds into evaluation? It will come if there is a lobby of breeders and biotechnologists for such work, and cooperative networks will be essential to represent the needs of developing countries. On this point, many developing countries need to lead first with strong, aggressive plant introduction programs and the training of breeders before initiating genetic resources conservation programs; such needs have tended to be overlooked in recent years.

Appendix *table 4*. Maize (*Zea mays*) in genebanks.

No. of accessions	Type of storage	Institution	Location
19,858	M, L	VIR	Leningrad, USSR
12,500	M	CIMMYT	El Batan, Mexico
12,500	L	NSSL	Fort Collins, CO, USA
8,000	M	IMR	Belgrade, Yugoslavia
8,000	M	INIA	Chapingo, Mexico
4,890	S	ICA	Medellin, Colombia
4,225	M	ISU	Ames, IA, USA
3,444	M	UNA	Lima, Peru
3,200	S	RICTP	Fundulea, Romania
3,000	M, L	INTA	Pergamino, Argentina
3,000	S	CNU	Daejeon, Rep. of Korea
2,654	M, L	NIAR	Tsukuba, Japan
2,220	M	CIFEP	Cochabamba, Bolivia
1,700	L	PGRO	Ottawa, Canada
1,571	M, L	IARI	New Delhi, India
1,500	M	NARS	Kitale, Kenya
1,368	M	CRIFC	Sukamandi, Indonesia
1,306	M	MRI	Trnava, Czechoslovakia
1,300	M, L	IPB	Los Banos, Philippines
1,218	M, L	INIA	Madrid, Spain
1,000	M	MI	Braga, Portugal

Appendix *table 5*. Sorghum (*Sorghum bicolor*) in genebanks.

No. of accessions	Type of storage	Institution	Location
24,600	M, L	ICRISAT	Hyderabad, India
14,000	L	NSSL	Fort Collins, CO, USA
9,815	S	SRPIS	Experiment, GA, USA
9,615	M, L	VIR	Leningrad, USSR
5,000	M, L	PGRC	Addis Ababa, Ethiopia
4,900	S	RICTP	Fundulea, Romania
4,610	S	USSCFS	Meridian, MS, USA
4,000	S	MITA	Mayaguez, Puerto Rico
4,000	S	ASP	Tihama, Yemen
3,000	M, L	CGI	Beijing, P.R. China
3,000	M	INIA	Chapingo, Mexico
2,700	S	INTA	Cordoba, Argentina
2,626	M	ORSTOM	Bondy, France
2,072	M, L	IPB	Los Banos, Philippines
2,000	M, L	IARI	New Delhi, India

Appendix *table 6*. Millets in genebanks.

No. of accessions	Type of storage	Institution	Location
PEARL MILLET (*Pennisetum typhoides*)			
18,148	M, L	ICRISAT	Hyderabad, India
12,431	S	AICMIP	Poona, India
2,737	M	IRSTOM	Bondy, France
1,776	L	PGRO	Ottawa, Canada
1,000	L	NSSL	Fort Collins, CO, USA
FOXTAIL MILLET (*Setaria italica*)			
3,588	S	AICMIP	Poona, India
3,226	M, L	CGI	Beijing, P.R. China
1,429	S	AICMIP	Bangalore, India
1,260	M, L	ICRISAT	Hyderabad, India
FINGER MILLET (*Eleusine coracana*)			
2,960	S	AICMIP	Bangalore, India
2,944	S	AICMIP	Poona, India
1,863	M, L	ICRISAT	Hyderabad, India

Research needs

Although the processes and needs for collecting, conservation, characterization, evaluation, and documentation of germplasm accessions are now widely understood in the plant genetic resources community, what has been outlined above does indeed imply that not all processes are being pursued properly. Against this background, there is also a need to recognize that the knowledge base on which the work is founded is extremely slender. Accordingly, IBPGR has emphasized strategic research to secure the fundamental advances necessary before applied research can be undertaken effectively. The research agenda of IBPGR comprises a number of initiatives, examples of which follow.

Seed conservation technology still requires research, especially to find newer, more cost-effective methods of conservation, *e.g.*, the long-term storage of very dry seed (< 5% moisture) at room temperature. The genetic stability of such stored seeds has hardly been researched. Despite progress with orthodox seeds, problems remain with recalcitrant seeds. One new

Appendix *table 7*. Legume crops genebanks.

No. of accessions	Type of storage	Institution	Location
Soybean (*Glycine max*)			
10,200	M, L	AVRDC	Shanhua, Taiwan, China
8,350	L	NSSL	Fort Collins, CO, USA
3,500	M, L	NIAS	Tsukuba, Japan
3,000	M, L	VIR	Leningrad, USSR
3,000	S, M	OBCI	Wuhan, P.R. China
2,900	S, M	SAA	Shadong, P.R. China
2,900	M	CRIFC	Sukamandii, Indonesia
2,337	M, L	IARI	New Delhi, India
1,500	M	INIA	Chapingo, Mexico
1,359	M, L	IITA	Ibadan, Nigeria
1,339	M, L	IPB	Los Banos, Philippines
1,000	S, M	LAA	Harbin, P.R. China
1,000	L	PGRO	Ottawa, Canada
Common bean (*Phaseolus vulgaris*)			
30,790	M, L	CIAT	Cali, Colombia
9,321	M	WRPIS	Pullman, WA, USA
8,900	M	INIA	Chapingo, Mexico
5,000	M	CU	Cambridge, UK
4,456	L	NSSL	Fort Collins, CO, USA
4,202	S	EMBRAPA	Goias, Brazil
3,851	M, L	ZGK	Gatersleben, German D.R.
3,109	S	ICA	Bogota, Colombia
2,679	S	INIPA	Lima, Peru
2,627	M, L	NIAVT	Tapiozele, Hungary
2,575	M	NRPIS	Geneva, NY, USA
2,420	M, L	IPIGR	Sadovo, Bulgaria
2,000	S	NU	Nairobi, Kenya
2,000	S	MU	Lilongwe, Malawi
1,800	L	CENARGEN	Brasilia, Brazil
1,712	S	THRS	Thike, Kenya
1,700	S	IIHR	Bangalore, India
1,500	S	UNA	Lima, Peru
1,242	S	ICCPT	Fundulea, Romania
1,162	S	FONAIAP	Caracas, Venezuela
1,042	L	INIA	Madrid, Spain
1,000	L	CGN	Wageningen, Netherlands
1,000	S	NIAR	Butare, Rwanda

Appendix *table 7*. (Continued).

No. of accessions	Type of storage	Institution	Location
Lima bean (*Phaseolus lunatus*)			
3,846	M	NBI	Bogor, Indonesia
2,527	M, L	CIAT	Cali, Colombia
Groundnut (*Arachis hypogaea*)			
11,641	M, L	ICRISAT	Hyderabad, India
6,274	S	NRCG	Junagadh, India
6,133	S	SRPIS	Experiment, GA, USA
4,229	L	NSSL	Fort Collins, CO, USA
4,200	M	NCSU	Raleigh, NC, USA
4,000	M	OSU	Stillwater, OK, USA
2,712	M	INTA	Manfredi, Argentina
2,496	M	TAMU	Stephenville, TX, USA
1,800	L	IAC	Campinas, Brazil
1,730	M	CRIFC	Sukamandi, Indonesia
1,500	M	UF	Mariana, FL, USA
1,200	M	UF	Gainesville, FL, USA
1,053	M, L	VIR	Leningrad, USSR
900	S	ISRA	Bambey, Senegal
Cowpea (*Vigna unguiculata*)			
12,000	M, L	IITA	Ibadan, Nigeria
3,930	M	NBI	Bogor, Indonesia
2,537	M, L	CENARGEN	Brasilia, Brazil
1,616	M, L	IARI	New Delhi, India
1,386	M, L	IPB	Los Banos, Philippines
1,150	S	SRPIS	Experiment, GA, USA
1,050	M, L	VIR	Leningrad, USSR
Chickpea (*Cicer arietinum*)			
14,361	M, L	ICRISAT	Hyderabad, India
5,585	M	ICARDA	Aleppo, Syria
3,396	M	WRPIS	Pullman, WA, USA
2,584	L	PARC	Islamabad, Pakistan
2,031	L	NSSL	Fort Collins, CO, USA
2,000	M, L	IARI	New Delhi, India
1,685	M, L	VIR	Leningrad, USSR

Appendix *table 7*. (Continued).

No. of accessions	Type of storage	Institution	Location
Pigeonpea (*Cajanus cajan*)			
10,818	M, L	ICRISAT	Hyderabad, India
1,500	M, L	IARI	New Delhi, India
Lentil (*Lens culinaris*)			
6,086	M	ICARDA	Aleppo, Syria
1,197	M, L	IARI	New Delhi, India
Faba bean (*Vicia faba*)			
3,293	M	ICARDA	Aleppo, Syria
2,525	L	VIR	Leningrad, USSR
1,469	M, L	IG	Bari, Italy
1,136	L	FAL	Braunschweig, F.R. Germany
Mungbean (*Vigna radiata*)			
5,736	M, L	UPLB	Los Banos, Philippines
5,483	M, L	AVRDC	Shanhua, Taiwan, China
3,000	S	PAU	Ludhiana, India
2,172	M	CRIFC	Sukamandi, Indonesia
2,100	S	UM	Columbia, MO, USA
1,850	S	IARI	New Delhi, India
Pea (*Pisum sativum*)			
4,090	L	IG	Bari, Italy
3,062	L	NGB	Lund, Sweden
3,058	M	ICARDA	Aleppo, Syria
2,694	M	NRPIS	Geneva, NY, USA
2,067	M	JIT	Norwich, UK
1,818	L	ZGK	Gatersleben, German D.R.
1,429	L	NSSL	Fort Collins, CO, USA
1,296	L	FAL	Braunschweig, F.R. Germany
1,223	L	NIAVT	Tapiozele, Hungary
1,011	L	PGRC	Addis Ababa, Ethiopia

Appendix *table 8*. Root and tuber crops in genebanks.

No. of accessions	Type of storage	Institution	Location
Potato (*Solanum* spp.)			
9,435	M, L	VIR	Leningrad, USSR
6,500	M, L	CIP	Lima, Peru
5,000	M, L	CENARGEN	Brasilia, Brazil
4,286	M	UB	Birmingham, UK
4,000	S	INIA	Toluca, Mexico
2,800	M, L	IRPIS	Sturgeon Bay, WI, USA
2,605	M	CPRI	Simla, India
2,370	M, L	FAL	Braunschweig, F.R. Germany
1,282	M, L	AVRDC	Shanhua, Taiwan, China
1,250	M	SCRI	Edinburgh, UK
1,000	S	ICA	Tibaitata, Colombia
Sweet potato (*Ipomoea batatas*)			
1,243	M, L	CIP	Lima, Peru
1,200	M	NBI	Bogor, Indonesia
1,200	S	KNAES	Kagoshima, Japan
1,200	M, L	AVRDC	Shanhua, Taiwan, China
1,000	M, L	IITA	Ibadan, Nigeria
Cassava (*Manihot esculenta*)			
3,700	M, L	CIAT	Cali, Colombia
1.829	M, L	IITA	Ibadan, Nigeria
1,327	M, L	CTCRI	Kerala, India

approach in this area is the examination of embryo technology; in some cases it seems that embryos may be ‘orthodox’ and more easily conserved that the entire seed.

The potential of *in vitro* conservation has attracted the genetic resources community for some time, but current methods are generally at the developmental stage and, where available, are only reasonably reliable for medium-term storage of a limited number of crops. The Centro Interncional de Agricultura Tropical (CIAT, Colombia) and IBPGR are testing the methodology for medium-term conservation using cassava as a model. For long-term conservation, research on cryopreservation appears to offer the best potential. The long-term part is under investigation, again using cassava. Much more research is needed on other crops [2]. One complication is that *in vitro* culture technology can generate variation (somaclonal variation). Although this may be useful for breeding, for conservation the release of such variation must be minimized. This area is being researched in association with the development of culture technologies.

One spin-off of IBPGR *in vitro* research has been the development of low-technology collecting techniques for embryos or vegetative materials difficult to handle in the field. These techniques have been researched for cocoa and are now being used in the field by several national programs for cassava, perennial cotton species and coconut. Techniques for sterile or shy-seeding African forages and species of *Musa* are being tested.

For most germplasm maintained *in vitro,* disease indexing is an essential prerequisite to moving materials among countries. Such work needs to be accelerated, and IBPGR is currently testing a new concept of totally enclosed indexing that should provide relief to overworked or poorly functioning quarantine services. In the same vein, quarantine and pathology aspects of movement of seed of wild germplasm, which often is available only in small quantities, have largely been overlooked. Adequate checks are needed, but the excessive use of seed in destructive methods of testing might be avoided if new indexing methods are developed.

A major constraint facing all programs in the past five years in implementing collecting of diversity from the wider genepools has been the lack of knowledge on the distributions of wild species, breeding systems and species relationships in complex genepools. Newer biochemical and molecular techniques are proving to be of great value in this area of applied research on diversity. IBPGR is researching a number of species complexes in primary, secondary, and tertiary genepools, the extent depending on the level of sophistication of breeding of specific crops, but a much wider effort is required to cover immediate practical problems. This would improve our understanding of the ranges of variation in collections and aid in the identification of collecting gaps.

References

1. Williams J.T. 1984. A decade of crop genetic resources research. In: J.H.W. Holden and J.T. Williams (eds.) Crop Genetic Resources: Conservation and Evaluation pp. 1–17. George Allen and Unwin, London.
2. IBPGR. 1987. Annual Report for 1986. International Board for Plant Genetic Resources, Rome, Italy.
3. Plucknett D.L., Smith N.J.H., Williams J.T. and Anishetty M. 1987. Genebanks and the World's Food. Princeton University Press, Princeton. 247 pp.

Program for plant germplasm preservation in the United States: the U.S. National Plant Germplasm System

HENRY L. SHANDS[1], PAUL J. FITZGERALD[2] and STEVE A. EBERHART[3]

[1] *National Program Leader for Germplasm, National Program Staff, Agricultural Research Service, U.S. Department of Agriculture, Rm. 140, Bldg. 005, BARC-W, Beltsville, Maryland 20705, U.S.A.*
[2] *Agricultural Science Advisor for Plant Germplasm, Agricultural Research Service, U.S. Department of Agriculture, Department of Agronomy, Purdue University, West Lafayette, Indiana 47907-7899, U.S.A.*
[3] *Director, National Seed Storage Laboratory, Agricultural Research Service, U.S. Department of Agriculture, Colorado State University, Fort Collins, Colorado 80523, U.S.A.*

Abstract The United States' National Plant Germplasm System (NPGS) coordinates Federal, State, and private sector efforts to collect, maintain, and preserve plant germplasm for potential use in the improvement of crops used in agriculture and industry. The NPGS accomplishes its functions by use of specialized units organized to provide operational, advisory, and administrative support. Base and working collections of all important crop plants in the U.S. are maintained as seed, whole plants, or plant parts in special facilities. These collections are evaluated, enhanced, and distributed under policies that assure accessibility, genetic integrity, and long-term security. The NPGS is supported by research programs to enhance use of genetic diversity and to develop safe and economical methods of alternative preservation. The Germplasm Resources Information Network, a centralized national information repository, is a source of readily available information about NPGS collections. The NPGS maintains active relations with the international plant germplasm system through exchanges of genetic materials and serves as a backup repository for international collections of maize, rice and other crops.

Key words: germplasm, genebank, GRIN database, national plant germplasm system, National Seed Storage Laboratory, regional plant introduction station, national clonal germplasm repository

Mission

The National Plant Germplasm System (NPGS) is comprised of a network of cooperating institutions, agencies, and research units in the Federal, State, and private sectors. Participants at the National Workshop on Plant Germplasm held at Peoria, Illinois, April 9-12, 1981 developed a statement

L. Knutson and A.K. Stoner (eds.). Biotic diversity and germplasm preservation, global imperatives.
 ISBN 0-7923-0178-1

of the mission of the NPGS which is just as meaningful today:

> 'The National Plant Germplasm System (NPGS) provides the genetic diversity necessary to improve crop productivity and to reduce genetic vulnerability in future food and agriculture development, not only in the United States but for the entire world. The NPGS acquires, maintains, evaluates, and makes readily accessible to plant scientists a wide range of genetic diversity in the form of seed and clonal germplasm of crops and potential new crops.'

These activities form a continuous flow of genes from source to end use, a continuum that keeps high-yielding varieties on the market; improves the quality of agricultural products; minimizes production costs; reduces dependence on pesticides, thus enhancing the quality of the environment; and minimizes the vulnerability of agriculturally important germplasm to pests and environmental stresses [1–3].

Structure

The NPGS is organized into three functional elements: operations, advisory, and administrative (Fig. 1). Each element has a distinct but interacting

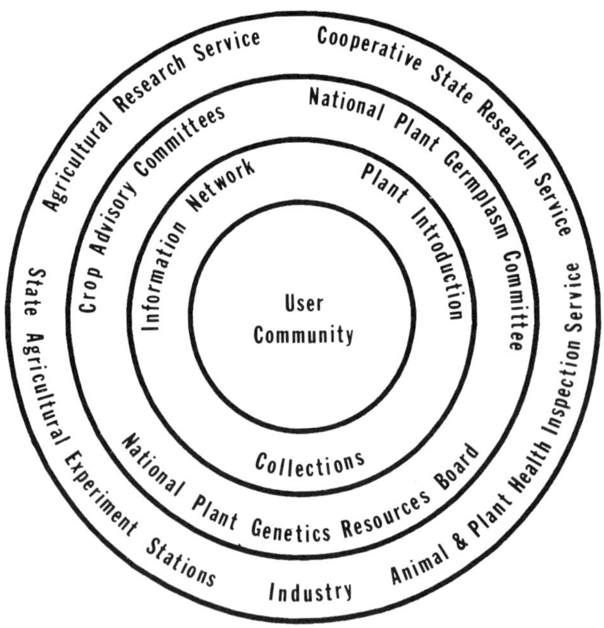

Figure 1. Schematic representation of the National Plant Germplasm System showing relationship of the operations, advisory, and administrative elements and their focus on the user.

Table 1. Common abbreviations used by the U.S. National Plant Germplasm System. May 1988.

APHIS	–	U.S. Department of Agriculture (USDA) Animal and Plant Health Inspection Service – enforces national quarantine regulations including issuance of certificates for introduction and export of plant materials.
ARS	–	USDA Agricultural Research Service.
ASTA	–	American Seed Trade Association – national organization of private firms involved with seed trade issues, including germplasm.
CAC	–	Crop Advisory Committee – composed of Federal, State, and private industry members of many disciplines who provide crop-specific technical advice to the National Plant Germplasm System.
CGIAR	–	Consultative Group for International Agricultural Research – headquartered in Washington, DC, and involved with management of agricultural development programs at thirteen international centers.
CSRS	–	USDA Cooperative State Research Service – supports research cooperatively between the Federal Government and the State Agricultural Experiment Stations.
GRIN	–	Germplasm Resources Information Network database of the Germplasm Services Laboratory, Plant Sciences Institute, ARS, Beltsville, Maryland.
GSL	–	Germplasm Services Laboratory, Plant Sciences Institute, ARS, Beltsville, Maryland – houses GRIN, Plant Exploration Office, and Plant Introduction Office.
IARC	–	International Agricultural Research Center – managed by CGIAR and the Food and Agriculture Organization, World Bank, and donors.
IBPGR	–	International Board for Plant Genetic Resources – one of the thirteen IARC's; dedicated to the international acquisition and preservation of germplasm resources; headquartered in Rome, Italy.
IR	–	Inter-Regional research project of CSRS – notably IR-1 potato project, Sturgeon Bay, Wisconsin) and IR-2 (virus-free fruit stocks, Prosser, Washington); authorized by 1946 Agricultural Marketing Act.
NCCPB	–	National Council of Commercial Plant Breeders – a component of the ASTA which deals with commercial breeding interests including plant property rights and germplasm issues.
NCGR	–	National Clonal Germplasm Repository – preserves clonally propagated fruit and nut crops and initially developed by ARS and CSRS joint funding, now supported by ARS operations funding.
NPGC	–	National Plant Germplasm Committee – national advisory committee to the National Plant Germplasm System; Federal, State, and industry membership.
NPGQC	–	National Plant Germplasm Quarantine Center, Beltsville, Maryland – jointly managed by APHIS and ARS; plant virus indexing work conducted by NPGQL personnel of ARS.

Table 1. (Continued).

NPGQL	–	National Plant Germplasm Quarantine Laboratory, ARS Plant Sciences Institute, Beltsville, Maryland – supplies personnel to conduct indexing and research at the NPGQC.
NPGRB	–	National Plant Genetic Resources Board – advisory to the Secretary of Agriculture on policy matters; Federal, State, and industry membership.
NPGS	–	National Plant Germplasm System – the total set of operatives relating to plant germplasm involving Federal, State, and private institutions.
NPL	–	National Program Leader – on ARS National Program Staff; NPL for Germplasm coordinates the activities of the NPGS.
NPS	–	National Program Staff of ARS.
NSGC	–	National Small Grains Collection – housed at Aberdeen, Idaho.
NSSL	–	National Seed Storage Laboratory – houses the base collection at Fort Collins, Colorado.
PEO	–	Plant Exploration Office of the GSL, Beltsville, Maryland.
PGOC	–	Plant Germplasm Operations Committee – comprised of curators of major collections and service components concentrating on operational efficiency of the NPGS.
PIO	–	Plant Introduction Office of the GSL, Beltsville, Maryland.
RPIS	–	Regional Plant Introduction Station(s) – four stations, authorized under the 1946 Agricultural Marketing Act.
RTAC	–	Regional Technical Advisory Committee – comprised of Regional Coordinator, Administrative Advisor, and state representatives to regional committee associated with RPIS.
SAES	–	State Agricultural Experiment Station(s) – act in regional associations to coordinate regional and national activities.
TAC	–	Technical Advisory Committee – advisory to a National Clonal Germplasm Repository site, specializing in those commodities maintained there.
USDA	–	U.S. Department of Agriculture – Assistant Secretary for Science and Education is responsible for ARS and CSRS; Assistant Secretary for Marketing and Inspection Services administers APHIS.

relationship with other elements. It is important to recognize that while many distinct elements represent the various components of the NPGS, these elements generally work in harmony because they are held together by common interests, values, and objectives [4–5]. Common abbreviations associated with various elements of the NPGS are provided in Table 1.

Operations

The operations element involves the acquisition and preservation of plant germplasm and the distribution of information and seed. The three major operational functions are the working (active) collections, the base collection, and germplasm services. Operations are funded and staffed by the Agricultural Research Service (ARS) of the U.S. Department of Agriculture (USDA) with strong support, both financial and personnel, from State universities.

The working collections are held primarily in four Regional Plant Introduction Stations (RPIS) authorized under the Agricultural Marketing Act of 1946. The stations, all of which receive joint Federal-State support, are located at Griffin, Georgia; Ames, Iowa; Geneva, New York; and Pullman, Washington. In earlier years the RPIS's were used primarily as repositories for exotic germplasm, but they now maintain a mixture of exotic, domestic, and elite germplasm collectively known as 'working collections' of the species for which each station has responsibility (Table 2). Each RPIS is staffed with a curator; from two to four scientists representing agronomy, horticulture, pathology, and entomology; research associates where feasible; and technical support. The RPIS's increase and distribute seed to users, evaluate germplasm, and conduct other research activities [6]. As needed, seed is also grown at the RPIS's to maintain adequate sample sizes of accessions held in the base collection.

Working collections of several important crops are held in special commodity collections at other U.S. locations (Table 3). Special genetic stock collections are identified in Table 4. Accessions in NPGS seed working collections total 300,970.

The genetic integrity of some economically important U.S. crops cannot always be preserved as seed. Many of the fruit and nut species fall into this category. That germplasm is preserved in the National Clonal Germplasm Repositories (NCGR). Representatives of each important species are maintained at the NCGR's as living plants in an environment relatively free of diseases and insect pests [7]. Since 1980, a number of clonal repositories have been established (Table 5). At present, the NCGR's serve as both working and base collections, although seed backups are available for certain species. As advances in preservation technology are made, other economically feasible methods of storage may permit the separation of the working and base collections. Accessions in the NCGR's and in quarantine at Beltsville, Maryland, total 24,720.

Duplicate samples of material in the working and commodity collections are held in a base collection at the National Seed Storage Laboratory (NSSL), Fort Collins, Colorado [3, 5]. The NSSL maintains this base

Table 2. Major plant genera and number of accessions at Regional Plant Introduction Stations. May 1988.

Ames, Iowa:		
Zea	5,929	
Cucumis	3,581	
Helianthus	3,059	
Amaranthus	2,541	
Brassica	1,533	
Other	9,134	25,777
Pullman, Washington:		
Phaseolus	10,231	
Medicago	3,795	
Cicer	3,758	
Lens	2,244	
Carthamus	1,624	
Other (Many grasses)	16,370	38,022
Geneva, New York:		
Lycopersicon	5,163	
Pisum	2,876	
Trifolium	1,865	
Brassica	1,290	
Cucurbita	674	
Other (Ornamentals)	2,904	14,772
Griffin, Georgia:		
Sorghum	14,625	
Vigna	8,172	
Arachis	7,797	
Cajanus	4,089	
Capsicum	3,115	
Other	20,023	57,821
	Total accessions	136,392

collection for the U.S. and – through formal and informal cooperation – for the global network of genetic resources centers administered by the International Board for Plant Genetic Resources [5]. In operation since 1958, the NSSL's principal mission is long-term preservation of valuable plant germplasm at low temperatures ($-20°C \pm 2°$) [7] with 85 percent or higher viability. NSSL seed is withdrawn only when all other sources of seed have been exhausted or when germination of the stored seed declines to a level that indicates that regeneration is necessary.

Table 3. National Plant Germplasm System holdings in major commodity collections. May 1988.

Soybean collection (Urbana, Illinois and Stoneville, Mississippi):			
Urbana, Illinois	Short season	8,950	
	Perennial	522	
Stoneville, Mississippi	Long season	2,995	12,467
Cotton collection (College, Station, Texas):			
College Station, Texas	Various species	2,916	
Phoenix, Arizona	*G. barbadense*	1,000	
Stoneville, Mississippi	*G. hirsutum*	1,590	5,506
Small Grains collection (Aberdeen, Idaho):			
Triticum		40,925	
Hordeum		25,649	
Avena		20,353	
Oryza		17,984	
Secale		2,304	
Tritico Secale		932	
Aegilops		617	
Other		169	108,933
Grass collections (Sites listed):			
Byron, Georgia	Bamboo	80	
Brookings, South Dakota	Native grasses	1,280	
Logan, Utah	Forage & Range grasses	1,770	
Tifton, Georgia	Pearl millet	4,287	
Tifton, Georgia	Wild *Pennisetum*	507	7,924
Other collections (Sites listed):			
Bushland, Texas	*Helianthus* species	15	
Davis, California	*Lycopersicon* species	950	
Fargo, North Dakota	*Linum*	2,617	
Lexington, Kentucky	*Trifolium* species	1,500	
Raleigh, North Carolina	*Tripsacum*	353	
Salinas, California	Leafy vegetables	2,434	
Stillwater, Oklahoma	*Arachis* species	4,005	
Sturgeon Bay, Wisconsin	*Solanum*	3,824	15,698
	Total accessions		150,528

The accessions in the NSSL base collection total 232,000 (Table 6) and include 396 genera and 1,848 species. Present categories of seed stocks in storage include basic plant introductions, recently released and obsolete cultivars, open-pollinated cultivars, parental lines, genetic stocks, differen-

Table 4. Special genetic stock collections of the National Plant Germplasm system. May 1988.

Barley	Fort. Collins, Colorado	3,000
Brassica	Madison, Wisconsin	450
Maize	Urbana, Illinois	5,000
Pea	Geneva, New York	3,000
Tomato	Davis, California	2,000
Wheat	Columbia, Missouri	600
	Total accessions	14,050

tial host and virus indicator stocks, and type specimens for future reference.

Seed accepted for storage at the NSSL is handled under four general categories:

1. The base collection,
2. Quarantined germplasm samples,
3. Genetic stocks and special collections, and
4. Plant Variety Protection Act voucher samples.

The NSSL also preserves certain genetic stocks including genetic marker lines, aneuploid stocks, etc., that merit preservation because of their genetic characteristics or scientific value. These stocks are exempted from minimum seed and germination requirements, as the condition of the material is the responsibility of the curator of the genetic stocks.

Scientists in the NSSL also conduct long-range studies on biochemical, physiological, and genetic changes in seed during storage and effects of seed moisture content, storage environment, and storage containers on seed longevity. The overall objective of these studies is to improve technologies for storing seed. Laboratory procedures for accurately monitoring the viability of stored seed are established on a crop-by-crop basis. NSSL is also responsible for conducting basic research on the cryopreservation of clonal germplasm, the objective of which is to develop technologies that could lead to the establishment of a base collection for this germplasm at the NSSL.

Electrophoresis and high performance liquid chromotography technologies are expected to be useful in comparing suspected duplicate samples and in evaluating genetic diversity among accessions of certain species. The goal is to ensure the preservation of a sample of each accession so as to represent as fully as possible the genetic diversity in each species.

With more than 230,000 accessions plus backup germplasm from international collections already in storage, the NSSL's seed vaults are expected to be filled to capacity within the next two years. Congress has appropriated funds to plan and design the NSSL's expansion, and constructon funds have been included in the FY 1989 Administration budget request to Congress.

Table 5. Holdings at National Clonal Germplasm Repositories. May 1988.

Brownwood, Texas:	*Carya*		181
Corvallis, Oregon:	*Pyrus*	2,034	
	Rubus	802	
	Vaccinium	499	
	Mentha	480	
	Frageria	470	
	Other	354	4,639
Davis, California:	*Prunus*	1,272	
	Vitis	1,267	
	Ficus	272	
	Juglans	180	
	Pistacia	145	
	Other	609	3,745
Geneva, New York:	*Malus*	2,500	
	Vitis	1,050	3,550
Hilo, Hawaii:	Tropical species		491
Miami, Florida:	*Saccharum*	2,000	
	Persea	418	
	Mangifera	277	
	Theobroma	269	
	Coffea	191	
	Other	4,121	7,276
Mayaguez, Puerto Rico:	*Theobroma*	282	
	Musa	47	
	Other	95	424
Orlando, Florida:	Citrus		400
Riverside, California:	*Citrus* spp.		789
	Citrus relatives	153	942
Brawley, California:	*Phoenix*	72	72
Beltsville, Maryland:			
Beltsville Agricultural Research Center-East			
National Plant Germplasm Quarantine Center:			3,000
Total clonally propagated fruit and nut accessions			24,720

The Board on Agriculture of the National Academy of Sciences has appointed a committee to assist in the development of specifications. Seed vaults approximately four times the size of current vaults have been requested to provide both conventional refrigerated storage at $-20\,^{\circ}C$ to $-36\,^{\circ}C$ and cryogenic storage at $-196\,^{\circ}C$. These vaults will also ensure protection against such natural disasters as floods and tornados as well as

Table 6. Base collection plant germplasm at the National Seed Storage Laboratory, Fort Collins, Colorado. May 1988.

Genus	Species	Common name	Number
Triticum	*aestivum*	Wheat	29,082
Triticum	*durum*	Durum wheat	5,463
Sorghum	*bicolor*	Sorghum	18,971
Zea	*mays*	Corn	17,070
Glycine	*max*	Soybean	10,802
Oryza	*sativa*	Rice	10,762
Avena	*sativa*	Oat	9,912
Hordeum	*vulgare*	Barley	9,435
Lycopersicon	*esculentum*	Tomato	6,439
Arachis	*hypogaea*	Peanut	5,234
Phaseolus	*vulgaris*	Bean	5,018
Solanum	*tuberosum*	Potato	3,398
Cicer	*arietinum*	Chickpea	2,793
Gossypium	*hirsutum*	Cotton	2,543
Linum	*usitatissimum*	Flax	2,492
Pisum	*sativium*	Peas	2,145
Medicago	*sativa*	Alfalfa	1,545
Carthamus	*tinctorius*	Safflower	1,499
Secale	*cereale*	Rye	1,243
Eleusine	*coracana*	Finger millet	1,212
Cyamopsis	*tetragonoloba*	Guar	1,164
Andropogon	*gerardii*	Big bluestem	1,068
Cucumis	*melo*	Muskmelon	1,027
Dactylis	*glomerata*	Orchardgrass	1,018
Ricinus	*communis*	Castorbean	997
Sesamum	*indicum*	Sesame	995
Cucurbita	*pepo*	Squash	939
Lens	*culinaris*	Lentil	884
Lactuca	*sativa*	Lettuce	869
Citrullus	*lanatus*	Citron	841
Brassica	*oleracea*	Cabbage	759
Vigna	*unguiculata*	Cowpea	742
Trifolium	*pratense*	Red clover	706
Capsicum	*annuum*	Pepper	666
Eragrostis	*curvula*	Weeping lovegrass	551
Helianthus	*annuus*	Sunflower	542
Festuca	*arundinacea*	Tall fescue	541
Setaria	*italica*	Italian millet	507
Lolium	*perenne*	Perennial ryegrass	438
Miscellaneous			69,688
		Total accessions	232,000

against vandalism. Expanded facilities are needed to initiate research on the development of techniques for early detection of declining seed viability and to find improved methods for preserving plant tissue of clonal germplasm including pollen, buds, apical meristems, and plant and cell tissue cultures.

The Germplasm Services Laboratory at Beltsville, Maryland, provides a number of important supporting elements that are essential to the effective functioning of the NPGS:

The Plant Exploration Office maintains information on previous plant explorations; provides valuable assistance in planning and facilitating new explorations; maintains important contacts in many countries where the U.S. has germplasm interests; and fosters joint studies with other nations on the adaptation and distribution of germplasm as a basis for future explorations.

The Plant Introduction Office serves as the key link between the U.S. and other nations with regard to the movement of plant germplasm. This office maintains current inventories of all U.S. collections; facilitates the addition of new accessions to U.S. collections by obtaining necessary quarantine clearances and assigning appropriate Plant Introduction numbers; works closely with USDA's Animal and Plant Health Inspection Service (APHIS) to prevent potentially damaging plant diseases and/or pests from entering the U.S. inadvertently on plant germplasm; and maintains contact with counterpart offices in other countries to facilitate the smooth and safe flow of germplasm in both directions [3].

The National Plant Germplasm Quarantine Center (NPGQC) inspects, indexes, and certifies that germplasm entering the U.S. is free of economically important diseases or insects. The greatest challenge to the NPGQC staff is the identification of virus and viruslike entities that are carried by vegetatively propagated germplasm of stone fruits (peach, plum, cherry, etc.), small fruits, vegetatively propagated grasses, and tuberous and root crops. Current diagnostic procedures make this task extremely difficult, slow, and costly. The NPGQC is jointly operated by APHIS and ARS.

The Germplasm Resources Information Network (GRIN) database links germplasm sites and accumulates information about plant germplasm from a wide range of sources. Curators and other authorized personnel only are permitted to enter inventory and evaluation data into GRIN, but these data are accessible to any ARS scientist who has access to a computer and a modem. GRIN is also accessible to State and private germplasm workers as well as to scientists in Canada, Mexico, and the International Agricultural Research Centers (IARC). Generally, common descriptors of both national and international collections are used to provide compatibility between systems and to enhance accessibility.

GRIN has become much more than a supporting element in the NPGS; GRIN binds the NPGS operational units together. It provides an automated retrieval capability for the collection and dissemination of plant germplasm information. It provides data processing expertise to support the production of databases consisting of some 450–500,000 discrete accessions. GRIN allows scientists and breeders in the U.S. and throughout the world to stay in communication. It aids in locating the plant material needed for research or breeding and facilitates rapid exchange. GRIN provides a capability to manage germplasm information and programs in a manner not previously possible [9].

Additional operational management functions are embodied in the Plant Germplasm Operations Committee (PGOC), the Regional Technical Advisory Committees (RTAC), and the Technical Advisory Committees (TAC).

PGOC membership is comprised of the key operators at the program level; they include the director of the NSSL, the curators of the RPIS's and NCGR's, and others with similar levels of responsibility. The PGOC addresses concerns that have been identified by the advisory groups and, where appropriate, develops new procedures and approaches to improve the effectiveness of the NPGS.

The RTAC's serve an important liaison function between the Directors of the State Agricultural Experiment Stations (SAES) and the RPIS's [6]. RTAC membership includes a representative from each State in the region and representatives from USDA's Soil Conservation Service, Forest Service, and Cooperative State Research Service (CSRS). Additional members may be included from specialized research laboratories such as the ARS Northern Regional Research Center. ARS is represented *ex officio* by the director of the regional RPIS. Each RTAC functions under the guidance of an Administrative Advisor appointed by the Association of State Agricultural Experiment Station Directors for the region. RTAC's maintain oversight of regional research projects and recommend changes as needed to address the highest priority needs. Annual budget requests are reviewed, and SAES Directors are advised of recommendations. Regional research projects are approved and funded annually by action of the Director's Association within each region.

TAC's perform a similar function for the NCGR's as do the RTAC's for RPIS's. TAC membership is not based on individual State representation but rather on bringing to the TAC the special expertise and experience relative to the crops for which the repository is responsible [7]. TAC's advise curators on technical matters relating to the development and operation of the clonal repositories. TAC's do not recommend levels of support for NCGR operations, but they do identify priority program needs that may justify budgetary requests. In view of the availability of Crop Advisory

Committees (CAC) for individual crops or crop groups, it may be necessary to reexamine the role of TAC's at the point when the NCGR's are well established and operating in a maintenance – rather than a developmental – mode.

Advisory

The NPGS draws upon many sources of advice and counsel to improve its operational effectiveness and efficiency. Primary among these are the National Plant Genetic Resources Board (NPGRB), the National Plant Germplasm Committee (NPGC), and more than three dozen Crop Advisory Committees (CAC). The ARS National Program Leader (NPL) for germplasm coordinates the activities of the many NPGS operational units and committees. The NPL also ensures the development of appropriate policies and long-term plans for the total system.

The NPGRB was established in 1975 to advise the Secretary of Agriculture about the assembly, description, maintenance, and effective utilization of living resources represented by crop cultivars and by primitive and wild forms of our crops [2]. The NPGRB is comprised of about 20 members and is perhaps the most influential body in the development of national germplasm policy. Members are appointed by the Secretary of Agriculture to a two-year term but may serve as many as three consecutive terms; they represent both the public and private sectors and are chosen on the basis of their diverse capabilities and are distinguished by their knowledge and interest in the management of plant genetic resources. NPGRB members are cognizant of domestic and international activities to minimize genetic vulnerability of crops. They formulate recommended actions and policies on the collection, maintenance, and utilization of plant genetic resources; recommend actions to coordinate plans relative to plant genetic resources of domestic and international organizations; recommend policies to strengthen plant quarantine and pest monitoring activities; and advise on new and innovative approaches to plant improvement. The NPGRB carries out its duties under the guidance and chairmanship of the Assistant Secretary of Agriculture for Science and Education.

The NPGC was established in 1974 after the 1972 ARS reorganization, which altered working relations with State cooperators [1, 4]. The NPGC resulted from the restructuring of the National Coordinating Committee for New Crops, created in 1949 by SAES Directors. The majority of members represented the State and Federal sectors, but industry and professional societies were also represented. Membership was altered again in 1981 when ARS replaced its representatives from the program operations level with administrators with area, regional, or national responsibilities comparable with State Administrative Advisors representing the RPIS's. Oriented more toward operations than policy, the NPGC coordinates research and service efforts of Federal, State, and industry units engaged in the introduction, preservation, evaluation, and distribution of plant germplasm through representation of all views of its members; recommends policies for conducting the national plant germplasm program and for its relationships to international plant germplasm programs; develops research and service proposals and justification for adequate funding of regional and national plant germplasm activities; and actively advocates mutually agreed-upon programs with SAES associations and USDA agencies. The NPGC serves as the principal forum in which SAES interests can be presented and harmonized with Federal interests at a technically informed level. The committee of 15–20 members normally meets twice a year.

CAC's, as currently structured and utilized, are a new and important addition to the NPGS. Each CAC is composed of a national working group of specialists who provide analysis, data, and advice about germplasm within a specific crop or group of related crops of current or future economic importance. CAC's are interdisciplinary peers of Federal, State, and industry scientists. They advise the States and USDA on all aspects of germplasm relating to the particular commodity, and they prioritize commodity needs in acquisition, maintenance, evaluation, and enhancement.

The CAC's began as commodity or crop improvement committees and remained functional as such until about 1975, when the newly created NPGRB established representative peer groups of scientists to address the status and needs of about a dozen major U.S. crops [2, 4]. Committee numbers expanded as more crops were added. The CAC further expanded when the Germplasm Resources Information Project, the forerunner of GRIN, asked these crop experts to assist in developing crop descriptors for uniformly recording germplasm evaluation data. Today, close to 40 CAC's are organized and functional. As with GRIN, the CAC's provide a matrix that gives the NPGS flexibility, structure, and stability. They are organized to function with relevance and at an acceptable level of uniformity while maintaining some degree of autonomy in their operations. They are expected to provide a ready source of technical information and advice necessary for the effective planning and operation of the NPGS.

Administrative support

National systems require administrative support and a source of funding, and the NPGS is no exception. The USDA establishes national policy for the NPGS. The ARS provides national leadership for the various phases of germplasm activities which range from acquisition to the development and release of improved cultivars [10]. Federal funding of the NPGS, at $26.4 million in FY 1988, is provided in the ARS appropriation. Program leadership and coordination is vested in the NPL for Germplasm, ARS National Program Staff. This NPL is responsible for program planning, resource allocation, and coordination of plant germplasm research. The ARS provides resources to establish and maintain facilities for germplasm collections and for research that supports the national germplasm program, a substantial portion of which is conducted cooperatively with other Federal agencies, the States, and – to a much smaller degree – with the private sector. The USDA can provide the necessary continuity of programs and funding stability that is not possible by individual or groups of States or by the private sector.

The USDA's CSRS serves as an important liaison between the Federal Government and the SAES's for research programs of national scope. Until FY 1987, CSRS received direct appropriations for clonal germplasm research. Beginning in FY1987, however, these funds were added to the ARS appropriation for plant germplasm. The CSRS currently supports plant germplasm activities through the allocation of Hatch and Evans-Allen (1890 institutions) funds to experiment stations and universities and through competitive grants.

The more than 50 SAES's have a unique partnership with the Federal Government in the conduct of the NPGS. They provide substantial land and laboratory and office space for their own plant germplasm scientists as well as for many Federal scientists. The SAES's often provide technical, clerical, and other administrative support services to the NPGS. Collectively, SAES contributions to the NPGS are estimated at $15–20 million, but this contribution could be much larger if more research on basic genetics and early stages of plant breeding were to be included as germplasm activities. Historically, Federal-State cooperation in plant germplasm has been strong and productive. Coordination of the State inputs to the NPGS remains a challenge to the leadership of the NPGS.

APHIS is charged with the legal responsibility for establishing regulations and procedures to minimize the risk of entry into the U.S. of serious disease, insect, and weed pests. APHIS and ARS work closely to assure the continued introduction of plant germplasm free of disease and insects.

Although some 98 percent of germplasm moving across State and U.S. borders moves freely without detention, some plant species – especially those in the genus *Prunus* – are known to be carriers of virus and viruslike entities. Thus careful attention must be given to these species. Foremost in this prohibited group are the stone fruits (peach, plum, cherry, etc.), which must be indexed in quarantine facilities before APHIS can certify them as safe for release for domestic use. At present, indexing is costly and time-consuming, but it is a necessary process if we are to ensure continued access to foreign germplasm. ARS has assumed the responsibility for indexing prohibited plants. These procedures are designed to determine the presence of disease agents and may require growing clones of the test plant in protected nursery gardens until they bear fruit. If, after appropriate testing, the introduced germplasm remains free of disease symptoms, APHIS will certify that the germplasm is free of exotic disease and will approve its release by the NPGQC to the germplasm user. For many other kinds of germplasm, however, it is only necessary that APHIS issue a phytosanitary certificate that will permit movement of these materials across international borders.

Industry is an important partner in the NPGS as a user translating germplasm developments to the farmer and consumer. Industry, because of its many and diverse components, also becomes involved in establishing national priorities for plant germplasm. As a practical matter, the NPGS looks to the National Council of Commercial Plant Breeders and the American Seed Trade Association (ASTA) to represent industry's viewpoints. However, ASTA membership includes some 600–800 companies ranging from large corporations to family-owned seed companies, and their views on some issues are diametrically opposed.

Nevertheless, the NPGS attempts to consider the views of all parties in conducting germplasm programs. Industry provides some support – primarily financial – on a case-by-case basis, but a better mechanism is needed for receiving and utilizing its support on a continuing and broader program basis. Industry also provides a valuable service to the NPGS through its membership on the CAC's and other advisory committees by offering objective suggestions and criticisms through program reviews and by speaking out to the public and to Congress on behalf of the NPGS.

Policy of the NPGS

Over the years, a sizable body of policies has been developed by organizations and institutions to guide and control germplasm operations at every level of activity [3–4]. These policies have contributed significantly to the effective functioning of the NPGS. However, the NPGS policy that supersedes all others in importance is that of free and unrestricted exchange of germplasm with all nations: to permit access to U.S. collections by any *bona fide* scientist or breeder, domestic or international. This policy has grown out of the belief that germplasm is a heritage of mankind and is not owned by any individual or nation. Germplasm is a heritage to be freely shared for the benefit of all mankind. Any other approach to the holding and distribution of germplasm can only result in diminished benefits to all nations. No nation is self-sufficient in genetic resources. While some nations are more sufficient than others, all nations are interdependent with regard to plant germplasm. It is in everyone's best interests to refrain from building barriers to unrestricted exchange.

Status and management of the collections

After several years in preparation, an operations manual was developed and distributed to the clonal repositories early in 1986 [11]. In September of that year, the PGOC set out to develop an operations manual covering various activities at the seed repositories. This activity is under way, but progress is slow because of the many unique situations generated by the diverse collection of plants, locations, and management structures within the NPGS. A few examples of protocols for regeneration procedures include:

1. Population size needed to maintain diversity,
2. Insect cage specifications and use and management of insect pollinators,
3. Field crossing techniques that involve all individuals in the population such as chain sibbing or mixed pollen procedures, and
4. Isolation requirements.

Among the scientific issues confronting the NPGS are the identification of genetic diversity, storage of genes or genotypes, development of gene pools, practical use of a core collection concept, and confirmation and removal of duplicative or nonuseful materials. Research studies and evaluations are being undertaken on several of these problem areas to help determine appropriate management protocols.

Curators of working collections will continue to upgrade their storage facilities to keep pace with technological advances in seed preservation. Highest priority for all working collections is maintenance of clean, disease-free seed. Proper seed drydown and seed moisture in storage are essential for maintaining longevity. Once seed is dried properly, storage life in the working collections can be extended significantly by reducing the relative humidity in the above-freezing temperatures at which most seed is stored. Where practical, curators will use freezers and construct cold rooms with efficient insulating characteristics, refrigeration, and dehumidification equipment to preserve material as long as possible. The costs of regeneration – whether measured in terms of mechanical/labor inputs or in loss of genetic diversity through seed inviability and genetic drift – will eventually be quite substantial and will have to be balanced against energy and facility costs.

NCGR sites have received clear direction for reducing the risk exposure and field costs of the base and working clonal collections. It is appropriate that seeds and pollen of the vegetatively propagated crops be conserved to preserve their genes, but many of these materials must be examined for tolerance to standard seed moisture/temperature conditions. Research is in progress at the NSSL to investigate pollen cryopreservation of these clonal collections and of species with recalcitrant seeds. Cryopreservation of vegetative parts of clonally propagated materials may be of use in the future to provide a base collection of this germplasm.

We have discussed the various NPGS collections and the number of samples being maintained. In 1986, the PGOC identified the need to reconcile the lists of the NPGS working and base collections. This has been accomplished by GRIN and at the collection sites. Without the GRIN database, this reconciliation would not have been possible. Congress has provided funding in the FY 1988 budget increases for regeneration of samples needed to provide fresh seed to both working and base collections. As a result of the GRIN reconciliation study, several classes of seed have been identified for regeneration:

Seed in working and/or base collections. For many years before the base collection concept was developed, NSSL stored cultivars and seed for programs being terminated, whereas the RPIS's stored plant introductions. No mandate existed for the RPIS's to send seed to NSSL or for NSSL to send seed to the RPIS's and other working collections. With a few exceptions as noted previously, this duplicative storage is now normal operating procedure.

IBPGR samples that are to be incorporated into the NPGS collections. Until recently protocols were unclear for managing the IBPGR samples, and it was assumed that they were to be backup base collection samples. That issue has been clarified, and the NPGS is moving forward to regenerate appropriate samples as quickly as possible. The large number of IBPGR sorghum lines held in quarantine at NSSL pose a formidable problem for NPGS, as do all quarantined materials. However, growout procedures are being developed for this germplasm.

Samples with insufficient or low viability. Rescued samples entering the NPGS with low viability or those that have gone out of condition slowly over the years are to be regenerated as soon as possible. Rescue operations at the NSSL and the RPIS's, based on research at NSSL, will be used to maintain as much genetic diversity as possible. Protocols for each crop group will be presented to the CAC's for review to assure the best job possible in managing these and regular growouts.

Summary

We are confident that the NPGS has instituted a program of management oversight and review which will ensure the best possible preservation of the valuable germplasm under its jurisdiction. Also, the related research is designed to address the problems and questions affecting the operational protocols for each crop. In these ways, the NPGS will become stronger and provide to the user more germplasm, properly cared for and documented.

References

1. Science and Education Administration. 1977. The National Plant Germplasm System, USDA Program AID No. 1188. U.S. Dept. Agric. Washington, DC. 12 pp.
2. National Plant Genetic Resources Board. 1979. Plant Genetic Resources: Conservation and Use. 20 pp.
3. Jones, Q. and S. Gillette. 1982. The NPGS: An Overview. Diversity, Spec. Rep. 1. Lab. Info. Sci. Agric., Fort Collins, Co. 7 pp.
4. Murphy, C.F.(ed.). 1981. The National Plant Germplasm System. Sci. Educ. Admin., U.S. Dept. Agric., Washington, DC. 204 pp.
5. Plant Germplasm Preservation and Utilization in U.S. Agriculture. 1985. Coun. Agric. Sci. Tech. Rep. 106. 35 pp.
6. National Program for Conservation of Crop Germplasm. 1971. A progress report on Federal/State cooperation. State Agric. Exp. Sta. Dir. Assoc. North Cent., North East., Southern, and West. Reg. of U.S., and Agr. Res. Ser., U.S. Dept. Agric., Washington, DC, 73 pp.
7. Brooks, H.J. and D.W. Barton. 1977. A plan for national fruit and nut germplasm repositories. Hort. Sci. 12(4): 298–300.
8. Bass, L.N. 1973. Controlled atmosphere and seed storage. Seed Sci. Technol. l(2): 463–492.
9. Germplasm Resources Information Network. 1985. A Manual for Public Users. Plant Genet. Germplasm Inst., Agr. Res. Ser., U.S. Dept. Agric., Washington, DC. 208 pp.
10. USDA. 1983. Agricultural Research Service Program Plan. Pp. 27–30 in U.S. Dept. Agric., Agric. Res. Serv., Misc. Publ. 1429.
11. USDA. 1986. Operations Manual for National Clonal Germplasm Repositories. U.S. Dept. Agric., Agric. Res. Serv., Misc. Publ. Unnumbered. 150 pp.

Development of research on crop germplasm resources in China

ZHANG YIHUA and DONG YUSHEN
Institute of Crop Germplasm Resources, Chinese Academy of Agricultural Sciences, Beijing, People's Republic of China

Abstract During the past 40 years, increased attention has been given by Chinese scientists to their country's rich plant germplasm resources. Research to collect, evaluate, preserve, and use plant germplasm is carried out at the national level by the Chinese Academy of Agricultural Sciences, Institute of Crop Germplasm Resources, and other crop-specific institutes within the Academy. Germplasm research is also conducted by the provincial agricultural academies. Germplasm collected in China now numbers about 300,000 accessions. During the 1980's, modern facilities have been constructed in China for low temperature seed storage. Also, a systematic evaluation of the collected germplasm has begun. The international exchange of germplasm resources is recognized as having contributed significantly to the development of many crops in China and throughout the world. A catalog of Chinese plant germplasm available for exchange with foreign countries has been published. Sequels to the catalog are planned.

Key words: germplasm exchange, germplasm evaluation, gene banks

China is an ancient agricultural country with a long history, vast territory, complicated terrain, and a wide variety of cultivation systems. Since many major field, vegetable, cash, and fruit crops originated in China, it is particularly rich in plant germplasm resources.

Germplasm resources in China have attracted the attention of agricultural experts for many centries. An ancient and famous book on agriculture entitled 'Qinming Yiaoshu', published 1500-odd years ago, states that '...grain crops have the characters of maturing earlier or later, status higher or shorter, seeds less or more, eating quality good or bad, tolerance stronger or weaker, smell bad or fragrant and grain inflated or disinflated'. This quotation shows that variations of different characters for crops were noted even at that long ago time. The book 'Dao Pin' (Rice Varieties), written by Huangshen in the Ming Dynasty 400-odd years ago, is the earliest compendium of rice germplasm resources, recording all morphological and agronomical characters of 38 rice varieties in the Taihu Lake area of China. Since the 20th century, with the advancement of modern agricultural science in China, increasing attention has been paid by scientists to China's germplasm resources. However, with very little research being conducted, there were no professional research units to study germplasm resources. Thus the

L. Knutson and A.K. Stoner (eds.). Biotic diversity and germplasm preservation, global imperatives.
 ISBN 0-7923-0178-1

abundant germplasm resources were not collected and preserved for research and use. Since the founding of the People's Republic of China in 1949, much attention has been paid by the State to research on germplasm resources. During 1956–1958, the government organized a nationwide collection of germplasm resources. Since then, units have been established in many agricultural institutions to conduct research on germplasm resources of various crops.

In the last 10 years, research on germplasm resources has advanced to a new stage. In the Chinese Academy of Agricultural Sciences (CAAS) in 1978, the Institute of Crop Germplasm Resources (ICGR), Beijing was established as the national research center for crop germplasm resources. Research sections for germplasm resources also have been set up in other institutions of the CAAS, such as the Institute of Cotton, the Institute of Oil Crops, the Institute of Bast-Fiber Plants, the Institute of Vegetable Crops, and the Institute of Pomology. Institutes or sections for research on crop germplasm resources also have been set up in most provincial agricultural academies. All of these units conduct research on the collection, identification, preservation, and use of crop germplasm and are gradually forming a working system of germplasm resources throughout the country.

The working system of crop germplasm resources in China

A) The Institute of Crop Germplasm resources of the CAAS is the national research center for germplasm resources. Its tasks are as follows:
   1. To organize and coordinate the national research work for crop germplasm resources,
   2. To manage the exchange of germplasm resources with the rest of the world,
   3. To manage the long-term preservation of germplasm resources for the whole nation, and
   4. To collect, identify, and evaluate grain and legume crops and provide medium-term preservation of germplasm for exchange both at home and abroad.

B) The tasks of the germplasm resources research sections of other research institutes of the CAAS are as follows:
   1. To organize and coordinate the important research projects on related crop germplasm resources,
   2. To provide medium-term preservation of germplasm resources for related crops, and
   3. To collect, identify, and evaluate germplasm resources for related crops.

C) The tasks of the institutes or sections of the provincial academies of agricultural sciences are as follows:
   1. To collect and provide medium-term preservation of indigenous germplasm resources,
   2. To collect, identify, and evaluate local germplasm resources in the province, and
   3. To undertake projects organized by the Institute of Crop Germplasm Resources and other professional institutes of the CAAS.

The research network for the national germplasm resources of various crops has been formed. The responsible institutes for research work are listed in Table 1.

The system of crop germplasm resource research stations in China is not yet complete. Eventually, we plan to set up experiment stations to study crop

Table 1. Institutes responsible for research on germplasm resources of various crops in China.

Crop	Institute
Rice	Institute of Crop Germplasm Resources of the Chinese Academy of Agricultural Sciences (CAAS) and China National Rice Research Institute (CNRRI)
Wheat	Institute of Crop Germplasm Resources of the CAAS
Corn	Institute of Crop Germplasm Resources of the CAAS
Millet	Institute of Crop Germplasm Resources of the CAAS
Edible beans	Institute of Crop Germplasm Resources of the CAAS
Sorghum	Institute of Crop Germplasm Resources of the CAAS and Liaoning Academy of Agriculture Sciences
Barley & broad bean	Qinghai Academy of Agricultural Sciences and Zhejiang Academy of Agriculture Sciences
Oat & buckwheat	Regional Academy of Agriculture Sciences of Inner Mongolia Autonomous Region
Panic crops	Shanxi Academy of Agriculture Sciences and Yikezhaomeng Agricultural Institute of Inner Mongolia
Sweet potato	Xiuzhou Agricultural Institute of Jiangsu Academy of Agricultural Sciences
Potato	Keshan Agricultural Institute of Heilongjiang Province
Cotton	Institute of Cotton of the CAAS
Oil crops	Institute of Oil Crops of the CAAS
Bast-fiber	Institute of Bast-fiber of the CAAS
Sugar Beet	Institute of Sugar Beet of the CAAS
Vegetable crops	Institute of Vegetable crops of the CAAS
Fruit trees	Xingcheng Institute of Pomology and Citrus Institute of the CAAS
Tea	Institute of Tea of the CAAS
Mulberry	Institute of Sericulture of the CAAS
Herbage	Institute of Grassland and Institute of Animal Science of the CAAS

varieties in different ecological regions, to study more intensively the regional germplasm resources, and gradually to provide for medium-term preservations.

Collection of crop germplasm resources

Since the founding of the People's Republic of China, the collection of crop germplasm resources has been carried out mainly in two ways: a) by mobilizing the masses and relying on administrative authorities and b) by organizing the investigation and collection of crop germplasm resources of priority crops by research units in priority regions.

In the early 1950's, the Ministry of Agriculture organized each province and region to collect crop germplasm resources. In 1958, 200,000 accessions of germplasm of more than 50 field crops were collected, some in duplicate. Except for some remote or inaccessible areas, all local varieties of the main crops in China were collected. Those materials are preserved in appropriate institutions in each province, with some preserved in appropriate CAAS institutes. In the 1960's, some of the materials were lost, but 150,000 accessions of germplasm remain.

In 1978, the same year the CAAS Institute of Crop Germplasm Resources (ICGR) was established, the Ministry of Agriculture and the State Science and Technology Commission jointly ordered each province in China to make supplemental collections of germplasm on a large scale with ICGR in charge of all organization and coordination. The collection program lasted for five years (1979–1984), with scientists, rural technicians, and masses of farmers participating. About 100,000 accessions of germplasm (some in duplicate) representing more than 50 crops were collected, while some previously lost materials were found and some local varieties in imminent danger of extinction were rescued.

About the same period (1975-1985), several investigations of crop germplasm resources were organized, such as the exploration and collection of crop germplasm resources in Yunan and Tibet, the nationwide exploration and collection of germplasm resources of wild rice and soybean, and the exploration and collection of germplasm resources of forage crops in temperate and sub-tropical zones. Some provinces also organized their own explorations for crop germplasm resources. From these explorations and investigations, about 50,000 accessions of germplasm were collected. We are still investigating germplasm resources in the mountainous area of western Hubei Province and in the Hainan Islands. At the same time we are investigating the germplasm resources of the Triticeae throughout China with the support of the International Board for Plant Genetic Resources.

Collected germplasm material throughout China now numbers about 300,000 accessions (with some duplication). We now are organizing the management of germplasm resources in China, organizing collaborating groups by crops, standardizing data collection protocols, consolidating data, etc. After examining the agronomic characters and morphological features of the collected germplasm, we plan to discard duplicate materials, arrange for the increase of seeds, and eventually transfer them into the National Germplasm Bank for long-term preservation and character identification. On the basis of preliminary observation and evaluation, we already have produced annals of various kinds, such as the 'Variety Annals of Wheat in China', 'Variety Annals of Millet in China', 'Variety Annals of Peanut in China', 'Variety Annals of Tobacco in China', etc.

Conservation of crop germplasm resources

Before the 1970's, China lacked modern storage facilities for crop germplasm resources. Germplasm was conserved mainly by traditional methods in various provincial and national agricultural research institutes. Workers in some units placed sun-dried seeds in large jars, sealed them, then kept them in relatively cool places for conservation. Some units sent their crop seeds to Xinjiang Autonomous Region, Qinghai Province, etc., where the climate is dry, and entrusted the local agricultural units with conservation of their crop seeds. To preserve the vitality of these germplasm resources, we frequently had to replant them for rejuvenation.

Since the 1980's, many modern facilities have been constructed for low temperature seed storage (i.e., low temperature germplasm banks). The CAAS/ICGR national germplasm bank (the No. 1 Bank) was constructed through government investment and put into operation in 1985. The temperature inside the bank is 10 °C, and the holding capacity is 80,000 accessions. To preserve the large number of germplasm resources in China properly and indefinitely, the Rockefeller Foundation and the International Board for Plant Genetic Resources (IBPGR), with Dr. T. T. Chang as Technical Advisor, provided financial aid to construct a new, larger national germplasm bank (the No. 2 Bank) in the ICGR. The floor area of this bank is 3000 m^2, and is divided into three parts–the experimental area, the seed treatment area, and the storage area. The 300 m^2 low-temperature storage area is maintained at − 18 °C with a relative humidity below 50%. This area is divided into two rooms with movable shelf units used for seed storage. The shelves can store 400,000 accessions of seeds of various crops. Besides research on the physiology of conserved crop seeds at the new national germplasm bank, we are carrying out studies on preservation of stem tips of vegetatively propagated crops (e.g., sweet potato and potato) and on preservation of pollen and plant tissue at ultra low temperatures (in liquid nitrogen).

Conditions of crop germplasm conservation in other professional institutes and provincial academies are gradually being improved. Some units already have built relatively modern low-temperature structures for medium-term preservation of germplasm.

Evaluation and identification of crop germplasm resources

Some agronomic and morphological characters of the most collected crop germplasm resources in China have been observed and recorded, but comprehensive, systematic evaluation of characteristics of the germplasm resources has just begun. The priority for evaluation is aimed at characters most needed in breeding work, such as resistance to diseases and insect pests, and tolerance to drought, cold, and saline-alkaline conditions. At the same time, the quality of the agricultural products is evaluated. Resistance characters are determined primarily in view of special needs in various places, but they also have been carried out jointly by organized specialists of related disciplines in different ecological areas of China since 1978. For instance:

1. Joint identification of the disease and insect pest resistance of foreign rice varieties was conducted from 1978 to 1984. The evaluation work was presided over by the ICGR with 15 scientific research units in 12 provinces participating; 3,510 accessions of rice introduced from foreign countries were evaluated for resistance to rice blast, rice bacterial leaf blight, rice yellow dwarf virus, brown planthopper, and rice leafroller. Among these accessions, 186 were determined to be resistant to rice blast, 161 to rice bacterial leaf blight, 39 to rice yellow dwarf virus, 142 to brown planthopper, and 6 to rice leafroller. These results along with the agronomic characters of each variety have been published.

2. Cold tolerance was evaluated in rice germplasm resources from 1979 to 1985. The research was presided over by the ICGR; 13 units in 8 provinces participated. Evaluation of cold tolerance was conducted at 6 growing stages, and more then 2,700 accessions of materials from China and abroad were studied. The growing stages are: budding, seedling, tillering, meiotic division, blooming, and filling. The varieties found to have outstanding cold tolerance are: The Japanese variety ‘Zongxin 134’, which has cold tolerance in 5 growing stages; ‘Heixuan No. 5’, from Yunnan Province which can bloom normally at 15 °C, or 3 °C lower than the temperature generally needed by Japonica rice; and ‘Ce 21’, from Zhejiang Province

which has strong tolerance for cold, as do most of the varieties developed from it.

3. Resistance to rust, powdery mildew, and yellow dwarf virus in wheat was evaluated by the CAAS Institute of Plant Protection. Altogether, 9,552 accessions were evaluated between 1983 and 1987. These included 3,952 Chinese varieties, 717 kindred plants, and 5,243 foreign varieties. The evaluation was conducted in 5 nurseries – stripe rust nursery, leaf rust nursery, stem rust nursery, powdery mildew nursery, and yellow dwarf virus nursery. All varieties were evaluated using the physiological race of the fungus from China. Many rust-resistant wheat accessions were identified; more *Aegilops* germplasm was found to be resistant to powdery mildew than in other genera, and more *Agropyron* and *Leymus* germplasm was found to be resistant to yellow dwarf virus than in other genera.

Exchange of germplasm resources with the rest of the world

International exchange of germplasm resources has promoted crop improvement and agricultural development in various countries. The abundant plant germplasm resources in China have contributed a great deal to the development of many crops (such as soybean and rice) throughout the world. In the same way, the introduction of crop varieties from foreign countries has played an important role in improving various crops in China. For example, in recent decades, all of the rust-resistant wheat varieties used in China have been derived from foreign varieties. During the 1950's, the varieties used were 'Mentana' from Italy, 'Quality' from Australia, and some from the U.S. During the 1960's, the varieties used were 'Funo' and 'Abondan' from Italy. During the 1970's, the variety used was 'Lovrin' from Romania. The maize hybrid 'Zhongdan No. 2', which is a cross between inbred 'Mol7' from the U.S. and a Chinese inbred, has been planted on 2 million ha and has become the hybrid maize most commonly grown in China. Rice varieties 'IR 24', 'IR 26', and 'IR 661', etc., introduced from the International Rice Research Institute (IRRI), are good restorer lines of hybrid rice in China, and they have played an important role in the development of hybrid rice in China. Therefore, we recognize that the international exchange of germplasm resources is beneficial to all parties involved. In recent years, the exchange of germplasm with foreign countries has developed continuously until we now exchange germplasm with more than 90 countries and regions on 5 continents.

From 1972 to 1986, China provided 16,786 accessions of crop germplasm resources to foreign countries, as shown in Table 2.

Table 2. Plant germplasm distributed from China to foreign countries between 1972 and 1986.

Crop	Accessions
Food crops	12,650
Cash crops	199
Oil crops	1,437
Fruit trees	685
Vegetables	1,323
Forage and green–manurecrops	340
Others	152
	16,786

From 1972 to 1986, the crop germplasm resources introduced from foreign countries numbered 92,791 accessions, as shown in Table 3.

China's exchange of germplasm resources with foreign countries is conducted in many ways, such as cooperative agreements between governments, cooperative agreements in science and technology between scientific research institutions, mutual exchange by scientists, and collection through visits and explorations.

In recent years, we also have had a cooperative agreement with the IBPGR. China is entrusted with the conservation of crop germplasm resources of certain species (e.g., Chinese cabbage) in the national germplasm bank. The IBPGR supports our propagation of seeds and evaluation of germplasm resources for certain crops (e.g., barley, wheat) in China.

For the convenience of exchanging germplasm resources with foreign countries, we published 'A Catalogue of Chinese Crop Germplasm for

Table 3. Plant germplasm introduced into China from foreign countries between 1972 and 1986.

Crop	Accessions
Food crop	75,635
Cash crops	2,351
Oil crops	4,851
Fruit trees	1,717
Vegetables	5,948
Forage and green-manure crops	1,696
Others	593
	92,791

Exchange' in 1987, in cooperation with the IBPGR. There are 4,403 accessions of 10 crops listed in this catalog. We are planning to publish sequels to the catalog continuously. All of the crop seeds included in the catalog can be provided to foreign countries. To provide seeds not included in the catalog, the approval of the Chinese government is needed.

A good beginning has been made in joint surveys of Chinese germplasm resources by Chinese and foreign scientists. Since 1986, we have conducted an expedition on the Triticeae in the Inner Mongolian Autonomous Region, Sichuan Province, and other areas with scientists from France and Sweden. We hope that such investigations can be continued. We also wish to continue to study crop germplasm resources in China and in the rest of the world with scientists of various countries. We welcome foreign scientists who bring germplasm with them to China to conduct research with their Chinese collegues. We also hope that foreign laboratories will welcome Chinese scientists and invite them to travel abroad with germplasm to conduct research. Joint training of graduate students is also a good way to foster international cooperation. All of these endeavors should be conducted only after careful planning.

We sincerely hope that cooperation in germplasm resources between China and various countries will continue to develop further and, moreover, that through research, crop germplasm resources throughout the world can be induced to contribute even more to the needs of mankind.

Management and use of plant germplasm collections

TE-TZU CHANG[1], SHERL M. DIETZ[2] and MELVIN N. WESTWOOD[3]

[1] *International Rice Research Institute, Los Baños, Philippines.*
[2] *Western Regional Plant Introduction Station, Agricultural Research Service, U.S. Department of Agriculture, Pullman, Washington, U.S.A.*
[3] *National Clonal Germplasm Repository-Corvallis, Agricultural Research Service, U.S. Department of Agriculture, Corvallis, Oregon, U.S.A.*

Abstract Sound management of sizeable plant germplasm collections requires a systems approach, with well-defined objectives, adequate physical facilities, competent staff, and sustained administrative and financial support. Management practices should cover not only preservation, research, multiplication/propagation, documentation, distribution, and rejuvenation, but also collection or re-collection, evaluation, communication, promotion of conserved germplasm use, public information, and training. Appropriate maintenance is a function of the scope of the collections, the genetic structure of the populations, their storage behavior, the capacity of storage facilities, and ease of rejuvenation. Components of management are interrelated and, for efficacy and security, should be integrated into a national plant germplasm system and international networks.

Examples are drawn from:

1. The base collection of rice at the International Rice Research Institute in the Philippines,
2. The multicrop collection at the U.S. Western Regional Plant Introduction Station at Pullman, Washington, and
3. The National Clonal Germplasm Repository for Small Fruits, Pears and Nuts at Corvallis, Oregon.

Crop seeds are generally adapted to cold storage, while the clones are kept as living trees, rootstocks, and tissue cultures. The required facilities are discussed.

Conservation programs must be justified by their effective use. To ensure active use of germplasm, seed/plant/tissue materials and full information on all accessions should be freely and readily available to a broad segment of scientists and growers. Examples of highly successful use of plant germplasm collections are described. Areas for future research and operational linkages are outlined.

Biotechnology is expected to be complementary and synergistic with conservation. Public awareness and sustained support from all sectors of society are needed to protect irreplaceable plant germplasm and to make it available to benefit future generations.

Key words: seed preservation, clonal germplasm, genetic integrity, data management, crop advisory committees, genebank management, training, international collaboration, use of germplasm, genetic diversity

L. Knutson and A.K. Stoner (eds.). Biotic diversity and germplasm preservation, global imperatives.
 ISBN 0-7923-0178-1

Introduction

The appropriate management system of a genetic resources center is determined by its primary mission and the types of germplasm being conserved. Management practices should aim for maximum security and cost effectiveness. The ultimate goal of a management system is to preserve the complete genetic diversity to maximize the use of plant collections for research and agricultural production.

An integration of clearly defined objectives, well designed physical facilities, trained staff of adequate size, and sustained administrative and financial support is needed to enable a center to execute its mandate. To show the myriad dimensions and contrasts, the contents and use of the plant germplasm collections of three centers are depicted in this paper:

1. The single-crop, large and rich base collection of rice at the International Rice Research Institute (IRRI), Los Baños, Philippines,
2. The multicrop active collection at the U.S. Western Regional Plant Introduction Station (WRPIS) supported by the U.S. Department of Agriculture (USDA), Agricultural Research Service (ARS), State Agricultural Experiment Stations and Washington State University at Pullman, Washington, and

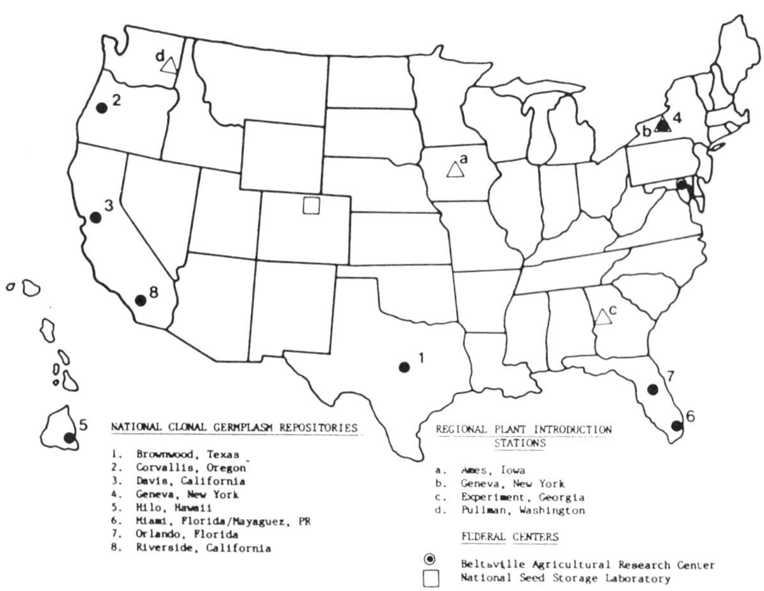

Figure 1. The eight National Clonal Germplasm Repositories and four Regional Plant Introduction Stations in the U.S. National Plant Germplasm System.

3. The National Clonal Germplasm Repository for Small Fruits, Pears, and Nuts (NCGR-Corvallis) supported by USDA-ARS in cooperation with Oregon State University at Corvallis, Oregon.

Although the WRPIS and the NCGR-Corvallis are separate units with their own facilities, they operate as components of the U.S. National Plant Germplasm System (NPGS) (Fig. 1). Their primary mandate is for introduction, increase, preliminary evaluation, documentation, maintenance, and distribution of plant germplasm. The NPGS components have a great variety of international linkages.

IRRI is one of the international agricultural research centers (IARCs) in the Consultative Group for International Agricultural Research (CGIAR) system. IRRI has a global mandate for the conservation, distribution, evaluation, and use of rice germplasm. IRRI has formalized links with many national rice research centers, regional genebanks, another IARC (International Institute of Tropical Agriculture [IITA]), and the International Board for Plant Genetic Resources (IBPGR).

This paper also aims to review and assess constraints in managing large plant germplasm collections and to suggest areas for future endeavor.

Management issues

Management is the vital link between conservation on one hand and germplasm availability on the other. Sound management of large plant germplasm collections requires a systems approach to architectural and engineering designs and flow of operations based on well defined objectives. Adequate and functional physical facilities, sufficient and competent staff, and sustained administrative and financial support are imperative to good management practices.

The scope of management practices should cover not only preservation, research on conservation methods, seed multiplication or plant propagation, documentation and database management, distribution of seed or plant material, and regeneration of aged or exhausted stocks but also should include field collection or re-collection in target areas, linkage with plant quarantine services, multidisciplinary evaluation, international and inter-institutional linkages, free communication, promotion of conserved germplasm use, public information, and human resource development.

Physical requisites for seed banks

The site. The choice of a site is crucial to efficient and secure operations. The site should have:

1. Ample space for buildings, auxilliary structures, and fields,
2. Ready access to irrigation water,
3. Reliable supply of electricity,
4. Ample and inexpensive supply of labor,
5. Freedom from flooding, and
6. Proximity to a research-education complex that can enhance the technical capacity of the center.

The representativeness of the site to the climatic-edaphic adaptation of the conserved materials is biologically important. In recent years, local costs in land acquisition, construction costs, and daily wages also have become important considerations.

Structural design of the building. The building housing the storage rooms and associated equipment should have:

1. A strong foundation to withstand earthquakes,
2. Devices to insulate against water infiltration from the outside,
3. A strong structure to cope with powerful storms,
4. Security devices against natural hazards, electrical failures, or unexpected accidents, willful damage, and
5. Measures to ensure workers' safety.

Component areas. Well-maintained cold and dehumidified storage rooms form the heart of most seedbanks. Other facilities and working areas associated with seed storage may be grouped under:

1. Processing: temporary holding, cleaning, fumigation, registration, drying, preparation, and packaging,
2. Storage: rooms with different temperatures and relative humidity levels for various kinds of seeds,
3. Viability monitoring and rescue: germination rooms, growth chambers, and greenhouse,
4. Database management: records and computer rooms,
5. Research laboratories on seed storage and viability, and
6. Refrigeration equipment and backup components: electric power center, compressors, and dehumidifier assembly, cooling tower, and emergency generator.

When cryopreservation is included, a separate and well ventilated room for placing the liquid nitrogen (LN) tanks and an adjacent area for LN storage are necessary.

The working areas mentioned above should be placed according to the functional flow of operations. Moreover, dry areas should be separated from wet rooms. A fumigation room should have effective insulation and aeration. Drying and low-temperature areas may require the installation of airlocks to minimize air exchange. Cold rooms should have efficient and durable insulation so as to minimize operating costs. The floor plan of the International Rice Germplasm Center (IRGC) at IRRI is provided to illustrate the system of reinforced insulation and the graduated steps of cooling and dehumidification (Fig. 2).

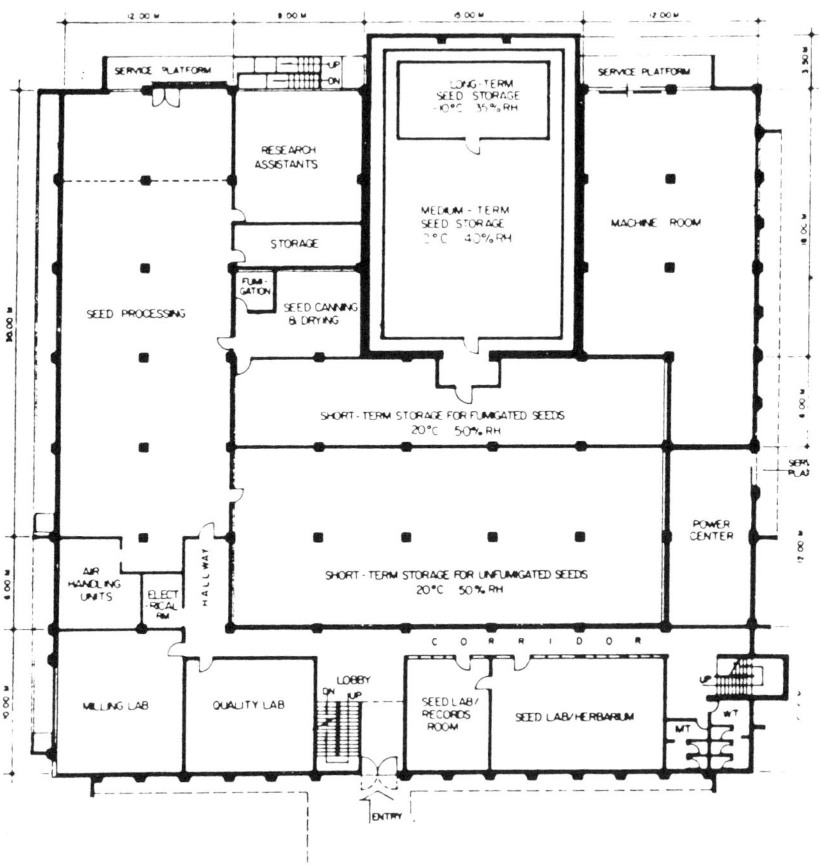

Figure 2. Floor plan of the Rice Genetic Resources Laboratory building at the International Rice Research Institute (IRRI), showing the multi-tiered system of cooling and dehumidification from seed processing area, through seed canning and drying room, fumigated short-term storeroom and medium-term storeroom to the long-term storeroom.

Within the seedbank complex, other essential rooms or areas are spaces for administrative officers, offices/labs for scientific staff, a reception area for visitors, a seminar room combined with a library area, a training room, and a tool shop. A herbarium is necessary for storage of plant specimens, especially of the wild taxa.

A well designed physical facility and rationally placed working areas are essential to an effective flow of operations and economic use of energy and water [1]. Seed storage in the humid tropics requires additional precautions in design and equipment [2].

Seed storage rooms and shelving. Modern storage rooms make use of prefabricated insulation panels, assembled in modular form, and which are more efficient in heat insulation and maintainance than are solid-walled rooms. For use in subfreezing temperatures, the insulation panels should have sufficient thickness and thermal stability. The minimum thermal conductivity should be within 0.017–0.023 W/m °C and should last for at least 30 years [3].

Large storerooms are less costly to build and to maintain on a per accession basis than are small walk-in cabinets [4].

For storing seeds, mobile (compactile) shelves increase storage space within a room and thus lower energy costs, although the initial costs of such shelves are high.

Seed containers. Durable and airtight seed containers not only prolong the shelf life of stored seed but also cut down the cost of viability monitoring and the need for frequent rejuvenation. Frequent cycles of rejuvenating seed in small populations have several disadvantages:

1. Liable to lead to mixtures and errors,
2. May involve undetected genetic drift within the population,
3. Possible loss of seed because of adverse weather and pests,
4. Added workload in sorting, labelling, storage, and inventory, and
5. More costly to perform a cycle of regeneration than to store under 'good' conditions [5, 6].

Among various kinds of seed containers, non-rusting metal cans and glass bottles are superior to plastic materials in durability and reliability. Lowering of seed moisture content via a slow, low-temperature drying process and the exclusion of air prior to packing would aid markedly in extending seed longevity. For bottles and metal cans, the choice of a reliable gasket for the cap or lid also is crucial. Natural rubber or solvent-based rubber gaskets are superior to paper, plastic, or water-based rubber seals.

Table 1. Size of germplasm collections and strorage conditions for seed stocks maintained at the International Rice Research Institute (IRRI) and the Western Regional Plant Introduction Station (WRPIS), Pullman, Washington, and the total holdings of four Regional Plant Introduction Stations of the National Plant Germplasm System (NPGS).

Number of genera	Number of species	Total number of accessions	Storeroom conditions	Seed container(s)
IRRI				
5	28	83,000	Short-term: 19°C, 50% RH	Paper bags in plastic trays
			Medium-term: 2°C, 40% RH	Aluminium cans with partial vacuum; glass jars containing silica gel (old system)
			Long-term: −10°C, 37% RH	Aluminium cans with partial vacuum
Western Regional P.I. Station				
213	1,722	38,611	3°C, 30% RH	Metal cans with screw-top lids, or Manila envelopes placed in steel drawer cabinets
Four Regional P.I. Stations combined				
816	4,426	136,689*	(similar to the above)	Same types as above or in glass canning jars with screw caps and rubber gaskets

* Totals include only samples having a. P.I. no., sufficient seed for seed distribution, and entered in U.S. Department of Agriculture Germplasm Resources Information Network.

2.1.1.6. *Duplicate storage.* Every base collection should have a backup site for storage of duplicate seed samples. For instance, the U.S. National Seed Storage Laboratory (NSSL) at Ft. Collins, Colorado, provides space for storing a duplicate set of the rice collection of IRRI and the maize germplasm of the Centro Internacional de Mejoramiento de Maiz y Trigo (CIMMYT). The active/working collections of the four Regional Plant Introduction Stations and some others, such as the Small Grains Collection at Aberdeen, Idaho are also backed up at the NSSL (Shands *et al.*, this volume).

Auxilliary facilities for seed increase and regeneration. When seed multiplication and regeneration are a function of a seedbank, a complete array of facilities such as fields, isolation areas, irrigation system, isolation glasshouses, screened nurseries, virus testing facility, farm machines, repair shop, and threshing, cleaning, and drying equipment will be needed.

Seed stores of IRRI and the four Regional Plant Introduction Stations. The size of collections and storage conditions of the IRRI and the four Regional Station are enumerated in Table 1. Seed stores should be upgraded when further advances in refrigeration technology and additional funding are available.

Physical requisites for clonal repositories

Site. The recent development of U.S. clonal fruit repositories to maintain the genetic diversity of both domestic cultivars and their wild relatives is without precedent. Many previous collections contained domestic forms only. Some of the wild species are found in botanic gardens and arboreta, but no collection is complete. Since 1980, eight National Clonal Germplasm Repositories (NCGR) have been established at various sites (Fig. 1). Each NCGR is located where the climate is conducive to growing the designated crop species. Consideration also was made to place each repository at or near a major research facility of a State Agricultural Experiment Station (SAES) or of the USDA, ARS. Specific NCGR sites are given in Table 2.

In addition to the above eight sites, a station at Sturgeon Bay, Wisconsin, maintains the common potato collection.

Because the field collections of clonal plants also become the base collections for these crops, it is important that the plants be grown where they will survive winter freezes, winds, and other environmental hazards, and where diseases and pests will not threaten them. In some cases, where the security of the field collections might be threatened, duplicate collections are grown in screenhouses or glasshouses [7].

Table 2. NCGR sites.

Repository site	Established	Plant genera maintained
Corvallis, Oregon (Oregon State Univ.)	1980	*Corylus* (hazelnut), *Pyrus* (pear) and related genera, *Fragaria* (strawberry), *Rubus* (raspberry, blackberry), *Vaccinium* (blueberry, cranberr), *Ribes* (currant, gooseberry), *Mentha* (mint), *Humulus* (hops).
Davis, California (Univ. of California)	1981	*Prunus* (stone fruits), *Juglans* (walnut), *Vitis* (grape), *Pistacia* (pistachio nut), *Ficus* (fig), *Olea* (olive), *Actinidia* (kiwi), *Diospyros* (persimmon), *Punica* (pomegranate), *Morus* (mulberry).
Miami, Florida/ Mayaguez, Puerto Rico (USDA Plant Introduction Station)	1984	*Mangifera* (mango), *Musa* (banana, plantain), *Persea* (avocado), *Bertholletia* (Brazil nut), *Coffea* (coffee), subtropical *Zizyphus* (jujube).
Brownwood, Texas (USDA-ARS Research Station)	1984	*Carya* (pecan, hickory), *Castanea* (chesnut).
Geneva, New York (New York Agric. Exp. Station)	1985	*Malus* (apple), *Vitis* (American grapes and their hybrids).
Hilo, Hawaii (Univ. of Hawaii)	1986	*Macadamia* (macadamia nut), *Ananas* (pineapple), *Psidium* (guava), *Passiflora* (passion fruit), *Carica* (papaya), *Arctocarpus* (breadfruit), *Anacardium* (cashew), *Malpighia* (acerola cherry).
Orlando, Florida (USDA-ARS Research Station)	1986	*Citrus*.
Riverside/Brawley, California (Univ. of California/USDA Station)	1987	*Citrus* and related genera, *Phoenix* (date).

Structural design of the building. The main building for a clonal repository should be built to withstand the natural hazards of the area (e.g., wind, rain, snow buildup, and so forth), and the germplasm and its information should be protected from fire damage. It should contain the offices, computer and information center, library, herbarium, tissue culture room, cryopreservation facility, and research labs.

Component areas. When field plantings are vulnerable to losses, screenhouses or greenhouses are needed to maintain a duplicate set of clones. These structures must be provided with temperature controls so the collection will not be subjected to lethal cold or heat. The plants, usually grown in pots, also must be irrigated frequently. Ideally, an automatic watering system is provided. Other clonal facilities include cool rooms for temporary storage of propagules, sample processing and propagation area (headhouse and greenhouse complex), virus indexing rooms and thermotherapy chambers for virus elimination, tissue culture room, LN cryopreservation facility for meristem and pollen storage, research laboratory for germplasm identification and evaluation, freezer storage ($-20\,^{\circ}\mathrm{C}$) for seed of wild species, equipment repair shop, and an emergency generator to support the entire repository electrical system. Specific needs of a repository depend upon the climate (tropical, subtropical, cold-temperate) and upon the nature of the clonal crops maintained, e.g. strawberry *vs.* walnut.

Clonal maintenance. Field-grown collections are fenced to exclude large herbivores and vandals. Each tree or plant contains a permanent label that includes a unique identifying number and preferably a unique bar code that can be read by a portable electronic unit, an extension of the computer. When necessary, the plants are irrigated and treated with pesticides. Clonal germplasm is distributed as budwood or dormant scions removed from an established plant, or as sterilized *in vitro* microplants and meristem cultures. These last may be distributed to countries where quarantine regulations prevent the importation of scion wood. Unlike seed crop regeneration, new shoots and meristems of the clones grow out each year to renew the germplasm. Thus, there is no problem of genetic drift, low seed viability, and so forth. However, preserved clones need to be examined from time to time for possible somatic mutations that may occur infrequently.

Duplicate preservation. Because of the high cost of maintaining clonal plants, the NCGR is both the base collection and the working/active collection. Clonal cultivars of fruits and nuts must be asexually propagated to retain their genetic integrity. The cultivars are propagated asexually by rooting of stem cuttings, grafting scions to a rootstock, mound layerage, meristem culture, crown separation, etc. This propagation, plus virus elimination and long-term maintenance of growing plants, requires more people, resources, and specialized facilities than does seed storage. As stated earlier, any necessary duplication is done at the site, *e.g.*, by *in vitro* microplants or protected potted plants in a screenhouse.

From the above, it is clear that the cost of maintaining a clone is much higher than maintaining a seed-propagated cultivar. The annual cost per clone at Corvallis is $106, but the cost may vary depending upon the crop and the degree of genetic evaluation being performed at the genebank.

Size and scope of U.S. clonal repositories. The holdings of the NCGR at Corvallis are shown in Table 3; those of the other seven NCGRs are given in Table 4. It should be noted that the oldest of these repositories is but eight years old; thus, few of the collections are complete at this time [7].

Requisites for maximizing longevity of conserved materials

The goal in conservation is to maximize the viability of conserved materials at minimal cost; for materials to be distributed, maintaining the health of the plant material is equally important.

Seed materials

Biological factors of importance to longevity are proper physiological maturity, healthy state, freedom from mechanical injury, non-disturbance of seed dormancy, and recognition of interspecific and intervarietal variation in seed longevity. At the IRRI, every seed is visually and manually selected for purity, full development, and freedom from discoloration.

Physical factors of equal importance are the threshing process and speed, seed moisture content prior to drying, temperature and drying rate, sorting of seed by specific gravity, and exclusion of moisture and oxygen prior to packing. Additional details on seed processing were provided by Chang [8, 9].

Table 3. List of clonal crop accessions at National Clonal Germplasm Repositories-Corvallis, Oregon (1988).

Genus	No. of accessions		
	Seed	Clones	Total
Corylus (hazelnut)	3	108	111
Fragaria (strawberry)	47	304	351
Humulus (hops)	2	115	117
Mentha (mint)	13	475	488
Pyrus (pear)	224	1,013	1,237
Ribes (currant)	136	125	261
Rubus (raspberry, blackberry)	436	332	768
Vaccinium (blueberry, cranberry)	262	272	534
Related genera	535	183	718
Totals	2,658	2.927	4.585

Table 4. Clonal accessions of U.S. repositories at Brownwood, Texas; Davis, California; Geneva, New York; Hilo, Hawaii; Miami, Florida; Orlando, Florida; and Riverside, California (1988).

Genus	Number of accessions
Actinidia (Chinese gooseberry, kiwi)	199
Anacardium (cashew)	NA*
Ananas (pineapple)	161
Arctocarpus (bread fruit)	NA
Bertholletia (Brazil nut)	NA
Carica (papaya)	50
Carya (pecan)	241
Castanea (chesnut)	NA
Citrus (orange, lemon, etc.)	NA
Coffea (coffee)	NA
Diospyros (persimmon)	7
Ficus (fig)	135
Juglans (walnut)	1,239
Macadamia (nut)	22
Malpighia (acerola cherry)	NA
Malus (apple)	2,743
Mangifera (mango)	NA
Morus (mulberry)	NA
Musa (banana, plantain)	NA
Olea (olive)	106
Passiflora (passion fruit)	NA
Persea (avocado)	NA
Phoenix (date)	NA
Pistacia (pistachio)	219
Platycarya (nut)	1
Prunus (stone fruit)	1,177
Psidium (guava)	26
Pterocarya (nut)	43
Punica (pomegranate)	5
Vitis (grape)	3,866
Zizyphus (jujube)	NA

* NA – Not available – not yet in U.S. Department of Agriculture Germplasm Resources Information Network database.

Drying is a crucial step in preparing seed for long-term storage. The IBPGR Working Group on Seed Storage [10] has recommended a combination of aerating seeds at 15 °C and 10–15% RH for safe drying. This slow process may suit seeds of vegetables and seeds of high moisture content. The drying process used for lowering moisture content rice seed to 6% at the IRRI consists of heating seed of 12% moisture content in 38 °C ovens where the incoming air is first dehumidifed to 8% RH and then cooled to 30°C. The drying process is completed within 20 hours. For medium-term storage, a satisfactory process used earlier was to place 250-g seedlots in paper bags inside an airtight glass jar along with a liberal amount of activated silica gel. The dehydrating agent was replaced two weeks later with a fresh lot and the seed moisture content was maintained slightly above 8%. The dry state of the stored seed is readily monitored by the blue color of the indicating type of silica gel inside the jars. A high level (95–96%) of seed viability thus has been maintained for two tropical rice cultivars throughout a 24-year period [11].

Clonal materials

For fruit trees, either the grafted trees or young plants need protection from harmful or lethal diseases, damaging viruses, injurious insect pests, killing frosts, and other weather hazards. For potted plants, screenhouses or greenhouses provide the protective structure. Given this protection, a clonal plant may live for 25 to 100 years. For tissues maintained *in vitro*, a minimum nutrient medium and a reduced temperature may prolong the two- to four-year shelf life of stored materials. Growth retardants also may be added to the culture media to extend shelf life, but the techniques need further refinement [see 12, 13, 14].

Because of the long life of clonal plants, they can become infected (via insect vectors, pollen transmission, or grafting) with viruses, mycoplasmas, or other viruslike systemic pathogens. It thus is important that clonal genebanks develop routine procedures to index for and eliminate viruses. Viruses in symptomless clones are detected by inoculating sensitive indicator plants with tissue from the clone or by use of the serological enzyme-linked immunosorbent assay (ELISA). Once detected, a virus may be eliminated from a plant by thermotherapy, in which the plant is placed in a heat chamber at 38°C. This process slows the multiplication of the virus but permits rapid shoot growth, so that the tip grows away from the virus. The tip meristems are then removed and grown *in vitro* or are micrografted to a virus-free stock plant. After the heated meristem is re-established, it must be re-indexed to determine whether or not the virus was eliminated.

Maintenance of genetic integrity and reducing redundancy

Maintenance of genetic integrity of accessions

Maintaining the genetic integrity of an accession is second in importance to security. The potential hazards of frequent rejuvenation have been mentioned as one of the factors that could affect the genetic structure of self-fertilizing populations. For heterogeneous or heterozygous accessions, attention should be given to the appropriate means of preserving the original population structure as much as possible.

For a multicrop collection, increasing the seed or plant of each species requires an art and skill of its own.

The genetic integrity of clonal crops is easily maintained because they are asexually propagated. Infrequent somatic mutations occur, however, so accessions should be verified from time to time.

For sexually propagated plants, the usual mode of reproduction will determine the appropriate choice or choices. Normally, self-fertilized plants can be reproduced by a reasonable level of separation or isolation. For example, harvests of rice panicles are transported and dried inside cloth bags to minimize mechanical mixing during handling. Individual panicles of wild species of rice are bagged to collect the highly shattering grains.

Often cross-pollinated plants need to be bagged. Naturally cross-pollinated plants may require sib-pollination, caging, or even adding insect pollinators inside a cage. Alfalfa plants are grown under screened cages and pollinated by leaf-cutter bees between sister plants. Certain beans and many sorghums that are sensitive to long photoperiod need to be increased in a subtropical/tropical site. Compositing the plants in a race of maize is an alternate approach. Harvesting and keeping the seeds in bulk and in sufficient quantity is a necessary measure.

Primitive cultivars (land races) and wild species are best regenerated in environments similar to their natural habitats. The sample being grown should be of adequate size to ensure representativeness of the genetic variation of the original population [see 15 for rice].

For perennial and rhizomatous plants, some of which are clones, measures should be taken to prevent rhizomes of one accession from invading the adjacent plot of another accession or from becoming a noxious weed. Soil of potted plants should not be reused unless sterilized by steam or fumigation. Fields should be allowed to fallow for a certain period so that the dropped seeds have time to sprout. Spaced planting also aids identification of contaminants of native grasses in the seed increase fields.

Wind-pollinated grasses grown in blocks are separated by use of other plants as barriers.

Biennial plants such as head cabbage require overwintering in a cool room to initiate flowering.

It is a common practice for rice and bean workers to have a seed file consisting of a small sample of original seeds. In the case of rice, samples in the seed files are frequently used for comparative purposes prior to storage, distribution, or regeneration. Plants being rejuvenated are checked against characterization data obtained earlier. For wild species, herbarium specimens are kept at the time of the first seed increase [see 5].

For critical experiments, a purified strain to be used as a tester or differential strain should be maintained as such with a new accession number or a suffix to the original number.

Reducing redundancy in a collection

Hawkes [16] has indicated that 250 to 500 cultivars of a clonal crop, properly selected, should provide the genetic diversity (plus the most important gene combinations) of that crop. This number of cultivars provides the genetic diversity of the crop but does not attempt to provide all possible combinations of these genes. If there existed only 10,000 allelic pairs of the 100,000 or so genes it takes to 'run' a plant, it still would be impossible to obtain all possible combinations. The same is true for seed crops, even though storage costs are very much lower. But consider the redundancy in the wheat genebanks with 400,000 samples, rice with more than 200,000, maize with 100,000, and common bean with 57,000 [17]. As far as we know, nearly all of the genes are there, so a prudent strategy might be to reduce the collecting activity and increase the genetic evaluation of what we have. Plucknett *et al.* [17] stated, 'Germplasm collections need to be much more extensively evaluated and documented so that biotechnologists (and breeders) can readily obtain what they need.'

It is inevitable that duplication in a collection will grow as accessions are received in batches from different sources over a period of time. In the process of international exchanges, some accessions may have acquired modified or new names. Some accessions become ecostrains of the original sample after being dispersed to and grown at different locations. Some others may have lost their true genetic identity after repeated regeneration or faulty handling of seed. Many others may have been contaminated and become mechanical mixtures of different genotypes [see 5].

Because of the high cost of preservation, redundancy should be reduced as much as possible by a variety of means. Accession name, pedigree history, geographic origin, and seed source are useful criteria at the stage of registration. Comparison of seed and plant morphology is the next step. Characterized data can be retrieved and compared statistically to identify effectively obvious or probable duplicate accessions. Moreover, electrophoretic techniques and use of restriction fragment length polymorphism (RFLP) markers can add effective means toward this goal. It is essential that redundancy of accessions in a genebank be minimized in such a hierarchial manner. However, distinct ecostrains having the same accession name may be kept as separate accessions, since ecoedaphic differentiation under a new environment can readily lead to permanent genotypic changes [see 5].

Database management

It is imperative that all the documentation aspects of a germplasm center are computerized with respect to passport data, acquisition files, characterization data, rejuvenation records, evaluation data, seed or clonal inventories, viability records, distribution records, and various kinds of research data. The different component files should be integrated into a 'user-friendly' database system, which is linked to those of other genebanks and major user-organizations. Some of the essential components for the establishment of such a system are a uniform recording system, standardized descriptors and descriptor-states, and compatible hardware and software across different components of the global database system.

The Genetic Evaluation and Utilization (GEU) data management system of IRRI consists of the following files:

1. GB-basic – 5 identification characters and 45 morpho-agronomic descriptors provided by the germplasm center staff,
2. GB-GEU traits – 38 descriptors of biotic and ecologic responses and grain quality provided by research staff of various disciplines,
3. History of crosses – hybridization records of crosses made by the breeders,
4. Nurseries and breeding records – nursery records, progeny evaluation data and yield trial data, and
5. International Rice Testing Program (IRTP) nurseries.

The files are integrated in a database that is used frequently by IRRI scientists. Moreover, searches for accessions in the germplasm bank that satisfy combinations of desirable traits are frequently made to fill seed requests, both within IRRI and from national workers [see 18, 19].

With the increasing availability and improvement of microcomputers in recent years, dependence on mainframe computers has been greatly reduced. Most of the data-processing chores in a genebank can be handled by microcomputers. During 1988, the IRRI initiated a database management training course for plant genetic resources (PGR) workers of national genebanks. The course contents are based on microcomputer use.

The various seed and clonal plant germplasm units and their collections in the U.S. National Plant Germplasm System (NPGS) are now all linked together by the Germplasm Resources Information Network (GRIN) via a Prime computer in the National Agricultural Library in Beltsville, Maryland. A user-friendly version is open to public users, and steps have been taken to broaden the use of GRIN by some IARCs. However, international linkage with other collections, such as the Chinese clonal collections, and with other data networks, such as the IBPGR network, needs to be developed further.

Operational linkages

A plant germplasm collection cannot operate in isolation. Although the mandates of each of the three institutions represented in this paper vary, each center serves as one of the links in national/international networks. The organization of the U.S. National Plant Germplasm System, involving both state and federal inputs, has been discussed in these proceedings by Shands *et al.* and will not be described here. It is sufficient to say that the responsibilities for maintaining the large number of crops/plants involved are shared by the National Seed Storage Laboratory, the four Regional Plant Introduction Stations, the eight Clonal Repositories and other specialized agencies or programs in the NPGS such as the USDA Small Grains Collection, Soybean Collection, Cotton Collection, Potato Collection, World Rice Collection, and special genetic stocks of tomatoes, peas, barley, etc. [see 20, 21]. It should be noted that the 38 Crop Advisory Committees (CACs) form the grassroots of the NPGS. The CACs have been actively pursuing:

1. Assessing the genetic vulnerability of their respective crop,
2. Determining gaps in the present germplasm collections,
3. Recommending future exchanges of and explorations for germplasm to fill these gaps,
4. Defining the evaluation of specific characteristics used to describe the collections, and
5. Identifying uniform rating systems for each characteristic.

Cooperation has to stem from the heart of the crop scientists.

All components of the NPGS have operational linkages with numerous national PGR centers and the IARCs either in formalized collaboration or scientist-to-scientist cooperation. The areas of cooperation are in exploration and introduction, germplasm exchange, regeneration and duplicate storage, multi-site evaluation, third-country quarantine service, and sharing of information on germplasm. The WRPIS, for instance, has close ties with CIAT (Colombia) on beans and with ICARDA (Syria) on chickpeas and lentils. Several IARCs obtained large segments of crop germplasm from U.S. sources that were the early stocks of their holdings.

For other large and germplasm-rich countries, the national plant germplasm network of India has been described by Paroda and Arora [22]. For China, a national scheme has been developed (see Dong and Zhang, this volume).

A well-organized international network for rice has been in operation since 1978 [23]. Regional responsibilities for collection and preservation were defined, and back-up storage was agreed upon. The network is schematically illustrated in Fig. 3. Periodic conferences or small group discussions provide continuity in international and interinstitutional cooperation. The International Rice Germplasm Center at IRRI has an advisory committee of international composition to guide its operational direction.

At the global level, since 1987 the International Agricultural Research Centers (IARCs), under the Consultative Group of International Agricultural Research (CGIAR) system, have taken steps to intensify inter-

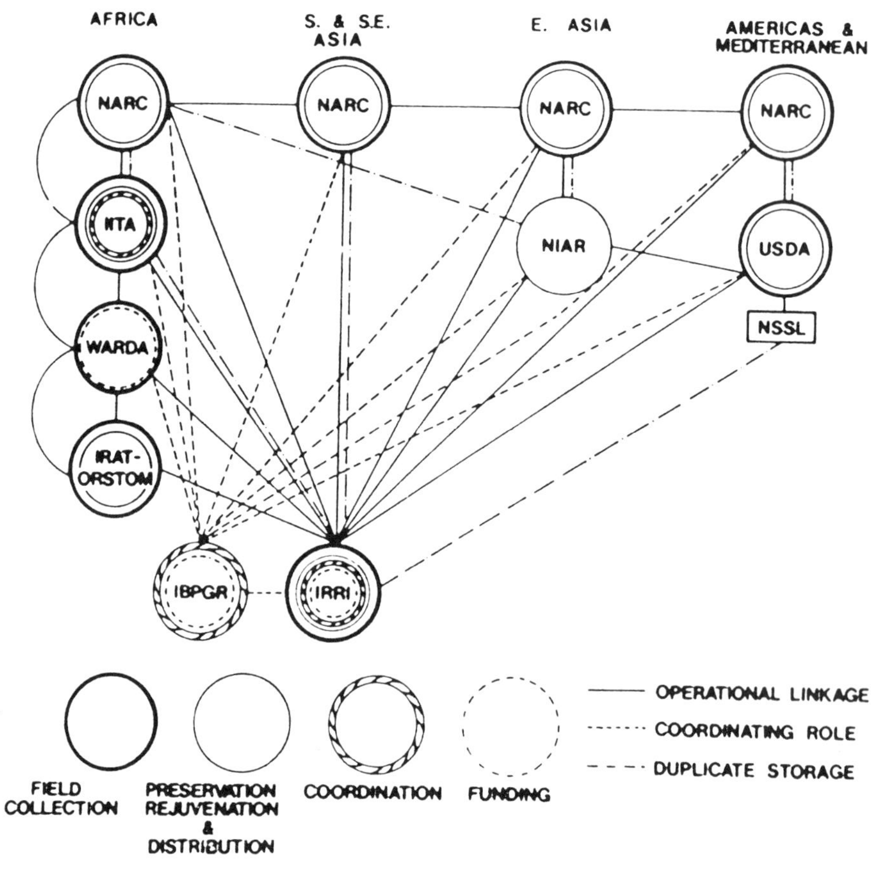

Figure 3. The global and four regional networks of linkages in rice germplasm conservation: field collection; preservation, rejuvenation, and distribution; regional coordination; and funding of projects. Acronyms: NARC – National Agricultural Research Centers; WARDA = West African Rice Development Association; IRAT = Institut de Recherches Agronomiques Tropicales et des Cultures Vivrieres (France); ORSTOM = Office de la Recherche Scientifique et Technique d'Outre-Mer (France); NIAR = National Institute of Agrobiological Resources (Japan); IBPGR, IITA, IRRI, NSSL, and USDA are explained in the text.

center cooperation through periodic meetings of the heads of their plant genetic resources programs.

While the exchange of seed/plant materials has attained high levels of mutual convenience in recent years, the linkages on data exchange have lagged behind. Efficient database management, common language in computer programs (software), and telecommunication facilities will be needed to attain a state of free and speedy exchange.

Operational linkages at all levels could be further strengthened by more frequent meetings and periodic follow-ups. This Symposium is a step in that direction.

The human factor

While plant germplasm activities have been initiated or markedly expanded in many countries in recent years, the development of trained human resources has greatly lagged behind the needs. Moreover, the scientific or technical competency of the germplasm workers has not kept pace with recent advances in preservation technology, biosystematics, cell biology, and potential contributions from biotechnology.

What are the qualifications of a well-rounded curator (manager) and of a middle-level germplasm scientist? In our assessment, the curator should be an applied biologist by training, well versed in the knowledge of the plants under his/her care and their biological diversity, able to motivate staff members of diverse disciplines to work as a team, and capable of communicating with a wide array of people in administration, physical support services, computer services, public media, and fellow scientists in other disciplines. The scope of a curator's mission is so broad and deeply involved with diverse groups of people that a prerequisite is a unique love for the plants in relation to human welfare. The scientific staff should be competent in their areas of specialization, vigilant about the security of the conserved materials, and willing to respond to changes in scientific direction. Above all, every worker should be keenly interested in the plants, conservation-oriented in their scientific outlook, and dedicated to serving other workers.

Plant conservation biology had not been taught in any university until the University of Birmingham began a graduate program in plant genetic resources (PGR) in 1969. However, the Birmingham programs are not closely affiliated with practical agricultural research and plant germplasm conservation programs. Many other short-term PGR training courses in the past (largely funded by the IBPGR) dealt with field collection, genebank operations, and *in vitro* techniques. The IRRI, recognizing the need for broad and practical training, has implemented training programs on the comprehensive package of rice conservation [11] and more recently on PGR database management. We hope that other large crop genetic resources centers also will provide such training programs in cooperation with agricultural colleges. The IBPGR is in the process of publishing a genebank handbook for clonal germplasm. When complete, this should serve as a useful text for training germplasm workers. The U.S. National Plant Germplasm Committee (NPGC) also has published an *Operations Manual for National Clonal Germplasm Repositories* [7].

During the summer of 1985, the IBPGR organized and funded an international short course on Clonal Genebank Management. This training course was presented jointly by NCGR-Davis and NCGR-Corvallis and covered crop collection criteria, collecting techniques, establishment of collections, virus testing and plant health, computer use, *in situ* reserves, and germplasm use. Twelve countries participated, including some from eastern Europe, southern Europe, Asia Minor, Central Asia, and East Asia.

Meanwhile, the plant germplasm conservation centers need to continue on-the-job training for their new staffs or for young staff members of other, smaller centers. Human resources development should be an integral part of the functions of a well-organized PGR center [3].

In retrospect, the large plant collections of the past were well managed largely by those plant scientists who were deeply involved in the exploration, introduction, evaluation, maintenance, or breeding facets from the beginning of the projects. The collections were a part of the workers' active lives, and this accounted for their intense dedication to the mission. As these workers retired or disappeared from the scene, the continuity of sustaining the well-tended programs became adversely affected, and the usefulness of the plant collections was greatly lowered. For many PGR centers, successors should be recruited in advance and nurtured in a staff development program.

Moreover, germplasm workers in many national and state centers have been continuously stressed by excessive workloads in proportion to the available facilities, funds, and supporting staff. The morale of such service-oriented workers has been lowered further by their inability to conduct research experiments that would lead to publishable findings and opportunities for career advancement and professional recognition. Until this situation is remedied, well-trained and talented young scientists will not choose the curatorial mission of PGR as a promising career.

To sum up, the human resources development facet has been and will likely remain as the most serious constraint in PGR programs since conservation-oriented activities began in a serious manner as late as the mid 1960s [24]. Intensified efforts to fill the needs will be pivotal to the future success of PGR programs.

Utilization of germplasm

Our focus on the use of plant germplasm is directed to the ways and means by which PGR workers can contribute to the full and effective use of the conserved germplasm rather than to the process of using germplasm. The latter subject is discussed by other authors in these proceedings.

Prerequisites

The prerequisites for effective use of germplasm on a scientific level are:

1. Availability and easy access to the conserved materials and the scientific findings derived from research with germplasm,
2. Communication between PGR workers and cooperating scientists in the user community – the latter include biological researchers, plant breeders, agricultural engineers, food processors, and growers as well as anthropologists, historians, economists and other social scientists, and
3. The service capability of the PGR centers.

The above facets require adequate administrative support and funding. Moreover, the level of service spirit of the PGR workers and their collaboration with the users will determine the extent of use. On the international level, political barriers in germplasm exchange between not-so-friendly countries have been largely circumvented by the IARCs [25]. International testing nurseries, such as those coordinated by the International Rice Testing Program at IRRI, have further intensified the exchange of both unimproved and elite rice germplasm [see 26].

For conserved clonal germplasm to be useful to plant improvement activities, the choice of plant materials and accompanying data is crucial. The choice of domestic cultivars is based upon a set of priority gene descriptors determined by the CAC or other crop specialists. Cultivars thus are selected on a specified genetic basis and are not simply a collection of all known varieties. For wild species, populations of all known primary species, sub- species, and ecotypes should be collected as seed because most of the genetic diversity of a genus is found in these wild relatives. From these seed samples, a few clones of each are grown out and maintained for species verification and for ready availability for genetic evaluation. Rootstock clones are selected by completely different criteria than are cultivars.

The passport data and other information known about incoming clones are extremely important, because it is likely to be the only genetic information that will be available to users for several years. For wild species, this includes site data such as latitude, elevation, soil type, rainfall, winter temperatures, pests present, etc. [16].

For large collections, systematic evaluation, efficient documentation, and effective dissemination of evaluation results are a vital link between conservation and use. Much of the evaluation results by numerous plant scientists in the past were widely scattered and unavailable to interested workers. Recently established database management systems in major genebanks constitute a giant step toward the efficient use of accumulated data. Experiences gained by IRRI staff indicate that in addition to a database system, communication among various segments of the international scientific community through workshops, monitoring tours, and newsletters such as the *International Rice Research Newsletter* has further enhanced cooperation among workers.

Similar experiences in the U.S. have shown that GRIN has greatly enhanced communication, brought workers closer together, and allowed quicker analysis of all types of data. The GRIN database management team has provided numerous training workshops at the curatorial site and at headquarters in Beltsville so that the system can be better monitored. Enhanced cooperation can be found among workers at all levels. The *Diversity* journal also significantly adds to national and international communication.

Evaluation and data dissemination

Evaluation provides the essential link between conservation and use. The accompanying database management system should encompass all facets of conservation. Efficient documentation and database management are particularly imperative to the use of evaluated germplasm because the availability or unavailability of evaluation data will determine the users' interest in accessing the seeds or plant materials. Much of the evaluation data made by numerous plant researchers in the past remains as manual records in widely scattered locations that are not available to interested and potential users.

Systematic evaluation of rice germplasm at IRRI has been discussed elsewhere in detail [19, 27] and the companion documentation system has been described [18, 19, 27]. Extensive use of the evaluated germplasm stem from the readily accessible data files.

Crop Advisory Committees in the U.S. recently have set up specific evaluation files for wheat, alfalfa and 36 other crops in line with the GRIN documentation system.

Under the NPGS data files of the plant collections of eight National Clonal Germplasm Repositories, the four Regional Plant Introduction Stations, and USDA/ARS research scientists stationed in different states are linked to each other and to the entire NPGS by the GRIN. Any interested user can access the stored data by communicating with the appropriate crop curator (see Mowder and Stoner, this volume).

However, a huge backlog of evaluation data accumulated by numerous USDA, state, and private industry workers in the past needs to be located, interpreted, transcribed, and entered into the GRIN files. We recognize that with the constraints of time and space and the passing of earlier workers, a substantial portion of the invaluable data may never be made available to other interested workers.

Examples of successful use of exotic germplasm

The costly investments in plant germplasm collections must be justified by profitable returns through use and further enhancement. The ultimate criterion for assessing the success of plant germplasm conservation programs is the benefits derived from the conserved materials. Among different categories of plant materials, introductions/exchanges and their progenies following hybridization have given the highest returns. This is followed by varietal improvement based on the use of land races. The wild species generally remain the least conserved and little exploited category, although impressive advances in breeding have been attained in several crops such as wheat, rice, tomato, sugarcane and tobacco, with genes contributed by the wild taxa [28].

Space limitations allow us to discuss only the beneficial use of germplasm conserved by our centers and collaborators, although the impact of useful germplasm on successes in plant breeding is becoming increasingly apparent.

International Rice Research Institute (IRRI) and collaborating national centers

The 'Green Revolution' in rice started with the high–yielding semidwarf varieties having the *sd*-1 gene, which originated in China. Taichung Native 1 and IR8 led the way in the subtropics and the humid tropics [29]. Semidwarfs derived from these varieties now are planted on more than 40 million ha or nearly 42% of the irrigated rice fields in South and Southeast Asia. The area planted to the semidwarfs in all of Latin America is about 26%, but the proportion jumps to 70% when Brazil is excluded. About 5% of the rice lands in Africa are planted to the high-yielding varieties [30]. Practically all the *indica*-type varieties (about 20 million ha) on the China mainland are semidwarfs sharing the same gene [31]. The hybrid rices now grown on nearly 11 million ha in China also carry the *sd*-1 gene in their parentage. The sterile cytoplasm that made hybrid rice a reality, and a second Green Revolution in rice, came from a wild relative found on Hainan Island of China [32]. The multiple resistance to diseases and insect pests in the more recently developed IRRI varieties were derived from land races of tropical Asia. A strain of *O. nivara* collected from northern India has furnished the only source of resistance to grassy stunt virus biotype 1 [33]. The estimated contributions of the high-yielding varieties to increased rice production in eight Asian countries from 1965–80 amounted to 23% of the increase. The varietal factor was estimated at $4.5 billion [34]. Further details may be found in papers by Barker *et al.* [35] and Dalrymple [30].

Western Regional Plant Introduction Station and collaborators
The breeding of the short-strawed wheat cultivar 'Gaines' may illustrate the contributions from conserved wheat germplasm. Gaines was widely grown in the Pacific Northwest and has contributed substantially to increased wheat yields. The high-yielding and short-statured Gaines wheat developed by Orville A. Vogel came through a long process of repeated crossing and selection over a period of three decades. All of its parents were foreign introductions into the U.S. (S.M. Dietz, pers. comm.).

Another unique example is the wheat accession PI 178383, which was collected from Turkey by Jack Harlan in 1948. This accession has multiple disease resistance to common bunt, dwarf bunt, and some tolerance to stripe rust, snow mold, and mildew. It took a number of years to unveil its outstanding characteristics for disease resistance. Several high yielding varieties of the Pacific Northwest derived from this introduction [36].

The introductions from Turkey were reassessed for their geographic origin and usefulness. This study led to re-collection trips to eastern Turkey, during which additional sources of resistance were obtained (S.M. Dietz, pers. comm.).

National Clonal Germplasm Repositories and collaborators
Successful uses of clonal germplasm have been attained in:

Strawberries. – Resistance to soil-borne *Verticillium* wilt, red stele, *Rhizopus* root rot, *Botrytis* fruit mold, powdery mildew, two-spotted mite, and the strawberry aphid were derived from *Fragaria chiloensis* Duchesne and *F. virginiana* Duchesne and bred into high-quality and productive (8-fold increase in California from 1940 to 1980) varieties [37],

Grape. – Resistance to high lime soils and to the *Phylloxera* root louse were obtained from wild rootstock species [38],

Apple. – Resistance to apple scab was obtained from several wild species of *Malus*, and resistance to wooly aphid and feeding by mice were also derived from other wild species [39], and

Pear. – Resistance to the root aphid, pear psylla, *Phytophthora* root rot, nematodes, blister mite, pear decline, and fire blight were obtained from wild *Pyrus* germplasm; tolerance to adverse soils was also incorporated from the wild relatives [40].

By use of wild species as rootstocks, desirable genes often can be used directly without long years of breeding. For example, Comice pear, a low yielding cultivar, was grafted to nine related wild species and grown for 10 years. Amazingly, use of some rootstocks enabled a 10-fold increase in fruit yield efficiency [41].

Additional references on the successful use of land races and wild taxa maybe found in papers by Stalker [42], Harlan [43], Oldfield [44], Rick [45], and Chang [4].

Reinstating genetic diversity in major crops

Success in plant breeding inevitably leads to a narrowing of the genetic base in a crop. A monumental example may be drawn from the case of southern leaf blight associated with the Texas type *cms* of maize [46]. Potential dangers in other crops have been pointed out by many authors and summarized by Day [47]. The situation is especially vulnerable in rice because of:

1. The common *sd*-1 compound-complex gene used by numerous rice breeders. Semidwarf rices sharing the *sd*-1 gene now also occupy about one-fourth of the rice acreage in the U.S. [30],
2. The Cina cytoplasm present in many HYVs, and
3. The Wild Abortive *cms* in Chinese hybrid rice [see 48].

The semidwarfing genes in the high-yielding wheats are slightly broader than those of rice. The commonly used *Rht* genes may be traced back to four: *Rht*-l, *Rht*-2, *Rht*-8 and *Rht*-9, of which the first two came from Norin 10. These two genes were widely used in the Mexican (CIMMYT), Indian, U.S., and Japanese programs [49]. This is another case where genetic diversity needs to be reinstated.

Germplasm workers, including Crop Advisory Committees and key scientists, should monitor the genetic diversity of principal cultivars in a major crop and help breeders diversify and broaden the genetic base. At IRRI, we have made special efforts to identify new sources of semidwarfism in the rice collection by making hundreds of allelic tests with the well-used *sd*-1 gene [50]. We also collect as many *cms* sources as possible to build up the genetic stocks.

PGR worker's role in promoting the use and security of germplasm

The PGR scientists and advisors play a crucial role in determining the extent and depth of using conserved germplasm because gaps exist in informing interested users of the potentials of little-known germplasm and especially because of the reluctance of plant breeders to experiment with agronomically poor germplasm. Constraints in expediting the use of germplasm have been discussed in detail [28].

We consider it the primary duty of PGR workers to accelerate and intensify the use of the germplasm collections under their charge in a variety of ways:

1. Through person-to-person communication with users, initially through the crop advisory committees, on a comprehensive array of topics including field collection and re-collection,
2. Through broad dissemination of information on conserved and evaluated materials using catalogs, publications, newsletters, and computer printouts,
3. Through feedback to administrators about the dissemination and use so that sustained or increased support can be assured,
4. Through dialog with the public sector by receiving visitors, especially civic groups, journalists, and school teachers, and by preparing and releasing news/teaching materials, and
5. Through interacting with socio-political activists who could be persuaded to support PGR programs.

The conservationist also should serve as an apostle in fostering the reinstatement of genetic diversity in major crop plants. The CAC members and the crop curator comprise the few knowledgeable sources of information on the extent of genetic erosion in specific germplasm groups. PGR programs can benefit from the awareness and support from all sectors of society to justify the seemingly endless demand for inputs [25, 51].

As a custodian of many national collections, the International Rice Germplasm Center (IRGC) at IRRI has returned entire national collections to several donor-countries when called upon to do so. The WRPIS and other units of the NPGS also have returned lentils to Turkey, the entire 'teff' collection to Ethiopia, and safflower land races to India. The bean collection of CIAT received seeds from WRPIS to initiate its operations. The capability to restore germplasm to national genebanks can avert the criticism that genebanks have 'confiscated' the land races of developing countries.

Linkage with biotechnology

Biotechnology depends on germplasm collections for experimental materials – genes and gene combinations. Up to this stage of development, germplasm workers have been performing the support and service role that fuels experiments in genetic engineering. Their assistance includes the know-how on growing the plant materials and on their genetic compositions. The bulk of the rice materials demanded by biotech workers from the IRGC is exotic germplasm (temperate zone rice cultivars and wild relatives), which is difficult to multiply. The same is true for wild species of chickpeas, some of which have never before produced seed under cultivation except with special care at Pullman, Washington; and for the conversion of tropical maize, sorghum, disease cleanup, etc., for U.S scientists.

Meanwhile, conventional plant breeders and their associates are apprehensive that biotechnology presents a serious competition for available funds, facilities, and talented young scientists [52, 53]. Moreover, the rapid decline of U.S. universities in training plant breeders will lead to serious gaps in crop improvement.

Germplasm workers should recognize the potential areas of cooperation and assistance that biotechnology could offer to genetic conservation activities [see 54]. The most promising areas are in micropropagation, elimination of viruses in tissue cultures, provision of molecular diagnostic techniques, *in vitro* storage and cryopreservation, and research on germplasm. It becomes obvious that genetic conservation and biotechnology can be complementary and synergistic [55].

Future perspectives

Research directions

Looking ahead at the never-ending task of conserving plant germplasm collections for productive use, a number of necessary research areas are looming on the horizon. Further research to improve preservation techniques in the interest of cost efficiency ranks high. Drying and storage conditions for high-oil or high-protein seeds should be refined. For recalcitrant seeds and many difficult-to-preserve plants, new techniques need to be developed with the collaboration of different disciplines. To minimize the destructive sampling of stored seed for viability monitoring, indirect testing techniques should be explored. Sequential testing has been proposed as an initial step in saving seeds during conventional viability tests [56]. The genetic stability of conserved materials over time is such an important aspect that it certainly deserves more in-depth studies; this need is probably of greater concern for *in vitro* cultures than for seeds. A more thorough understanding of the germplasm collections being conserved in relation to the total range still existing in nature would lead to more meaningful re-canvassing and re-collection activities. The scope of genetic diversity in the major food crops should be monitored continually, both spatially and temporally, and remedial measures formulated. The situation is particularly worrisome in rice, wheat, and in the rice-wheat cropping sequence.

While collection size may be a favorable factor in ensuring a broad genetic base, obvious redundancy must be reduced by improved techniques, such as molecular diagnostics, to make maintenance costs more effective. Consolidation within a large collection and between collections is necessary.

Germplasm workers also should play a more active role not only in supplying materials for evaluation but also in systematic studies so that the users can be assisted by the knowledge and expertise of the former on the germplasm materials. Detailed analysis of evaluation results could form the basis for re-collection or for assessing genotype x environment interactions [15]. Again, the Crop Advisory Committees appear to be the appropriate venue for a closer collaboration.

Collaborative ventures

Since germplasm knows no political boundaries or linkages on international, interinstitutional, or interdisciplinary levels, collaborative ventures offer the only avenue to protect the dwindling genetic resources that are the only biological heritage mankind can draw upon to feed the continually expanding population. Linkages already exist to a certain extent in field collection, *ex situ* conservation, duplicate storage, evaluation, research, exchange, and use, but the efforts still fall short of the real needs to ensure improved security and full use. Germplasm workers themselves have to exert a more vigorous role in making the services more user-oriented, complementary, and supportive of enhanced use.

While recent litigations and intergovernmental confrontations about the ownership of germplasm have temporarily hampered the free exchange of plant germplasm among certain countries, the situation can be expected to improve when communication is broadened. Through open dialogue, decision-makers and activists of both industrialized and less-developed countries can appreciate the complementary roles that each nation can play in safeguarding the available germplasm. Moreover, cooperation cannot be brought about by statutory means alone – genuine collaboration must stem from the hearts of working scientists who share a common interest in germplasm [25].

It is imperative that a free exchange of germplasm materials and related information is available to all interested users. This will insure the users' trust in the value of the germplasm collections and enhance collaborative germplasm research.

Improvements in other areas

Genetic conservation of plants is a relatively young endeavor. Its future growth will benefit from concurrent growth in allied fields. An expanded concept of conservation biology has been presented by Soulé [57] and will serve as a useful reference for PGR workers.

Meanwhile, PGR workers should make use of advances in preservation technology, refrigeration engineering, computer technology, and telecommunications to upgrade the physical facilities and laboratory equipment continually. Investments in more efficient facilities and equipment will yield handsome dividends in the long run.

Plant quarantine measures for pre- and post-entry operation at genebanks need to be reassessed and improved. Rice workers have recently taken such a step to insure the health of germplasm being exchanged [58].

Constraints

No plant germplasm collection can fare well without well trained and dedicated workers. When a highly knowledgeable leader retires or passes from the scene, the security of the irreplaceable collection is threatened. In the absence of well established post-graduate programs on PGR, different institutions must collaborate and pool their resources to supply the future cadres who can take on the complicated and often distressing mission. Similarly, the training of competent plant breeders should not be neglected in the current race to capitalize on biotechnology.

While the important contributions of rice germplasm to the Green Revolution in rice and of the multitude of foreign introductions that appear on the food tables of every American family are increasing the public awareness of germplasm, the constraints of germplasm workers in discharging their heavy responsibilities have not been widely recognized. One major challenge is the continuous growth in the size of collections. For instance, since 1972, the rice collection of IRRI has grown by more than four thousand samples per year. Meanwhile, the size of the distributions of the conserved materials from all three centers involved in this paper also has been expanding during the past decades.

The magnitude of seed distribution by the WRPIS has grown from an annual average of 3,420 during 1952–64 to 21,603 in 1986, while distribution to foreign countries has grown from an annual average of 4,364 during 1964–79 to 9,916 in 1986. Similarly, rice seed dissemination by the IRGC to both IRRI researchers and foreign scientists reached 55,000 in 1977 and leveled off to about 45,000 during the last two years, but the number of requests keeps growing [59]. Both maintenance and distribution phases involve physical facilities (fields, labs, greenhouses, growth chambers, tissue culture facility, and storerooms), human inputs, and related expenses in handling incoming and outgoing samples. Unless concurrent increases are provided to cope with their ever increasing growth, germplasm managers will find it impossible to cope with the continuing expansion, while maintaining the level of service desired. It is essential that CACs and germplasm leaders establish upper limits to the number of accessions for each crop.

We sincerely hope that a greater awareness is being generated across all sectors of human society in recognizing the arduous and unglamorous task of genetic conservation so that sustained support is ensured. Assured continuity will enable this vital hub of biological activities to fulfill its mission of providing food to the teeming human populace which is now slightly over five billion and to reach about 6.5 billion by the turn of the century.

References

1. Chang, T. T. 1986a. Genebank design. Pp. 224–235 in K. L. Tao, R. Ellis, P. Stanwood and T. T. Chang (eds.), Genebank Management and Seed Storage (in Chinese). International Board for Plant Genetic Resources-Chinese Academy of Agricultural Sciences, Beijing, China. [A reprint in English is available from the author, upon request]. 238 p.
2. Chang, T. T. 1983. Guidelines for the cold storage of orthodox crop seeds in the humid tropics. IBPGR Regional Committee for SE Asia Newslett. 7(2/3): 23–26.
3. National Research Council. 1988. Expansion of the U.S. National Seed Storage Laboratory: Program and Design Considerations. National Academy Press, Washington, DC. 16 p.
4. Chang, T. T. 1989a. The case for large collections. Pp. 123–135 in A.H.D. Brown, O.H. Frankel, D.R. Marshall and J.T. Williams (eds.), The Use Of Plant Genetic Resources. Cambridge University Press, Cambridge. 383 p.
5. Chang, T. T. 1976. Manual on the Genetic Conservation of Rice Germplasm for Evaluation and Utilization. International Rice Research Institute, Los Baños, Philippines. 77 p.

6. Chang, T. T. 1981. The rice genetic resources program of IRRI and its impact on rice improvement. pp. 85–106 in Rice Improvement in China and Other Asian Countries. International Rice Research Institute, Los Baños, Philippines. 307 p.
7. National Plant Germplasm Committee (NPGC). 1986. Operations Manual for National Clonal Germplasm Repositories. U.S. Department of Agriculture, Beltsville, MD. 261 p.
8. Chang, T. T. 1985c. Preservation of crop germplasm. Iowa State J. Res. 59(4): 365–378.
9. Chang, T. T. 1986b. Genebank management. Pp. 214–219 in K. L. Tao, R. Ellis, P. Stanwood and T. T. Chang (eds.), Genebank Management and Seed Storage (in Chinese). International Board for Plant Genetic Resources-Chinese Academy of Agricultural Sciences, Beijing, China. [A reprint in English is available from the author, upon request]. 238 p.
10. Cromarty, A. S., R. H. Ellis, and E. H. Roberts. 1985. The Design of Seed Storage Facilities for Genetic Conservation. International Board for Plant Genetic Resources, Rome. 100 p.
11. International Rice Research Institute. 1987. Annual report for 1986. Pp. 3–5. International Rice Research Institute, Los Banos, Philippines. 639 p.
12. Withers, L. A. 1984. Germplasm conservation *in vitro*: present state of research and its application. Pp. 138–157 in J. H. W. Holden and J. T. Williams (eds.), Crop Genetic Resources: Conservation and Evaluation. George Allen and Unwin, London. 296 p.
13. Bajaj, Y. P. S. 1986. *in vitro* preservation of genetic resources techniques and problems. Pp. 43–57 in Nuclear Techniques and *in vitro* Culture for Plant Improvement. International Atomic Energy Agency, Vienna. 529 p.
14. Withers, L. A. and J. T. Williams. 1986. *In vitro* Conservation. IBPGR Research Highlights 1984–1985. International Board for Plant Genetic Resources, Rome. 21 p.
15. Chang, T. T. 1985b. Collection of crop germplasm. Iowa State J. Res. 59(4): 349–364.
16. Hawkes, J. G. 1980. Crop Genetic Resources Field Collection Manual. International Board for Plant Genetic Resources, Rome. 37 p.
17. Plucknett, D. L., N. J. H. Smith, J. T. Williams, and N. M. Anishetty. 1983. Crop germplasm conservation and developing countries. Science 220: 163–169.
18. Gomez, K. A., D. Tuazon, and N. E. Nano. 1979. Germplasm bank information retrieval system. International Rice Research Institute Res. Pap. Ser. 45. 24 p.
19. Chang, T. T. 1985d. Evaluation and documentation of crop germplasm. Iowa State J. Res. 59(4): 379–398.
20. National Plant Genetic Resources Board (NPGRB). 1981. The National Plant Germplasm System. U.S. Department of Agriculture, Washington, DC. 12 p.
21. Jones, Q. and S. Gillette. 1982. The National Plant Germplasm System Special Report No. 1. Diversity Special Issue. 7 p.
22. Paroda, R. S. and R. K. Arora. 1986. Plant Genetic Resources – An Indian Perspective. National Bureau of Plant Genetic Resources, New Delhi, India. 34 p.
23. International Rice Research Institute (IRRI). 1978. Proceedings of the Workshop on the Genetic Conservation of Rice. International Rice Research Institute, Los Baños, Philippines. 54 p.
24. Chang, T. T. 1985a. Principles of genetic conservation. Iowa State J. Res. 59(4): 325–348.
25. Chang, T. T. 1987a. Saving crop germplasm. SPAN 30(2): 62–63.
26. Chang, T. T., D. V. Seshu, and G. S. Khush. 1988. The rice seed exchange and evaluation programs of IRRI. Pp. 21–32 in Rice Seed Health. International Rice Research Institute, Los Baños, Philippines. 372 p.
27. Chang, T. T. 1985f. Evaluation of exotic germplasm for crop improvement. Pp. 95–103 in K. L. Mehra and S. Sastrapradja (eds.), Proc. of the International Symposium on Southeast Asian Plant Genetic Resources. Lembaga Biologi Nasional-LIPI, Bogor, Indonesia. 211 p.

28. Chang, T. T. 1985e. Germplasm enhancement and utilization. Iowa State J. Res. 59(4): 399–424.
29. Chang, T. T. 1979. Genetics and evolution of the Green Revolution. Pp. 187–209 in R. de Vicente (ed.), Replies from Biological Research. Consejo Superior de Investigaciones Cientificas, Madrid. 418 p.
30. Dalrymple, D. G. 1986a. Development and Spread of High-Yielding Rice Varieties in Developing Countries. U.S. Agency for International Development, Washington, DC. 117 p.
31. Shen, J. H. 1980. Rice breeding in China. Pp. 9–30 in Rice Improvement in China and Other Asian Countries. International Rice Research Institute, Los Baños, Philippines. 307 p.
32. Lin, S. C. and L. P. Yuan. 1980. Hybrid rice breeding in China. Pp. 35–52 in Innovative Approaches to Rice Breeding. International Rice Research Institute, Los Baños, Philippines. 182 p.
33. Khush, G. S. 1984. IRRI breeding program and its worldwide impact on increasing rice production. Pp. 61–94 in J. P. Gustafson (ed.), Gene Manipulation in Plant Improvement. Plenum Publishing Corporation, New York. 668 p.
34. Herdt, R. W. and C. Capule. 1983. Adoption, Spread and Production Impact of Modern Rice Varieties in Asia. International Rice Research Institute, Los Baños, Philippines. 54 p.
35. Barker, R., R. W. Herdt, and B. Rose. 1985. The Rice Economy of Asia. Resources for the Future, Washington, DC. 324 p.
36. Burgess, S. (ed.). 1971. The National Program for Conservation of Crop Germplasm. University of Georgia, Athens, Georgia. 73 p.
37. National Council on Gene Resources. 1982. Strawberry Genetic Resources. California Department of Food and Agriculture, Berkeley. 252 p.
38. Howell, G. S. 1987. *Vitis* rootstocks. Pp. 451–472 in Rom and Carlson (eds.), Rootstocks for Fruit Crops. John Wiley and Sons, New York. 494 p.
39. Thompson, M. M. 1981. Utilization of fruit and nut germplasm. HortScience 16(2): 132–135.
40. Westwood, M. N. 1982. Pear germplasm of the new National Clonal Repository: its evaluation and uses. Acta Horticulture 124: 57–65.
41. Westwood, M. N. and P. B. Lombard. 1982. Rootstocks for pear. Proc. Oregon Hort. Soc. 73: 64–79.
42. Stalker, H. T. 1980. Utilization of wild species for crop improvement. Adv. Agron. 33: 111–147.
43. Harlan, J. R. 1984. Evaluation of wild relatives of crop plants. Pp. 212–222 in J. H. W. Holden and J. T. Williams (eds.), Crop Genetic Resources: Conservation and Evaluation. George Allen and Unwin, London. 296 p.
44. Oldfield, M. L. 1984. The Value of Conserving Genetic Resources. U.S. Department of Interior, National Park Service, Washington, DC. 360 p.
45. Rick, C. 1984. Conservation and use of exotic tomato germplasm. Pp. 121–131 in Conservation and Utilizatioin of Exotic Germplasm to Improve Varieties. Report of 1983 Plant Breeding Research Forum, Pioneer Hi-Bred International, Des Moines, Iowa. 252 p.
46. National Academy of Sciences. 1972. Genetic vulnerability of major crops. National Academy of Sciences, Washington, DC. 307 p.
47. Day, P. R. 1973. Genetic variability of crops. Annu. Rev. Phytopathol. 11: 293–312.
48. Chang, T. T. 1984. Conservation of rice genetic resources: luxury or necessity? Science 224: 251–256.
49. Dalrymple, D. G. 1986b. Development and Spread of High-Yielding Wheat Varieties in

Developing Countries. U.S. Agency for International Development, Washington, DC. 99 p.

50. Chang, T. T., C. Zuno, A. M. Romena, and G. C. Loresto. 1985. Semidwarfs in rice germplasm collections and their potentials in rice improvement. Phytobreedon. 1(1): 1–9.
51. Chang, T. T. 1987b. The availability of crop germplasm. Pp. 225–231 in Crop Exploration and Utilization of Genetic Resources. Taichung District Agricultural Improvement Station, Changhua. Taiwan. 285 p.
52. Sprague, G. F., D. E. Alexander, and J. W. Dudley. 1980. Plant breeding and genetic engineering: a perspective. BioScience 30: 17–21.
53. Borlaug, N. E. 1983. Contributions of conventional plant breeding to food production. Science 219: 689–693.
54. Goodman, R. M., H. Hauptli, A. Crossway, and V. C. Knauf. 1987. Gene transfer in crop improvement. Science 236: 48–54.
55. Chang, T. T. 1988. Conservation of gene pools. COGENE Symposium: Genetic Experimentation and Evolutionary Change. Jan. 21–23, 1988. Basel, Switzerland. (Abstract).
56. Ellis, R. H., E. H. Roberts, and J. Whitehead. 1980. A new, more economic and accurate approach to monitoring the viability of accessions during storage in seed banks. Plant Genet. Resour. Newslett. 41: 3–18.
57. Soule, M. E. 1985. What is conservation biology? BioScience 35(11): 727–734.
58. International Rice Research Institute. 1988a. Rice Seed Health. International Rice Research Institute, Los Baños, Philippines. 372 p.
59. International Rice Research Institute. 1988b. Annual report for 1987. International Rice Research Institute, Los Banos, Philippines. 640 p.

Population genetics and the dynamics of conservation*

GENE NAMKOONG

USDA Forest Service, Southeastern Forest Experiment Station, Genetics Department, Box 7614, North Carolina State University-Raleigh, North Carolina 27695-7614, U.S.A.

Abstract Dynamic evolution of natural populations implies that collections made at any one time and statically maintained cannot capture all potentially useful alleles or efficiently save useful alleles. The structure of mating demes and the presence of heterogeneous selection force species to evolve new alleles, traits, and populations continuously, some of which may be useful for breeding. Breeding also is a dynamic system, and breeders can develop different populations and use new gene effects to create selective responses to new economic demands or to new requirements for ecological adaptability. Static collections, however, are not necessarily easy to use. The genetic management of both natural and breeding populations can involve single or multiple populations selected for homogeneous or diverse objectives. Through collection and maintenance programs that mediate between evolution and breeding, germplasm conservation can include the active development of collections and the creation of diversity.

Key words: evolution, breeding, population genetics, selection, diversity

Introduction

By the evolution of populations and species we traditionally mean changes in population sizes, including founding and extinction, as well as changes in the sizes and forms of phenotypic and genotypic variability caused by selection, mutation, drift, and migration. Populations are thus viewed as dynamic systems. Some genetic elements change in response to genetic (epistatic), physical, and biotic elements that in turn respond to other external forces and changing elements of the interactive system. If threatened, the evolving system may require protection. However, it is often convenient to consider populations as static assemblages of genotypes which, if threatened by loss or deleterious change, may require preservation. In that case, we use population genetics to help understand how widely or narrowly to distribute our sampling to collect and store populations more efficiently. Sampling designs within and among more or less widely dis-

* Paper No. 11574 of the Journal Series of the North Carolina Agricultural Research Service, Raleigh, NC 27695-7601

L. Knutson and A.K. Stoner (eds.). Biotic diversity and germplasm preservation, global imperatives.

ISBN 0-7923-0178-1

persed populations have been recommended for this purpose [1–6] and analyses of ecogeographic surveys have helped to describe population structure [7].

Efforts to capture the extant variation have focused on ensuring that all potentially useful alleles are preserved and cataloged in some secure collection, and amassing ever larger collections has become a major operational objective. Particularly with the new interest in biotechnology and genetic engineering, and the hunt for quantitative trait loci (QTL), germplasm conservation is sometimes thought to be reducible to storage technology. As expressed by Lewis [8], 'An adequate gene resource conservation program is to genetic engineering as a library is to knowledge. They both are sources of the past, the present, and the future, and both are equally essential to the development of their counterpart.' Breeding may then be reduced to simply accumulating good genes lexicographically. If the library functions well, breeders simply pull the right genes off the shelf to construct new varieties. Thus, population genetics is reduced to collecting pieces of DNA from a global biotic machine that may be losing useful pieces of itself, and breeding is reduced to tinkering with those parts to reconstruct or repair the organic mechanisms.

Aside from the difficulties of ever knowing as much as would be needed to understand the functioning of many genes or their interactions, there are fundamental problems with that mechanistic vision. While collections are clearly necessary, they have not been developed to be sufficient for our needs. As stated by Plucknett, *et al.* [9], 'Simply storing material, as in a museum, is not enough...', and several authors in this volume address important issues in breeding that indicate a far more dynamic view of organisms.

From selection studies, it is clear that germplasm is not static in its development or its ultimate uses. There is a continuous creation and erosion of genetic variability as demonstrated by continued responses to recurrent selection, due to release of cryptic variation [10] or to new variations created by mutation [11]. In addition, feedback mechanisms exist in natural as well as breeding populations. Since selection effects are at least partially dependent on what has already been selected, the alleles, traits, and even the gene actions developed are dependent on an interaction or feedback between different parts of the sequence of populations. The specific alleles that are effective at specific loci thus depend on many other alleles in the populations in which they occur and in which they are to be used. Furthermore, the external forces of the economic and ecological milieu for breeding are constantly shifting. The breeder must be able to respond to or anticipate shifting goals. Therefore, breeding is not reducible to merely accumulating 'good' alleles at QTL; it involves a dynamic manipulation of genes and populations. While good collections are necessary for such endeavors, they are not sufficient for agricultural productivity.

If the uses of collections are not static, collections that are static at the time of sampling may not be the most useful assemblage of genotypes for breeding, and they may not include all potentially useful alleles. Only if sampled populations are themselves static can collections reflect a stable distribution of genetic variation if that is desired. For such populations, genetic variations may not be dynamic responses to evolutionary forces. They may merely be random noise that we collect to avoid the genetic erosion that occurs around the ragged edges of evolutionary stable states. If populations, genes, and traits are changing in an evolutionary dynamic, however, any collection can only represent a sample of transient allelic states. Its preservation is an arbitrary fixation of a dynamic system that never includes all potentially useful allelic states. Rational germplasm conservation, therefore, requires meshing a dynamic evolutionary system with a dynamic breeding system. Under these circumstances, it may not be necessary to conserve all populations, but it is necessary to understand their dynamics.

In this paper, some elements of the dynamics of population genetics will be examined for their effect on genetic distributions, biotic diversity, and germplasm conservation. Then, the breeding objectives of conservation will be discussed, and some suggestions for relating evolutionary diversity to breeding use will be offered.

Natural population genetics

Most species are subdivided to some extent, with different alleles and allele frequencies existing among subpopulations. Whenever a species occupies a large enough area that random mating among all parts of its range is untenable, the possibility of subdivision exists, even if no selective differences exist. Since such areas can be very small, especially if inbreeding is common or if either male or female gametes have any migratory constraints, many population differences can persist. If selective differences are at all correlated with boundaries of migratory restrictions, interpopulational differences per se can be important considerations in germplasm conservation.

Consider first, genetic loci that have no average effects of selective importance. If strict mating barriers exist, populations will differentiate at a rate strictly determined by the effective population size of each 'island' population. If mating is random within populations and each 'island' has had the same number of equally fertile founders, then the variance among populations is simply:

$$V(p_i) = [\bar{p}1 - \bar{p})/2N_e] , \quad (1)$$

when $\bar{p}$ is the species-wide average allele frequency, and N_e is the variance effective number. However, even with many hundreds of individuals within the boundaries of 'islands', N_e can be very small. In most plant and animal communities, effective reproduction is limited to a relatively few individuals, population sizes tend to fluctuate, and, for perennial or iteroparous organisms, repeated reproductive dominance and parent-offspring mating can occur. All of these events depress N_e well below any census or density estimates and inflate $V(p)$. If populations remain segregated, the variance increases from generation to generation.

If migration occurs, the rate of differentiation is slowed and even low-frequency migrants can have strong effects on the distribution of allele frequencies and their probabilities of loss or fixation. With some variations for different types of species structure, only a few migrants per generation per population are generally required to allow an equal probability of all allele frequencies to exist among populations [12]. While this condition substantially reduces the probability of losing alleles in any of the populations of a species, it still allows substantial differentiation to exist among them. Similarly, the probability distribution of the allele frequencies can be nearly flat. If mutation introduces new alleles into the populations, and if the size of each population is approximately equal to the reciprocal of the mutation rate, all allele frequencies can be equally likely to occur. Thus, both mutation and migration inhibit the rate of allelic losses, but they cannot prevent interpopulational differentiation. Sufficiently large N_e is required to ensure the persistance of alleles in some populations, however.

The question for conservation then is the extent to which populations differentiate as a result of these evolutionary factors and how we can estimate any past or anticipated divergences. For nonselectively affected loci, we can consider a model species in which no migration occurs among populational mating demes. We define an 'effective number of populations' (N_p) as an analogue to the more familiar effective population size (N_e). We consider that a species, with total population size NT, is apparently composed of P populations, each of size N_e, and $NT = PN_e$, and we assume that P, and N_e can be estimated from field observations. If the populations have no migration, the distribution of average allele frequencies is reflected in the variance effective number, N_e, and the $V(p_i)$. If the populations are connected by migrants, however, the allele frequencies will not be as strongly differentiated, and the population will behave as if N_e was larger than estimates made on the basis of within-population observations. Hence, for the same total population size NT, a large N_e implies the existence of a smaller effective number of populations such that $N_p \leq P$. At the limit, $N_p = 1$, a full exchange of migrants is implied such that the whole species is essentially panmictic without genetic subdivision. In that case, sampling neutral genes can ignore any apparent ecological boundaries and collecting within one area is as good as sampling among broadly distributed collection sites. At the other extreme, $N_p = P$ and unique or 'private' alleles [13] may be present in some populations, if N_e is small. Broad interpopulational sampling will then be needed to ensure sampling in those populations where p_i is high. If $1 < N_p < P$, some degree of drift will differentiate populations, but not at the same rate as for P independent populations. Thus, it is more important to know N_p than P – to understand the number of different constituent populations that may exist and the nonstatic nature of species composition and evolution, even in the absence of diversifying selection.

Selection

If we add an element of selection to this system, even one that is static over time but differs among parts of the species' range, then variations in selectively affected alleles can be substantially more complex. Even in the absence of any boundaries to random mating, directional selection in some parts of the species' range does not necessarily lead to allelic fixation [14]. Since migration often is correlated with distance, physical barriers, and environmental differences, there is often some association between the boundaries of mating and selection demes. In cases where population differentiation is reinforced, extreme differences can be generated quickly, even if selection is effective on only one sex [15, 16]. Allelic variations are much easier to maintain, and population differentiations for different alleles are easier to maintain. In such cases private alleles also are more likely to exist, and sampling must be dispersed. If boundaries are related to ecological or geographic distance variables, targeted sampling can be more easily designed. We again must conclude, however, that while it is important for sampling purposes to know how selection can differentiate populations, selection itself can rarely be considered a static process.

In particular, selection in competitive ecosystems or in host-pathogen systems can generate permanent nonequilibrium states such that neither phenotypic nor genotypic stasis is ever reached. First, consider a simple case where complete random mating exists for a species with genetic variation in susceptibility to a pathogen or pest. If the susceptibility reaction is such that one homozygote genotype may be more vigorous but becomes more susceptible as its frequency in the field increases, its relative fitness declines when it reaches high frequency. Similarly, the other homozygote genotype may also become more susceptible to attack if it reaches high frequency. In that situation, an intermediate equilibrium frequency might be expected.

Contrary to such expectations, however, research has shown that such frequency-dependent systems can generate cyclic behavior in relative representation of the genotypes – that there never is a stable equilibrium [17, 18].

If we further consider the case where two species are competing and one of them has genetic variation in its capacities to withstand intra- and interspecific competition, even more complex behavior can be generated [19]. For example, assume that one homozygote (aa) is a rapid early grower, is not suppressed by intraspecific competition, but is susceptible to interspecific competition. If the 'a' allele is fixed, the species might be locked into a competitive exclusion contest with another species. Sometimes one species would eliminate the other, and at other times the roles would be switched, depending on initial species densities. An alternate allelic homozygote (AA) might be a better interspecific competitor than an intraspecific one, and, if the 'A' allele is fixed, both species could coexist. Where both alleles are present in a random mating population, some surprising results ensue. If there is an element of frequency dependence in either gene effects or interspecies competitive interactions, stable and persistent cycles of allele frequencies and sizes of both species can appear. Even in the absence of frequency dependence, however, the genetically invariant species can never exclude the genetically variable one. The allele (a) that would allow competitive exclusion to occur would sometimes allow the invariant species to increase. That increase, however, would decrease its own species population size and would then favor alleles (A) with good interspecific competitive ability. Then, either that allele would tend to be fixed and competitive coexistence would result, or the genetically variable population size would increase fast enough that the allele (a) favoring competitive exclusion would be fixed, but only if the other species goes to extinction. With any frequency dependence or with mating barriers such that both alleles are protected, permanent cycling of population sizes and allele frequencies can ensue. Thus, even without external environmental factors selectively maintaining nonequilibrium states, the biotic environment can be a major selective agent generating permanent disequilibria.

Biotic factors such as pests and pathogens are obvious generators of both equilibrium and nonequilibrium coevolution because genetic variations often exist in virulence and resistance types of interspecies effects, as well as in intraspecies competition and fecundity. With genetic and ecological variations both generating response dynamics, it becomes less likely that a species will have one strongly attractive equilibrium. In fact, even without interspecific density effects on probabilities of successful attack by pathogens, or on resistance by hosts, genetic variations in virulence and susceptibility generate permanent cyclic changes in allele frequencies in both host and pathogen populations [20]. The addition of density dependence in such factors as birth rates, likelihood of successful attack, and host escape or frequency dependence virtually guarantees very rich dynamics. The dynamics of such systems can generate several stable and unstable equilibria. Hence, a species exposed to one kind of physical environment can exist in a stable equilibrium composed of several different states. In fact, it can exist in a chaotic dynamic, never even settling into cyclic behavior [21]. Hence, biotic selection factors themselves can generate permanent instabilities, and the natural system need never achieve any maximum fitness for any of the participating species. Such interactions at the genetic and ecological levels may protect species from extinction in complex ecosystems, but it is not clear that genetic complexity generally leads to ecosystem stability. It is also obviously not true that ecosystem stability leads to genetic complexity.

Models of biotic selective factors present a highly complex view of selection because they are derived assuming complete random mating, no mutation, and no other selective factors such as climate or soil that induce variations in the interaction coefficients. In coarse–grained environments, environmental forces can create situations where qualitatively different dynamics are simultaneously forcing populations into different evolutionary paths. If selection factors are reinforced by mating preferences, populations may respond to strongly diversifying pressures that are not obviously related to any single cause and may not be directed to any obvious equilibria. Furthermore, since the selection effects of biotic factors are nonlinear and depend on the initial state of the populations, the changes in species and allele frequencies in any one community are not understandable without knowledge of its prior states.

Diversity

For alleles and loci that had or now have selectively significant effects, patterns of heterogeneity in allelic frequencies among populations are not easily predicted. Nevertheless, high levels of genetic diversity at many loci exist in many species. They continue to be generated by both neutral and selective natural processes. For alleles and loci that have not been subject to selection or carried in linkage disequilibrium with selected alleles, the distribution will be dictated by mutation and migration patterns of ancestral relationships. If the species is structured simply, as in one large random mating population, then any pattern of sampling is equally likely to capture these alleles as is any other. In fact, if several such populations exist in isolation but all are large, or only recently separated, they would all share the same allelic distributions and any sufficiently large sample of a few thousand individuals [22, 23] would capture all but the rarest alleles with high probability, even those that are deleterious but maintained by recurrent mutation rates of 10^{-6}.

In Table 1 [from 5], sample sizes necessary to ensure low probabilities of missing alleles are given. For each sample site, there is a Bernoulli trial for each of s independent loci with a probability of loss that can be computed as a function of the sample size and the inbreeding coefficient as

$$Q = [(1 - q)^2 + Fq(1 - q)]^n$$

for allele frequency q and for n independent samples. The expected number of allele losses is sQ. Therefore, the risks can be expressed in terms of the expected number of alleles lost. If Table 1 is used, an average loss of one allele at one of the s loci is arbitrarily chosen as the criterion of an acceptable loss rate.

A 10-fold increase in the numbers of loci considered doubles the number of sampled plants required, while a similar decrease in average allele frequency increases the sample size by 10-fold. Thus, sample size is more strongly affected by the rarity of the alleles designated to be saved than by the number of loci. Nevertheless, if many loci affect traits of interest, we may not be as concerned with all of the rarest alleles at each. Hence, sample sizes of several hundred or a few thousand will likely suffice.

Increasing the number of alleles at each locus for 10 loci increases the sample size by approximately 62% for all frequencies. The sample size increase needed to compensate for any increase in number of loci is identical regardless of whether 1 or 4 alleles are involved. Therefore, the general levels for requisite sample size remain intact.

When sampling a species containing subdivided populations, one approach is to disperse samples into N_p local populations, with n genotypes per sample site. If a single allele of interest exists at frequency q in only certain populations and those populations exist at frequency p, then the probability of loss for a single site sample would be:

$$Q = [(1 - p) + p(1 - q)^{2n}],$$

and for N_p site samples, each with n genotypes,

$$Q = [(1 - p) + p(1 - q)^{2n}]N_p ,$$

Table 1. Minimum number of genotypes required for an average loss of one allele at any of the loci with four rare alleles per locus.

No. of loci (s)	Frequency of rare alleles at each locus			
	0.05	0.01	0.005	0.001
10	72	367	735	3681
50	104	528	1057	5295
100	117	597	1196	5988

If some or all of the natural populations have been persistently small and isolated, differentiation may have occurred, and the smaller N_e within isolates is, the more likely are large differences in allele frequencies. As previously suggested, the joint effects of small N_e and low interpopulation migration must combine to generate much chance for private alleles to exist which requires sampling many populations for their capture. In those cases, dispersed sampling among the N_p populations is needed, but the number of populations needed to sample for private alleles depends on the acceptable risk of missing any unique populations [5]. Any knowledge of patterns of allele dispersal should be used to maximize the probability of sampling the populations that contain rare alleles. Sampling should be dispersed to include peripheral populations or sites with marginal species fitness in which rare alleles that may be useful in the future are likely to be more frequent. For alleles that are uniformly dispersed throughout the population, p may be close to 1, making any pattern of sampling equally effective. Thus, dispersing samples does not diminish the effectiveness of sampling for homogeneously dispersed alleles, but does increase the effectiveness of sampling the heterogeneously distributed alleles.

Under conditions of selection or hitchhiking, with a single large population and wide intermating, a single dynamic evolutionary system exists though it may well not be stable. With mating subdivision but with no difference in environmental selective pressures, the populations may be moving to different equilibria because the ecological and genetic effects are complex. If environmental factors also generate different dynamics, and if mating barriers exist, entirely different dynamics can be occurring simultaneously in different parts of the species range. In such cases, sampling to capture alleles may require additional population samples. Although the distribution of alleles is not random, it cannot be predicted without detailed knowledge of population genetic dynamics, which is rarely available. If some estimate of N_p is available, however, distributing samples among the smaller and more environmentally extreme populations can be expected to provide a probability of missing alleles of low frequency that is no greater than for neutral alleles sampled from random populations. Since these sample population sizes can be just a few thousand, the possibilities of collecting for purposes of capturing extant alleles seems quite feasible. While such sampling is not yet done for many species, population genetics indicates no great theoretic problems in achieving that objective.

If a major objective of collecting is for future uses, and the alleles themselves are changing and new allelic combinations are being generated, any static collection can only freeze the genotypic state at an arbitrary time. For populations in static equilibrium, a static collection is sufficient, but for evolving species or for uses that require further evolution, it is insufficient.

One could envision continuous collections that ultimately include infinitely large collections of the infinitely many alleles. Obviously, however, not all allelic variants are either functionally effective or valuable. We can limit attention in conservation programs to preserving or using any newly arisen useful alleles. Current research on gene structure seems to indicate that relatively few high-frequency 'allelomorphs' and many low-frequency unique amino acid sequence 'alleles' exist in the coding regions of transcribed genes. A relatively high frequency of an allele may reflect significance in natural selection and may, therefore, also be of some future use in natural or cultivated ecosystems. Alleles that have a major favorable effect on physiology, morphology, resistance, etc., would be expected to have higher frequency than alleles carried only by mutation-selection balance. However, newly arisen or low-frequency alleles that have an effect of no present value, but possible future value, are at risk of loss unless selected for. For individual qualitatively significant alleles at low frequency, selection in populations of a few thousand can rapidly increase their frequency and drastically reduce the probability of loss. For quantitative traits with small effects at many loci, the selection effect per locus is small and low-frequency alleles are easily lost, but with many loci the population sizes needed to obtain continuous responses to selection can be on the order of tens or a few hundreds [11]. Thus, for trait responses that use particular qualitatively effective, but low-frequency alleles, or those that are quantitative in effects, useful variations are more dependent on selection in large populations than on collections of large populations. Clearly, large initial population sizes are needed to capture existing low-frequency alleles, but these need only be on the order of a few thousand to retain those that are maintained by recurrent mutation even though they are deleterious in the wild. Static collections will not capture newly arisen mutations, and if they are maintained without selection, any effective and useful alleles have a greater chance of loss than if they are managed with selection.

The major problem with selection is to decide what directions, traits, adaptabilities, or other criteria to use. Any one set of objectives would impose positive selection pressure on some alleles, but negative pressure on others. If those selected against occur at low frequency, the probabilities of their loss are high. Stabilizing or frequency-dependent selection can save alternate alleles and protect alleles from extinction. Alternatively, maintaining multiple populations and selecting for different traits or adaptabilities can increase the frequency of different alleles in the separate populations. Since selection can be effective in natural or managed populations, the utility of the set of populations maintained depends on how well selection can provide allelic variation for present and future needs. Regardless of the specific measures taken, multiple populations managed for genetic differentiation can clearly generate more evolutionary diversity than can be contained in a single population, whether or not that single population is kept static. Management of populations will depend on the objectives of the conservation program and how the conserved germplasm can be used.

Breeding population dynamics

Populations from which the raw breeding materials are to be drawn may be unstable and evolving new alleles and trait expressions. Uses of genetic variability in environmental or economic adaptability and in responses to evolving biotic challenges are also evolving. Germplasm conservation, therefore, must involve the active or passive management of two dynamic systems. Physically, it is improbable that collections alone can ever capture all potentially useful variations. Theoretically, it seems unlikely that we can limit ourselves to the one gene management technique of seed or clonal storage. This is not to deny the vital function that collections can fulfill in ensuring the viability of sets of populations. In maize collections, for example, it is 'the acquisition, increase, and safeguarding of collections already on hand that must take immediate precedence' [24]. Attention must be focused on how germplasm is or can be used in breeding systems and how germplasm conservation programs may be designed to serve these systems.

Breeding systems

By the term 'breeding system', I mean the whole array of activities involved in producing useful, genetically altered (and presumably improved) clones, lines, varieties, populations, etc., from plants or animals in some originally less useful state. The sequence of activities includes selecting materials found in the wild or in primitive cultivation, various types of observation and testing, preliminary breeding with unadapted materials, all forms of enhancement breeding, and developing primary breeding stocks for the final development of released elite materials.

These systems are perhaps most advanced in the grain crops (rice, maize, wheat, barley, sorghum, oats) and a few other food species such as potatoes, tomatoes, and soybeans. For these plants, long traditions of development have created their own dynamics. The utility of new introductions now depends very much on the elite lines or varieties already developed, and any newly adapted releases affect what may be subsequently selected and developed. New opportunities for breeding are created by external factors such as new demands for climatic or edaphic adaptation, and by new cultivation and biotic challenges. Since these forces are partially affected by the varieties previously used, there are further feedback mechanisms involved in the breeding system dynamics. The dynamic nature of breeding is also reflected in some germplasm banks where new materials are constantly being added as new varieties and populations are developed [25].

In one sense, these new varieties are the modern-day equivalents of more ancient traditional breeding practices that evolved crops for adaptability to local conditions and food preferences. The diversification of rice into its modern forms, for example, is largely a result of human activites and disruptive selection for different adaptabilities [26]. However, partly as a result of breeders' past success in creating elite lines, it has become very difficult to introduce new unimproved germplasm into commercial use. The efforts required to introduce genetic materials from unadapted germplasm into elite lines are substantial and usually involve several generations of segregation following hybridization and backcrossing [27]. An alternative method of bringing whole new populations into breeding stocks requires special selection procedures. In maize, for example, recurrent selection in exotic elite populations or in hybrid base populations can be successful [28]. Investments in such programs have been low, and while the economic impact of particular introductions has been large, the actual infusion of exotic germplasm has been small [29]. In that sense, the evolution of advanced breeding populations has been channeled more than when more and different source populations were available for infusing intrapopulation genetic variability, or for substituting competing populations. Since most of the breeding materials in new varietal releases come from elite x elite crosses, it would be useful to estimate the genetic diversity in the new acquisitions of germplasm banks and the structure of the genetic variation being stored. A dynamic germplasm storage system might be better served by developing improved varieties from independent source materials. Also, deliberate development of independent populations for use in a storage system, or for use as primary materials for breeding stocks, might be more cost effective than storing derivative elite lines.

For less intensively developed species, the ancestral line from primitive races or wild varieties to materials released from breeders stocks is much shorter. For crops such as forest trees where the generation cycle is very long and selective breeding has only recently been practiced, the differences between wild stock and improved varieties is not very great at present, but it is rapidly increasing. Intensive selection can increase yield at a rate of 1–2% per year even in species with long generation cycles (e.g. 10–20 years) [30]. To provide genetic variations to accomodate future needs, however, either very large populations must be carried, or selection intensity must be reduced. Alternatively, intensive selection can be used in multiple independent populations to create a Multiple Population Breeding System (MPBS) with a broader differentiation among populations both in their source germplasm and in their traits and environmental adaptabilities [31]. By developing independent populations instead of relying on any one particular 'elite' population, the breeder can continually develop alternate breeding stock, any one or subset of which could be used for contemporaneous varietal development or release. On the order of 20–50 breeding individuals or pairs per population are sufficient to generate continuous gain from selection in quantitative traits of moderate heritability, based mainly on the gene effects of newly arisen mutants [11]. For species with long generation cycles, the obvious advantage of such a multiple population system is that the long development time for introducing new genes or traits is eliminated. For species that are just beginning to be developed in breeding programs, it costs no more to develop multiple small populations than single large ones. For species for which breeding systems are already established, the advantage is that entirely new populations and alleles are introduced to the system. However, to establish populations of more independent origin requires enhancement efforts such as Goodman's [28] exotic maize breeding program. This step requires comparable effort to the backcrossing systems often used for other grains [27, 29], but promises to use new alleles more effectively.

Some method of prebreeding is needed when new materials are incorporated into an already bred population. For species with relatively low present investment in breeding, the problem of developing populations for future breeding use may not be qualitatively different from that for heavily bred species with elite lines. In either case, the development of multiple populations from independent source materials to establish closer but broader ancestral linkages to the total germplasm of any species may require more rapid and expanded prebreeding techniques and more precisely controlled ecogeographic surveys and testing than presently used. Investments in research on these breeding methods could broaden effective use of germplasm collections and ultimately provide the foundations for continually evolving breeding stocks.

In the multiple population breeding system (MPBS), emphasis is placed on the control of interpopulational variablity both in the diversity of its sources and in the traits or environmental adaptabilities that define selection objectives. Selection is assumed to improve performance in multiple traits across a span of environmental variables. Hence, for long-term breeding, value is not a one-dimensional concept. To satisfy an array of future demands, the MPBS simultaneously develops breeding stock for any of several possible futures by recurrent selection within each population [32]. In contrast, a breeding system such as HOPE (Hierarchical OPen Ended system) can be organized as a set of hierarchical populations as described by Kannenberg [33]. A very large base population is maintained and is open to new materials. Higher level populations are more stringently selected and tested up to the development of a high level gene pool for extracting elite lines. It was intended for hybrid line development in a type of modified reciprocal recurrent selection, but it could serve as a model for systems based on simple recurrent selection. It was also intended to assist when more population introductions are needed than can be worked on or even be effectively controlled. One contrast between the two systems is in whether introduced materials are incorporated into the single base population of HOPE or into one of possibly several populations in the MPBS. Another is whether complex selection is attempted in a single population or simple recurrent selection is used in two or more populations that use different alleles or select for divergent objectives, or both.

Appropriate collection and maintenance for a germplasm conservation system depend upon the number and organizational relationship of possible breeding populations. For species with fairly stable markets and uniform cultural systems, a single, best population might be selected, and breeding would be intended only to modify its basic traits such as with new resistances or special adaptabilities to particular environments. In such cases, replicate populations [34] may be advantageous if different alleles can be used to increase yield or other trait performances. Alternatively, Goodman [28] suggests starting with different, but elite, source populations and selecting for adaptability such that performances can be improved with different sources. Thus, either by arbitrary subdivision of original breeding stock or by using naturally subdivided original stocks, breeding options are increased by use of independent populations.

Since there is always some shift in economic or ecological requirements, some continual shifting of the optimum population will always be needed. With sufficiently slow changes in requirements, it may always be possible to shift the same population, but there may also exist some practical limit to how far it can shift before some other population with different origin is better to use. The strategic question for breeders is how long to continue with the one population before switching, and how and when to prepare alternate populations.

When greater uncertainty exists about future markets for any of the constituent traits determining crop value, or when present and anticipated environmental variations are broader than can be optimally accommodated by sources from a single population, then more than one population is needed to maximize breeding value. In spite of selection for uniformly high stable yield, it is assumed that limits to adaptability will persist. With more than one population tracking changes in requirement or bracketing expected needs, it has been shown that multiple populations always generate higher value than can a single population [35]. If population yield or adaptability is of a Gaussian form and the environmental distribution is normal, the breeding value of two populations always exceeds one, three exceed two, etc., and $n + 1$ always exceeds n, but at a decreasing marginal value. The marginal value depends on the homeostasis of yield relative to the variance of the environmental distribution. Value is large when homeostasis is low relative to the anticipated environmental variance. However, the optimal interpopulational differentiation increases along with anticipated variation, and tends toward a stable portion of that environmental variance. Thus, for highly uncertain environments with a very large variance of anticipated possible futures, the more widely should populations be differentiated. The optimal differentiation among breeding populations also is large for widely homeostatic population performances, but the marginal value of additional populations is smaller than for narrowly adapted populations. Hence, there is always a finite limit to the number of populations that are economically feasible to breed. For wide ranges in the ratio of population homeostasis to environmental variance, this number can be quite small (e.g. less than 5). Therefore, for breeding purposes, an optimum number of populations can be defined that maximizes a utility measure for a set of populations. For a finite number of trait or environmental dimensions, this N_p might be on the order of a few tens.

In some extreme cases where needs for adaptations or trait performances not now available in any breeding population are anticipated, selection for more extreme trait expressions or for adaptation to extreme environmental conditions may be the only way to save alleles or populations that might be useful in the future. In such cases, breeding to extend extremes rather than to maximize value under constrained conditions may be appropriate. This approach maximizes what may be termed 'inclusiveness' in contrast to the 'utility' of a set of populations. Inclusiveness is more likely to be important for species with less intensively cultivated environmental controls such as forest trees, but may be an important objective for some breeding agencies even with grain crops. In fact, for most agencies involved in germplasm conservation, some mixture of objectives is likely to exist. If inclusiveness is important, both the initial collections and their subsequent management in selection programs can affect conservation objectives by directing goals towards extremes rather than towards clear and present utility.

For some species, the objectives of breeding are clear, stable, and relatively homogeneous over space and time, and breeding methods are well established. For the major grain crops yield *per se* is rarely an objective of breeding with exotics. Elite breeding stocks normally are modified only to add special traits such as resistances [29]. However, even for some of these species and certainly for many others that are not so uniform and stable in their trait breeding objectives or environmental adaptability, breeding alternate populations can be economically useful [28]. For species with little likelihood of developing narrowly defined, stable adaptabilities, prebreeding for multiple populations can maximize the value of breeding and allow for nondisruptive spatial and temporal substitutions of one breeding population for another. For agencies concerned with developing the broadest economic and environmental adaptabilities, populations should be differentiated as much as possible. If particular, known variations in adaptability are required, more narrowly adapted sets of populations can be bred. The only question for breeders is how many, and how well differentiated, the breeding populations should be. Collections of breeding populations can thus be viewed as a controlled evolutionary system. Some species may be maintained in a single homogeneous random mating structure with uniform selection. Other species may be differentiated into separate populations with different selection pressures, and still others may be subjected to homogeneous selection but maintained in separate mating demes.

Conservation dynamics

The above very sketchy descriptions of the dynamics of natural populations and dynamic breeding systems might lead one to view germplasm conservation as taking materials from the natural system for use in the breeding system, while the systems otherwise function independently. Since advanced breeding sometimes leads to losses of genetic variability, however, it seems obvious that the two systems are linked in more complex ways than one serving the other. Germplasm conservation might be viewed as the coordination of the genetic dynamics of more natural populations with the needs of evolving breeding populations. In general, it would not be involved in ecosystem conservation or advanced breeding, but might be involved in ecogeographic surveys and testing to identify populations for direct infusion into breeding stocks. The collection and maintenance of populations would then depend on the distribution of genetic variations in the more natural populations and in breeding populations. While independent populations with private alleles may exist in some species, they need not be maintained in any structure but can merely be mixed into a single, large, base breeding population. For identification and testing purposes, and for ultimate use of many disease and insect resistance alleles, maintaining separate populations would be most useful. Species that exist as a single large interbreeding population may nevertheless contain genes for particular adaptabilities or extreme trait expressions. With such species, conservation may be accomplished best by disruptive selection and restructuring into multiple breeding populations. In any case, the collection and maintenance programs would be affected by the structure of the natural population and the structure of the breeding population.

First consider a worst-case scenario in which all populations that are not stored suddenly become extinct or otherwise unavailable. In that event, any present collections would represent an initial population, and we would hope that population contains a large set of genotypes such that N_e is large. The germplasm manager might choose to maintain that closed set as a single interbreeding population or to split it into multiple populations with or without selection. If neutral genes are randomly distributed in this initial population, their frequencies among and within populations would remain nearly constant. For alleles that can be made subject to selection, any that are newly arisen or otherwise exist at low frequencies can be more probably saved if selected for. Hence, to save alleles, some stabilizing or disruptive selection would be useful. For extending extremes in trait expressions, divergent selection in separate populations can be effective, even when sizes per population are relatively small. For increasing the future utility of breeding stock, selection in multiple populations for traits and adaptabilities dispersed around average anticipated needs would be optimal. Thus, for saving alleles, increasing diversity, and increasing utility, some selection program would be useful. The details of how many populations and how different they should be would vary widely among species and how they are managed. As demands of breeders shift, the population array of these breeding stocks would also shift. Hence, the utility of the collected populations would be improved by shifts that tracked anticipated needs. In that case, some initial populations would be discarded and replaced by derivative populations or by competitor populations. Neither the seeds nor the active breeding populations of discarded populations would have to be maintained. If such a program also maintained pairs of more broadly inclusive diversified populations, there would be no loss of alleles or trait expressions suffered by discarding obsolete populations. Thus, collection and maintenance in germplasm conservation would involve the prebreeding of multiple populations that track shifting demands. Maintaining any initial state of the species, which in this case was an arbitrary collection, would not be necessary. This could be a risky alternative if not done with considerable caution, but can be designed even for highly risk–averse objectives.

In more typical cases where more or less 'natural' populations remain, other opportunities for conservation are available. We can more completely sample the extant distribution and possibly enrich currently stored populations, and we can use 'natural' populations for maintenance or for generating new variations. If the 'natural' populations are stable in demographics and allelic distribution, and N_e is greater than a few thousand, then several alleles at each of many loci at frequencies of 0.001 could be retained for many generations. If there is also no danger of their extinction or destabilizing change, then there is no need for any *ex situ* collection and maintenance. Only if there is danger of population loss or if it is convenient to do so, is it necessary to maintain *ex situ* collections. In that case, a collection of a few thousand is necessary to capture alleles and would have to be subdivided if the natural distribution is subdivided. Once the species has been collected, however, its natural structure does not have to be maintained, but may be further subdivided if diversifying selection is needed for breeding utility or inclusiveness. Alternatively, well-differentiated populations may be mixed into a single population as suggested for HOPE [33]. For stable populations, the selection-mutation-migration balance would presumably continue and the distribution of alleles would be largely dependent on the migration-inbreeding structure. For highly inbred species or those that exist in many populations with little migration, highly dispersed sampling is needed. For widely outcrossing species, even if selection is strongly diversifying, alleles will be widely distributed and the number of different populations needed to sample can be small. If collections have been made of sufficient size and distribution, or if diversified and dynamic populations have been created by prebreeding, however, there is no breeding need for conserving natural populations. This stage may have already been reached or may be easily reached for some species where there may be more useful and more diverse genetic variation in the breeding stock populations than may naturally exist. This is not to deny that there are many other reasons for conserving natural populations, but, in such cases, breeding utility is not one of them.

If natural populations are unstable and are evolving new alleles and interlocus combinations, and if genotypic frequencies and traits are responding to external forces or internal ecosystem dynamics, the utility of their collection and maintenance depends on the maintenance of genetic structures in the evolving populations and how that evolution can be used in the breeding system. For use in breeding systems that have stable populations of breeding stock and uniform objectives, the temporal and spatial heterogeneity of natural populations requires dispersed sampling in both time and space to collect potentially useful genotypes and populations. If the natural populations are following a single dynamic system and are differentiated only because of their initial conditions, only a finite number of stable points or orbits may be useful to sample. In that case, a finite number of populations, possibly only one, is needed because all populations will eventually arrive at those points or orbits. If they are contained within a dynamical collection, again no breeding need exists for conserving the natural system. Broad population sampling might still be useful, however, if it takes a long time to reach stable states, or if breeding can use other than stable populations. Then one or multiple *in situ* conservation populations that are permitted to evolve can be a useful supplement or substitute for any static or dynamic *ex situ* collections. These *in situ* populations must be secure and continuous monitoring and sampling must be possible. If not, *ex situ* collections would have to be allowed to evolve.

For breeding systems that use multiple populations, natually differentiated initial populations seem reasonable. They may constitute replicate populations or may serve as initial differentiation states to the extent that natural differentiation correlates with breeding objectives. This approach might be particulary advantageous for the genetic management of host-pathogen systems where populations develop different resistance mechanisms over generations. Since the utility of natural evolution for generating populations for sampling and testing depends on the correlation between natural and breeding selection objectives, it could be useful for the breeder to consider means of influencing the 'natural' evolutionary system within managed ecosystems. For many traits such as resistances and environmental stress tolerances, the breeder might select from targeted ecosystems in which 'natural' selection processes are directed by environmental management. This might be conducted in conjunction with ecosystem management in 'managed resource areas'. Such seminatural populations might then be equivalent to enhanced breeding populations that would otherwise be developed *ex situ.* They might also be viewed as improved 'semiwild' populations for sampling and testing. For traits and species that are managed or behave similarly in their wild and cultivated states, such management interventions for genetic objectives would be highly useful. They might also be useful in either natural or breeding populations for increasing differentiation to widen the inclusiveness of alleles and trait expressions.

Distinctions have thus been blurred between conservation and breeding, between natural and breeding populations, and between '*in situ*' and '*ex situ*' techniques. For any one species, collection can mean random sampling from wild but stable populations or saving and discarding breeding populations, and maintenance can mean *in situ* conservation stand management or breeding for differentiation. In this dynamic view of population genetics, neither the 'natural' system nor the breeding system is static, and germplasm conservation cannot be viewed as merely collecting mechanical parts for warehousing. Germplasm conservation is a broad and dynamic part of evolutionary biology.

References

1. Allard, R.W. 1970. Population structure and sampling methods. pp. 97–107 in O.H. Frankel and E. Bennett, Eds. Genetic Resources in Plants: their Exploration and Conservation (IBP Handb. No. 11), Oxford University Press, Oxford, England.
2. Chesser, R.K. 1983. Isolation by distance: relationship to the management of genetic resources. pp. 66–77 in C.M. Schonewald-Cox, S.M. Chambers, B. MacBryde, and W.L. Thomas, Eds. Genetics and Conservation. Benjamin/Cummings, Menlo Park, California.
3. Liu, E.H. and M.J.W. Godt. 1983. The differentiation of populations over short distances. pp. 78–95 in C.M. Schonewald-Cox, S.M. Chambers, B. MacBryde, and W.L. Thomas, Eds. Genetics and Conservation. Benjamin/ Cummings, Menlo Park, California.
4. Marshall, D.R. and A.H.D. Brown. 1975. Optimum sampling strategies in genetic conservation, pp. 53–80 in O.H. Frankel and J.G. Hawkes, Eds. Crop Genetic Resources for Today And Tomorrow (International biological programme 2), Cambridge University Press, Cambridge, England.
5. Namkoong, G. 1988. Sampling for germplasm collections in Proc. XXIInd International Horticultural Congress, University of California at Davis, California. August 11–20, 1986. HortScience 23: 79–81.
6. Shonewald-Cox, C.M. 1983. Conclusions: guidelines to management: a beginning attempt. pp. 414–445 in C.M. Schonewald-Cox, S.M. Chambers, B. MacBryde, and W.L. Thomas. Eds. Genetics and Conservation. Benjamin/ Cummings, Menlo Park, California.
7. Zeuli, P.L.S. and C.O. Qualset. 1987. Geographical diversity for quantitative spike characters in a world collection of durum wheat. Crop Sci. 27: 235–241.
8. Lewis, L.N. 1985. Genetic engineering: one leg of a three-legged stool. Calif. Agric. 39: 2.
9. Plucknett, D.L., N.J.H. Smith, J.T. Williams, and N.M. Anishetty. 1987. Gene banks and the world's food. Princeton University Press, Princeton, New Jersey.
10. Dudley, J.W. 1977. 76 generations of selection for oil and protein percentage in maize. pp. 459–474 in Proc. of the International Conference on Quantitative Genetics, August 16–21, 1976, Iowa State University. The Iowa State University Press, Ames, Iowa.
11. Hill, W.G. and J. Rasbash. 1986. Models of long-term artificial selection in finite population with recurrent mutation. Genet. Res. Camb. 48: 125–131.
12. Crow, J.F. and M. Kimura. 1970. An introduction to population genetics theory. Harper & Row, New York.
13. Slatkin, M. and N. Takahata. 1985. The average frequency of private alleles in a partially isolated population. Theor. Popul. Biol. 28: 314–331.
14. Hedrick, P.W. 1986. Genetic polymorphism in heterogeneous environments: a decade later. Ann. Rev. Ecol. Syst. 17: 535–566.
15. Gregorius, H.-R. and G. Namkoong. 1983. Conditions for protective polymorphism in subdivided plant populations. Theor. Popul. Biol. 24: 252–267.
16. Namkoong, G. and H.-R. Gregorius. 1985. Conditions for protected polymorphism in subdivided plant populations. 2. Seed versus pollen migration. Am. Nat. 125: 521–534.
17. Selgrade, J.F. and G. Namkoong. 1984. Dynamical behavior of differential equation models of frequency and density dependent populations. J. Math. Biol. 19: 133–146.
18. Namkoong, G. and J.F. Selgrade. Frequency-dependent selection in logistic growth models. Theor. Popul. Biol. 29: 64–86.
19. Selgrade, J.F. and G. Namkoong. 1986. Mathematical analysis of density dependent coevolution with interspecific competition. Math. Biosci. 82: 201–220.
20. Leonard, K.J. and R.J. Czochor. 1980. Theory of genetic interactions among populations of plants and their pathogens. Ann. Rev. Phytopath. 18: 237–258.

21. Schaffer, W.M. and M. Kot. 1986. Differential systems in ecology and epidemiology. pp. 158–178 in A.V. Holden Ed. Chaos. Princeton University Press, Princeton, New Jersey.
22. Gregorius, H.-R. 1980. The probability of losing an allele when diploid genotypes are sampled. Biometrics 36: 643–652.
23. Namkoong, G. 1980. Genetic considerations in management of rare and local tree populations. pp. 59–66 in Proc. Conference on Dendrology in the Eastern Deciduous Forest Biome. Publ. #FWS-2-80, School of Forestry and Wildlife Resources, VA Polytechnic Institute and State University, Blacksburg, Virginia.
24. Goodman, M.M. 1984. An evaluation and critique of current germplasm programs. pp. 197–216 in The 1983 plant breeding research forum 'Conservation and Utilization of Exotic Germplasm to Improve Varieties.' Pioneer Hi-Bred International, Des Moines, Iowa.
25. Chang, T.T. 1985a. Germplasm enhancement and utilization. Iowa State J. Res. 59: 399–424.
26. Chang, T.T. 1985b. Crop history and genetic conservation: Rice–A case study. Iowa State J. Res. 59: 425–458.
27. Frey, K.J., T.S. Cox, D.M. Rodgers, and P. Bramel-Cox. 1984. Increasing cereal yields with genes from wild and weedy species. pp. 51–68 In: V.L. Chopra, B.D. Joshi, R.P. Sharma, H.C. Bansal, Eds. Genetics: New Frontiers, Proc. of the XV International Congress of Genetics, December 12–22, 1983, New Delhi, India, Vol. 4, Appl. Genet. Mohan Primlani, Oxford and IBH Publ., New Delhi, India.
28. Goodman, M.M. 1985. Exotic maize germplasm: status, prospects, and remedies. Iowa State J. Res. 59: 497–528.
29. Cox, T.S., J.P. Murphy and M.M. Goodman. 1988. The contribution of exotic germplasm to American agriculture. pp. 114–144 in J. Kloppenberg, Jr., Ed. Seeds and Sovereignty, Duke University Press, Durham, North Carolina.
30. Zobel, B.J. and J. Talbert. 1984. Applied Forest Tree Improvement. J. Wiley & Sons, New York, New York.
31. Namkoong, G., R.D. Barnes, and J. Burley. 1980. A philosophy of breeding strategy for tropical forest trees. Tropical Forestry Papers No. 16, University of Oxford, Oxford, England.
32. Namkoong, G. 1983. Strategies for gene conservation in forest tree breeding. pp. 79–92 in C.W. Yeatman, D. Kafton and G. Wilkes, Eds. Plant Gene Resources: A Conservation Imperative. AAAS selected symposium 87. Westview Press, Boulder, Colorado.
33. Kannenberg, L.W. 1983. Utilization of genetic diversity in crop breeding. pp. 93–110. C.W. Yeatman, D. Kafton and G. Wilkes, Eds. Plant Gene Resources: A Conservation Imperative. AAAS selected symposium 87. Westview Press, Boulder, Colorado.
34. Baker, L.H. and R.N. Curnow. 1969. Choice of population size and use of variation between replicate populations in plant breeding selection programs. Crop Sci. 9: 555–560.
35. Roberds, J.H. and G. Namkoong. 1986. Maximization of expected value for a trait in an environmental gradient. pp. 179–186 In: C.G. Tauer and T.C. Hennessey, Eds. Physiologic and Genetic Basis of Forest Decline, Proc. ninth north American forest biology workshop, June 15–18, 1986, Oklahoma State University, Stillwater, Oklahoma. Oklahoma State University, Publ. # AI-6154-0686, RO.

SECTION THREE

Frontiers in Germplasm utilization

Sampling, evaluation and utilization of animal genetic resources

KEITH E. GREGORY and GORDON E. DICKERSON

P.O. Box 166, U.S. Meat Animal Research Center, Clay Center, Nebraska 68933, U.S.A.

Abstract No program exists in the U.S. to sample, evaluate, and use optimally the genetic diversity present in animals important as food and fiber worldwide. Yet animal products provide U.S. consumers with substantial amounts of essential nutrients. Animals and animal products constitute about 50% of total farm sales in the U.S. Furthermore, all domesticated food and fiber animals now in the U.S. originated in, or were introduced from, other parts of the world. Because the U.S. has a limited sample of animals important as food and fiber worldwide, there is a need to develop a comprehensive program of systematically sampling, evaluating/characterizing, and developing breeding systems to use these resources in an optimal manner. Such an effort could provide the foundation for an organized program of acquiring and maintaining the genetic variation needed in our populations of food and fiber animals to achieve and maintain optimum adaptation to continuing changes in our diverse production-marketing ecosystems. An organized program of animal genetic resources management is needed to provide the framework for a comprehensive program of animal breeding research and development. The development of procedures for movement of animal genetic resources with minimal risk to animal health should receive high priority. The level of genetic diversity needed is greatest and perhaps easiest to maintain for the extensively managed food and fiber animal species (e.g., beef cattle, sheep, goats) because of their variable, dynamic production environments and marketing situations. Continuing optimum adaptation of global genetic stocks to varied production-marketing ecosystems is required for most effective use of U.S. animal production resources and to insure an adequate supply of reasonably priced animal products to meet the nutritive requirements of the U.S. population in the 21st century and beyond.

Key words: optimum adaptation, production-marketing ecosystems, germplasm

Introduction

Animal products provide U.S. consumers with 70% of their protein, 35% of their energy, 80% of their calcium, 60% of their phosphorous, 40% of their iron, nearly all of their vitamin B-12, and substantial amounts of other essential vitamins and minerals [1, 2]. Animals and animal products constitute about 50% of total farm sales in the U.S. [3]. Thus U.S. programs to ensure efficient animal production are essential both to producers and consumers of animal products. All domesticated food and fiber animals now

L. Knutson and A.K. Stoner (eds.). Biotic diversity and germplasm preservation, global imperatives.
 ISBN 0-7923-0178-1

used in the U.S. originated in, or were introduced from, other parts of the world. Even turkeys were introduced from the Americas into Europe, where they were domesticated, then later reintroduced into the U.S. Because the U.S. has a limited sample of animals important as food and fiber worldwide, it is in our immediate and long-term interests to create a program of systematically sampling, evaluating, and developing breeding systems that will optimally use global animal genetic resources. Such a program is necessary: (1) to identify the animal genetic resources that are best adapted to specific production-marketing situations or ecosystems and (2) to organize the research efforts needed to develop breeding systems to achieve maximum bioeconomic efficiency or maximum conversion rates of production resources into animal products of greatest economic value [4, 5]. This approach can be the foundation for an organized program of managing animal genetic resources to meet current and future U.S. needs for animal genetic diversity.

Rationale for a managed program supported by the public

Most of the current food and fiber animal genetic resources in the U.S. were introduced by the private sector. Furthermore, except for recent efforts with beef cattle [6, 7, 8] and sheep [9], organized and publicly supported efforts in importation and evaluation/characterization have been limited. Because of this lack of a comprehensive program of germplasm sampling, evaluation/characterization, and developing breeding systems for optimal utilization, it is unlikely that present U.S. animal genetic resources provide an adequate base to achieve maximum bioeconomic efficiency in the long term under the dynamic livestock production-marketing ecosystems that likely will exist in the U.S. in the future.

An organized program of sampling, evaluating, and developing breeding systems for optimum utilization of animal genetic resources is highly technical and complex. It requires collaboration among governments when animals are imported from other countries. Because of uncertain profit potential, the private sector has been understandably reluctant to make the investments required to sample, evaluate/characterize, and develop breeding systems to use promising animal genetic resources from other countries in an optimal manner. This stalemate is not likely to change because genetic resources for extensively managed species (e.g., beef cattle, sheep, goats) are highly decentralized and controlled primarily by private farmers/ranchers with limited resources [4]. Because an organized program of sampling, evaluating, and developing breeding systems to use genetic resources optimally would lead to more effective use of animal production resources, in the form of more desirable animal products at lower prices, it is clearly in the public interest for a government organization to assume leadership for such a program.

A high return would be expected from investment of public resources in an organized program to allow matching biological characteristics of commercial breeding stocks to production resources and to market requirements for animal products. This expectation is illustrated by the favorable results observed for some of the cattle breeds introduced to the U.S. from continental Europe via Canada [6,7,8] and for sheep imports [9].

Reasonable questions may be 'What is the potential loss to the livestock industry and to the general public of our leaving largely to chance the animal genetic resources that are introduced?' and 'What is the potential expense of 'trial and error' by the livestock industry of finding out *what* genetic resources are useful and *how* to use them most effectively in each 'production-marketing ecosystem?' Such an 'ad hoc' approach to the use of genetic differences among breeds to optimize additive genetic (breed) composition and to use nonadditive genetic variation (heterosis) would be highly ineffective and costly both to the livestock industry and the public. Existing research and extension programs of the U.S. Department of Agriculture (USDA) and the Land Grant University System provide a partial answer now. However, a more organized, strategic, cooperative effort is needed.

Factors involved in a program of animal genetic resources management

Characterization of germplasm and ecosystems

An organized program of sampling and evaluating global animal genetic resources for bioeconomic potential and of developing breeding systems for optimal use is a long-term effort. Priorities must be established with adjustments made on a continuing basis.

First priority should be assigned to evaluating the sustainable production capabilities of major U.S. ecosystems, including ceilings and constraints, and to assessing global livestock genetic resources for bioeconomic potential within each major ecosystem. Such information is required to develop highly adapted livestock populations for the most efficient food and fiber production within each major ecosystem. Selective U.S. participation in global animal germplasm evaluation efforts should be cost-effective and can be justified to allow preliminary screening of animal genetic resources, thus increasing accuracy in sampling the most promising genetic material for importation, further evaluation, and optimal use in the U.S.

Maintenance of global genetic diversity

Livestock production systems and marketing situations in the U.S. and globally differ greatly because of differences in climate, exposure to diseases and parasites, nutritive environment favored by technological and economic factors, and uses of animal products. To the degree that livestock producers in each of these diverse ecosystems can be effective in developing the animal genotypes best adapted in bioeconomic efficiency to their production-marketing situation, a large global reservoir of genetic diversity would be maintained [5]. Such ecosystem diversity indirectly would maintain high levels of animal genetic diversity among different global ecosystems. To be fully useful, this reservoir of genetic stocks would need to be readily available for transport between countries and among ecosystems, surmounting animal health and trade barriers.

Organized national, regional, and international programs will be required to avoid potentially important losses of U.S. and global genetic variation in livestock species [4,10]. Risk of such losses arises from inadequately planned or ill-advised crossing of adapted local breeds with potentially higher producing but unadapted exotic stocks [4]. A large gain in performance is often observed from heterosis in first crosses of adapted native breeds in hostile environments with higher performing exotic breeds imported from more productive temperate regions. Part of such gains from heterosis is lost, however, when the first crosses are used to produce new composite populations or are backcrossed to increase the percentage of the exotic breed. An organized program of animal genetic resources management can develop breeding systems to use indigenous genetic stocks effectively in combination with 'improved' introduced stocks to minimize the loss of heterosis while retaining the optimum contribution of the indigenous stocks needed for adaptation [11].

It is in the long-term interests of the U.S. to participate selectively in global development of breeding systems for optimal use of germplasm resources and selection criteria to maximize bioeconomic efficiency of breeding stocks within each major global ecosystem. This approach should help to ensure a high level of global genetic diversity by maintaining populations with optimum adaptation to each of the varied global production-marketing ecosystems. Such potential reservoirs of increased genetic diversity could be a continuing source of germplasm contributing to optimum adaptation of genetic resources to dynamic U.S. production-marketing ecosystems.

Maintenance of genetic diversity through economically viable populations

The diverse and changing nature of agricultural development favors management of animal genetic resources primarily as economically viable commercial populations [5, 12]. Inability to predict long-term technological and economic changes in production and marketing situations emphasizes the need for continuing selection to maintain optimum adaptation (i.e. maximum bioeconomic efficiency) of livestock species within each of the diverse global livestock production-marketing ecosystems. Such efforts would help to maintain a high level of genetic diversity among ecosystems for future U.S. and global needs.

Although agricultural development causes ecosystems to change much faster than normal evolutionary rates, the risk of failing to maintain optimum adaptation of livestock populations within ecosystems can be reduced through national, regional, and international programs focused on:

1. Evaluating production capabilities and identifying constraints of major ecosystems,
2. Characterizing bioeconomic traits of animal genetic stocks within major ecosystems, and
3. Developing breeding systems for optimal use and further improvement of existing animal genetic stocks either as purebreds or as contributors to new composite breeds within each major ecosystem.

Changes in production and marketing situations for each major ecological zone are likely to be gradual enough to permit continuing adaptation of genetic stocks within each ecosystem from a combination of intra-population selection and, if major adjustments in germplasm are needed, by selection among populations from different ecosystems [5]. This approach could be the basic element of a program of animal genetic resources management to maintain the genetic diversity needed for continuing adaptation to regional changes in U.S. livestock production resources and markets for animal products. It would use an optimum combination of more recently introduced and traditional genetic stocks under private sector management, with breeding goals focused on maximum bioeconomic efficiency and with selection criteria adjusted to be in harmony with each dynamic production environment and marketing situation. Such management of animal genetic resources for optimum adaptation within each production-marketing ecosystem of the U.S. would permit sustained, efficient conversion of production resources into animal products of greatest economic value.

Sampling global animal genetic resources

A comprehensive, continuing U.S. program of sampling global genetic resources for evaluation/characterization and potential use is suggested for extensively managed beef cattle, sheep, and goats and the more intensively managed poultry, swine, and dairy cattle. Some highly fecund swine populations in the People's Republic of China may be a useful source of germplasm to improve reproductive rates in U.S. swine populations [13, 14, 15]. Preliminary investigations in France indicate that F1 Chinese x western crossbred females reach puberty 100 days earlier, produce three more pigs per litter, but have less favorable carcass composition and a lower rate of growth than do indigenous French stocks [16]. Some European swine populations have a more favorable lean-to-fat ratio than do many U.S. breeds of swine and may be useful for improving carcass composition of U.S. swine. The Belgian Pietrain and the Belgian Landrace breeds of swine have been selected intensively for a high lean-to-fat ratio. Crosses with breeds of higher prolificacy and maternal ability produced pigs with a more favorable carcass composition than general purpose genetic stocks [17, 18, 19, 20].

For sheep, access is needed to breeds in other countries that have higher fecundity, such as the recently imported Finnsheep, Booroola Merino, and Romanov [9, 21, 22], greater milk production needed to support higher fecundity (e.g., East Friesland), leaner carcasses produced by the Texel [23, 24, 25, 26], and longer breeding season. In comparison with Suffolk rams in topcrosses with Finnish Landrace x Dorset Horn ewes, Texel rams sired lambs with equal growth rate, heavier carcasses, higher dressing percentage, 2.9% more lean, and 3.8% less total fat [27, 28].

Some *Bos indicus* breeds of cattle in east Africa (e.g., Boran), southern Africa (e.g., Tuli), Pakistan and India (e.g., Sahiwal and Nellore) might well contribute tolerance to heat and nutritional stress and additional heterosis to U.S. beef production systems. Also, some tropically adapted breeds of *Bos taurus* cattle (Criollo) in several South American countries may have potential for beef production programs in the southern U.S. [29]. A very high level of heterosis has been observed in breed crosses of *Bos taurus* with *Bos indicus* cattle [30, 31, 32]. Even though European *Bos taurus* breeds of cattle have been sampled extensively (primarily via Canada) in the last two decades, others not yet sampled (e.g., Greyvieh in Austria) may have potential to make a contribution to improving maternal traits. Evaluation/characterization efforts at a number of research locations in the U.S. and in Canada have shown large differences in most bioeconomic traits among recently introduced European breeds and traditional British breeds of beef cattle [6, 7, 8].

Data on registration numbers show that breeds introduced into the U.S. from Europe via Canada in the late 1960's and early 1970's accounted for almost 25% of the beef cattle registered in U.S. breed association herdbooks in 1984, compared with less than 3% two decades earlier [33]. Although the percentage of commercial beef cattle in the U.S. affected by these recently introduced breeds is unknown, approximately 80% of the cattle fed for slaughter show evidence of crossbreeding. Because these recently introduced

European breeds are used extensively in beef crossbreeding programs, it is likely that as much as 50% of the cattle fed for slaughter have a contribution from one or more of these breeds.

Mohair fiber is produced mainly in central Texas from a narrowly based Angora goat population. Sampling from larger populations in Turkey and the Republic of South Africa would broaden the genetic base and thus offer increased opportunity for genetic improvement. Current use of goats for U.S. meat production is relatively unimportant, but access to more productive types (e.g., Boer goat in the Republic of South Africa) may permit meat goats to become important for specific production environments and marketing situations in the U.S.

Evaluation/characterization of animal genetic resources

A comprehensive program of evaluating promising animal genetic resources in U.S. production-marketing ecosystems is fundamental to effective selection among breeds/populations and for the development of breeding systems for their optimal use. Characterization of major bioeconomic traits for individual and maternal average and heterotic genetic effects would be ideal for making decisions on use of specific breeds or populations [3, 4]. However, such a complete genetic characterization generally is not economically feasible, and information on specific breed cross heterosis is not as important as average breed individual and maternal genetic effects in deciding:

1. What breeds to use,
2. Where to use them, and
3. How to use them.

A logical sequence for evaluation of genetic stocks for possible importation and use would be:

1. Use performance information that may be available in the country of origin to identify candidate breeds or populations judged to possess bioeconomic traits of potential importance for some U.S. production-marketing ecosystem,
2. Depending on the adequacy of the prior evaluation, either collaborate with appropriate interests in the country of origin in further evaluation or import a representative sample for more comprehensive evaluation/characterization efforts in appropriate U.S. production-marketing ecosystem(s),
3. For the slower reproducing cattle, sheep, and goats, the first priority in breed evaluation can be based on introduced males (semen) for use in topcross matings with appropriate indigenous female stocks to estimate

breed average individual and maternal genetic effects for major bioeconomic traits. More complete evaluation of heterotic as well as breed average individual and maternal genetic effects on bioeconomic traits is feasible in diallel crossing experiments with imported stocks of swine and poultry because of their higher reproduction rate.

However, diallel crossing experiments require adequate female populations of the introduced stocks. Importation of females as well as males also permits faster development of the introduced genetic stocks than would be possible through backcross use of males alone. When only topcross evaluations are feasible, average estimates of individual and maternal heterosis effects may be useful. For example, the level of heterosis is similar among crosses of different breeds of *Bos taurus* cattle [35].

Utilization of animal genetic resources
Alternate methods for commercial utilization of breeds are:

1. Expanding superior breeds,
2. Systematic crossbreeding to optimize breed and heterotic effects, and
3. Development of new, composite populations by continuing selection within a multi-breed foundation [34, 36].

Each of these approaches has merit, either singly or in combination, depending upon the genetic evaluation information. Formation of composite breeds is a simpler breeding system than systematic crossbreeding and allows optimal joint use of both additive (breed differences) and nonadditive (heterosis) sources of genetic variation in developing general purpose or specialized maternal and paternal (terminal sire) crossing stocks [34, 36]. Flexibility in developing composite breeds or crossbreeding combinations best adapted to specific production- marketing situations requires continued availability of a wide range of characteristics among the purebred stocks [34]. This procedure may be used to combine the special adaptive characteristics of native breeds and the higher output capability of introduced breeds to obtain increased bioeconomic efficiency [11]. Several composite breeds of beef cattle have been developed in the U.S. from crosses of *Bos indicus* and *Bos taurus* breeds of cattle to combine the tolerance to climatic and nutritive stress of the *Bos indicus* breeds with the higher output capability of *Bos taurus* breeds [31].

The genetic theory on which simultaneous use of both additive and nonadditive sources of genetic variation through the formation of composite breeds/populations, based on a multi-breed foundation, is not new [37]. More recently this genetic theory was developed further and adapted for the specific purpose suggested here [34, 36].

The heritability of breed differences in fecundity is high even though it is low among individuals within breeds, simply because breed means are based on a large number of individuals. Also, the heritability of growth rate and composition of gain is much higher than the heritability of fecundity among individuals [38, 39, 40, 41, 42]. Thus the logical strategy for utilization of breed or population differences in the formation of composite populations or new breeds is to select among breeds to make major changes, particularly for bioeconomic traits of low heritability and where large breed/population differences exist, or for bioeconomic traits for which a genetic intermediate is optimum. For example, a genetic intermediate is likely to be optimum in some major beef cattle production/marketing ecosystems [11] for such major bioeconomic traits as growth rate and mature size, milk production, carcass composition, and age at puberty in which large breed differences exist [6, 7].

Breed evaluation can be used to predict optimum additive genetic (breed) composition for composite populations. Composite populations provide a convenient procedure for using heterosis [6]. Retention of initial increases in crossbred heterozygosity after subsequent random (*inter se'*) mating within the crossbred population is proportional to $1 - \sum_i^n P_i^2$, where P_i is the fraction of each of n breeds in the pedigree of a composite population [34, 36, 37]. Heterozygosity is lost between the F_1 and F_2 generations. Even though retention of heterozygosity increases with the number of breeds in the crossbred foundation of a composite breed, the number of breeds contributing to the foundation should be balanced by possible losses of average genetic merit resulting from inclusion of a larger number of breeds.

Experimental results with beef cattle suggest that retention of heterosis is proportional to retention of heterozygosity in rotational crossbreeding systems [35]. In an experiment involving purebred, F_1, backcross, and F_2 and F_3 *inter se'* combinations in the Angus and Hereford breeds of cattle, retention of heterosis was proportional to retention of heterozygosity for all major bioeconomic traits except survival and pregnancy, and loss of heterosis for postweaning gain and carcass weight in the F_3 was less than expected on the basis of reduction in heterozygosity [43]. Thus early results from beef cattle breeding research suggest that breed evaluation can be used to predict optimum breed composition of composite breeds/populations for such bioeconomic traits as growth rate and mature size, milk production, carcass composition, and age at puberty and provide a convenient and useful procedure for use of heterosis. Subsequent selection opportunity in composite populations can be used for fitness (e.g., fecundity and survival) and other bioeconomic traits that may be at a sub-optimal level [6, 11]. In a review paper reporting results from 12 experiments with sheep, Young [44] reported inconclusive results; in some experiments less heterosis was retained than was expected from retained heterozygosity, and in other experiments more heterosis was retained than was expected on the basis of heterozygosity.

Females produced from crosses of highly fecund pigs from the People's Republic of China with western European breeds produce litters that are about three pigs larger than litters from crosses among western European breeds. However, progeny from the Chinese x western crosses have less favorable carcass composition and slower growth rate [16]. In sheep, the fecund Finnsheep and the Romanov produce about one more lamb per parturition than most of the traditional breeds. They also reach puberty before 6-7 months of age and produce their first lambs as yearlings. But the Finnsheep and the Romanov have sub-optimal rates of growth, higher than optimum fat levels on high energy diets, and wool of low quality [42]. Formation of composite breeds is a method for combining the high fecundity of the Chinese pigs with swine breeds that excel in lean tissue growth rate. Similarly, the high fecundity and early puberty of the Finnsheep and Romanov could be combined with the superior lean tissue growth rate of other breeds (e.g., Texel). In either case, subsequent selection within the composite would result in further improvement.

Perspective

Although agricultural development causes ecosystems to change much faster than do normal evolutionary rates, the risk of failing to maintain superior adaptation of livestock populations within ecosystems can be reduced through national, regional, and international programs focused on:

1. Evaluating production capabilities and identifying constraints of major ecosystems,
2. Characterizing bioeconomic traits of animal genetic stocks within major ecosystems, and
3. Developing breeding systems for optimal use and further improvement in adaptation (e.g., bioeconomic efficiency) of existing animal genetic stocks either as purebreds or as contributors to new breeds within each major ecosystem.

To achieve and maintain optimum adaptation of genetic stocks within ecosystems and the consequent genetic variation among diverse ecosystems, the appropriate national, regional, and international organizations should focus on acquiring and disseminating the information needed for effective use of breed differences and heterosis to achieve optimum initial additive genetic composition, and for continuing intra-population selection in harmony with changing requirements for optimum adaptation in each ecosystem. The U.S. interest in global animal genetic diversity justifies a national program, in collaboration with programs of other nations. National programs provide the basis for effective regional and international programs if effective coordinating leadership is provided at the international level. In fact, strong national programs serving enlightened long-term national interests are essential for an effective international program.

Accessibility of world animal germplasm resources

Animal health constraints

A comprehensive national program of sampling, evaluating, and utilizing global genetic resources requires development of minimal-risk methods for importing genetic stocks of cattle, sheep, goats, and swine that can contribute to improved productive efficiency and increased desirability of U.S. animal products.

Most of the present U.S. animal germplasm resources were imported many years ago when animal health management was primitive or nonexistent. Greatly improved animal health programs have been highly successful in eradicating and in preventing the entry and establishment of exotic animal diseases and the consequent large potential losses to the livestock industry and the public. The excellent health maintained in U.S. animal industries has contributed greatly to efficient production and to availability of animal products to consumers at reasonable prices with a high level of safety assurance [4].

Officials of the Animal and Plant Health Inspection Service (APHIS), USDA are charged with the major responsibility of preventing introductions of exotic animal diseases. Thus the policies and procedures developed to provide protection are, quite naturally, intended to maintain near-maximum levels of protection [4]. However, failure to introduce animal germplasm that could increase the efficiency of U.S. animal agriculture and thus reduce costs to consumers of animal products also is a potential cost to the public. Numerous genetic stocks of cattle, swine, sheep, and goats that could make important contributions to U.S. animal production systems are not currently available in the U.S. because of health requirements and/or prohibitive quarantine costs. Hence it is important to develop and implement less costly, minimal-risk methods of importing exotic animal germplasm [4].

The three methods available for transporting animal germplasm are:

1. Postnatal animals,
2. Frozen semen, and
3. Frozen embryos.

Technology for cryogenic preservation of semen has been developed for all livestock species, and cryogenic preservation of embryos is possible for all food and fiber animal species except poultry and swine [45]. High priority should be given to development of technology for cryogenic preservation of swine embryos to reduce the health hazard and the cost of shipping swine germplasm. Because of the reduced exposure to specific pathogens, animal health authorities generally agree that transport is often safer for embryos and semen than for postnatal animals in species where technology permits [46]. Costs of transport also favor semen or embryos over postnatal animals. Recent results suggest that washing protocols can be effective in removing specific viral pathogens from cattle and sheep embryos while the zona pellucida is still intact [47]. Preliminary results even suggest that the natural uterine environment may offer embryos a high level of protection from specific pathogens [48]. The use of this phenomenon together with an embryo-washing protocol could provide a low-risk method of germplasm movement. Further development of less expensive and more sensitive diagnostic tests for specific pathogens is also expected [4]. As embryo preservation technology and procedures for reducing animal health hazards develop further, animal health regulations are expected to be adjusted accordingly.

These advances in animal health technology indicate that a positive program of sampling and evaluating promising foreign animal germplasm under health supervision of Veterinary Services of the APHIS/USDA can be managed without increasing present hazards to animal health. Health hazards actually may be reduced by development of technology for the minimal-risk movement of embryos at costs much below those required for the long-term quarantine of postnatal animals. Continuing the present prohibitive and/or expensive import policies likely will encourage smuggling of animal germplasm as frozen semen or embryos, to the potential detriment of animal health in the U.S. [4].

Trade barriers

Trade barriers also may affect the availability of exotic germplasm in the U.S. and may directly or indirectly affect both animal export and import policy. For example, a 'perceived' animal health risk may be used indirectly to 'protect' a segment of a domestic industry from exotic imports. However, it is unlikely that trade barriers represent a major constraint on imports into the U.S. In general, bilateral negotiations to achieve reciprocity should result in equitable solutions that are mutually advantageous. For example, mutually satisfactory arrangements for the importation of highly fecund pigs from the People's Republic of China appear feasible because China has shown a continuing interest in importing U.S. swine germplasm. Samples of U.S. swine germplasm already have been exported to the People's Republic of China.

Future genetic diversity in United States food and fiber animal species

Dependence of U.S. on global genetic diversity

The U.S. is dependent upon access to adequate global genetic variation in biological characteristics of food and fiber animal species to allow most rapid adaptation to current and changing production environments and market requirements for animal products in each major ecological zone of the country. Modification of major components of a physical environment to the degree required to achieve efficient production from unadapted genotypes generally is not economically nor technologically feasible. Thus the practical breeding goal is to match the biological characteristics of the life cycle of available genetic stocks with the most technologically and economically feasible production environment and marketing situation of each major ecosystem to approach near optimum adaptation [11].

On a global basis, animal production ecosystems differ widely and are changing continuously in response to climatic, economic, technological, social, and political factors. It is in the interest of the U.S. to cooperate in international programs designed to maintain high levels of genetic diversity among ecosystems through selection toward optimum adaptation or for maximum bioeconomic efficiency in each ecosystem. Such maintenance of global genetic diversity would enhance continuing availability of the variety of adaptive characteristics that may be required by unforeseen changes in U.S. production environments and marketing situations. The range of animal genetic potentials that will be needed among ecosystems depends upon the range of production and marketing situations among ecosystems, which differs for the intensively managed dairy cattle, swine, and poultry versus more extensively managed beef cattle, sheep, and goats.

Species differences in ecosystem and genetic diversity

New technologies in production and processing of cereal grains, oilseed legumes, pastures, and forage crops have been primary factors contributing to increased availability of animal products, particularly in the U.S. and other developed countries. Improved efficiency of animal production has been another essential technological component contributing to availability of high quality animal products. Species of food and fiber animals differ in the global range of production environments and in the approaches needed to maintain genetic variation, primarily because of differences among livestock species in the feed resources for which they are competitive and in the variety of production systems favored by economic and technological factors. The increasingly intensive systems of production that have evolved in the poultry industry during the past 40 years are beginning to evolve in the swine industry. Such intensive production systems differ less among ecological zones than do extensive systems and thus generally favor greater uniformity of breeding stocks for commercial production and thus greater risk of losing genetic variation [4].

The rapid growth in U.S. egg, broiler, and turkey production has been feasible because of relatively low prices for cereal grains and legume oilseed meals resulting from new agronomic production technologies. The lower feed costs have led to relatively lower prices for eggs and poultry meat. Similarly, abundant supplies and resulting lower prices for feed grains have stimulated the feedlot finishing of most beef cattle on mixed diets of high cereal grain and oilseed meal content [4]. Even so, feed grains and oilseed meals provide no more than 15 to 20% of the total nutrients for beef cattle production in the U.S. [49]. In the dairy industry, economic efficiency has favored feeding large quantities of cereal grains and oilseed meals to achieve near maximum output of fluid milk per cow, because of the favorable price relationship of cereal grains relative to forages and milk volume. The dairy cow produces a unit of milk protein for about one-fifth of the feed energy required to produce a unit of beef protein [50]. Thus the high producing dairy cow is a relatively strong competitor for use of high energy or high protein plant resources, in contrast to other ruminants, where the primary output is meat and/or fiber. In all food and fiber animal species, but particularly in poultry, swine, and dairy cattle, the feeding of more cereal grains in developed countries has increased efficiency of production by increasing output and reducing the proportion of the total nutrient intake required for animal maintenance [4]. These more intensive production systems differ less among global geographic regions and require less diversity of genetic stocks than do extensive production systems. In dairy cattle, the additional factor of product pricing that favors maximum output of fluid milk per cow has led to a dairy industry in the U.S. in which more than 90% of current fluid milk supply is produced by a single breed, the Holstein-Friesian [51].

Rate of reproduction has a major impact upon life cycle ratio of costs to output value for different meat animal species and thus upon the competitiveness of a species for different types of production resources [52, 53]. The average beef cow is capable of producing only about 0.7 of her body weight per year in progeny market weight; the comparable value is about 8 for pigs, more than 70 for meat chickens, and more than 1,000 for some fish species [4, 54]. More than 50% of the total nutrients used by the beef cattle industry of the U.S. are needed just to meet maintenance requirements of the reproducing female population, whereas this percentage is less than 3% in meat chickens [55]. The relatively low life-cycle biological efficiency of beef cattle, sheep, and goats in converting high energy and high protein plant resources into meat and fiber relegates them largely to the consumption of high-fiber plant products not directly usable by humans nor economically competitive for use by the more efficient, specialized dairy ruminants and the nonruminant species with high fecundity [4].

Increased demand for cereal grains and oilseed meals for direct use in diets of humans is a reasonable long-term expectation as world population increases relative to arable land. Ruminant meat and fiber animals will not be strong competitors for use of scarce and expensive cereal grains and oilseed meals, except to the degree that appropriate supplementary feeding of such concentrate feeds continues to enhance the utilization of forage feed resources by increasing total output and reducing the cost per unit of output on a life-cycle basis. Because of relatively high efficiency in converting feed energy and feed protein into milk protein, the specialized dairy cow may continue to be a relatively strong competitor for use of limited concentrate feed resources [4].

No major changes in composition of feed resources are expected for swine and poultry diets as human competition for cereal grains increases. However, the economic feasibility of using large quantities of cereal grains and oilseed meals for swine, poultry, and dairy production will decline unless new production technologies or additional production areas result in compensating increases in cereal grain and oilseed meal production [4]. Thus the limited range of intensive commercial poultry and swine production environments may be expected to continue.

Approximately 40% of all nonfederal rural land in the U.S. is used mainly for production of pasture and forage, because of either terrain or moisture limitations [56]. The percentage of nonarable land usable only for pasture and forage production is even greater on a global basis. Thus, long-term continuation of ruminant animal production under a wide range of production environments is assured for meat, milk, and fiber [4], although total output will be reduced if supplementary feeding of cereal grains and oilseed meals is limited.

Species differences in genetic resource management

Genetic diversity is likely to be more important and easier to maintain for beef cattle, sheep, and goats than for poultry, swine, or dairy cattle because the relatively low life-cycle biological efficiency of ruminant meat and fiber animals makes them much more dependent upon adaptation to a wide range of climatic conditions and less favorable nutritive environments than is the case for the more intensively managed dairy cattle and nonruminant animal species with high fecundity.

Continuation of current trends toward greater intensification and higher levels of environmental control in production systems for poultry and swine tends to reduce the expected short-term importance of maintaining high levels of genetic diversity in these species [5]. However, genetic diversity in these species may still be important as insurance against potential outbreaks of new diseases or unexpected changes in production environments or in market preferences.

Several approaches can help to ensure adequate continuing genetic variation for dairy cattle, poultry, and swine to augment commercially viable living systems in which genetic diversity has been lost in these intensively managed species [4]. In dairy cattle, the Holstein breed is clearly superior to other breeds in volume of whole milk production per cow. However, this advantage is dependent upon a pricing policy that inadvertently favors low milk solids content and a high level of concentrate feeding. Thus a shift to a pricing policy based on milk solids or a marked increase in the price of concentrate and by-product feeds relative to forages or whole milk could make other breeds economically competitive with the Holstein breed [4]. Because of convenience and relatively low cost, long-term cryogenic preservation of additional semen samples from sires in active artificial insemination use is suggested for a broad genetic sample of Holstein-Friesian sires and a similar sample of sires from several of the less popular dairy breeds. This would be inexpensive insurance to maintain genetic variation both within and among breeds of dairy cattle.

The commercial poultry breeding industry is currently investing considerable resources in exploring the potential of recombinant DNA technology (gene transfer) because of evidence for specific genes with major effects on disease resistance/susceptibility [57]. This suggests that management of poultry genetic resources should include the live population or cryogenic preservation of unique genetic stocks that are not economically viable, if they carry genes with major effects on bioeconomic traits that could be useful when incorporated into commercially competitive stocks. The poultry industry also has long maintained composite, random-bred 'control' populations for use in comparing different experimental and commercial stocks and in estimating response from selection programs for major bioeconomic traits. These random-bred control stocks provide a valuable reservoir of genetic variation and should be continued for future contingencies. Furthermore, some experimental populations of chickens and turkey have been maintained by State Agricultural Experiment Stations. Though not commercially competitive, these populations should be conserved as a possible source of genetic variation for improving commercial stocks.

Although the swine industry is moving toward a breeding structure similar to that of poultry, and although several large commercial swine breeding organizations have been developed, swine breeding stock still is managed predominantly by a large number of farmer breeders [4]. Genetic variation for most bioeconomic traits is considerable in the currently used swine breeding stocks. Also, sampling, evaluation, and optimal use of swine germplasm from other parts of the world can add needed genetic variation in fecundity and carcass composition. Thus concern for future genetic diversity in swine does not require major conservation efforts beyond continuing to monitor the situation and applying effective genetic selection for maximum bioeconomic efficiency within the relatively narrow range of U.S. production systems and marketing situations. However, as in poultry, conservation of unique low fat, high fat, disease resistant, or other stocks is encouraged, as living populations or as frozen semen or embryos (if/or when feasible).

For beef cattle, sheep, and goats, a general program of global sampling, evaluating, and developing breeding systems for optimal use in diverse U.S. ecosystems and selecting for increased bioeconomic efficiency of live populations within each ecosystem appears adequate to maintain needed future genetic variation.

References

1. Johnson, A. O. 1978. Food Consumption, Prices and Expenditures. Economics, Statistics and Cooperative Service, U.S. Dept. Agric., Agric. Econ. Rep. 138, 1977 Supplement.
2. Marston, R. M. and B. B. Peterkin. 1980. Pp. 21–25 in Nutrient Content of the National Food Supply. NFR-9, Econ. Stat. Coop. Serv., U.S. Dept. Agric., Nat. Food Rev., Winter.
3. U.S. Department of Agriculture. 1982a. Agricultural Statistics. U.S. Government Printing Office, Washington, DC.
4. Council for Agricultural Science and Technology (CAST). 1984. Animal Germplasm Preservation and Utilization in Agriculture. Council for Agricultural Science and Technology, Ames, IA. Rep. 101.
5. Gregory, K. E. 1986. Preservation and management of genetic resources: Maintaining adaptation and diversity. Pp. 492–499 in G. E. Dickerson and R. K. Johnson (eds.), Proc. 3rd World Congress on Genetics Applied to Livestock Production, Management of Genetic Resources, XII. Univ. Nebraska, Lincoln.
6. Gregory, K. E., L. V. Cundiff, and R. M. Koch. 1982a. Comparison of crossbreeding systems and breeding stocks used in suckling herds in continental and temperate areas. Proc. 2nd World Congress on Genetics Applied to Livestock Prod. V: 482–503
7. Cundiff, L. V., K. E. Gregory, R. M. Koch and G. E. Dickerson. 1986. Genetic diversity among cattle breeds and its use to increase beef production efficiency in a temperate environment. Pp. 271–282 in G. E. Dickerson and R. K. Johnson (eds.), Proc. 3rd World Congress on Genetics Applied to Livestock Production IX. Univ. Nebraska, Lincoln.
8. Cundiff, L. V., K. E. Gregory, and R. M. Koch. 1989. Productivity of large sized cattle breeds in beef cow herds in the temperate zones of North America. Proc. 3rd World Congress on Sheep and Beef Cattle Breeding. Paris. (In press).
9. Dickerson, G. E. 1977. Crossbreeding evaluation of Finnsheep and some U.S. breeds for market lamb production. No. Cent. Reg. Publ. 246. Univ. Nebraska, Lincoln.
10. Hodges, J. 1986. Animal genetic resources in the developing world: goals, strategies, management and current status. Pp. 474–485 in G. E. Dickerson and R. K. Johnson (eds.), Proc. 3rd World Congress on Genetics Applied to Livestock Production, Management of Genetic Resources XII. Univ. Nebraska, Lincoln.
11. Gregory, K. E., J. C. M. Trail, R. M. Koch, and L. V. Cundiff. 1982b. Heterosis,

crossbreeding and composite breed utilization in the tropics. Proc. 2nd World Congress on Genetics Applied to Livestock Production VI: 279–292.

12. Hickman, C. G. 1982. Administrative and rational methods of livestock conservation. Proc. 2nd World Congress on Genetics Applied to Livestock Production VI: 129–135.
13. Zhang, Wen-Chen, J. S. Wu, and W. E. Rempel. 1983. Some performance characteristics of prolific breeds of pigs in China. Livestock Prod. Sci. 10: 59–68.
14. Cheng, Pei-Lieu. 1983. A highly prolific pig breed of China. The Taihu pig. Pig News and Information. Commonw. Agric. Bur. 4(4): 407–425.
15. Cheng, Pei-Lieu. 1984. Livestock breeds of China. Chap. 10, pp. 159–194 in Food and Agriculture Organization. Animal Prod. Hlth. Pap. 46.
16. Gianola, D., C. Legault, and J. C. Caritez. 1982. Pp. 39–55 in Evaluation of Chinese swine germplasm in France: Preliminary Results. Proc. Pig Industry Conf. Univ. Illinois, Urbana.
17. Howard, A. N. and W. C. Smith. 1976. The Belgian Pietrain as a sire of crossbred pigs slaughtered at 64 kg live weight. 1. Performance and carcass characteristics. Anim. Prod. 23: 389–393.
18. Howard, A. N., W. C. Smith, D. Lesser, and E. M. Carpenter. 1976. The Belgian Pietrain as a sire of crossbred pigs slaughtered at 64 kg live weight. 2. Butcher and consumer reactions to meat from the crosses. Anim. Prod. 23: 395–401.
19. Smith, W. C. and C. R. Toft. 1978. A note on the comparative performance of pigs sired by Pietrain x Large White and Large White Boars. Anim. Prod. 121–124.
20. Kalm, E. 1986. Evaluation and Utilization of Breed Resources: As Sire Lines in Crossbreeding. Pp. 35–44 in G. E. Dickerson and R. K. Johnson (eds.), Proc. 3rd World Congress on Genetics Applied to Livestock Production, Management of Genetic Resources X. Univ. Nebraska, Lincoln.
21. Gol'tsblat, A. I., A. D. Shatskii, and V. M. Vaevodova. 1982. Comparative productivity of Romanov and Finnsheep sheep. Soviet Agric. Sci. 10: 55–73.
22. Vesley, J. A. and E. E. Swierstra. 1986. Reproductive parameters of crossbred ewe lambs sired by Romanov, Finnish Landrace, Dorset and Western Range Rams. J. Anim. Sci. 62: 1555–1562.
23. Wolf, B. T., C. Smith, and D. I. Sales. 1980. Growth and carcass composition in the crossbred progeny of six terminal sire breeds of sheep. Anim. Prod. 31: 307–313.
24. Cameron, N. D. and D. J. Drury. 1985. Comparison of terminal sire breeds for growth and carcass traits in crossbred lambs. Anim. Prod. 40: 315–322.
25. Kempster, A. J., D. Craston, D. R. Guy and D. W. Jones. 1987. Growth and carcass characteristics of crossbred lambs by ten sire breeds, compared at the same estimated carcass subcutaneous fat proportion. Anim. Prod. 44: 83–98.
26. Craston, D., A. J. Kempster, D. R. Guy and D. W. Jones. 1987. Carcass composition of crossbred lambs by ten sire breeds compared at the same carcass subcutaneous fat proportion. Anim. Prod. 44: 99–106.
27. Latiff, M. G. A. and E. Owen. 1979. Comparison of Texel and Suffolk-sired lambs out of Finnish Landrace x Dorset Horn ewes under grazing conditions. J. Agric. Sci. Camb. 93: 235–239.
28. Latiff, M. G. A. and E. Owen. 1980. A note on the growth performance and carcass composition of Texel- and Suffolk-sired lambs in an intensive feeding system. Anim. Prod. 30: 311–314.
29. de Alba, J. 1987. Criollo cattle in Latin America. Animal Genetic Resources. Strategies for improved use and conservation. Anim. Prod. Hlth. Pap. 66. Food and Agriculture Organization – United Nations.
30. Cartwright, T. C., G. F. Ellis, Jr., W. E. Kruse, and E. K. Crouch. 1964. Hybrid vigor in Brahman-Hereford crosses. TX Agric. Exp. Sta. Tech. Monogr. 1.

31. Koger, M. 1980. Effective crossbreeding systems utilizing Zebu cattle. J. Anim. Sci. 50: 1215–1220.
32. Koger, M., F. M. Peacock, W. G. Kirk, and J. R. Crockett. 1975. Heterosis effects on weaning performance of Brahman-Shorthorn calves. J. Anim. Sci. 40: 826–833.
33. Koch, R. M., J. B. Gibb, and J. A. Gosey. 1986. Evaluation of beef cattle industry breeding programs. Pp. 398–409 in G. E. Dickerson and R. K. Johnson (eds.), Proc. 3rd World Congress on Genetics Applied to Livestock Production, Management of Genetic Resources IX. Univ. Nebraska, Lincoln.
34. Dickerson, G. E. 1969. Experimental approaches to utilizing breed resources. Anim. Breed. Abstr. 37: 191–202.
35. Gregory, K. E. and L. V. Cundiff. 1980. Crossbreeding in beef cattle: Evaluation of systems. J. Anim. Sci. 51: 1224–1241.
36. Dickerson, G. E. 1973. Inbreeding and heterosis in animals. Pp. 54–77. Proc. of Animal Breeding and Genetics Symp. in Honor of Dr. Jay L. Lush. American Society of Animal Science, Champaign, Illinois. 104 pp.
37. Wright, S. 1922. Effects of inbreeding and crossbreeding on guinea pigs. III. U.S. Dept. Agric. Bull. 1121.
38. Johnson, R. K. 1980. Heterosis and breed effects in swine. No. Centr. Reg. Publ. 262. Univ. Nebraska, Lincoln.
39. Hanrahan, J. P. 1982. Selection for increased ovulation rate, litter size and embryo survival. Proc. 2nd World Congress on Genetics Applied to Livestock Prod. V: 294–309.
40. Hill, W. G. and A. J. Webb. 1982. Genetics of reproduction in the pig. Pp. 541–563 in D. J. A. Cole and G. R. Foxcraft (eds.), Control of Pig Reproduction. Butterworths, London.
41. Wolf, B. T. and C. Smith. 1983. Selection for carcass quality. Pp. 493–514 in W. Haresign (ed.), Sheep Production. Butterworths, London.
42. Land, R. B. and D. W. Robinson (eds.). 1985. Genetics of reproduction in sheep. Butterworths, London.
43. Koch, R. M., G. E. Dickerson, L. V. Cundiff, and K. E. Gregory. 1985. Heterosis retained in advanced generations of crosses among Angus and Hereford cattle. J. Anim. Sci. 60: 1117–1132.
44. Young, L. D., G. E. Dickerson, T. S. Chang, and R. Evans. 1986. Heterosis retention in sheep crossbreeding. Pp. 497–508 in G. E. Dickerson and R. K. Johnson (eds.), Proc. 3rd World Congress on Genetics Applied to Livestock Production IX. Univ. Nebraska, Lincoln.
45. U.S. Congress, Office of Technology Assessment. 1987. Technologies to. Maintain Biological Diversity, OTA-F-330. U.S. Government Printing Office, Washington, DC.
46. Office International Des Epizooties. 1985. May, 1986 Update. Pp. 401. International Zoo – Sanitary Code 12, Reu de prony 75017. Paris, France. 5th ed.
47. Singh, E. L. 1987. The disease control potential of embryos. Theriogenol. 27(1): 9–20.
48. McVicar, J. W., E. L. Singh, C. A. Mebus, and W. C. D. Hare. 1986. Embryo transfer as a means of controlling the transmission of viral infections. XIII: Failure to detect foot and mouth disease viral infectivity associated with embryos collected from infected donor cattle. Theriogenol. 26: 595–601.
49. Byerly, T. C. 1975. Feed use in beef production. A review. J. Anim. Sci. 41: 921–932.
50. Reid, J. T., O. D. White, R. Anrique, and A. Fortin. 1980. Nutritional energetics of livestock: Some present boundaries of knowledge and future research needs. J. Anim. Sci. 51: 1393–1415.
51. U.S. Department of Agriculture. 1982b. Participation Report. Dairy Herd Improv. Newsl. ARS 58(1). September 1982.

52. Notter, D. R., J. O. Sanders, G. E. Dickerson, G. M. Smith, and T. C. Cartwright. 1979. Simulated efficiency of beef production for a mid-western cow-calf-feedlot management system. III. Crossbreeding systems. J. Anim. Sci. 49: 92–102.
53. Tess, M. W., G. L. Bennett, and G. E. Dickerson. 1983. Simulation of genetic changes in life-cycle efficiency of pork production. J. Anim. Sci. 56: 336–379.
54. National Cattlemen's Association. 1982. The Future for Beef. Beef Business Bulletin. National Cattlemen's Association, March 5.
55. Gregory, K. E. 1982. Breeding and production of beef to optimize production efficiency, retail product percentage and palatability characteristics. J. Anim. Sci. 55: 716–726.
56. Fox, A. S. and K. C. Clayton. 1981. Agriculture's Production Potential. Agricultural – Food Policy Review 4. Perspectives for the 1980's. Econ. Stat. Serv., U.S. Dept. Agric.
57. Crittenden, L. B. and J. S. Gavora. 1986. Genetic resistance to virus diseases. Pp. 624–634 in G. E. Dickerson and R. K. Johnson (eds.), Proc. 3rd World Congress on Genetics Applied to Livestock Production, Genetics of Reproduction, Lactation, Growth, Adaptation, Disease and Parasite Resistance, XI. Univ. Nebraska, Lincoln.

Why are there so many species of mollicutes? An essay on prokaryote diversity

ROBERT F. WHITCOMB and KEVIN J. HACKETT
Insect Pathology Laboratory, Plant Sciences Institute, USDA-Agricultural Research Service, Beltsville Maryland 20705, U.S.A.

Abstract Microbial species and genes are a natural treasure. Proximate factors determining their diversity include substrate, temperature, pressure, pH, atmosphere, osmolarity, redox, moisture and light, spatial and temporal features, and competitors. Interaction of these factors with microorganisms generates genetic and phenotypic diversity. Historical (evolutionary) factors, now being rapidly revealed by analysis of highly conserved macromolecules such as rRNA, determine diversity of cellular design and organization. The prokaryote class Mollicutes (mollicutes) can be used to illustrate these processes and relationships. These wall-less organisms, which are obligately associated with cells of eukaryotes, include commensals, mutualists, and pathogens, many of which are fastidious or noncultivable. Host ecologies organize proximate factors. Historical factors may reflect evolutionary origin at the time of radiation of the insect orders, and subsequent colonization of flowering plants and vertebrates. Based on current discovery curves, the mollicute genus *Spiroplasma*, which is obligately associated with arthropods, may be one of the largest genera of organisms. In this genus, a combination of nutritional and microhabitat specializations and ecological factors operating on both regional and local levels, are important determinants of diversity. Although most microbial taxa, like Mollicutes, are poorly known, current emphasis on microbial phylogeny should broaden our knowledge of microbial structure and function. Specialist microorganisms, whether primitive habitat-specific autotrophs, or symbiotes of eukaryotes, in their aggregate, have potentialities that vastly exceed those of eukaryotes, and merit attention and preservation.

Key words: mollicutes, diversity, phylogeny, ecology, evolution, prokaryote, spiroplasma, acholeplasma, mycoplasma, classification

Introduction

The world is essentially 'a microbial world.' Although we may pretend to be eukaryotes, we are impostors. Of our 10^{14} cells, only 10% are eukaryotic [1], and even these contain mitochondria – organelles that are in fact, from an evolutionary perspective, prokaryotic [2, 3]. In a sense, therefore, to discuss the determinants of prokaryote diversity is to discuss the determinants of all life.

In this paper, we cite some data and pose two riddles. One of the members of the prokaryote class Mollicutes, the genus *Spiroplasma*,

L. Knutson and A.K. Stoner (eds.). Biotic diversity and germplasm preservation, global imperatives.
 ISBN 0-7923-0178-1

although discovered a mere 17 years ago, may be the largest genus of any kind on Earth and may contain as many as a million species. Why are there so many species of *Spiroplasma*? And, what does mollicute diversity reveal about microbial diversity in general?

Microbial diversity

Microbial diversity can be thought of in different ways. At a superficial level, diversity is usually represented by formal taxonomic schemes based on phenotypic characters. However, analysis of the microbial genotype is now revolutionizing bacterial systematics by facilitating construction of taxonomies that use information derived at the most basic level of the cell, the genome.

Taxonomic diversity

In conventional taxonomic schemes [4, 5], the prokaryotes are divided into four divisions on the basis of their cell wall content. Organisms with a typical gram-positive cell wall are placed in the Firmacutes, classical gram-negative organisms in the Gracilicutes, organisms completely lacking a cell wall in the Tenericutes, and those without peptidoglycan (including the archaebacteria) in the Mendosicutes.

Within each division or phylum, prokaryotes are segregated into class, order, family, genus, and species by criteria that were developed one-by-one over more than a century of bacterial discovery and description. For this reason, few general principles exist for assignment of prokaryotes to higher taxa. Murray [5], while arguing that the current arrangement of higher taxa is useful, nevertheless considers this a time of 'taxonomic transition', during which conventional classifications are 'on hold', in anticipation of important new genomic and biochemical information.

It is important to recognize that the biological species concept [6], as applied to sexually reproducing populations of eukaryotes, is technically inapplicable to prokaryotes. Nevertheless, the taxonomic level 'species' still has meaning. As discussed later, bacterial species can be viewed (perhaps oversimplistically) as assemblages of nonbreeding, geographically universal, long-lived clones [7]. Fortunately for the concept of bacterial species, numbers of such clones are generally limited. And the partitioning of extrachromosomal and euchromosomal functions during the course of evolution – with the former involved with cell survival and reproduction in the short–term and the latter in the long–term [8, 9] – enables the use of euchromosomal genes in taxonomy and phylogeny [10]. For taxonomic purposes, it has been proposed [11, 12] that prokaryote species be considered assemblages of closely related clones, none of which show more than 70% DNA-DNA homology with other named species.

As of 1980 [13], more than 1,700 species, 282 genera, 65 families, 21 orders, and 8 classes of eubacteria had approved names. Although many more names have since been validly published, it is unlikely that current classification accurately represents underlying diversity. Future changes based on the results of genomic analyses and breakthroughs in the discovery of large groups of previously unknown bacteria are likely to be profound. The discovery of mollicutes in insects [14–16] alone promises to significantly increase the number of taxa at several hierarchical levels.

Phenotypic diversity

The most conventional view of microbial diversity is that revealed by study of the phenotype – the expression of the organisms' genes acting as a whole. This perspective often emphasizes properties of single clones, particularly as they are observed in the laboratory. The observed phenotypic characters are essentially either morphological or physiological.

As reviewed by Starr and Schmidt [17], prokaryotes exhibit significant morphological variation. For example, although many cells are spherical, subspherical or filamentous, some are flat and square, or even more unusual in shape. Numerous other adaptations, including modified cell walls or sheaths, add to diversity. Various types of resting stages are produced. Many of these adaptations reflect existence in environments where substrates, temperature, pH, and (particularly) water, are intermittently available or variable in quantity or quality.

Prokaryote cell size ranges from less than 100 nm to 100 μm [17]. Since cytoplasmic streaming does not occur in prokaryotes [18, 19], upper size may be defined by limits of diffusion; lower size may be determined by spatial needs for cellular components. While prokaryotic cells always contain ribosomes and a circular DNA genome, they may also contain extrachromosomal DNA, inclusions or organelles such as thylakoids, or sequential enzyme apparati [20, 21] that organize cellular metabolism.

Physiological diversity often is intimately correlated with morphological diversity. For example, coenosis allows sharing of metabolites, and some cells may be sacrificed altruistically as holdfasts or protective structures [17]. Other examples are flagellae or helicity, adaptations that allow microbes to obtain nutrients or optimize their gaseous environment [17] through taxes and motility [22, 23]. Many other physiological adaptations have no obvious morphological expression. This is especially true for metabolic pathways.

In general, particular phenotypes are consistent for a species, its life stage [24], nutrition, antibiotic exposure, growth temperature, or many other factors [17]. By studying the phenotypic diversity of microbes, much can be inferred about the selective pressures exerted – in both contemporary and evolutionary time – on the organisms. However, since each microbe must coordinate a great number of life processes, responses to a single determinant often may result in unique adaptations. For instance, the importance of the cell wall in protecting bacteria from osmotic stress [25] suggests that this factor likely has been an important selective force; however, another response to osmotic stress, one apparently employed by mollicutes, is avoidance of hypotonic conditions by development of life cycles in which transmission among hosts occurs only in osmotically acceptable fluids such as nectar or mucous. Or similar adaptations might have different causes. For example, the flexibility/rigidity and motility of a microbe may be [17], alternatively, a response to physicochemical conditions, prey, or the presence or absence of light.

Genotypic diversity

In prokaryotes, the genetic complement is the circular DNA of the genome and extrachromosomal elements [26] such as plasmids [27]; transposons are also present [7]. The genetic complement, particularly the chromosome, can be used to measure the diversity of a) species or higher level taxa, or b) populations. In this section we will consider two aspects of population genetics: a) allelic or clonal diversity, and b) underlying evolutionary mechanisms.

Population genetics. Biochemical methods of measuring genetic variation in populations have been applied extensively to eukaryotes since the 1960s and are now being applied to prokaryotes [7]. Until recently, analyses of bacterial diversity relied almost exclusively on phenotypic data such as serological or physiological characters. In the vast majority of cases, these properties were not linked to allelic substitutions at individual gene loci. In fact, some characters turned out to be determined extrachromosomally rather than euchromosomally [7].

As a result of breakthroughs in analysis of the prokaryotic genome, we are now beginning to obtain reliable quantitative information on allelic diversity. This approach, which started with Milkman [28, 29], has continued to gain momentum since 1980 [7]. The diversity of alleles in prokaryotes, most of which are believed to be selectively neutral [7], has been found to be extensive. For example, while surveying research on the population genetics of *E. coli* (perhaps the most intensively studied bacterium), Selander *et al.* [7] estimated 3.4–9.3 alleles per locus for various enzyme loci, with mean allelic diversity per locus (electrophoretic types) equal to 0.343–0.542, compared to values generally less than 0.15 for eukaryotes

[30], and only 0.063 for 50 enzyme loci in human populations [31]. According to Selander *et al.* [7], *E. coli* also has more polymorphic loci (94%) than eukaryotes (33%). And in 49 characters studied, *E. coli* exhibited great interlocus variance; diversity ranged from zero (one allele) to 0.82 (almost two dozen alleles), depending on the enzyme, biotype, or antibiotic sensitivity studied. Whereas *E. coli* was found to be more variable than eukaryotes and some bacteria, it was found to be less variable than certain other bacteria that are naturally competent for transformation.

Generally low recombination rates and extinctions of unfit clones have led to strong linkage disequilibria in many bacteria [7, 32, 33]. Thus, clonal diversity of bacteria is limited [7, 10, 34]. Consequently, although geographic dispersal is variable, and standing bacterial populations may be enormous, in some cases a single clone may make up 90% of all isolates of a species worldwide (R. Selander, pers. comm.). Bacterial species are therefore often assemblages of geographically universal [7, 35], (nearly) nonbreeding, long-lived clones [36], about 10^2 to 10^3 clones per population [7].

Evolutionary mechanisms. While we are beginning to understand how new genes may arise (e.g., through duplication of old genes, with subsequent divergence [37, 38]), the relative evolutionary importance of underlying genetic processes is not well known. Mutation is certainly the source of much genomic diversity, and important events in bacterial evolution may have resulted from successive single base-pair changes alone. Nevertheless, many discontinuous processes, e.g., gene rearrangements (inversions, additions, deletions, substitutions, and transpositions), might also play a major role in evolutionary change [37]. And bacteria are not entirely nonbreeding because of the presence of transformation, transduction, and conjugation. While gene transfer by these mechanisms is of low frequency (about the same as the mutation rate [39]), these factors may nevertheless be significant long-term evolutionary determinants [40]. For example, Selander *et al.* [7] summarize one study of *E. coli* lipopolysaccharides in which recombination, and not mutation, was the major determinant of genetic diversity.

Although prokaryotes have smaller and simpler genomes than eukaryotes, it has been beyond the means of science to relate bacterial genetic transformations to their evolution. Even the evolutionary quantum in prokaryotic evolution is, as is true for eukaryotic evolution, a matter of debate. In some organisms, the evolving unit may not be the entire genetic complex (euchromosomal, extrachromosomal, or both), but may instead be the gene. In fact, Dawkins [41] suggests that the gene, rather than the organism, is the only functional evolutionary unit. Although this view seems extreme, some 'selfish' genes [42] may even have escaped phenotypic selection [43, 44]. Nevertheless, we believe it more useful, at present, to view bacterial evolution first at the level of the clone. Consideration of the clonal nature of bacteria may, in fact, be the key to understanding their evolution. As summarized by Selander *et al.* [7], evolution of bacterial species, at least over the short term, is likely to proceed by clonal selection processes that include periodic selection of mutants (or possibly transformants) conferring fitness. The diversity of existing clones, although increased by genetic drift, is limited by competitive or stochastic extinctions [7]. Since the number of clones is environmentally determined and finite, it is appropriate, as with the metazoa, to think in terms of prokaryotic diversity at the species level. However, while we know much about the determinants of microbial diversity, the processes of speciation in prokaryotes are understood only superficially. Even in eukaryotes, which have been studied more intensively, and for which a much better fossil record exists, debates continue [45–47] as to whether evolution has involved periods of relative stasis followed by punctuated diversification, *e.g.*, as a consequence of species selection [48], or whether it has occurred by gradual phyletic transformation of phenotype in response to selective pressures [49]. Because of the paucity of the bacterial fossil record and the lack of discernable taxonomic characters, bacterial evolutionists have until now been spared embroilment in this controversy. Recent advances in rRNA analysis may soon change this. However, although the relative importance of various possible micro- as compared to macroevolutionary processes in bacterial evolution is a subject for future investigations, there is no evidence at present that these processes are decoupled. Because it is likely that evolution of clones is a stepwise process, it is reasonable to assume that phyletic gradualism may adequately explain bacterial speciation as well as observed diversities in extant lineages.

Finally, efforts have been made recently to estimate the rate of adaptation of prokaryote species. By the time *E. coli* has divided 30 times, requiring about 10 hours under optimal conditions [50] to 30 days under natural gut conditions [7], 1.5% of the cells may be mutants [50]. Milkman [51] predicted that a new adaptive genetic change in *E. coli* would require simultaneous incorporation of 3 or more changes at the codon level, with major episodes of periodic selection in *E. coli* occurring every 30,000 years. Certainly not all bacterial species are similarly constrained with respect to mutation or have equal opportunity for exploitation of emerging niches. Major episodes of evolution in other prokaryote lines may occur at much different periodicities.

Determinants of microbial diversity

Three classes of factors determine prokaryote diversity. The first is historical and can be expressed in terms of phylogenetics. Classifications based on well-founded phylogenies, whether or not couched in the formal language of the Code for Nomenclature [52], are the best predictors of unknown phenotypic or genotypic characters of any microbial taxon. The second class of determinants is proximate – these are the ecological factors recognized by classical microbial ecology. A final class of determinants can be described simply as stochastic 'drift.'

Historical (phylogenetic) determinants

Bacterial taxonomy, inheriting classification strategies originally applied to eukaryotes, has traditionally relied on phenotypic characteristics. However, Woese [53] recently proposed a new approach to taxonomy and phylogeny based on analysis of rRNA, molecules that best fulfill suggested criteria for molecular chronometers [54]. Important rRNA properties are:

a) universality,
b) intertaxal consistency in function,
c) lack of horizontal transfer,
d) experimental convenience, and most importantly,
e) high conservation of the primary sequence of the molecule, a feature that provides a measure of change over a broad evolutionary time scale.

Two means have been used to analyze the highly conserved 16S rRNA molecule. One of these methods [55] involves cataloging the oligonucleotides produced by enzymatic digestion of the rRNA. Oligonucleotides of six or more bases are then fractionated electrophoretically and sequenced. This method permits analysis of about one-third of the total number of bases in the entire molecule, but a significant fraction of the molecule yields oligonucleotides that are too short to be meaningfully analyzed. A method that permits study of essentially the entire molecule is the complete sequencing of 16S rDNA [56]. Other molecules have also been analyzed. Complete sequence analysis of 5S rRNA, although not providing as much information as analysis of 16S rRNA, is much less difficult and has been performed for a number of mollicutes and related gram-positive bacteria [57–61]. Sequencing of tRNA [58, 61–63], although laborious, has provided valuable ancillary data.

Using these technologies, Woese [53] and others [58, 59, 64, 65] have presented impressive evidence that all life forms derived from a 'universal ancestor', the seed of a single evolutionary tree, rooted billions of years ago. This ancestor gave rise to three urkingdoms of life – archaebacteria, eubacteria, and eukaryotes [53].

Archaebacteria. Phenotypically, archaebacteria range from anaerobic thermoacidophilic sulfur metabolizers to methanogens and aerobic non-methanogenic extreme halophiles [53, 66]. They are as different from eubacteria in genotype and phenotype as bacteria are from eukaryotes, a separation that apparently has been reinforced through evolutionary time. 'It is as if they (archaebacteria and eubacteria) had lived in two different worlds' [53]. Their differences notwithstanding, shared characteristics of extant members of archaebacteria [53] and eubacteria [67] suggest a common beginning. Primitive members of both groups are thermophilic, presumably reflecting an origin in a hot environment, perhaps less than 800 million years after the formation of Earth.

Woese [53] postulates that primitive archaebacteria have evolved slowly and may therefore more closely resemble the hypothesized 'universal ancestor' than do eubacteria or eukaryotes. Nevertheless, the long since extinct hypothetical ancestor (which Woese [53] termed a 'progenote') must have differed substantially from any cells on earth today. And, although extant prokaryotes, and particularly archaebacteria, may share many characteristics of primitive organisms, they should not be considered to be 'less highly evolved' than eukaryotes. For example, not only do eukaryotic cells have many primitive features, but prokaryotes are often more metabolically sophisticated. Many characteristics of eukaryotic cells (e.g., absence of H^+-translocating ATPase, lack of rigid cell wall, and DNA-binding proteins) that are present in archaebacteria might arguably [68] have also been selected for in primitive hot, acidic environments, where resident microorganisms may have been primarily lithotrophic. And in prokaryotes many pathways have evolved that are unknown in eukaryotes, such as those by which *E. coli* uses methionine [69].

Eubacteria. The eubacteria, best known by their common representatives, have a diversity that has made possible the creation of quite distinct upper level taxa. They can be divided into ten phyla: purple bacteria (including typically gram-negative bacteria); gram-positive bacteria (including actinomycetes [high G + C] and clostridia and their relatives [low G + C]; cyanobacteria and chloroplasts; spirochetes and leptospires; green sulfur bacteria; bacteroides, flavobacteria, and relatives; planctomyces and relatives; chlamydiae; radioresistant micrococci and relatives; and green non-sulfur bacteria and relatives [53]. In addition, an extant thermophile, *Thermotoga maritima*, may represent an early divergence from the eubacterial tree, and may offer an important clue about the habitat and perhaps the era of divergence of eubacteria [53, 67].

Woese [53] has used phylogenetic analysis of 500 species of eubacteria to elucidate the origin of common phenotypes among the groups, and inferentially, determinants of phenotypic diversity. For instance, the mollicutes are seen not as primitive bacteria, as had been suggested [70], but as highly simplified descendants of gram-positive bacteria. Thus one expects and finds that the mollicutes are phenotypically more similar to gram-positive bacteria than they are to other prokaryote lineages [53, 66, 67]. Furthermore, other previously used criteria for constructing higher level taxa, including photosynthesis, trophic strategies, and oxygen use or sensitivity, are scattered across the bacterial phyla [53]. For example, photosynthesizers are found in a number of different phylogenetic groups, many of which have given rise to nonphotosynthesizers. Heterotrophs and autotrophs are similarly intermixed. Overall, these phenotypic traits can be explained by the adaptation of the major eubacterial lineages in the course of evolution of Earth [53, 71–77], from;

a) its anoxic inorganic beginnings (anaerobic chemolithotrophs and eventually anoxygenic photosynthetic autotrophs) to
b) the accumulation of organic substrate (anaerobic heterotrophs), and
c) the evolution of free oxygen by oxygenic photosynthesizers, a scenario that led to
d) aerobic metabolizers.

Eukaryotes. A portion of the microbial world is eukaryotic; the studies of Woese [53], in fact, have focussed attention on primitive eukaryotes, of which microsporidia may be among the earliest [78, 79]. Interestingly, because extant members of this group are parasites, the history of parasitism may be traceable to earlier stages of life on the planet than had been previously supposed.

Proximate (ecological) determinants

Although earth provides exploitable resources on a grand scale, microbial niches are comparatively tiny patches on a vast quilt created by the weavings of substrate, redox, pH, water, and light. In this sense, diversity simply reflects permutations and combinations of these critical factors, combined with the ability of microorganisms to disperse to microhabitats, exploit the resources that they find there, and avoid extinction [7, 17, 53, 80, 81].

A consideration of non-host prokaryote habitats [81] reveals the importance of substrate distribution in determining diversity. For example, microbes have evolved various mechanisms for dispersing to and remaining at natural interfaces (sites of nutrient accumulation), whether water-solid, water-oil, or water-air. These include the development of motilities (by cilia or air vacuoles), chemotaxes (to substrates, or from toxins), or surface tension or adherence adaptations. Other potential determinants of diversity range from the gradients that form from the leaching of organic matter into streams to the immense archipelagos of planktonic islands in the sea.

Differences in types of substrate and energy resource are also important determinants of prokaryote diversity [74, 75]. Little has been overlooked by microbes. Not only are virtually all natural products consumed by prokaryotes, even the rocks and light have not escaped. Spectacular exploitation strategies are common (reviewed by Gottschalk [74] and Schlegel and Jannasch [81]). For example, nitrogen-fixing bacteria use nitrogen directly from the atmosphere; other bacteria reduce sulfur, while some can even use methane, which is not used by any eukaryote. And mycobacteria produce iron phorins that capture insoluble iron [82]. In their aggregate, the metabolic capabilities of prokaryotes vastly exceed those of eukaryotes. This ability is understandable from a consideration of the major ecological differences between early prokaryotes and eukaryotes. The former were primarily producers of biomass; the latter, presumably protozoa, were consumers – as predators or parasites. Predation or parasitism likely required a simpler metabolism – one that involved degradation of bacteria and use of component monomers, rather than *de novo* monomer synthesis. Metabolic evolution in the prokaryotes was no doubt facilitated by their small size (and their related ability to extract energy from small quantities of substrate), large populations, short generation times, lack of pressure to coordinate metabolism with complex morphologies and behaviors, and tolerance of reducing environments.

Of the physicochemical conditions determining diversity, oxygen may have played a pivotal role [53, 71, 77]. Pressure to cope with the variety of redox potentials found in nature, from anaerobic muds and animal rumens to aerobic sites on the leaves of plants, is certainly one of the key determinants of microbial diversity. In fact, many responses to redox are evident in bacteria [83]. Oxygen-tolerant fermentative anaerobes, for example, have well-developed cytochrome oxidases capable of capturing oxygen as an electron acceptor, reducing the level of toxic hydroxyl radicals [81, 84]. Redox potential can itself be considered to have diversified as an interaction of the abiotic and biotic components that make up Earth. The evolution of Earth, its atmosphere, hydrosphere, and biosphere, can be traced from the (often postulated) anaerobic primordial soup to the accumulation of oxygen, a result of the oxygenic photosynthesis of cyanobacteria (and later algae and plants), and finally to the use of oxygen by heterotrophic prokaryotes and eukaryotes [53, 76]. Prokaryotes are much better adapted than eukaryotes to grow anaerobically; in fact, few eukaryotic organisms (exceptions include a few protozoa, fungi and helminths) can grow indefinitely without oxygen [81].

Microbes can also tolerate extreme pH values. For example [81], the pH in coal mine drainage, the habitat of some archaebacteria [85, 86], can reach 0.7 as a result of the metabolism of thiobacilli; even the pH of abundant habitats (e.g., pine forest soils) may be as low as 3.0–5.5. At the other extreme, prokaryotic residents of alkaline springs may tolerate pHs of 11 or more; although more typical values are pH 8–9 (for example, for the Szik [salt and alkali] soils of the Hungarian lowlands [81]). Some bacteria have even been cultured at pH 11.4 [87].

Tolerance for extreme pH and salt is tied intimately to water needs [88]. In this sense, prokaryotes, with their demands for aqueous environments, have narrower tolerances than eukaryotes. According to Schlegel and Jannasch [81], most prokaryotes do not tolerate water activity [a_w] of less than 0.98, the a_w of sea water. Halophilic bacteria are an exception, with a_w tolerances of 0.75; some molds can withstand an a_w of 0.6.

The history of temperature on Earth provides a further important clue to determinants of microbial diversity. The earliest prokaryotes were no doubt thermophiles [53, 89]. Since primitive members of all three urkingdoms [53] have adaptations consistent with life in hot environments, it is in fact likely that these taxa diverged before the planet had cooled appreciably. Even today, extant thermophiles may grow at temperatures near the boiling point of water [81]. Mesophily and particularly psychrophily (growth at temperatures as low as $-18°C$ [81]) must be comparatively recent adaptations.

Temperature extremes do not constitute the only thermal selective pressure. Although some microbes live in thermally insulated habitats such as oceans (90% of which are 5°C or cooler), soil (about 12°C in temperate zones [81]), tropical environments, or within hosts (particularly homeothermic ones), other organisms experience significant temperature fluctuations. Many mesophiles appear to be extremely adaptable with regard to temperature fluctuations, and can tolerate changes of 30°C or more [81].

In addition to challenges posed by the availability of adequate substrate and physicochemical conditions, the presence of other life has also been an important determinant of phenotypic diversity. Like eukaryotes, bacteria have developed different survival strategies, depending on competition. As shown experimentally [90], microbial species may even respond to *r* (density-independent) or *K* (density-dependent) selective pressures, depending on the type of environmental challenge.

In many cases, cooperative strategies with the same or other prokaryotes or with single-celled eukaryotes have proved to be competitively effective [17, 81]. The association of prokaryote cells, *e.g.*, in pairs, sheets, or fruiting bodies, may be the result of needs for production of extracellular enzymes or phorins, or may afford protection from desiccation. Many symbiotic intracellular associations of bacteria with protozoa also have been described [91], and some bacteria have formed symbioses with fungi, *e.g.*, lichens.

However, perhaps no event has influenced the diversity of life on Earth so much as the rise of multicellular eukaryotes and the potential internal and external microhabits associated with them or that result as a consequence of their existence.

In summary, then, it is the living and nonliving ecological diversity of the Earth that provides the driving force that has molded bacterial phenotypes. Selection at the phenotypic level provides, with stochastic processes, the driving forces that mold the bacterial genotype. Taxonomic diversity reflects these interactions.

Determinants of diversity in the Mollicutes

The wall-less prokaryotes (class Mollicutes) are strict commensals or parasites of eukaryotic hosts. Because many cause disease in animals (man and livestock), plants (citrus and corn), and beneficial insects (honey bees), and others are associated with important pest insects, they have attracted considerable attention [92, 93].

Does their host dependency act as a constraint on their diversity? At one time it may have seemed so, but that was before the astounding species richness of spiroplasmas (15) was appreciated. The discovery of this group has led instead to a new question: why are there so many species of mollicutes?

Phenotypic diversity

Much of the rich morphological diversity of prokaryotes, particularly that manifested by cell wall modification, is absent in the mollicutes, which show no trace of cell wall or, in fact, any vestige of the peptidoglycan pathway [94]. This does not mean that mollicutes completely lack morphological diversity. Remarkably, even in the absence of peptidoglycan, many species have achieved structural differentiation [95].

Metabolic diversity is also reduced. Mollicutes apparently do not produce complex substances such as photosynthetic pigments. In fact, they appear to produce few metabolically unique products of any kind. None, for example, are known to produce antibiotics or any other kind of antimicrobial defense substance. They lack cytochromes and, while possessing some Krebs cycle enzymes, lack the complete pathway [96]. In general, mollicutes are mundane in their catabolic capabilities (*e.g.*, ability to use urea, glucose, or arginine as energy sources [97, 98]), or anabolic diversity (*e.g.*, the synthesis of carotenoids by some acholeplasmas [99, 100].

The lack of diverse phenotypic characters in mollicutes presents a dilemma for the study of mollicute taxonomy and genetics, in that few mutable characters are available for analysis. At the end of this section, we discuss the phenotypic modifications that characterize mollicute genera.

Taxonomic and genotypic diversity
The mollicutes are currently classified as the sole class of a new bacterial division (Tenericutes), parallel [4, 5] to Firmacutes, Gracilicutes, and Mendosicute bacteria [4, 5]. The class is divided at present into 3 orders, 4 families, and 6 genera. For significant properties of these taxa, see Table 1.

The diversity of mollicutes can be best interpreted by considering their evolutionary roots. Although phylogenies obtained from rRNA analyses differ, depending on the rRNA moieties chosen for study, all reconstructions agree in the general conclusion that the most probable ancestral lineage for the mollicutes is a subtree (Fig. 1) of gram-positive bacteria that includes, in addition to the mollicutes, certain anaerobic clostridia with low G + C content, such as *Clostridium innocuum* or *C. ramosum* [53, 55–58]. All extant mollicute genera can be arranged as a monophyletic cluster (clade) within this subtree [58]. Interestingly, although the clostridial species of this subtree have often been isolated from infected human tissues, they (like many mollicutes [101]) are noted for their metabolism of trehalose [102], an extremely abundant dimer in the hemolymph of insects [16]. This is in accord with other evidence [16, 103, 104] suggesting that mollicutes may have arisen in, and almost certainly radiated extensively with, terrestrial arthropods.

It is important to emphasize that the 5S and 16S rRNA and tRNA molecules used in phylogenetic reconstructions are molecules that are highly conserved in size and sequence; in this respect they contrast with the DNA of the mollicute genome, which was dramatically reduced in size when the wall-less mollicutes evolved from their clostridial ancestors. Thus, the evolutionary stability of the rRNAs makes them particularly valuable as molecular chronometers [54]. Several detailed reviews of mollicute phylogeny are available [54, 105, 106].

In some mollicutes, the genome has been reduced to less than 25% of the ancestral (eubacterial) size of 3×10^9 daltons [61, 106]. This reduction appears to have been achieved in steps. In four of the mollicute genera – *Acholeplasma*, *Anaeroplasma*, *Asteroleplasma*, and *Spiroplasma* – the ancestral genome size is less than half of the primitive eubacterial size. In two genera – *Mycoplasma* and *Ureaplasma* – this value of 1×10^9 daltons was halved again, to reach a size of about 5×10^8 daltons, which may represent a lower limit for the size of the prokaryotic genome. Although the genetic composition has been determined for only a small fraction of the genome of any mollicute species, reductions appear to involve both deletion of genes and reduction of repetitive sequences and/or operon numbers. For

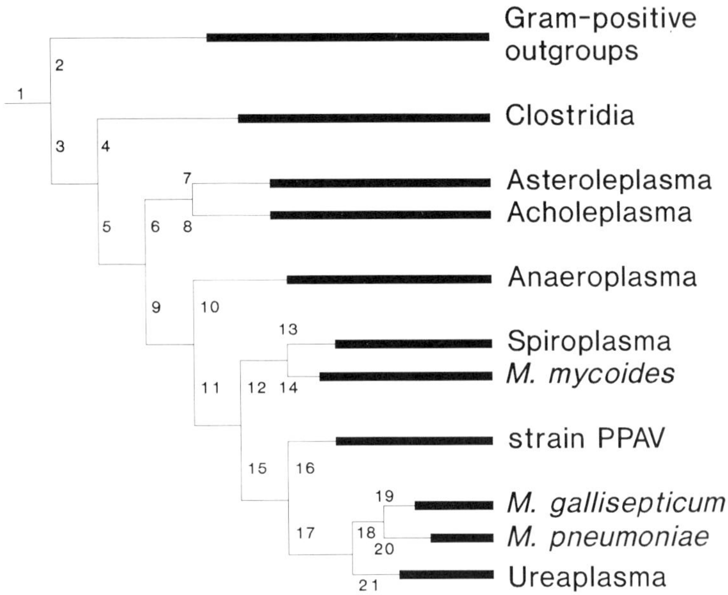

Figure 1. Phylogenetic reconstruction of mollicute phylogeny. The depicted nodes represent a consensus phylogeny from rRNA data [54, 106] and from other characters [61, 106]. In many cases, the character states for many species within clades are unknown; the cladogram, which was hand (rather than computer) generated, is therefore a preliminary iteration. In the following discussion, proposed apomorphies (a), synapomorphies (s), autapomorphies (au), and plesiomorphies (p) are identified. An initial divergence from an (1) anaerobic gram-positive ancestor is postulated. The proposed reconstruction requires that the gram–positive outgroups (2) including bacilli and lactobacilli, developed aerobic metabolism by an event independent of the emergence of the aerobic metabolism of the mollicutes. Cataloging and sequence analyses of rRNA have consistently identified certain *Clostridia* (*C. innocuum* and *C. ramosum*) as the closest gram-positive allies of mollicutes. These analyses are supported (3) by the sharing (s) of rifampicin sensitivity by the *Clostridia*-mollicute clade; presumably, the 5S rRNA molecule had been shortened (a) to 114 nucleotides in *Clostridia* (3). In the clostridial (4) lineage, (p) cell wall was retained, as was a (p) genome size of 3×10^9 Da. In the mollicute (5) lineage, (s) cell wall was lost, and (a) the genome was reduced to 1×10^9 Da. Also, (a) the number of rRNA operons may have been reduced to 2. In the *Asteroleplasma-Acholeplasma* (6) lineage, the ancestral (p) lack of sterol requirement was retained. In *Asteroleplasma* (p) obligate anaerobic metabolism was retained (7), whereas in *Acholeplasma* (s), aerobic metabolism and carotenoid synthesis evolved (8). The present-day habitats of acholeplasmas and anaeroplasmas (in each case the guts of vertebrate or invertebrate animals) suggest that the early divergences leading to Mollicutes may have occurred with this scenario. The development of a sterol requirement provides a synapomorphy defining the vast majority of mollicutes (9). In *Anaeroplasma*, (p) anaerobic metabolism was conserved (10), whereas in the remaining mollicutes, (s) oxygen tolerance emerged (11). Furthermore, it appears that at this node (11), there was a significant change in codon usage (a or s) in which UGA is read as tryptophan, rather than functioning as a stop codon. Also (11), (a) filamentous cell shape emerged. In the spiroplasma–*M. mycoides* lineage (12), the 5S rRNA was reduced to 107 nucleotides. It is likely that divergence of the *Acholeplasmas* and *Spiroplasma* lineages occurred after development of host association with arthropods. The *Spiroplasma* lineage (13)

example, the number of rRNA operons in mollicutes appears to have been reduced from as many as 10 in *Bacillus subtilis* to just 1 or 2 in mollicutes [61, 106, 107]. Although the value of 5×10^8 daltons may represent an approximate lower size limit to the prokaryote genome, there is some degree of phenotypic diversity among organisms in this range. For example, details of metabolic pathways differ considerably in urealytic, glycolytic, and arginolytic mollicutes [108, 109]. Thus, there is no single metabolic solution to the problem of minimal size.

Mycoplasma. Members of the genus *Mycoplasma* are largely colonists of vertebrate cell surfaces. Genome size estimates (obtained largely by renaturation kinetics) range from 3×10^8 to 7×10^8 daltons (61), but difficulties in experimental determinations are considerable, so that detailed statistical analyses will be required to assess apparent small differences in these values [110].

Metabolically, many *Mycoplasma* species fall into one of two categories [111]. More than half are strong fermenters of glucose; others do not ferment glucose but do hydrolyze arginine. Glucose, of course, is a primary energy reservoir in many vertebrate and invertebrate animal fluids and tissues. Arginine metabolism is characteristic of a large cluster of *Mycoplasma* species [112].

Although the two main substrate utilization strategies of *Mycoplasma* species appear to involve *either* glucose or amino acids, some species (e.g., *M. iowae*) utilize both of these substrates, while others (e.g., *M. bovis*) utilize neither [111]. The driving force in energy metabolism of species that use neither glucose nor arginine is particularly enigmatic.

Perhaps because of the ubiquity of sugars and amino acids in hosts,

---

is defined by two conspicuous synapomorphies–helicity and translational motility. In the *M. mycoides* lineage (14), the genome size (s) was reduced to about 5×10^8 Da. In the remaining mollicutes, we presume an ancestor (p), that was filamentous but not helical, with a genome size of 1×10^9 Da, 2 rRNA operons, and a 5S rRNA molecule of 107 or 108 nucleotides (15). The PPAV strain (16) resembles this ancestor, but (au) has lost an rRNA operon. The remaining mollicutes (17) are associated with vertebrates. Filamentous shape is reduced or lost (s). The genome has been reduced (s) to about 5×10^8, in an event independent to that occurring with the *M. mycoides* group. In one clade, cellular polarity developed (18) in the form of attachment organelles (a). In this clade, the 5S rRNA is 108 nucleotides in length. In *M. gallisepticum*, the organelles are bipolar (au). In *M. pneumoniae*, the blebs are unipolar, and are associated (20) with cells that are relatively filamentous, at least *in vivo*. The specific P1 attachment protein is an autapomorphy that defines *M. pneumoniae* (20). Specific details of membrane attachment proteins of most mollicutes are probably species-specific, and constitute autapomorphies that permit immunological identification of the species. All species of *Ureaplasma* (21) are defined by the conspicuous synapomorphic utilization of urea. The 5S rRNA of ureaplasmas is reduced to 104 nucleotides, confirming its highly apomorphic position in Mollicute phylogeny.

Table 1. Taxonomy and Properties of Mollicutes.[a]

Classification	No. Species	Genome Size (MDa)	G + C of DNA (mol %)	Sterol requirement	Distinctive properties	Hosts
Order I: Mycoplasmatales						
Family I: Mycoplasmataceae						
Genus I: *Mycoplasma*	80	500	23–41	+		Animals, Insects; Plants
Genus II: *Ureaplasma*	3	500	27–30	+	Urease +	Animals
Family II: Spiroplasmataceae						
Genus I : *Spiroplasma*	10	1000	25–31	+	Helicity	Arthropods, Plants
Order II: Acholeplasmatales						
Family I: Acholeplasmataceae						
Genus I: *Acholeplasma*	11	1000	27–36	–	Carotenoids	Animals, Insects; Plants
Order III: Anaeroplasmatales						
Family I: Anaeroplasmataceae						
Genus I: *Anaeroplasma*	4	1000	29–33	+	Obligate Anaerobiosis	Bovine-Ovine Rumen
Genus II: *Asteroleplasma*	1	1000	40	–	Obligate Anaerobiosis	Bovine-Ovine Rumen

[a] Modified from Razin and Freundt (1984).

carbohydrate or arginine substrate specializations have yet to be clearly correlated with mollicute niches *in vivo*. Biological fluids (e.g., plant cell sap, invertebrate hemolymph, vertebrate blood, and protoplasm) are rich in free and covalently bound amino acids. Arginolytic mycoplasmas colonize the same tissues as do fermentative mycoplasmas [113, 114]. A frequent inadvertent experiment of this kind is performed when tissue culture cells become contaminated with mycoplasmas [113, 114]. In such cases, both the arginolytic *Mycoplasma arginini* and the fermentative *M. hyorhinis* have occasionally been found to contaminate the same cell lines [113]. The most common portal of entry of these organisms into tissue culture systems is probably the animal serum component of media used to propagate the cells [113, 114].

But metabolic diversification in the genus *Mycoplasma* is by no means the only phenotypic manifestation of diversity. Despite the seeming simplicity of the limiting membranes of these 'minimal cells', some are morphologically differentiated, and some are motile. Cells of some species exhibit polarity in the form of attachment organelles [95]. In general, the host cell receptors to which mycoplasmas attach are glycoproteins such as sialoglycoconjugates [95]. Differentiation among domains of host cell membranes presumably permits discrimination among cellular receptor sites and, accordingly, permits partitioning of the host cell surface into a variety of attachment site types. An interesting example of this specificity involves two colonists of *Homo sapiens*. The membranes of *Mycoplasma pneumoniae*, the etiological agent of human primary atypical pneumonia [115], and *M. genitalium*, a resident of both urethral and respiratory tracts, have species-specific attachment proteins [116].

Whether partitioning is based on chemical attachment affinities, substrate availability, or simple opportunism in the absence of competing mollicute species, two or more mycoplasmas often parasitize the same vertebrate host. Thus it is not unusual to obtain a diversity of mollicute isolates from a single host animal [117].

In a minority of *Mycoplasma* species, there are suggestions of a third driving force in determination of phenotypic diversity. Although most mycoplasmas appear to be transmitted by direct animal-to-animal contact, there is some evidence that certain caprine and ovine mycoplasmas may be associated with arthropods [118, 119]. A putative *Mycoplasma* species has recently been isolated from the hemocoel of an insect [120], strengthening suggestions of an arthropod link. Although no arthropod-vertebrate cycle has been conclusively demonstrated in *Mycoplasma* or in any other mollicute genus, the future may very well reveal that such cycles exist.

For many years, some workers regarded the genus *Mycoplasma* as a polyphyletic cluster of mollicutes [121, 122]. Recent reconstruction of phylogenies using highly conserved rRNAs [53, 106] brought both good news and bad news to mollicutologists. The good news was that all of the wall-less organisms that they studied – except *Thermoplasma acidophilum*, which was conceded by all to be an archaebacterium: reviewed by Razin [106] – appeared to be monophyletic. The bad news was that the genus *Mycoplasma* itself appeared to be polyphyletic [58], in that it represented at least two and perhaps as many as three separate divergences from an ancestral stem with a genome size of 1 x 10^9 daltons. In particular, all analyses agree that *Mycoplasma mycoides* and related species are closely related to, and possibly diverged from, the *Spiroplasma* lineage [106]. Other mollicutes, including a newly discovered group (represented by the PPAV strain, isolated [123] from the surface-sterilized seed of an apple tree) appear [56] to belong to still other clade(s) (Fig. 1).

This discovery poses a great dilemma for taxonomists. Because *M. mycoides* is the type species of the genus [124], its generic name must stand intact. But to separate the unrelated *Mycoplasma* species from the type species would require changing the generic nomenclature of more than 70 species, many of which are commonly recognized as *Mycoplasma* species by medical and veterinary professionals. Understandably, the Subcommittee on Taxonomy of *Mollicutes* voted unanimously at its 1988 meeting to let some dust settle in the wake of rapidly changing concepts of mollicute phylogeny before contemplating drastic taxonomic change (Subcommittee minutes, in preparation). Hopefully, further phylogenetic analyses will facilitate resolution of this dilemma.

Ureaplasma. The genus *Ureaplasma* [125, 126] is a cluster of mollicutes, all of which have been shown to be associated with vertebrates. As in the case of most animal-associated mollicutes, their discovery came in the course of studies directed primarily to disease etiology. In *Homo sapiens*, they cause nongonococcal urethritis [127] and can be primary pathogens in immunocompromised patients [128]. Putative roles for these organisms are also being considered in endometriosis, amnionitis, postpartum fever, infertility, and spontaneous abortion [126]. Other vertebrate hosts have been identified, including the usual range of domestic animals (cattle, goats, cats, dogs), minks, racoon dogs, and camels, several domesticated bird species, and a variety of primates such as squirrel monkeys, talapoin, patras and cynomolgus monkeys, chimpanzee, marmoset, and green and rhesus monkeys [129]. This list, of course, betrays an almost exclusive bias for search in domesticated animals or primates, and little effort has been expended in examination of wild hosts, a bias that extends to all vertebrate-associated mollicutes.

Ureaplasmas appear to be primarily or exclusively residents of the urogenital tract, a habitat in which there is intense selective pressure for use of a dominant substrate – urea. Although the energy-yielding metabolic pathways of ureaplasmas are largely unknown, it appears that these organisms are not fermentative [108]. As with other mollicutes, they appear to lack cytochromes and, while possessing some TCA cycle enzymes, lack the intact TCA cycle [130]. Because urea is used in substrate quantities, the metabolic pathways by which it is metabolized are probably idiosyncratic [108] and will have to be studied on their own terms.

The ureaplasmas illustrate an important principle of mollicute diversity: that is, that some species track a superabundant substrate, whose exploitation permits occupation of a unique microbial niche. The uniqueness of this niche is demonstrated by the ability of other mollicutes such as *Mycoplasma hominis* to colonize the same tract and achieve the same titer without utilizing urea [131]. *Ureaplasma urealyticum* and *M. hominis* frequently coexist in the same tract with no suggestion of competitive exclusion.

Acholeplasma. In the genus *Acholeplasma*, the genome is twice that of *Mycoplasma*. One would expect such organisms, while not being as self-sufficient as their gram-positive bacterial ancestors, to have greater biosynthetic capabilities than the genetically simpler mycoplasmas. Evidence for this expectation, in terms of metabolic pathways, is now accumulating (D. Pollack, pers. comm.).

Unlike the situation in the genus *Mycoplasma*, diversification in *Acholeplasma* appears to be more metabolic than structural, and may be associated with dispersal into a wide variety of vertebrate/invertebrate hosts. For many years, because acholeplasmas had been isolated frequently from domestic animals [132], it had been assumed that their hosts were largely vertebrates. Within the last decade, however, this perception changed radically [120, 133, 134] with the discovery of acholeplasmas on plant surfaces. Subsequently, the plant association of acholeplasmas was explained by their discovery in many species of the four major holometabolous insect orders (Fig. 2). Acholeplasmas appear to be transmitted in relatively simple cycles from plant surface to insect gut to excrement to plant surface. Because this cycle is presumed to be primitive, the evolutionary date of its appearance could provide an estimate for the emergence of acholeplasmas. The existence of leaf-chewing insects from the early Permian era 275 million years ago is documented by fossil evidence [136].

Anaeroplasma and Asteroleplasma. Bovine and ovine rumens contain, in addition to many other microorganisms, a group of obligately anaerobic (oxygen-sensitive) wall-less prokaryotes [137]. This assemblage of orga-

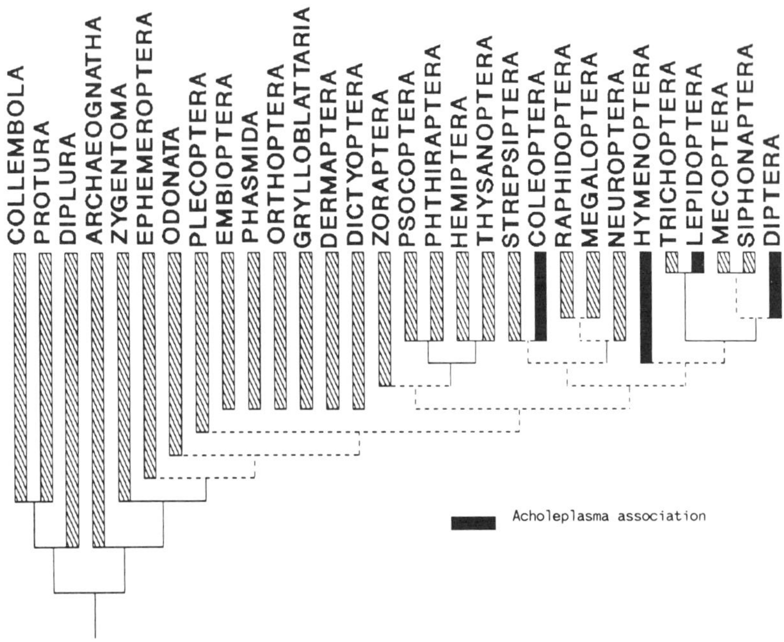

Figure 2. Acholeplasma-insect associations. Orders from which acholeplasmas have been isolated are indicated by solid black. Phylogenetic scheme from Kristensen [135].

nisms is remarkably diverse. Some species require sterols; others do not. Also, the G + C content of DNA ranges from 29 to 40. At present, five species are recognized; two genera (*Anaeroplasma*, in which sterols are required and *Asteroleplasma*, in which they are not required) and a new order (*Anaeroplasmatales*) have been created [138].

The affinities of *Anaeroplasma* are clearly indicated by properties of its highly conserved rRNA [58]. The genome size of 1×10^9 daltons suggests that this genus is primitive, an indication confirmed by the length of the 5S rRNA and by analysis of the 16S rRNA, which places it close to acholeplasmas on the mollicute tree [58].

If anaeroplasmas are truly primitive, their metabolic pathways, particularly those most closely associated with anaerobiosis, should have affinities with those of related clostridial species, such as *C. innocuum* or *C. ramosum*. Such affinities should be explored experimentally. A search for anaeroplasmas in insect guts also is clearly warranted.

Spiroplasma. These considerations bring us to the genus *Spiroplasma*. This genus, discovered a mere 17 years ago [139], could (arguably) be the largest genus on Earth. Although regarded at the time of their discovery as plant pathogens [139], spiroplasmas were subsequently found [14–16] to be associated principally – like acholeplasmas – with insects (Fig. 3).

The Spiroplasma phenotype. Spiroplasmas differ from *Mycoplasma* species in several profound ways, including retention of the ancestral genome size of 1×10^9 daltons. The most obvious difference is in their helicity and motility, which, unlike the similarly sized spirochetes, is accomplished without cell wall or axial filament [140]. But even within the general plan of helicity in spiroplasmas, there is room for specialization. Morphological diversity is apparent in the number of coils per organism, diameter of the organism or coil, and tightness of coil wrapping. And in contrast to other spiroplasmas, which have helices with coils of uniform amplitude, the Colorado potato beetle spiroplasma (CPBS) has coils of decreasing amplitude [15, 16], enabling remarkable translational motility. Further, in the host gut, the spiral cells of the CPBS become compressed into a single plane. These forms become coinlike and inert, requiring exposure to fresh media to reactivate the motile spiral form. This spiroplasma, unlike most other currently known cultivable species, appears to be a specialist, infecting only the Colorado potato beetle [15, 16].

Although morphological adaptations in *Spiroplasma* attributable to

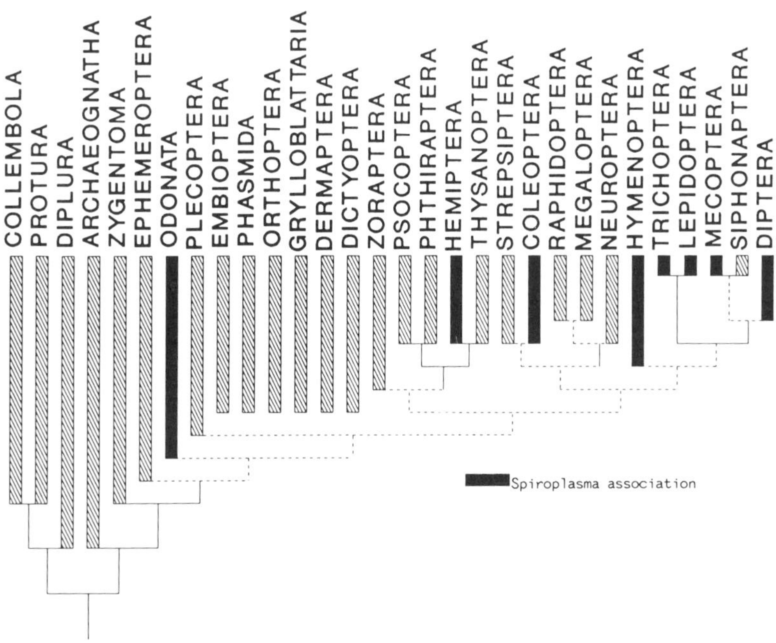

Figure 3. Spiroplasma-insect associations. Orders from which spiroplasmas have been isolated are indicated by solid black. Phylogenetic scheme from Kristensen [135].

specialization are uncommon, nutritional diversification may be universal. The CPBS and the *Drosophila* sex-ratio spiroplasma, both highly specialized to coexist within a single host, proved to be extremely fastidious and escaped cultivation for many years. Although these organisms have recently been cultivated [141, 142], and some of the factors involved in the narrow breadth of their trophic niche have been determined [143], we are only beginning to understand the underlying mechanisms that drive their nutritional specialization(s).

Spiroplasmas, in their differential utilization of substrates or in their choice of intrahost microhabitats, are no less diverse than members of the genus *Mycoplasma*. Translational motility may facilitate their passage of insect gut barriers, allowing them greater access to hemolymph, in which many can establish persistent infections [16]. Motility and capacity for hemocoel invasion also may enhance intracellular exploitation of their hosts, an event that could increase their pathogenic potential in the short term (contemporary time scale) but which could be a step in development of a symbiotic relationship in the long term (evolutionary time scale).

Such symbiosis is suspected in the case of *Drosophila* sex-ratio spiroplasmas [144]. These organisms exist, apparently stably, in a minority of individuals in their host population in nature. They were at first mistaken for a genetic factor because they are lethal to the male progeny. Since they are transovarially transmitted in the flies, it is presumed that they confer a selective advantage to their hosts [144].

Taxonomic diversity in Spiroplasma. Taxonomic diversity of spiroplasmas may be among the highest in the microbial world. Although the *Spiroplasma* inventory now includes only 10 described species [145, 146], twenty additional groups (putative species) or subgroups have been partially characterized [146]. But this existing inventory is probably only the tip of an iceberg. Initially, it was typical to find one new spiroplasma species in every ten new species of insects surveyed [16]. Excluding tabanid flies, overall isolation rates of about 6% have been obtained from holometabolous insect species [147]. In tabanid flies, isolation rates are nearly 50% of all individual flies examined, representing (so far) at least seven spiroplasma species or groups [148]. Given estimates that there may be more than 10 million insect species [149], these figures suggest that spiroplasma species richness may be legion. Recently, an ad hoc committee of the Subcommittee on Taxonomy of *Mollicutes* [150] proposed criteria for establishment of a group classification for spiroplasmas. Considering the probable species richness of this genus, such a group classification could remain, for many years at least, the effective official classification of the group.

Clonality. Issues concerning the clonality of microbes attain sharp focus in the genus *Spiroplasma*. In this genus, not only is there a seemingly endless array of species, but within many species it is common to find clusters of clones that show a high degree of intercluster heterogeneity [61, 146].

In group I spiroplasmas, the clones can be grouped into eight clusters of partially related subgroups [151]. Genomic differences among these subgroups are of a magnitude that permits them to be regarded as species or putative species. Four of these, all associated with plant or insect disease, have in fact been given species epithets [145, 152–154].

A less tractable array of clones is that of *Spiroplasma apis* (group IV). This species, as currently constituted, is represented by a large assemblage of isolates that differ markedly in their reciprocal serological cross-reactivities [155]. This was confirmed by 2-dimensional gel electrophoretic patterns, which, like the serological patterns, indicated substantial differences [61]. Limited DNA-DNA hybridization tests [61] suggest that the underlying genomic differences that generate the observed phenotypic diversity in group IV spiroplasmas are insufficient to warrant assignment of new species. At the time the group IV studies were performed, it was not known whether this case history would prove to be unusual. Today, however, there are substantial indications that diffuse patterns of clonal diversity may be extremely common, not only in spiroplasmas, but in mollicutes in general. For example, serological analyses of group XVI spiroplasmas [156] have suggested the existence of a diffuse clonal assemblage, a pattern that has been confirmed in a recent detailed study (M. Abalain-Colloc, pers. comm.). And in our own serological studies of spiroplasmas associated with tabanid flies [157, 158], we have found similar clone assemblages in group XXIII spiroplasmas. In these cases, the serological differences may be of comparable magnitude to those found in group IV.

Clone assemblages have been found in three other mollicute genera. DNA-DNA hybridization studies indicated that *Acholeplasma axanthum*, a species that is often isolated from plant surfaces and that is therefore suspected to be associated with insects, exists as a large array of partially related strains that form a homology continuum with no obvious gaps that would permit designation of subgroups [159]. Similarly, strains of *Mycoplasma hominis*, a colonist of the human urogenital tract, show relationships reflected by DNA-DNA homologies as low as 50%, with other values distributed continuously from the lowest level to levels indicating near identity [160]. And *Ureaplasma urealyticum*, which also colonizes the human urogenital tract, not only contains 14 serovars but is actually a complex of two discrete clone clusters [161]. These clusters show a pattern of partial DNA-DNA relatedness [40–50%] that may permit eventual designation of a second species [162].

There are exceptions to the apparent rule of clonal clustering. One of these is *Mycoplasma pneumoniae*, which shows tight patterns of clonality when examined by restriction enzyme analysis [163]. And isolates of group IV spiroplasmas from diseased honey bees (May Disease) in France [164], of the pathogen *Spiroplasma kunkelii* (agent of corn stunt disease) [153], and *Spiroplasma melliferum* (an agent of honeybee disease) [152] in the U.S., show very limited serological or electrophoretic variability.

These observations, which suggest that organisms from a single niche in a single host may be limited in their clonal diversity, strongly implicate host association as a selective force. Specialist microorganisms are subject to intense intraspecific pressure. As rapidly as mutants may arise in a host-specific organism, they must immediately compete with the wild–type clone from which they have been derived. If they manage to escape by transmission of a single mutant cell or clone in the absence of wild-type cells, they would still be subject to pressure from back-mutation. Under such conditions, there may be little opportunity for emergence of new clones.

This hypothesis can be accepted only if convincing explanations can be provided for the comparatively large clonal diversity of *M. hominis* [160] and *U. urealyticum* [161, 162]. We believe that fundamental differences in the transmission frequency, and perhaps especially intrahost site specificity, may explain clonal diversity of urogenital mollicutes. In their human host, these organisms colonize both urethral and vaginal tissues [126]. Moreover, even in the same individual, their infection is not limited to a single tissue [165]. No site-specific attachment mechanisms are known. And it is known that different strains have widely differing abilities to colonize various target tissues [166]. Within the host, these generalists are noted for persistence, even in the face of immunological challenge. It is easy to imagine a mutant of either of these genital mollicutes surviving by escaping to a new target tissue where its fitness would be greater than that of the wild-type. Transmission of this mutant to a new host in the absence of wild-type organisms would in effect spawn a new microbial clone, free to evolve in its own way. Whether or not this specific hypothesis of clonality is eventually substantiated, we feel that an understanding of clonality, and evolution of the mollicutes in general, depends strongly on understanding, in a quantitative way, the interactions of transmission dynamics, host colonization, and inter- and intraspecific competition. The importance of such factors and extrinsic ecological selective mechanisms renders searches for mechanisms of prokaryote evolution in sole terms of intrinsic genetic mechanisms predictably futile.

Determinants of Spiroplasma diversity. Other factors limit the species richness of spiroplasmas. One of these may be geographical (158). For example, preliminary data suggest that the number of spiroplasmas isolated in warm climates may be greater than in cool climates. Also, the number of spiroplasmas isolated in dry climates may be greater than that in wet climates. At a very superficial level, this may be explained by the possible adverse effects of precipitation on organisms that are transmitted on plant surfaces. However, transmission also may be favored in dry climates by a strong synchronization of floral blooms that attract large assemblages of flower-visiting insects (16). Even in most temperate climates, spiroplasma maintenance may be linked to seasonal flowering [167]. Colonization of seasonally synchronized floral blooms requires an interbloom reservoir, which for most mollicutes is almost certainly long-lived insect(s). At warmer latitudes, the life history strategy of some insect taxa (e.g., sap-sucking insects) shifts from an obligate egg diapause strategy – which is excellent for escaping cold winters – to a facultative adult diapause whose onset is determined by many factors, including temperature (degree-days), or, in the tropics, timing of wet/dry cycles. Because few spiroplasmas are known to be vertically transmitted, at northern latitudes the increased frequency of egg diapause may serve as a barrier to spiroplasma maintenance in such taxa. Of course, diapausing immature or adult insects can serve as overwintering spiroplasma hosts at any latitude. For example, wild bees that have an adult diapause organized by the bloom of their host flower have been found to be infected with spiroplasmas before they emerge from their overwintering cells [16]. Also, *Cotinus* (Coleoptera) larvae apparently carry group IX spiroplasma overwinter in the soil in Maryland [168].

Linkage of spiroplasma diversity to the nature of host reservoirs introduces another factor. Specialization, in an evolutionary context, is density-dependent. Dominant hosts attract specialists. For all the reasons outlined by MacArthur and Wilson [169] and later applied to insect evolution [170], host plant species that have been abundant in the Quaternary have accumulated specialists [171]. Actually, *contemporary* host plant dominance is one of the most powerful determinants of insect species diversity [172]. This relationship probably also governs spiroplasma occurrence; in general, spiroplasmas are often associated with dominant insects (including pests) – especially those that aggregate at some time in their life history. The net effect of dominance and gregariousness probably increases the microorganismal transmission rate over time.

Finally, there are evolutionary constraints on diversity. Spiroplasmas appear to be associated only with more recently evolved insect orders [103, 147; see also Fig. 3]. They have been searched for but not found in such primitive insect orders as Plecoptera, Ephemeroptera, and Orthoptera. In the single primitive order in which they have been found (Odonata), they may represent only a transient presence that could be explained simply by ingestion of infected prey.

Genotypic diversity

In Mollicutes, genetic diversity is presumably limited by the organisms' simplicity. Overall reduction in genome size has been accomplished both by gene deletion, or by reduction in operon size by reduction of numbers of redundant genes.

The mechanisms by which simplification has been achieved in Mollicutes may be of special interest to microbial geneticists. For example, in *Mycoplasma capricolum* the 'universal' stop codon UGA is read as tryptophan [173]. This codon usage, which has also been documented for *M. pneumoniae* [174], *S. citri* [61], and *S. melliferum* [61], may block expression of mollicute genes in recombinant bacterial clones. For example, the cloned gene for spiralin, the dominant membrane protein of *S. citri*, was fully expressed in *E. coli*, whereas the replicative form (RF) of the SpV-4 virus of *S. melliferum* was not. This result is readily explained by the lack of tryptophan in the spiralin molecule. This unusual codon usage, which was not observed in acholeplasmas [175], appears to be a significant apomorphy (or possibly even synapomorphy) that defines the *Spiroplasma-Mycoplasma* clade of the Mollicutes. Presumably, this usage, which has also been reported for various mitochondria [176–178], has arisen at least twice, since mitochondria, which appear to be most closely allied to gram-negative bacteria [106], do not appear to be members of any clade of the Mollicutes.

Study of gene modification and operon simplification may also lead to an understanding of bacterial evolution. For example, Hu *et al.* [174] recently studied codon usage in *M. pneumoniae*. They found that all 61 possible sense codons were utilized. These data contrasted with reported codon usage of ribosomal protein genes of *M. capricolum* [179], in which 13 of the 61 possible codons were not used. Also, in *M. pneumoniae*, there was no evidence of preferential base utilization in any position of 3086 codons analyzed. Thus, although AT-biased mutation pressure may have been a factor in gene evolution of *M. capricolum*, the origin of the $tRNA_{uca}$ gene in *M. pneumoniae* appeared to be unrelated to such pressure. Instead, it was proposed [174] that replacement of $tRNA_{cca}$ with $tRNA_{uca}$ genes in *M. pneumoniae* reflected a simplification of the codon reading system. Such simplification has also been proposed for *M. mycoides* [180] and for mitochondria.

In spiroplasmas, there is also evidence of genomic simplification. Both rRNA and tRNA operons are reduced in number in comparison to those of their eubacterial progenitors [61]. In the case of *S. melliferum*, there are 10 clustered tRNA genes [63]. The sequence upstream of the tRNA gene cluster shows a sequence very similar to that of eubacterial RNA polymerase promoters [62]. The sequence 3′-flanking to the tRNA gene cluster shows regions with dyad symmetry that can fold into stem and loop structures such as those involved in transcription termination [62]. If these sequences do indeed function as promoters and terminators, the gene cluster represents a single operon. This would contrast with the condition in *Bacillus subtilis*, in which 21 tRNA genes are organized into an operon distal from the rRNA gene set [181].

As much as studies of this kind may contribute to an understanding of the molecular basis for prokaryote (and therefore mollicute) evolution, they do not point to any special enabling mechanisms that enhance diversity. The explanation for high mollicute diversity must, presumably, be sought elsewhere.

Conclusion

The evidence that we have summarized enables us to address our riddles. In the spectacularly speciose genus *Spiroplasma*, diversity is determined by a combination of nutritional specialization, microhabitat specialization, and ecological factors that determine host ecology on both regional and local levels. But spiroplasma diversity also reflects evolutionary history, and particularly, the explosive rise of arthropods 300 million years ago. The increase in complexity and number of niches accompanying this event provided a statistically and qualitatively new world for microbes, a world that even today must be full of vacant niches awaiting innovative microbial strategies to exploit them. From a historical perspective, the origin of *Mollicutes* as host-dependent organisms led to the emergence of the spiroplasmas. Thus, the answer to one riddle is the answer to the other as well. The diversity of spiroplasmas and of mollicutes in general, as a result of their obligate or nearly obligate host dependence, is determined primarily by the spatio–temporal diversity of their hosts, both on contemporary and evolutionary time scales.

Acknowledgements

We acknowledge comments and critique of the manuscript by J. G. Tully. D. E. Lynn assisted with the graphics. Supported in part by BARD agreement US 1030-85 and USDA Competitive Grant 87-CRCR-1-2372.

References

1. Savage, D. C. 1977. Microbial ecology of the gastrointestinal tract. Annu. Rev. Microbiol. 31: 107–133.
2. Margulis, L. 1981. Symbiosis in Cell Evolution. Freeman and Company, San Francisco.
3. Ebringer, L. and J. Krajcovic. 1987. Prokaryotic character of choloroplasts and mitochondria–the present knowledge. Folia Microbiol. 32: 244–280.
4. Gibbons, N. E. and R. G. E. Murray. 1978. Proposals concerning the higher taxa of bacteria. Int. J. Syst. Bacteriol. 28: 1–6.
5. Murray, R. G. E. 1984. The higher taxa, or a place for everything? Pp. 31–34. In: Bergey's Manual of Systematic Bacteriology, vol. 1. Ninth Edition. N. R. Krieg and J. G. Holt (eds.). Williams and Wilkins, Baltimore.
6. Dobzhansky, T. 1937. Genetics and the Origin of Species. Columbia University Press, New York.
7. Selander, R. K., D. A. Caugant, and T. S. Whittam. 1987. Genetic structure and variation in natural populations of *Escherichia coli*. Pp. 1625–1648. In: *Escherichia coli* and *Salmonella typhimurium*: Cellular and Molecular Biology, vol. 2, F. C. Weidhardt (ed.). American Society for Microbiology, Washington, D. C.
8. Reanney, D. C. 1977. Genetic engineering as an adaptive strategy. Brookhaven Symp. Biol. 29: 248–271.
9. Campbell, A. 1981. Evolutionary significance of accessory DNA elements in bacteria. Annu. Rev. Microbiol. 35: 55–83.
10. Ochman, H. and A. C. Wilson. 1987. Evolutionary history of enteric bacteria. Pp. 1649–1654. In: *Escherichia coli* and *Salmonella typhimurium*: Cellular and Molecular Biology, vol. 2, F. C. Weidhardt (ed.). American Society for Microbiology, Washington, D. C.
11. Johnson, J. L. 1984. Nucleic acids in bacterial classification. Pp. 8–11. In: Bergey's Manual of Systematic Bacteriology, vol. 1, N. R. Krieg and J. G. Holt (eds.). Williams and Wilkins, Baltimore.
12. Wayne, L. G., D. J. Brenner, R. R. Colwell, P. A. D. Grimont, O. Kandler, M. I. Krichevsky, L. H. Moore, W. E. C. Moore, R. G. E. Murray, E. Stackebrandt, M. P. Starr, and H. G. Trüper. 1987. Report of the ad hoc committee on reconciliation of approaches to bacterial systematics. Int. J. Syst. Bacteriol. 37: 463–464.
13. Skerman, V. B. D., V. McGowan, and P. H. A. Sneath. 1980. Approved lists of bacterial names. Int. J. Syst. Bacteriol. 30: 225–420.
14. Clark, T. B. 1977. *Spiroplasma* sp., a new pathogen in honey bees. J. Invertebr. Pathol. 29: 112–113.
15. Clark, T. B. 1982. Spiroplasmas: diversity of arthropod reservoirs and host-parasite relationships. Science 217: 57–59.
16. Hackett, K. J. and T. B. Clark. 1989. The ecology of spiroplasmas. Pp. 113–199. In: The Mycoplasmas, vol. V, R. F. Whitcomb and J. G. Tully (eds). Academic Press, New York.
17. Starr, M. P. and J. M. Schmidt. 1981. Prokaryote diversity. Pp. 3–42. In: The Prokaryotes, vol. II, M. P. Starr, H. Stolp, H. G. Trüper, A. Balows, and H. G. Schlegel (eds.). Springer-Verlag, Berlin.
18. Morowitz, H. J. 1967. Biological self-replicating systems. Progr. Theor. Biol. 1: 35–58.
19. Pirie, N. W. 1973. On being the right size. Annu. Rev. Microbiol. 27: 119–132.
20. Srere, P. A. 1987. Complexes of sequential metabolic enzymes. Annu. Rev. Biochem. 56: 89–124.
21. Welch, G. R. and T. Keleti. 1987. Is cell metabolism controlled by a 'molecular democracy' or by a 'supramolecular socialism'? Trends Biol. Sci. 12: 216–217.

22. Daniels, M. J. and J. M. Longland. 1984. Chemotactic behavior of spiroplasmas. Curr. Microbiol. 10: 191–193.
23. Daniels, M. J., J.M. Longland and J. Gilbart. 1980. Aspects of motility and chemotaxis in spiroplasmas. J. Gen. Microbiol. 118: 429–436.
24. Baker, D. A. and R. W. A. Park. 1975. Changes in morphology and cell wall structure that occur during growth of *Vibrio* sp. NCTC 4716 in batch culture. J. Gen. Microbiol. 86: 12–28.
25. Koch, A. L. 1985. How bacteria grow and divide in spite of internal hydrostatic pressure. Can. J. Microbiol. 31: 1071–1084.
26. Davey, R. B. and D. C. Reanney. 1980. Extrachromosomal genetic elements and the adaptive evolution of bacteria. Evol. Biol. 13: 113–147.
27. Novick, R. P. 1980. Plasmids. Sci. Am. 243: 102–123.
28. Milkman, R. 1973. Electrophoretic variation in *Escherichia coli* from natural sources. Science 182: 1024–1026.
29. Milkman, R. 1975. Allozyme variation in *E. coli* of diverse natural origins. Pp. 273–285. In: Isozymes, vol. IV, C. L. Markert (ed.). Academic Press, New York.
30. Nevo, E., A. Beiles, and R. Ben-Shlomo. 1984. The evolutionary significance of genetic diversity: ecological, demographic, and life history correlates. Pp. 13–213. In: Evolutionary Dynamics of Genetic Diversity. Lecture Notes in Biomathematics, vol. 53, G. S. Mani (ed.). Springer-Verlag, Berlin.
31. Neel, J. V. 1984. A revised estimate of the amount of genetic variation in human proteins: Implications for the distribution of DNA polymorphisms. Am. J. Human Genet. 36: 1135–1148.
32. Ohta, T. 1982. Linkage disequilibrium with the island model. Genetics 101: 139–155.
33. Ohta, T. 1982b. Linkage disequilibrium due to random drift in finite subdivided populations. Proc. Natl. Acad. Sci. (USA) 79: 1940–1944.
34. Maruyama, T. and M. Kimura. 1980. Genetic variability and effective population size when local extinction and recolonization of subpopulations are frequent. Proc. Natl. Acad. Sci. USA 77: 6710–6714.
35. Ørskov, F., I. Ørskov, Jr., D. J. Evans, R. B. Sack, D. A. Sack, and T. Wadstrom. 1976. Special *Escherichia coli* serotypes among enterotoxigenic strains from diarrhea in adults and children. Med. Microbiol. Immunol. 162: 73–80.
36. Selander, R. K. and B. R. Levin. 1980. Genetic diversity and structure in *Escherichia coli* populations. Science 210: 545–547.
37. Riley, M. 1985. Discontinuous processes in the evolution of the bacterial genome. Evol. Biol. 19: 1–31.
38. Riley, M. and A. Anilionis. 1978. Evolution of the bacterial genome. Annu. Rev. Microbiol. 32: 519–560.
39. Levin, B. R. 1981. Periodic selection, infectious gene exchange, and the genetic structure of *E. coli* populations. Genetics 99: 1–23.
40. Lewontin, R. C. 1985. Population genetics. Annu. Rev. Genet. 19: 81–102.
41. Dawkins, R. 1976. The Selfish Gene. Oxford University Press, New York.
42. Orgel, L. E. and F. H. C. Crick. 1980. Selfish DNA: The ultimate parasite. Nature 284: 604–607.
43. Doolittle, W. F. and C. Sapienza. 1980. Selfish genes, the phenotype paradigm and genome evolution. Nature 284: 601–603.
44. Nur, U., J. H. Werren, D. G. Eickbush, W. D. Burke, and T. H. Eickbush. A 'selfish' B chromosome that enhances its transmission by eliminating the paternal genome. 1988. Science 240: 512–514.
45. Smith, J. M. 1987. Darwinism stays unpunctured. Nature 330: 516.

46. Smith, J. M. 1988. Punctuation in perspective. Nature 332: 311–312.
47. Gould, S. J. 1988. Punctuated equilibrium prevails. Nature 332: 211–212.
48. Eldredge, N. and S. J. Gould. 1972. Punctuated equilibria: An alternative to phyletic gradualism. Pp. 82–115. In: Models in Paleobiology, T. J. Schopf (ed). Freeman, San Francisco.
49. Darwin, C. 1859. On the Origins of Species. John Murray, London.
50. Reader, J. 1986. The rise of life: The first 3. 5 billion years. Alfred A. Knopf, New York.
51. Milkman, R. 1985. Two elements of a unified theory of population genetics and molecular evolution. Pp. 65–83. In: Population Genetics and Molecular Evolution. T. Ohta and K. Aoki (eds.). Japan Scientific Societies Press/Springer-Verlag, Tokyo and Berlin.
52. Lapage, S. P., P. H. A. Sneath, E. F. Lessel, V. B. D. Skerman, and H. P. R. Seeliger. 1975. International code of nomenclature of bacteria, 1976 revision. American Society for Microbiology, Washington, DC.
53. Woese, C. R. 1987. Bacterial evolution. Microbiol. Rev.51: 221–271.
54. Maniloff, J. 1983. Evolution of wall-less prokaryotes. Annu. Rev. Microbiol. 37: 477–499.
55. Woese, C. R., J. Maniloff, and L. B. Zablen. 1980. Phylogenetic analysis of the mycoplasmas. Proc. Natl. Acad. Sci. (USA) 77: 494–498.
56. Laigret, F., O. Grau, and J. M. Bové. 1989. Comparison of 16S rDNA sequences of various mollicutes. Art. Zbl. Bakt., I, Orig. Hyg. A (in press).
57. Walker, R. T., E. T. J. Chelton, M. W. Kilpatrick, M. J. Rogers, and J. Simmons. 1982. The nucleotide sequence of the 5S rRNA from *Spiroplasma* species BC-$_3$ and *Mycoplasma mycoides* sp. *capri* PG-3. Nucl. Acid Res. 10: 6363–6367.
58. Rogers, M. J., J. Simmons, R. T. Walker, W. G. Weisburg, C. R. Woese, R. S. Tanner, I. M. Robinson, D. A. Stahl, G. Olsen, R. H. Leach, and J. Maniloff. 1985. Construction of the mycoplasma evolutionary tree from 5S rDNA sequence data. Proc. Natl. Acad. Sci. (USA) 82: 1160–1164.
59. Hori, H. and S. Osawa. 1986. Evolutionary change in 5S rRNA secondary structure and a phylogenetic tree of 352 5S rRNA species. Biosystems 19: 163–172.
60. Dams, E., E. Huysmans, A. Vandenberghe, and R. De Wachter. 1987. Structure of clostridial 5S ribosomal RNAs and bacterial evolution. Syst. Appl. Microbiol. 9: 54–61.
61. Bové, J. M., P. Carle, M. Garnier, J. Renaudin, and C. Saillard. 1989. Molecular and cell biology of spiroplasmas. Pp. 243–364. In: The Mycoplasmas, vol. V., R. F. Whitcomb and J. G. Tully (eds.). Academic Press, New York.
62. Rogers, M. J., A. A. Steinmetz, and R. T. Walker. 1986. The nucleotide sequence of a tRNA gene cluster from *Spiroplasma melliferum*. Nucl. Acid Res. 14: 3145.
63. Rogers, M. J., A. A. Steinmetz, and R. T. Walker. 1987. Organization and structure of the tRNA genes in *Spiroplasma melliferum*. Isr. J. Med. Sci. 23: 357–360.
64. Fox, G. E., E. Stackebrandt, R. B. Hespell, J. Gibson, J. Maniloff, T. J. Dyer, R. S. Wolfe, W. E. Balch, R. S. Tanner, L. J. Magrum, R. Blakemore, R. Gupta, L. Bonen, B. J. Lewis, D. A. Stahl, and C. R. Woese. 1980. The phylogeny of prokaryotes. Science 209: 457–463.
65. Pace, N. R., G. J. Olsen, and C. R. Woese. 1986. Ribosomal RNA phylogeny and the primary lines of evolutionary descent. Cell 45: 325–326.
66. Achenbach-Richter, L., K. O. Stetter, and C. R. Woese. 1987. A possible link among archaebacteria. Nature 327: 348–349.
67. Achenbach-Richter, L., R. Gupta, K. O. Stetter, and C. R. Woese. 1987. Were the original eubacteria thermophiles? Syst. Appl. Microbiol. 9: 34–39.
68. Searcy, D. G. 1986. Some features of thermo-acidophilic archaebacteria preadaptive for the evolution of eukaryotic cells. Syst. Appl. Microbiol. 7: 198–201.

69. Bains, W. 1986. Evolutionary paradoxes and natural non-selection. Trends Biol. Sci. 12: 90–91.
70. Wallace, D. C. and H. J. Morowitz. 1973. Genome size and evolution. Chromosoma 40: 121–126.
71. Fridovich, I. 1975. Oxygen: boon and bane. Am. Sci. 63: 54–59.
72. Olson, J. M. 1978. Precambrian evolution of photosynthetic and respiratory organisms. Evol. Biol. 11: 1–37.
73. Schwartz, R. M. and M. O. Dayhoff. 1978. Origins of prokaryotes, eukaryotes, mitochondria, and chloroplasts. Science 199: 395–403.
74. Gottschalk, G. 1979. Bacterial Metabolism. Springer-Verlag, New York.
75. Lake, J. A., M. W. Clark, E. Henderson, S. P. Fay, M. Oakes, A. Scheinman, J. P. Thornber, and R. A. Mah. 1985. Eubacteria, halobacteria, and the origin of photosynthesis: The photocytes. Proc. Natl. Acad. Sci. (USA) 82: 3716–3720.
76. Prosser, C. L. 1986. Adaptational Biology: Molecules to Organisms. John Wiley and Sons, New York.
77. Ochman, H. and A. C. Wilson. 1987. Evolution in bacteria: evidence for a universal substitution rate in cellular genomes. J. Mol. Evol. 26: 74–86.
78. Vossbrinck, C. R., J. V. Maddox, S. Friedman, B. A. Debrunner-Vossbrinck, and C. R. Woese. 1987. Ribosomal RNA sequence suggests microsporidia are extremely ancient eukaryotes. Nature 326: 411–414.
79. Cavalier-Smith, T. 1987. Eukaryotes with no mitochondria. Nature 326: 332–333.
80. Atlas, R. M. and R. Bartha. 1981. Microbial ecology: Fundamentals and Applications. Addison-Wesley, London.
81. Schlegel, H. G. and H. W. Jannasch. 1981. Prokaryotes and their habitats. Pp. 43–82. In: The Prokaryotes, vol. II, M. P. Starr, H. Stolp, H. G. Trüper, A. Balows, and H. G. Schlegel (eds.). Springer-Verlag, Berlin.
82. Ratledge, C. 1976. The physiology of the mycobacteria. Pp. 115–244. In: Advances in Microbial Physiology, vol. 13, A. H. Rose and D. W. Tempest (eds.). Academic Press, New York.
83. Condon, S. 1987. Responses of lactic acid bacteria to oxygen. FEMS Microbiol. Rev. 46: 269–280.
84. Fridovich, I. 1976. Oxygen radicals, hydrogen peroxide, and oxygen toxicity. Free radicals in biology, vol. 1, W. A. Pryor, (ed.). Academic Press, New York.
85. Darland, G., T. D. Brock, W. Samsonoff, and S. F. Conti. 1970. A thermophilic acidophilic mycoplasma isolated from a coal refuse pile. Science 170: 1416–1418.
86. Woese, C. R. and R. S. Wolfe. 1985. The Bacteria, vol. 8. Archaebacteria. Academic Press, New York.
87. Souza, K. A. and P. H. Deal. 1977. Characterization of a novel extremely alkaline bacterium. J. Gen. Microbiol. 101: 103–109.
88. Natochin, Y. V. and R. G. Parnova. 1987. Osmolality and electrolyte concentration of hemolymph and the problem of ion and volume regulation of cells in higher insects. Comp. Biochem. Physiol. 88A: 563–570.
89. Brock, T. D. 1986. Notes on the ecology of thermophilic archaebacteria. Syst. Appl. Microbiol. 7: 213–215.
90. Luckinbill, L. S. 1978. r and K selection in experimental populations of *Escherichia coli*. Science 202: 1201–1203.
91. Preer, J. R., Jr. and L. B. Preer. 1984. Endosymbionts. Pp. 795–811. In: Bergey's Manual of Systematic Bacteriology, vol. 1, N. R. Krieg (ed.). Williams and Wilkins, Baltimore.
92. Razin, S. and M. F. Barile (eds.). 1985. The Mycoplasmas, Vol. IV. Academic Press, New York.

93. McGarrity, G. J. and D. L. Williamson. 1989. *Spiroplasma* pathogenicity *in vitro* and *in vivo*. Pp. 365–392. In: The Mycoplasmas, vol. V, R. F. Whitcomb and J. G. Tully (eds.). Academic Press, New York.

94. Bebear, C., J. Latrille, J. Fleck, B. Roy, and J. M. Bové. 1974. *Spiroplasma citri*: un mollicute. Colloq. Inst. Sante Rech. Med. 33: 35–42.

95. Razin, S. 1985. Mycoplasma adherence. Pp. 161–202. In: The Mycoplasmas, vol. IV, S. Razin and M. F. Barile (eds.). Academic Press, New York.

96. Manolukas, J. T., M. F. Barile, D. K. F. Chandler, and J. D. Pollack. 1988. Presence of anaplerotic reactions and transamination, and the absence of the tricarboxylic acid cycle in mollicutes. J. Gen. Microbiol. 134: 791–800.

97. Rodwell, A. W. and A. Mitchell. 1979. Nutrition, growth, and reproduction. Pp. 103–139. In: The Mycoplasmas, vol. I, M. F. Barile and S. Razin (eds.). Academic Press, New York.

98. Pollack, J. D. 1979. Respiratory pathways and energy-yielding mechanisms. Pp. 188–211. In: The Mycoplasmas, vol. I, M. F. Barile and S. Razin (eds.). Academic Press, New York.

99. Razin, S. and R. C. Cleverdon. 1965. Carotenoids and cholesterol in membranes of *Mycoplasma laidlawii*. J. Gen. Microbiol. 41: 409–415.

100. Tully, J. G., and S. Razin. 1968. Physiological and serological comparisons among strains of *Mycoplasma granularum* and *Mycoplasma laidlawii*. J. Bacteriol. 95: 1504–1512.

101. Chang, C. J. 1989. Nutrition and cultivation of spiroplasmas. Pp. 201–241. In: The Mycoplasmas, vol. V, R. F. Whitcomb and J. G. Tully (eds.). Academic Press, New York.

102. Smith, L. D. and G. Hobbs. 1974. Genus III. *Clostridium* Prazmowski 1880, 23. Pp. 551–572. In: Bergey's Manual of Determinative Bacteriology, 8th ed., R. E. Buchanan and N. E. Gibbons (eds.). Williams and Wilkins, Baltimore.

103. Hackett, K. J. 1989. Evolutionary considerations–arthropod and plant mollicutes. Art. Zbl. Bakt., I Orig. Hyg. A (in press).

104. Hackett, K. J. 1989. Adaptational biology and spiroplasmas. Art. Zbl. Bakt., I Orig. Hyg. A (in press).

105. Neimark, H. 1986. Origins and evolution of wall-less prokaryotes. Pp. 21–42. In: The Bacterial L-forms, S. Madoff (ed.). Dekker, New York.

106. Razin, S. 1989. Molecular approach to mycoplasma phylogeny. Pp. 33–69, In: The Mycoplasmas, vol. V, R. F. Whitcomb and J. G. Tully (eds.), Academic Press, New York.

107. Amikam, D., G. Glaser, and S. Razin. 1984. Mycoplasmas (mollicutes) have a low number of rRNA genes. J. Bacteriol. 158: 376–378.

108. Pollack, J. D. 1986. Metabolic distinctiveness of ureaplasmas. Pediatr. Inf. Dis. 5: S305–S307.

109. Pollack, J. D., M. V. Williams, M. C. McElwain, D. DeSantis, J. Manolukas, C. J. Chang, E. A. Clark, R. F. Whitcomb, and K. J. Hackett. 1988. The metabolism of spiroplasmas. Abstr. 7th. Congr. Int. Org. Mycoplasmol., Vienna, 1988. Abstr. No. 19.

110. Pyle, L. E., L. N. Corcoran, B. G. Cocks, A. D. Bergemann, J. C. Whitley, and L. R. Finch. 1988. Pulse-field electrophoresis indicates larger-than-expected sizes for mycoplasma genomes. Nucl. Acids Res. 16: 6015–6025.

111. Razin, S. and E. A. Freundt. 1984. The mycoplasmas. Pp. 742–770. In: Bergey's Manual of Systematic Bacteriology, vol. 1, N. R. Krieg and J. G. Holt (eds.). Williams and Wilkins, Baltimore.

112. Sjostrom, K. E., K. C. S. Chen, and G. E. Kenny. 1986. Detection of end products of the arginine dihydrolase pathway in both fermentative and nonfermentative *Mycoplasma* species. Int. J. Syst. Bacteriol. 36: 60–65.

113. McGarrity, G. J., V. Vanaman, and J. Sarama. 1984. Cytogenetic effects of mycoplasmal infection of cell cultures: a review. In Vitro 20: 1–18.
114. McGarrity, G. J. and H. Kotani. 1985. Cell culture mycoplasmas. Pp. 353–390. In: The Mycoplasmas, vol. IV, S. Razin and M. F. Barile (eds.). Academic Press, New York.
115. Cassell, G. H., and B. C. Cole. 1981. Mycoplasmas as agents of human disease. New England J. Med. 304: 80–89.
116. Morrison-Plumber, J., D. H. Jones, K. Daly, J. G. Tully, D. Taylor-Robinson, and J. B. Baseman. 1987. Molecular characterization of *Mycoplasma genitalium* species-specific and cross-reactive determinants: Identification of an immunodominant protein of *M. genitalium*. Isr. J. Med. Sci. 23: 453–457.
117. Tully, J. G., and R. F. Whitcomb (eds.). 1979. The Mycoplasmas, Vol. II. Academic Press, New York.
118. Shifrine, M., K. P. Bailey, and S. S. Stone. 1972. Contagious bovine pleuropneumonia: Isolation of *Mycoplasma mycoides* var. *mycoides* from ticks collected from infected cattle and infection attempts using these ticks. Bull. Epizoot. Dis. Africa (Special Issue) 20: 43–45.
119. Cottew, G., and F. R. Yeats. 1982. Mycoplasmas and mites in the ears of clinically normal goats. Austr. J. Vet. Sci. 59: 77–81.
120. Tully, J. G. 1989. Class Mollicutes: New perspectives from plant and arthropod studies. Pp. 1–31. In: The Mycoplasmas, vol. V., R. F. Whitcomb and J. G. Tully (eds.). Academic Press, New York.
121. Neimark, H. and J. London. 1982. Origins of the mycoplasmas: Sterol-nonrequiring mycoplasmas evolved from streptococci. J. Bacteriol. 150: 1259–1265.
122. Neimark, H. 1982. Evolution of mycoplasmas and genome losses. Yale J. Biol. Med. 56: 377–383.
123. Saillard, C., J. C. Vignault, F. Bonnet, D. L. Rose, J. G. Tully, P. Carle, and J. M. Bové. Isr. J. Med. Sci. 23: 696–698.
124. Borrel, A., E. Dujardin-Beaumetz, Jeantet, and C. Jouan. 1910. Le microbe de la péripneumoniae. Ann. Inst. Pasteur (Paris) 24: 168–179.
125. Shepard, M. C., C. D. Lunceford, D. K. Ford, R. H. Purcell, D. Taylor-Robinson, S. Razin, and F. T. Black. 1974. *Ureaplasma urealyticum* gen. nov., sp. nov.: Proposed nomenclature for the human T (T-strain) mycoplasmas. Int. J. Syst. Bacteriol. 24: 160–171.
126. Cassell, G. H., W. A. Clyde, Jr., G. E. Kenny, W. M. MacCormack, and D. Taylor-Robinson (eds.). 1986. Ureaplasmas of humans: with emphasis upon maternal and neonatal infection. Pediatr. Infect. Dis. 5: S221–S354.
127. Taylor-Robinson, D. 1985. Mycoplasmal and mixed infections of the human male urogenital tract and their possible complications. Pp. 27–63. In: The Mycoplasmas, vol. IV. S. Razin and M. F. Barile (eds.). Academic Press, New York.
128. Taylor-Robinson, D., P. M. Furr, A. D. B. Webster. 1986. *Ureaplasma urealyticum* in the immunocompromised patient. Pediatr. Infect. Dis. 5: S236–S238.
129. Tully, J. G., and D. Taylor-Robinson. 1986. Taxonomy and host distribution of the ureaplasmas. Pediatr. Infect. Dis. 5: S292–S295.
130. Davis, J. M., Jr., J. Manolukas, G. Capo, J. Espinoza and J. D. Pollack. 1988. Pyruvate metabolism and the absence of a TCA cycle in *Ureaplasma urealyticum*. 7th Congr. Int. Org. Mycoplasmol., Vienna, Austria, 1988. Abstr. P42.
131. Taylor-Robinson, D., and McCormack, W. M. 1979. Mycoplasmas in human genitourinary infections. Pp. 307–366. In: The Mycoplasmas, vol. II, J. G. Tully and R. F. Whitcomb (eds.). Academic Press, New York.
132. Tully, J. G. 1979. Special features of the acholeplasmas. Pp. 431–449. In: The Mycoplasmas. vol. I, M. F. Barile and S. Razin (eds.). Academic Press, New York.

133. Eden-Green, S., and J. G. Tully. 1980. Isolation of *Acholeplasma* spp. from coconut palms affected by lethal yellowing disease in Jamaica. Curr. Microbiol. 2: 311–316.
134. Tully, J. G., D. L. Rose, R. F. Whitcomb, K. J. Hackett, T. B. Clark, R. B. Henegar, E. Clark, P. Carle, and J. M. Bové. 1987. Characterization of some new insect-derived acholeplasmas. Isr. J. Med. Sci. 23: 699–703.
135. Kristensen, N. P. 1981. Phylogeny of insect orders. Annu. Rev. Entomol. 26: 135–157.
136. Smart, J. and N. F. Hughes. 1973. The insect and the plant: Progressive palaeoecological integration. Pp. 143–155. In: Symp. Royal Entomol. Soc. (London), Vol. 6: Insect/Plant Relationships. H. F. Van Emden (ed.). Blackwell Scientific, London.
137. Robinson, I. 1979. Special features of the anaeroplasmas. Pp. 515–528. In: The Mycoplasmas, vol. I, M. F. Barile and S. Razin (eds.). Academic Press, New York.
138. Robinson I. M., and E. A. Freundt. 1987. Proposal for an amended classification of anaerobic mollicutes. Int. J. Syst. Bacteriol. 37: 78–81.
139. Davis, R. E., R. F. Whitcomb, T. A. Chen, and R. R. Granados. 1972. Current status of the aetiology of corn stunt disease. Pp. 205–214. In: Pathogenic Mycoplasmas, Ciba Geigy Symp. Elsevier-Excerpta Medica-North-Holland, The Netherlands.
140. Tully, J. G., and R. F. Whitcomb. The genus *Spiroplasma*. Pp. 2271–2284. In: The Prokaryotes, vol. II, M. P. Starr, H. Stolp, H. G. Trüper, A. Balows, and H. G. Schlegel (eds.). Springer-Verlag, New York.
141. Hackett, K. J. and D. E. Lynn. 1985. Cell-assisted growth of a fastidious spiroplasma. Science 230: 825–827.
142. Hackett, K. J., D. E. Lynn, D. L. Williamson, A. Ginsberg, and R. F. Whitcomb. 1986. Cultivation of the *Drosophila* sex-ratio spiroplasma. Science 232: 1253–1255.
143. Hackett, K. J., D. E. Lynn, A. S. Ginsberg, S. Rottem, R. B. Henegar, J. Adams, D. L. Williamson, and R. F. Whitcomb. 1987. Cell-assisted culture of fastidious spiroplasmas: Initial analysis of growth factors. Isr. J. Med. Sci. 23: 667–670.
144. Williamson, D. L. and D. F. Poulson. 1979. Sex ratio organisms. Pp. 175–208. In: The Mycoplasmas, vol. III, R. F. Whitcomb and J. G. Tully (eds.). Academic Press, New York.
145. Saglio, P., M. L'Hospital, D. Lafleche, G. Dupont, J. M. Bové, J. G. Tully, and E. A. Freundt. 1973. *Spiroplasma citri* gen. and sp. nov.: a mycoplasmalike organism associated with 'stubborn' disease of citrus. Int. J. Syst. Bacteriol. 23: 191–204.
146. Williamson, D. L., J. G. Tully, and R. F. Whitcomb. 1989. The genus *Spiroplasma*. Pp. 175–208. In: The Mycoplasmas, vol. V, R. F. Whitcomb and J. G. Tully (eds.). Academic Press, New York.
147. Hackett, K. J., R. F. Whitcomb, R. B. Henegar, A. G. Wagner, E. A. Clark, J. G. Tully, F. Green, W. H. McKay, P. Santini, D. L. Rose, J. J. Anderson, and D. E. Lynn. 1989. Mollicute diversity in arthropod hosts. Art. Zbl. Bakt., I Orig. Hyg. A (in press).
148. French, F., R. F. Whitcomb, J. G. Tully, K. J. Hackett, E. A. Clark, R. B. Henegar, and D. L. Rose. 1989. Tabanid spiroplasmas of the Southeast USA: New groups and correlation with life history strategy. Art. Zbl. Bakt., I Orig. Hyg. A (in press).
149. May, R. 1986. How many species are there? Nature 324: 514–515.
150. Whitcomb, R. F., J. M. Bové, T. A. Chen, J. G. Tully, and D. L. Williamson. 1987. Proposed criteria for an interim serogroup classification for members of the genus *Spiroplasma*. Int. J. Syst. Bacteriol. 37: 82–84.
151. Bové, J. M., C. Mouches, P. Carle-Junca, J. R. Degorce-Dumas, J. G. Tully and R. F. Whitcomb. 1983. Spiroplasmas of group I: The *Spiroplasma citri* cluster. Yale J. Biol. Med. 56: 573–582.
152. Clark, T. B., R. F. Whitcomb, J. G. Tully, C. Mouches, C. Saillard, J. Bové, Wróblewski, P. Carle, D. Rose, R. B. Henegar, and D. L. Williamson. 1985. *Spiroplasma melliferum*, a new species from the honeybee (*Apis melliferum*). Int. J. Syst. Bacteriol. 35: 296–308.

153. Whitcomb, R. F., T. A. Chen, D. L. Williamson, C. Liao, and T. B. Clark. 1986. *Spiroplasma kunkelii* sp. nov.: Characterization of the etiological agent of corn stunt disease. Int. J. Syst. Bacteriol. 36: 170–178.
154. Saillard, C., J. C. Vignault, J. M. Bové, J. G. Tully, D. L. Williamson, A. Fos, M. Garnier, C. P. Gadeau, and R. F. Whitcomb. 1987. *Spiroplasma phoeniceum*, sp. nov., a new plant pathogenic species from Syria. Int. J. Syst. Bacteriol. 37: 106–115.
155. Tully, J. G., D. L. Rose, O. Garcia-Jurado, J. C. Vignault, C. Saillard, J. M. Bové, R. E. McCoy, and D. L. Williamson. 1980. Serological analysis of a new group of spiroplasmas. Curr. Microbiol. 3: 369–372.
156. Tully, J. G., D. L. Rose, E. A. Clark, P. Carle, J. M. Bové, R. B. Henegar, R. F. Whitcomb, D. E. Colflesh, and D. L. Williamson. 1987. Revised group classification of the genus *Spiroplasma* (Class Mollicutes), with proposed new groups XII to XXIII. Int. J. Syst. Bacteriol. 37: 357–364.
157. Clark, T. B., B. V. Peterson, R. F. Whitcomb, R. B. Henegar, K. J. Hackett, and J. G. Tully. 1984. Spiroplasmas in the Tabanidae. Isr. J. Med. Sci. 20: 1002–1005.
158. Whitcomb, R. F., K. J. Hackett, J. G. Tully, E. A. Clark, F. E. French, R. B. Henegar, D. L. Rose, and A. G. Wagner. 1989. Tabanid spiroplasmas as a model for mollicute biogeography. Art. Zbl. Bakt., I,Orig. Hyg. A (in press).
159. Aulakh, G. S., E. B. Stephens, D. L. Rose, and M. F. Barile. 1983. Nucleic acid relationships among *Acholeplasma* species. J. Bacteriol. 153: 1338–1341.
160. Christiansen, C., G. Christiansen, and O. F. Rasmussen. 1987. Heterogeneity of *Mycoplasma hominis* as detected by a probe for ATP genes. Isr. J. Med. Sci. 23: 55–70.
161. Razin, S., and D. Yogev. 1986. Genetic relatedness among *Ureaplasma urealyticum* serotypes (serovars). Pediatr. Infect. Dis. 5: S300–S304.
162. Christiansen, C., F. T. Black, and E. A. Freundt. 1981. Hybridization experiments with deoxyribonucleic acid from *Ureaplasma urealyticum* serovars I to VIII. Int. J. Syst. Bacteriol. 31: 259–262.
163. Razin, S., J. G. Tully, D. L. Rose, and M. F. Barile. 1983. DNA cleavage patterns as indicators of genotypic heterogeneity among strains of *Acholeplasma* and *Mycoplasma* species. J. Gen. Microbiol. 129: 1935–1955.
164. Mouches, C., J. M. Bové, J. G. Tully, D. L. Rose, R. E. McCoy, P. Carle-Junca, M. Garnier, and C. Saillard. 1983. *Spiroplasma apis*, a new species from the honey-bee (*Apis mellifera*). Ann. Microbiol. (Inst. Pasteur) 134A: 383–397.
165. McCormack, W. M. 1986. *Ureaplasma urealyticum*: ecological niche and epidemiological considerations. Pediatr. Infect. Dis. 5: S232–S233.
166. McCormack, W. M. 1972. Localization of genital mycoplasmas in women. Am. J. Obstet. Gynecol. 112: 920–923.
167. Hackett, K. J., A. L. Hicks, R. F. Whitcomb, J. E. Lowry, and S. W. T. Batra. 1984. Occurrence and frequency of subgroup I–6 spiroplasma in oldfields in Maryland and Virginia. Isr. J. Med. Sci. 20: 1006–1008.
168. Clark, T. B., R. F. Whitcomb, and J. G. Tully. 1982. *Spiroplasma* from coleopterous insects: New ecological dimensions. Microbial Ecol. 8: 401–409.
169. MacArthur, R. H., and E. O. Wilson. 1967. The theory of island biogeography. Princeton University Press, Princeton, New Jersey.
170. Opler, P. A. 1974. Oaks as evolutionary islands for leaf-mining insects. Am. Sci. 62: 67–73.
171. Southwood, T. R. E. 1973. The insect/plant relationship–an evolutionary perspective. Pp. 3–30. In: Insect/Plant Relationships, Symp. Royal Entomol. Soc. (London). vol. 6, H. F. Van Emden (ed.). Blackwell Scientific, London.
172. Strong, D. R.,Jr. 1979. Biogeographic dynamics of insect-host plant communities. Annu. Rev. Entomol. 24: 89–119.

173. Yamao, F., A. Muto, Y. Kawanchi, M. Iwami, S. Iwagami, Y. Azumi, and S. Osawa. 1985. UGA is read as tryptophan in *Mycoplasma capricolum*. Proc. Natl. Acad. Sci. (USA) 82: 2306–2309.
174. Hu, P. C., J. M. Inamine, S. Lechel, C. H. Huang, A. M. Collier, and W. A. Clyde, Jr. 1989. Analysis of the nucleotide sequence of the P1 attachment-protein operon of *Mycoplasma pneumoniae*: Codon usage and the emergence of $tRNA_{UCA}$. Art. Zbl. Bakt., I,Orig. Hyg. A (in press).
175. Citti, C., C. Saillard, and J. M. Bové. 1989. Art. Zbl. Bakt., I,Orig. Hyg. A (in press).
176. Barrell, B. G., S. Anderson, A. T. Bankier, M. H. L. deBruijn, E. Chen, A. R. Coulson, J. Drouin, I. C. Eperon, D. P. Nierlich, B. A. Roe, F. Sanger, P. H. Schreier, A. J. H. Smith, R. Staden, and I. G. Young. 1980. Different pattern of codon recognition by mammalian mitochondrial tRNAs. Proc. Natl. Acad. Sci. (USA) 77: 3164–3166.
177. Bonitz, S. G., R. Berlani, G. Coruzzi, M. Li, G. Macino, F. G. Nobrega, M. P. Nobrega, B. E. Thalenfeld, and A. Tzagoloff. 1980. Codon recognition rules in yeast mitochondria. Proc. Natl. Acad. Sci. (USA) 77: 3167–3170.
178. Heckman, J. E., J. Sarnoff, B. Alzner-DeWeerd, S. Yin, and U. L. RajBhandary. 1980. Novel features in the genetic code and codon reading patterns in *Neurospora crassa* mitochondria based on sequences of six mitochondrial tRNAs. Proc. Natl. Acad. Sci. (USA) 77: 3159–3163.
179. Muto, A., and S. Osawa. 1987. The guanine and cytosine content of genomic DNA and bacterial evolution. Isr. J. Med. Sci. 23: 334–341.
180. Samuelsson, T., Y. S. Guindy, F. Lustig, T. Boren, and U. Lagerkvist. 1987. Apparent lack of discrimination in the reading of certain codons in *Mycoplasma mycoides*. Proc. Natl. Acad. Sci. (USA) 84: 3166–3170.
181. Wawrousek, E. F., N. Narasimhan, and J. N. Hansen. 1984. Two large clusters with thirty-seven transfer RNA genes adjacent to ribosomal RNA gene sets in *Bacillus subtilis*: Sequence and organization of trrnD and trrnE gene clusters. J. Biol. Chem. 259: 3694–3702.

Microbial germplasm

SHUNG-CHANG JONG
American Type Culture Collection, 12301 Parklawn Drive, Rockville, Maryland 20852, U.S.A.

Abstract Microorganisms are interpreted here to include algae, bacteria, fungi, protozoa, and viruses. They are universal and fundamental elements of biotic diversity. Microbes are essential to the maintenance of stable ecosystems and the very existence of plant and animal life. Thus, on-site microbial diversity can only be sustained by maintenance of intact ecosystems. It is estimated that less than 20% of the world's microbes have been cultured and identified so far.

Microbial germplasm is extremely diversified. The genes that contribute to inheritable germplasm differ even from one strain to another within a species. Genetic variability within and among species is substantial. Species cannot be adequately studied or assessed without growing them in pure culture.

Microbes have an inherent tendency to mutate, which is accelerated under laboratory conditions. Among pure cultures, no cell population is genetically identical to the parent culture. Consequently, culture collections have been and are being established to conserve valuable germplasm important to economic, medical, agricultural, and scientific endeavors. Conservation technologies for microbial cultures are now an integral part of both basic and applied microbiology. Unfortunately, there is no method yet for the complete preservation of genotypes.

The worldwide program for conserving microbial gene pools and making them accessible to both the developed and developing countries has been launched through the establishment of the Global Network of Microbiological Resources Centers (MIRCENS). Today, this network program is comprised of 15 centers in various regions of the world.

No one knows exactly how many culture collections of microbial germplasms exist. The 1986 edition of the *World Directory of Collections of Cultures of Microorganisms* lists 327 collections in 55 countries. The world-wide collection contains on the order of 17,202 species or varieties of fungi (including lichens and yeasts), 7,455 bacteria, 2,348 algae (including blue-green algae), 314 protozoa, 2,259 animal viruses, 500 bacterial viruses, 307 plant viruses and 68 insect viruses. A total of 646,555 strains representing 51,453 species and 4,058 genera are conserved, consisting of 397,136 bacteria, 254,169 fungi, 11,800 algae, 6,939 animal viruses, 3,433 bacterial viruses, 1,444 protozoa, 1,091 plant viruses, and 80 insect viruses. Agriculture, medicine, and industry all have benefited significantly from products derived from these cultures. Any of these cultures offer new sources of biosynthetic products, new sources of genes for hybridization, or even sources of characteristics for genetic engineering.

In addition to the basic functions of acquisition, authentication, preservation, and distribution of cultures and associated information, culture collections of microbial germplasm provide other services such as identification and taxonomic analyses of cultures, safe deposit of proprietary cultures, contract services, consulting services, information services, patent depositories, publications, sponsoring of workshops and training courses, and accommodation of visiting investigators. For interactions among existing culture collections, many organizations promote administrative and scientific coordination and information exchange between different collections, or between collections and appropriate scientific communities at the international, regional, and national level.

L. Knutson and A.K. Stoner (eds.). Biotic diversity and germplasm preservation, global imperatives.
 ISBN 0-7923-0178-1

Key words: Microorganisms, microbes, algae, bacteria, fungi, protozoa, viruses, yeasts, microbial gene pools, pure cultures, culture collections, preservation techniques, biotechnology, genetic engineering

Introduction

Microorganisms (often referred to as microbes, meaning tiny living things) have an enormous impact on biotic diversity and germplasm as well as on global economic and social development. Microbes that are not typically classified as either plants or animals are interpreted here to include algae, bacteria, fungi, protozoa, and viruses. They not only cause diseases of humans, animals, and plants, but also enhance the quality of our daily life by their contributions to agriculture, food processing, medicine, industry, modern biotechnology, and genetic engineering. Microbes constitute about one-quarter of the total biomass on earth and occur in nearly all environments. The role of these organisms is fundamental to the proper functioning of ecological systems and to the very existence of plant and animal life [1]. The interdependency of this relationship is such that a loss of microbial species would result in a loss of animal and plant variety. The use of advanced agricultural and industrial technology, the human population explosion, and disturbances to ecosystems by resource exploitation already have resulted in both reduced microbial diversity and biomass. On-site (in nature) microbial diversity can only be sustained by maintenance of intact ecosystems [1–3].

In nature, microbes exist as large populations of individuals rather than as single entities. Such individuals, descendants of a single cell, are termed strains and are considered to be the basic units of microbial diversity. Strains sharing common biochemical, physiologic, and morphologic characteristics are recognized as species [1]. Genetic variability within and among species is substantial. The magnitude of the genes that contribute to inheritable germplasm differs even from one strain to another within a species. Culture collections have been and are being established to conserve germplasm important to economic, medical, agricultural, and scientific endeavors. These off-site (laboratory based) collections of microbial strains or isolates throughout the world represent only a small fraction of known and/or potential microbial species (on-site); nevertheless, they form a major genetic resource for all facets of biotechnology [4, 5].

Establishment of pure cultures

The biology of microbes cannot be adequately studied or assessed without growing them in pure (axenic) culture. Microbes grow only under favorable conditions, and there is no universal set of such conditions. A fairly thorough understanding of the biochemical factors that influence microbial growth can be a very useful guide to the establishment of pure cultures artificially in the laboratory. Nutrients, osmotic pressure, acidity or alkalinity, presence of oxygen, inhibitors, light, temperature and degree of agitation are primary factors influencing microbial growth. A pure culture of a microbial species rarely occurs in nature. Environmental factors determine which microbes will grow and predominate in the natural species mixed population (culture).

Since microbial germplasm is extremely diversified, there are various genetic and biochemical actions and reactions that maintain the stability of natural ecosystems, which in turn affect the growth and survival of microbes. These mutual relationships inherent in mixed populations are very difficult to understand. To date only a relatively small portion of microbes as compared to the total population in the environment have been successfully isolated into pure culture and identified [1]. For instance, it is estimated that less than 20% of extant species of bacteria and fungi are known.

Innumerable protocols for isolating pure cultures of microbes have been described in the literature [6–9]. Procedures vary from group to group, and sometimes a special method is required for a single species. The technique of choice may depend on the strains themselves, particularly in searches for microbes with special capabilities. Nutritional and physiological characteristics of microbes often dictate the procedure. The principles employed are those of selective enrichment or inhibition, which create environmental conditions favorable for growth of the desired microbes.

Techniques for isolating pure cultures were first developed for bacteria by Robert Koch in 1880 [5]. Such pure culture techniques were in turn applied to fungi, algae, protozoa, and viruses. All require sterile conditions, which involve the inactivation or physical elimination of all living cells from the culture environment. Pure culture techniques are used in isolation, identification, and propagation of microbes in the following disciplines: the chemical, brewing, pharmaceutical, and fermentation industries; agriculture, including biological nitrogen fixation, plant pathology, and dairying; food technology, biotechnology, and genetic engineering; and in environmental and ecological studies. Consequently the isolation and maintenance of microbial cultures has been an integral part of both basic and applied microbiology.

The growth and metabolic activity of microbes is based upon a very large store of genetic information. It is well known that microbes have an inherent tendency to mutate, a tendency that is accelerated under laboratory conditions. The genetic population often change in pure culture. Among pure cultures, no cell population is genetically identical to the parent culture. The increased percentage of any mutant in the total population is dependent upon culture conditions and the number of generations of growth permitted. Therefore, once pure cultures are established, they must be maintained in a genetically stable condition. Unfortunately, no method has yet been devised for the complete preservation of genotypes.

Technologies to conserve pure cultures

In keeping microbial cultures, conservation implies both maintenance and preservation. Maintenance means keeping the culture alive and pure whereas preservation means maintaining it with stable biological properties for a prolonged period. Therefore, a culture collection of microorganisms, in addition to maintaining culture viability, must keep the organisms pure and unchanged in character and appearance and preserve their physiological and genetic properties [10]. There is no universal technique for storing microbial cultures. Many methods have been developed for this purpose [8–14]. Selection of a method must be based on future use of the culture and the advantages or disadvantages of each method.

Subculturing

In a small collection, special requirements (media, light, temperature conditions, frequency of transfer) chosen to preserve the characteristics desired in individual strains can be accommodated with periodic subcultures. However, even this seemingly simple system has disadvantages. In every transfer is the danger of introducing contamination, of selecting cells without certain characteristics that may eliminate the 'normal' character of the strain, and of the gradual weakening and eventual death of the strain through continued growth under artificial conditions.

In larger collections, the problems associated with subculturing are multiplied significantly by the increased staff workload and the possible decrease in time available for individual attention to specific details.

The intervals between transfer of cultures for maintenance depend on the requirements of the microbe as well as on storage conditions. At room temperature the periods will vary from days to a few weeks to a few months. To avoid rapid drying of cultures on agar medium and to reduce the metabolic rate of the microbe, fresh mature cultures, which are not sensitive to cold, can be stored in a refrigerator at $-10°C$ to $-20°C$ [15, 16]. Under these low-temperature conditions, the viability period of most microbes is lengthened, and the interval between transfers can often be longer than at room temperature.

Immersion in mineral oil

The storage life of a fungal or bacterial culture can also be extended by covering a fresh, well-grown agar slant with sterile mineral oil at least 1 cm deep [17]. This depth will reduce dehydration of the medium and retard metabolic activity and growth of the culture [18]. Oil-covered slants are stored at room temperature or under refrigeration depending on the culture. Strains sensitive to cold will still be sensitive despite the reduced metabolic rate.

Medicinal-grade mineral oil, sterilized in flasks for 3 hours at 160–170 °C, is used. The oil is dispensed from a sterile, stoppered funnel. The test tube culture is held a certain distance below the dispenser to prevent contamination of the dispenser by any dry spores dislodged by the stream of oil. When only a few tubes are being processed, the oil can be dispensed with disposable pipettes.

Transfer of cultures under oil is messy, and regrowth is sometimes slow. As much oil as possible should be drained from the inoculum by touching it to the side of the test tube before it is removed to the new tube. It should be placed high on the new slant so that any oil remaining will drain away from the inoculum. Occasionally a second subculture will be necessary to regain normal growth.

Cultures preserved under oil grow at a reduced rate and will survive for years. As with most methods of preservation, much depends on the condition and vitality of the original isolate. If periodic transfers are made from the original tube and from a subculture, the longevity of the strain can be determined and intervals between transfers can be lengthened.

The oil overlay method is especially applicable to strictly mycelial or nonsporulating forms that cannot be freeze-dried successfully, or to small collections in which the installation of lyophilization or liquid nitrogen freezing procedures would not be economical.

Storage under oil will also deter mite infestation, and any mites already present in the tubes will be killed in 2–3 months [10].

Immersion in sterile distilled water

Storage in sterile distilled water is an alternative method for extending the shelf life of some microbe cultures. Castellani [19–23] first reported the storage and successful retrieval, without any change in morphology or physiological character, of human pathogenic fungi kept for 12 months in sterile distilled water. In fact, pleomorphic cultures so stored retained the same pleomorphic characteristics after retrieval. In his first trials, the glass tubes were sealed, but later plain cotton-plugged tubes were found to be satisfactory. Benedek [24] revised the system by using commercially available physiologic saline in small, screw-capped bottles. Sterile water storage has been used for human pathogens [25], plant pathogens [26, 27], root pathogens, and ectomycorrhizae of filamentous fungi, yeasts, and bacteria [28].

Microbial suspensions or agar blocks or disks cut from plates of actively growing cultures are placed into test tubes half filled with sterile distilled water and the caps screwed down tight to prevent evaporation. The tubes can be stored at room temperature or refrigerated, depending upon the thermal sensitivity of the organism. This method is simpler and less messy than the oil overlay method; however, cultures generally will survive longer under oil (reportedly 1–32 years) than in water storage (2–7 years) [14, 27]. Castellani suggested that longer survival might be possible by replenishing the water as it recedes, but he did not publish any results.

The storage of actively growing inocula and/or sufficient numbers of cells, spores, and hyphae are important factors in determining longevity in distilled water, but most results seem to indicate that survival of cultures by any preservation method is often strain specific.

Drying of cultures

Microbes stored at lowered temperatures, under oil, or in distilled water are in a state of reduced metabolism, which slows the production of metabolites and the staling of the medium. Microbes stored under dry conditions become dormant and will remain viable for many years. This is especially true of spore-forming bacteria and fungi and cyst-forming protozoa. For drying and ease of handling, spores have been stored in soil, sand, silica gel, and gelatin, or on filter papers and porcelain or glass beads.

Since 1918 scientists have stored fungi in sterilized soil for months or even years with little or no variability when compared with the same cultures maintained by serial subcultures. Bakerspigel [29] tested loam, clay, sawdust, peat, and silicaceous earth with a variety of fungi. His results indicated that viability and stability of character was best preserved in loam. This method of storage is used in many industrial collections to preserve viability and the characteristics of commercially important microorganisms [30–33]. In this procedure, moistened soil or sand is inoculated with a small amount of inoculum and incubated and grown before drying [31], or dried soil is inoculated with a large amount of cell or spore suspension and dried immediately [33].

In the first method, soil is placed in test tubes or bottles, moistened to 60% of capacity, and autoclaved at 15 lb pressure for 30 min on three alternate days. It is then inoculated and grown at the optimal temperature for the culture. The tube is allowed to dry out at room temperature and then stored in the refrigerator.

In the second method, soil is placed in the tubes but no water is added. The tubes are autoclaved as above and then heated in a drying oven. The soil tubes must be stored in a desiccator until used. A heavy suspension of cells or spores to be preserved is mixed into the soil. After air-drying, the tube is stored in the refrigerator. The culture is revived by sprinkling a few grains of the soil into broth or onto agar followed by incubation until grown.

The advantages of the soil culture method are:

1. Increased longevity of the culture,
2. Morphological changes are reduced or eliminated, and
3. A uniform inoculum is available for many years [31].

A method used by the Fungal Genetics Stock Center in the U.S. to preserve *Neurospora* spores and mycelia dried on anhydrous silica gel was described by Perkins [34, 35]. Cotton-plugged 13 × 100 mm tubes are filled to 65 mm with silica gel (Davison Refrigeration grade silica gel PA400, approximately 12–20 mesh size, without indicator), hot air sterilized at 180°C for 2 hours, then stored in a desiccator or moisture-proof box until used.

One week-old cultures are used to make conidial or mycelial suspensions in 0.5 ml cold, sterile water. This is diluted with 0.5 ml of cold, sterile, non-fat milk, mixed gently and pipetted drop-wise over the silica gel in a prelabeled tube. The tube is shaken to distribute the suspension throughout the crystals and placed in an ice bath for 15 min to absorb the heat produced when the silica gel is moistened. The tubes are allowed to dry at room temperature for 24 hours, sealed with a cap of Parafilm®, and stored at 5°C or −20°C in moisture-proof boxes. Cultures are revived by sprinkling a few crystals over an appropriate medium followed by incubation until grown. Ninety percent of 700 strains of genetic stock cultures, including nutritional and morphological mutants, survived for up to 5 years [33].

Since its publication, the method of Perkins [34] has been used with good results for many genera of insect and plant pathogens [36–39], yeasts [40], other selected fungi [41–43], and a range of bacterial species [44, 45]. Viability of non-conidiating strains was greater with this method (drying without freezing) than with lyophilization, in which the material is first frozen [34]. In some studies, viability was dependent on the strain used and on the medium on which it was grown before drying.

In 1956, Goldie-Smith [46] reported great success with preserving Blastocladieales by drying (pouring the water from established hemp seed cultures and allowing them to air dry). They were revived three years later by the simple addition of water. As a further refinement, she dried liquid *Allomyces* cultures on sterile filter paper squares. These were kept dry in envelopes and revived 17–18 years later. Members of the *Enterobacteriaceae*, as well as a variety of other bacteria, were also successfully preserved for several years in this manner [47]. Many heterotrophic bacteria can be preserved in dried gelatin drops or disks stored at $-20\,^{\circ}C$ [48]. Another convenient drying method recommended for most species of bacteria and fungal spores involves the use of porcelain beads [49, 50].

Freeze-drying (lyophilization)

Freeze-drying or lyophilization is an effective and economical means for the long term preservation of microbial cultures. It is a process of simultaneously freezing and drying microbes under vacuum. The freezing part of the process may be carried out separately from the drying. Drying a culture will reduce metabolic activity to a minimum by removal of water necessary for the function. Freezing a culture will accomplish the same purpose since water in the frozen state cannot be utilized for metabolism. Both of the methods, freezing and drying, are used for preservation, but both have disadvantages. Lyophilization was developed to try to solve these problems. However, lyophilization is disruptive to the cell, and mycelial cultures or cultures with fragile or elaborate spores usually do not survive.

Freeze-drying is a process of removing water from cells by sublimation, i.e., passing directly from the solid to the gaseous state. Sublimed water vapor is drawn off under vacuum, and the dried spores are sealed in ampoules for long-term storage. The materials to be dried must be solidly frozen before being subjected to the high vacuum. In this way, the solid particles of the cells are locked into place by an ice matrix and cannot interact. The ice matrix is removed by sublimation, and the dried material, being inert, theoretically maintains its original characteristics indefinitely, provided that re-entry of moisture is prevented.

In practice, however, there are limiting factors to the success of freeze-drying for long-term preservation. Fresh isolates are more likely to have longer survival times than cultures maintained for years in subculture before being freeze-dried. Cell suspensions must be sufficiently heavy to insure the survival of adequate numbers of cells from any freezing injury that may occur. A high percentage of viable cells is also necessary to minimize the selection of mutants. Some workers have found evidence of variability, especially among yeasts [51, 52], but others have reported no change at all or much less than would be found after use of serial subcultures for many years [53, 54].

Living cells will not survive freeze-drying without a protective agent. Many suspending media have been tested for their protective qualities, namely, to prevent total drying and to buffer the effects of electrolyte concentration during dehydration. Bovine serum was the earliest additive used;

others are sodium glutamate, peptone, various sugars or mixtures of sugars, and skim milk.

A slow freezing rate, about 1 °C/min, has been found to be important to lessened cell damage [55], and the material must remain frozen until it has been dried to at least 5% moisture. The final moisture content must be between 1% and 2% to prevent DNA damage and mutation or even death [14].

The rehydration procedure has been found to be an important factor in the viability of freeze-dried cultures. Freeze-dried cells must be allowed sufficient time to reabsorb moisture before transfer to a solid medium [56, 57].

In 1945, Raper and Alexander [58] reported good survival for microbes lyophilized at the Northern Regional Research Laboratory of the U.S. Dept. Agric. (USDA) at Peoria, IL, by a procedure now called the NRRL method. In this method, heavy suspensions of spores are made in sterile bovine serum or in dried skim milk reconstituted to double strength and sterilized. Approximately 0.5–1 ml of the suspension is dispensed into sterile, previously labeled, cotton-stoppered Pyrex® glass tubes. The cotton plugs are replaced and the exposed cotton is burned off. The remaining cotton is pushed down about 1/2 inch into the tube, and the tubes are inserted into rubber sleeves on a glass manifold attached to vacuum pump. The tubes are lowered into a bath of Dry Ice® and methyl Cellusolve® at −40 °C to −50 °C, which freezes the suspensions in a few seconds. The tubes are raised above the surface of the bath, maintained at a temperature of about −10 °C, and evacuated by means of a vacuum pump at about 200–500 mg of mercury. When the pellets appear to be dry, the coolant is removed and the evacuation is continued for 30 min at room temperature to insure complete desiccation. The tubes are sealed off under vacuum and stored at 4 °C.

Another method of lyophilization, the PRL or Prairie Regional Laboratory method, was found by Haskins and Anastosion [59] to be equally successful. It is similar to the double vial 'batch' method used by the American Type Culture Collection (ATCC) [33]. In the batch type method currently used at ATCC, the spore suspensions 'slow freeze' before primary drying at low temperatures, secondary drying at room temperature, and sealing off under a vacuum.

Heavy suspensions of mature spores or cells at least 10^6/ml are made in 20% (w/v) skim milk (Difco 0032) that has been sterilized at 116 °C for 20 min. Care must be taken to prevent carmelization of the milk by overheating. The cells of cultures grown in broth are centrifuged and resuspended in a solution of 12% sucrose plus half-strength, fresh growth medium. The sucrose is prepared as a 24% solution autoclaved at 120 °C for 20 min. Then

0.2 ml of the suspension is pipetted into a sterile, prelabeled, cotton plugged, 11 × 35 mm freeze-drying shell vial (VWR Scientific No. 27921-015). The filled vials are refrigerated immediately at 4 °C and processed within 2 hours to prevent the possible germination of spores [9, 10].

The vials are packed upright into a stainless steel container, which is placed in the bottom of a mechanical refrigerator at − 70 °C for 1 hour, which produces a freezing rate of about 1 °C/min. The container of vials is then attached to an Atmo-vac plate which is connected to a dry ice/methyl cellusolve-chilled condenser and a vacuum pump. Crushed dry ice is packed around the container, and the material is dried at a vacuum of 20–30 μm of mercury. The dry ice sublimes within two hours and the drying continues at room temperature overnight [9, 10].

Soft glass 'outer' shell vials (VWR Scientific 27921-015; 14.25 mm × 85.0 mm) are prepared in advance by covering the bottom with silica-gel granules (Fisher Scientific, grade 42, 6–16 mesh, 'Tel-Tale' brand). A small wad of cotton is inserted for cushioning, and the vials are heated at 100 °C overnight. The silica gel turns dark blue after heating and serves as a moisture indicator during storage. The vials are cooled in a dry cabinet (10% relative humidity). When the 'outer' vials have cooled, the vacuum line to the Atmo vac plate is clamped off and air is admitted from the dry cabinet through a silica gel column, which brings the pressure in the freezing-drying pan up to atmospheric pressure. The 'inner' vials are transferred to the dry cabinet, where each is inserted into a prepared outer vial. A wad of two layers of glass fiber paper is inserted above the cotton plug of the inner vial for insulation. The vials are removed from the dry cabinet, and the outer vial is constricted to a narrow capillary with an air-gas torch.

The vials are allowed to cool, attached to a 25-port manifold, evacuated to less than 50 μm of mercury, and hermetically sealed with a two way, air-gas burner. Freeze-dried preparations may be stored at room temperatures, but greater longevity is achieved by storage at 4 °C.

One sample of each culture freeze-dried is opened and checked for viability a few days after processing to determine the degree of any freezing injury. Once a culture has survived the freezing shock, it will probably retain its viability for years.

Liquid drying *in vacuo*, referred to as L-drying, is similar to freeze-drying in so far as the operation takes place in ampoules attached to a manifold and immersed in water at 20 °C The vacuum level is adjusted to allow evaporation without freezing. This method, originally described by Annear [60–62], has been improved and successfully used in preservation of various microbial cultures at the Institute for Fermentation in Osaka, Japan [63–70]. L-drying has the following advantages over freeze-drying: organisms are dried without freezing, cotton wool plugs (nonabsorbent) are used to prevent contamination, and the cells dry faster.

Cryogenic storage

Since the rates of mitotic recombination and mutation are likely to correspond to those of cell division and metabolic activity, an ideal system for microbial germplasm conservation would be to store living cultures in such a manner as to achieve a complete cessation of cell division and total arrest of metabolism while the cultures still remain viable. This may be accomplished by storage below $-139\,^{\circ}C$ where there is no growth of ice crystals and the rates of other biophysical processes are too slow to affect cell survival. Successful storage of a broad range of living materials such as algae, bacteria, blood, fungi, mycoplasmas, nematodes, plant and animal cells, protozoa, rotifers, semen, tissues, viruses, and yeasts has been achieved through the application of cryogenics (below $-100\,^{\circ}C$). The cryogenic storage temperatures now commonly used are those of liquid nitrogen ($-196\,^{\circ}C$) and liquid nitrogen vapor ($-150\,^{\circ}C$ and below) [9–12].

The critical periods in cryopreservation are during the freezing and thawing processes. As the temperature is lowered below the freezing point of water, ice crystals form, which can be damaging to cell membranes. Also the removal of water into ice can concentrate solutes into chemically damaging substances. Injury during thawing occurs as the reverse of freezing injury. As the temperature rises above $-139\,^{\circ}C$ water in the cells will crystallize, producing ice damage unless the temperature rises quickly enough to prevent it. In general, best results in the survival and recovery of most microbial cultures have been obtained using a slow cooling rate (*e.g.*, $1\,^{\circ}C$ per min) and rapid thawing.

At slow cooling rates, water in the suspending medium is frozen into ice. Intracellular water passes outward through the membranes as a result of the lowered vapor pressure of the extracellular water and the concentration of solutes. The cells shrink slowly due to dehydration produced by the loss of intracellular water, which prevents the formation of lethal intracellular ice crystals.

To reduce injury upon freezing and thawing, one feature common to most successful protocols for cryopreservation of living cells is the application of chemical cryoprotectants. The cryoprotectants are of two types:

1. Penetrating agents such as glycerol and dimethyl sulfoxide (DMSO), which readily pass through the cell membrane and exercise their protective effect within intracellular and extracellular conditions, and
2. Non-penetrating agents such as sucrose, lactose, glucose, mannitol, sorbitol, dextran, polyvinylpyrrolidone, and hydroxyethyl starch, which exert their protective effect external to the cell membrane.

By far, glycerol and DMSO have been proven most effective and are routinely employed at a concentration of 10% (vol/vol) and 5% (vol/vol), respectively. Glycerol is sterilized by autoclaving at 15 lb/in^2 for 15 min. DMSO is filter sterilized, collected in 15 ml quantities in test tubes, and stored in the frozen state at 5 °C (DMSO freezes at 18 °C). Because of the accumulation of oxidative breakdown products, an opened bottle of DMSO should not be used for more than 1 month. Cultures that do not survive freezing in glycerol are preserved with DMSO, which seems to be a better protectant than glycerol. DMSO is harmful if it comes in contact with the skin or eyes.

Cells, spores, or mycelial fragments are harvested by flooding slants or plates with 10% glycerol or 5% DMSO and gently scraping the surface. Then 0.5 ml of the suspension is pipetted into sterile, cotton-plugged, prescored borosilicate ampoules (T.C. Wheaton Co.) or 2.0 ml screw-cap polypropylene vials (Vangard International). Polypropylene vials have proven to be superior to borosilicate ampoules for freezing some strains of microbes [71]. Cultures producing only mycelia are grown on plates containing an appropriate medium. Three plugs of the advancing edge of the colony are removed with a 5-mm sterile cork borer and suspended in 0.4 ml of cryoprotective additive per vial or ampoule. Strains that yield colonies in liquid culture small enough to be pipetted are grown in flasks containing a suitable broth, either stationary or on a Gyrotory Shaker. Cultures in broth are fragmented in a sterile Waring Blender and suspended in equal parts of 20% glycerol and growth medium or equal parts of 10% DMSO and growth medium to give a final concentration of 10% glycerol or 5% DMSO, respectively. Some strains must be concentrated by centrifugation to obtain sufficient material for freezing. Human pathogens are not blended because of the hazard of aerosol dispersion [10].

Glass ampoules containing human pathogens are always maintained in the vapor-phase of the liquid nitrogen tank. They are not sealed so as to:

1. Eliminate the dangers involved in handling pathogens in the sealing process.
2. Prevent the possibility of explosion during thawing, and
3. Prevent the possibility of self-inoculation on opening the ampoules.

Ampoules stored in the liquid-phase of the liquid nitrogen must be completely sealed. An improperly sealed ampoule immersed in liquid nitrogen will permit entry of liquid and will explode at the time of thawing due to the sudden expansion of the nitrogen into gas, thus creating a serious hazard. The operator should wear a face mask to avoid danger from exploding ampoules. Polypropylene vials are stored in the vapor phase of liquid nitrogen because liquid nitrogen can enter the vial around the threads of the cap if immersed in the liquid phase.

Microscopic mounts are made during dispensing to ensure the adequacy of the suspension and the effect of the cryoprotectant on the microbe. Occasionally plasmolysis occurs. A prefreeze control inoculated onto agar is also made to determine any toxic effects of the cryoprotectant on the regrowth of the culture. A post-freeze viability check is run 3–4 days after freezing.

The filled vials or ampoules are placed onto pre-labeled aluminum canes (Nasco #A545) in racks that are then placed into the freezing chamber of a programmed freezer (Cryo-Med). The initial cooling rate is carried out at a rate of 1 °C/min from room temperature to − 40 °C then 10 °C/min to − 90 °C; subsequent cooling is rapid and un-controlled. The cooling rate is monitored by a Model D-5000 Strip Chart Recorder® (Houston Instrument). The vials or ampoules are then transferred to storage in liquid nitrogen at − 196 °C or liquid nitrogen vapor storage at a temperature range of − 150 °C to − 180 °C.

The level of liquid nitrogen in each refrigerator is maintained by an automatic controller that feeds in liquid nitrogen from a 3000-gallon storage tank. The flow of liquid nitrogen is controlled electronically by heat-sensitive thermistors. In case of malfunction or dangerously low liquid nitrogen levels, an alarm system is triggered.

To revive the cultures, the frozen ampoules or vials of cells are shaken vigorously in a 37 °C water bath until the last trace of ice is dissipated. This usually takes 40–60 seconds for the borosilicate ampoules and 60–120 seconds for polypropylene vials. The culture samples are aseptically transferred to an appropriate growth medium.

Microbiological resources centers (MIRCENS)

The need for conservation of the genetic uniqueness and diversity of microbes was specially emphasized in the principles and recommendations of the United Nations Conference on the Human Environment in Stockholm in 1972. As a result, the concept of a Global Network of Microbiological Resources Centers (MIRCENS) was proposed by M.K. Tolba, Executive Director of the United Nations Environment Program (UNEP) to a working group of the UNEP/UNESCO/ ICRO Panel on Microbiology in 1974. Since then UNEP and UNESCO have provided catalytic support for putting this concept into action. Today, the global MIRCEN network program, which embodies activities that are carried out within the framework of UNESCO's regular program activities in cooperation with the International Union of Microbiological Societies (IUMS), the International Cell-Research Organization (ICRO), the World Federation for Culture Collections (WFCC) and other bodies, is comprised of 15 centers (Table 1).

Eight of those (located in Japan, Brazil, Egypt, Guatemala, Kenya, Senegal, Sweden and Thailand) were identified and established within the framework of the now completed (1975–1985) UNESCO/UNEP project entitled 'Development of an Integrated Program in the Use and Preservation of Microbial Strains for Deployment in Environmental Management' [72, 73].

The worldwide program for conserving microbial gene pools and making them accessible to both the developed and developing countries has been

Table 1. Speciality and location of Microbiological Resources Centers (MIRCENS)

A. *World Data Center (WDC) MIRCEN*
   1) World Data Centre MIRCEN: The Life Sciences Research Information Section, RIKEN, 2-1 Hirosawa, Wako, Saitama 351-01, Japan

B. *Biological Nitrogen Fixation (BNF) MIRCENS*
   1) Beltsville MIRCEN: The Nitrogen Fixation and Soybean Genetics Laboratory, ARS, USDA, Beltsville, MD 20705, USA
   2) Bambey MIRCEN: The Centre National Recherches Agronomiques, de L'Institut Senegalais de Recherches Agricoles, B.P. 51, Bambey, Senegal
   3) Nairobi MIRCEN: The Department of Soil Science and Botany, University of Nairobi, P.O. Box 30197, Nairobi, Kenya
   4) Porto Alegre MIRCEN: The Institute de Pesquisas Agronomicas in Collaboration with the Universidade Federal do Rio Grande do Sul, Postal 776, 9000 Porto Alegre, Rio Grande do Sul, Brazil
   5) NifTAL MIRCEN: College of Tropical Agriculture and Human Resources, University of Hawaii, P.O. Box 'O', Paia, HI 96779, USA

C. *Biotechnology MIRCENS*
   1) Cairo MIRCEN: Ain Shams University, Faculty of Agriculture, Shobra-Khaima, Cairo, Arab Republic of Egypt
   2) Guatemala MIRCEN: Applied Research Division, Central American Research Institute for Industry (ICAITI), Ave, La Reforma 4-47 Zone 10, Apdo Postal 1552, Guatemala
   3) UK-IBS MIRCEN: The Institute for Biotechnological Studies, Research & Development Centre, The University of Kent, Canterbury, Kent CT2 7PD, United Kingdom
   4) Stockholm MIRCEN: Department of Bacteriology, Karolinska Institutet, Fack, S-10401 Stockholm, Sweden
   5) Bangkok MIRCEN: Thailand Institute of Scientific and Technological Research, 196 Phahonyothin Road, Bangken, Bangkok 9, Thailand
   6) Osaka MIRCEN: International Center of Cooperative Research In Biotechnology, University of Osaka, Suitashi 656, Osaka, Japan
   7) Tucuman MIRCEN: Planta Piloto de Procesos Industriales Microbiologicos (PROIMI), Avenida Belgrano y Paseje Caseros, 4000 S.M. de Tucuman, Argentina
   8) Guelph-Waterloo MIRCEN: University of Guelph, Guelph, Ontario N1G 2W1, Canada, and University of Waterloo, Ontario N2IK 3G1, Canada
   9) Maryland MIRCEN: Department of Microbiology, University of Maryland, College Park, MD 20742, USA

launched through the establishment of the world MIRCENS network, which is designed:

1. To provide the infrastructure for a world network, which would incorporate regional and interregional cooperating laboratories geared to the management, distribution, and utilization of the microbial gene pool,
2. To reinforce efforts relating to the conservation of microorganisms, with emphasis on *Rhizobium* gene pools in developing countries with an agrarian base,
3. To foster the development of new, inexpensive technologies native to the region,
4. To promote the applications of microbiology to strengthen rural economies, and
5. To serve as focal centers for the training of manpower and the diffusion of microbiological knowledge.

MIRCENS currently recognized are briefly classified as follows:

World Data Center MIRCEN

The World Data Center of Microorganisms (WDC) is an information center for culture collections. Users of the WDC realize quick and smooth exchanges of information and microorganisms to support research and development in life sciences and biotechnology. The WDC has published three editions of the *World Directory of Collections of Cultures of Microorganisms* and the *World Catalogue of Rhizobium Collections*, and the first edition of the *World Catalogue of Algae*. The WDC also has organized a number of training courses primarily on data management and computer usage in culture collections [73].

Biological Nitrogen Fixation (BNF) MIRCENS

Legumes are able to fix atmospheric nitrogen in root nodules formed by *Rhizobium* and to use these nitrogen compounds for growth. However, the production of legume crops often depends on the effective associations of host plant with appropriate strains of rhizobia that do not always occur naturally where the crop is to be grown. The benefits of the legume-*Rhizobium* symbiotic system versus the use of chemical fertilizers are well appreciated in farming systems among research workers, policy makers, and the general public. To optimize the benefits of biological nitrogen fixation and to widen the germplasm base for increased strain efficiency, collections of rhizobia have been established in many places throughout the world [73, 74].

Five MIRCENS are working effectively on the development of biofertilizers or *Rhizobium* inoculant material in cooperation with other international programs in various regions of the world. Over 5,000 strains of *Rhizobium* cultures are now maintained in the BNF MIRCENS collections (Table 2).

The USDA *Rhizobium* Culture Collection at the Beltsville MIRCEN is a major world germplasm resource of *Rhizobium* and *Bradyrhizobium*. Its major mission is to collect, evaluate, document, preserve, utilize, and supply the germplasm for basic research, commerce, and applied agriculture. With the aid of the U.S. Agency for International Development, the activities of the collection have extended to include:

1. Training of scientists from developing countries in *Rhizobium* technology and providing opportunities to participate in collaborative research.
2. Increasing the collection and evaluation of rhizobial strains for important legumes in developing countries, and
3. Providing advice, information, and germplasm to individuals and projects involved in nitrogen fixation research in developing countries.

The Bambey MIRCEN(MAO-MIRCEN) has focused on the collection of cultures of rhizobia encountered with legumes and nitrogen-fixing trees and of endomycorrhizae.

The activities of the Nairobi MIRCEN include collection and preservation of cultures; isolation, selection, and testing of *Rhizobium* strains; determinations of parameters affecting survival of rhizobia in soil; and evaluation of locally available materials as carriers for inoculants. This regional MIRCEN network now consists of 12 countries: Kenya, Sudan, Lesotho, Malawi, Mauritius, Rwanda, Tanzania, Uganda, Zambia, Zimbabwe, Ethiopia, and Madagascar.

The activities of the Porto Alegre MIRCEN (IPAGRO) are concentrated particularly in the fields of culture collection, serology, production and quality control of inoculants, identification techniques, greenhouse use, and field selection of rhizobia, and training in Latin America. The IPAGRO Culture Collection functions as a central rhizobia germplasm bank for all

Table 2. Holdings of *Rhizobium* culture collections at BNF MIRCENS

MIRCEN		Number of strains held
Bambey		100
Beltsville		2,334
NifTAL (Hawaii)		1,806
Nairobi		205
Porto Alegre		700
	Total	5,145

national institutions and centralizes the distribution of cultures to the inoculant industries. In 1987, 135 cultures were supplied for inoculant production and 405 cultures for research purposes.

The NifTAL MIRCEN is coordinating an Inter-MIRCEN project that involves collaboration of the MIRCENS located at Bambey, Bangkok, Beltsville, Cairo, Guatemala, Hawaii, Nairobi, and Porto Alegre. The objective is to enhance the capability of the MIRCEN network to contribute significantly to the deployment of microbial germplasm resources for international development purposes. The major activity of the NifTAL is to support development of appropriate technologies based on biological nitrogen fixation to enable small farmers in developing countries to produce increased amounts of food with reduced dependence on costly nitrogenous fertilizers. A comprehensive *Rhizobium* collection is assembled at NifTAL for legumes used at high or low elevations in the tropics for grain, forage, fodder, green manure, firewood, erosion control, shade, ground covers, and many other purposes [75].

Biotechnology MIRCENS

Biotechnology, in the contemporary sense, can be defined as the application of biological organisms, systems, or processes for production of goods and services. There can be little doubt that the origins of biotechnology can be identified in some of the traditional industrial processes using microbes. Microbial cultures have been developed and manipulated for the manufacturing of chemical, pharmaceutical, and agricultural products, and for the degradation of toxic and industrial wastes. Nine MIRCENS are strongly biotechnology oriented [73].

The Cairo MIRCEN for Egypt and the Arab States has affiliated laboratories in Sudan, Saudi Arabia, Syria, Iraq, Kuwait, Jordan, Qatar, Gadria, and other institutions in Egypt. The Cairo MIRCEN network is engaged in research on conservation and distribution of microbial cultures, execution and collaboration with different studies in the fields of biotechnology, genetic engineering, biofertilization, microbial control, utilization of agricultural and industrial wastes and by-products, and development of technical skills and scientific knowledge through training courses, symposiums, and conferences.

The Guatemala MIRCEN for Central America and the Caribbean possesses an established and reputed technical expertise in:

1. Food and industrial microbiology,
2. Biotechnology, bioengineering, and food engineering,
3. Wood and lignocellulose engineering and technology,
4. Agro–industrial processes, and
5. Analytical chemistry and biochemistry.

Such expertise is reinforced by technical contributions from affiliated laboratories in Colombia, Costa Rica, El Salvador, Guatemala, Honduras, Mexico, Nicaragua, Panama, Peru, Dominican Republic, Trinidad and Tobago, and Venezuela. The objective of the network is to provide a forum for technical discussions through workshops and seminars and to stimulate research creativity and alternative proposals for problem solving [76].

The UK Biotechnology MIRCEN offers considerable expertise in many of the facets of biotechnology through the Institute for Biotechnological Studies (IBS) and associated laboratories. The major research program has been designed to investigate generic problems associated with the extended use of immobilized biocatalysts.

The Stockholm MIRCEN has been developing a computer conferencing system called MIRCENET to link all the 15 MIRCENS networks electronically to facilitate and enhance communications among the member centers. It aims to accelerate information exchange and scientific interactions so as to bring a greater understanding and applications of biomass production, anaerobic digestion, and the bioconversion of lignocellulose.

The Bangkok MIRCEN serves microbiologists of Southeast Asia through the exchange of economically important microbial germplasm in the region and by offering training and fellowship programs and promoting research on utilization of microbial cultures for development of microbial products or processes. It has affiliated laboratories in the Philippines, Indonesia, Singapore, Malaysia, Hong Kong, and with other institutions in Thailand.

The Osaka MIRCEN is an International Center of Cooperative Research in Biotechnology (ICBiotech). Its major activities include conducting studies on microbial engineering in utilization of natural resources and promoting biotechnology with Southeast Asian countries by means of the exchange of scientists, cooperative research, and scientific seminars.

The South American Biotechnology and Applied Microbiology MIRCEN in Tucuman, Argentina, has more than 40 cooperating laboratories in Bolivia, Chile, Brazil, Ecuador, Colombia, Paraguay, Peru, Uruguay, and Argentina, each with different strengths and capabilities. It has similar goals as the Biotechnology MIRCEN for Central America and Caribbean.

The Guelph-Waterloo MIRCEN specializes in fermentation technology and diagnostic veterinary microbiology.

The Maryland MIRCEN is a Center of Marine Biotechnology, engaged in research on cloning of genes of ecological significance, characterization of marine bacterial polysaccharides, and establishment of a nucleotide and protein sequence data-base and analysis system. It has set up joint research projects with the Shangdon College of Oceanography in Quingao, China, and the Chinese University of Hong Kong.

World registry of culture collections

No one knows exactly how many culture collections of microbial germplasm exist. In the late 1960s, V.B.D. Skerman, Professor of Microbiology, University of Queensland, Brisbane, Australia, began compiling a register of existing culture collections of microbes maintained by various institutions throughout the world. The first *World Directory of Collections of Cultures of Microorganisms* was prepared by the World Data Center of Microorganisms in 1972 with support from United Nations Educational, Scientific and Cultural Organization, the World Health Organization, the Australian Commonwealth Scientific and Industrial Organization, and the Canadian National Research Council. The second edition of the *World Directory* was published in 1982 with additional financial assistance from the United Nations Environment Program (UNEP). The composition and activities of the collections listed in these two directories have been reviewed by Porter [72] and Sly [77], respectively.

The recent edition, published in 1986, listed 327 collections in 55 countries, with the names and addresses of the collections registered, their interests and functions, personnel, availability of cultures, catalogue holdings, etc. [78]. The directory is a valuable international document with an introduction and instructions for use in five languages. There can be little doubt that many existing collections are not included. For example, several important collections of microbial germplasm in the U.S., such as the *Bacillus* Genetic Stock Center, *E. coli* Genetic Stock Center, Yeast Genetic Stock Center (YGSC), Culture Collection of Algae at the University of Texas at Austin (UTEX), and the USDA Center for Forest Mycology Research Culture Collection are not registered with WDC.

The number of collections in various regions of the world and their sponsorship is listed in Table 3. The majority of the collections registered with the WDC are located in Western Europe, North America, Oceania, and Asia. Approximately 29% (94 collections) are in Western Europe, which includes Belgium, Denmark, Finland, France, Greece, Ireland, Italy, Norway, the Netherlands, Spain, Sweden, Switzerland, United Kingdom and West Germany. About 18% are in North America and Oceania, respectively. The former contains 59 collections housed in Canada and the U.S.; the latter has 58 collections housed in Australia, New Zealand, and Papua New Guinea. Collections are well distributed throughout most regions of Asia, which contain 17% of the total number. Asia includes China (PRC), Hong Kong, India, Indonesia, Japan, Malaysia, the Philippines, Singapore, Sri Lanka, Taiwan, and Thailand. In Eastern Europe, 26 collections are housed in Bulgaria, Czechoslovakia, East Germany, Hungary, Poland, Rumania, Turkey, U.S.S.R., and Yugoslavia. There are

Table 3. Culture collections in various regions of the world and their sponsorship

Region	No. of countries	No. of collections	Sponsor		
			University	Private Industry	Government
Africa	6	10(3%)	3	2	5
Asia	10	57(17%)	31	12	14
Australia – New Zealand – Papua New Guinea	3	58(18%)	17	26	15
Europe (Eastern) & USSR	7	26(8%)	7	4	15
Europe (Western)	18	94(29%)	29	37	28
Latin America	6	19(6%)	3	10	6
Middle East	3	4(1%)	0	1	3
North America	2	59(18%)	19	16	24
Total	55	327(100%)	109(33%)	108(33%)	110(34%)

19 collections in Latin America, housed in Argentina, Brazil, Chile, Colombia, Guatemala, Mexico, and Venezuela. Collections are sponsored evenly by universities (33%), private industries (33%), and governments (34%), respectively, indicating their value in education, research, and practical use as well as enormous financial investments by large academic, trade, or governmental agencies.

Collections vary greatly in scope, diversity, and size. Certain collections have broad general holdings, whereas others restrict their cultures to a specific field or taxonomic group. In many cases, bacteria and protozoa are not kept in the same collections. The major interests of various collections are summarized in Table 4, and the number of collections listing various holdings is in Table 5.

The worldwide collection contains about 17,202 species or varieties of fungi (including lichens and yeasts), 7,455 bacteria, 2,348 algae (including blue-green algae), 314 protozoa, 2,259 animal viruses, 500 bacterial viruses, 307 plant viruses, and 68 insect viruses (Table 6). A total of 51,453 species representing 4,058 genera or groups are preserved. However, these represent only a small portion of known and/or potential species of microbial diversity in nature.

The data in Table 7 show that a total of 646,558 strains are maintained in various collections in different regions of the world, consisting of 11,800 algae, 397,136 bacteria, 254,169 fungi, 1,444 protozoa, 6,939 animal viruses, 3,433 bacterial viruses, 80 insect viruses, and 1,091 plant viruses. Although

Table 4. Major interests of various collections in the world

Microbiological Interest	No. of collections
Agricultural	114
Environmental	17
Food	10
Forest products	30
General	132
Industrial	99
Insect	15
Medical	105
Veterinary	52

Table 5. Number of collections in the world listing various holdings

Kind of holding	No. of collections
Algae	43
Bacteria	244
Fungi (including lichens & yeasts)	280
Protozoa	38
Viruses:	
Animal	38
Bacterial	40
Insect	7
Plant	7

Table 6. Taxonomic range of various collections in the world

Kind of microbes	Genus	Species
Algae (including blue-green algae)	583	2,348
Bacteria (including valid & invalid names)	595	7,455
Fungi (including lichens & yeasts)	2,618	17,202
Protozoa	97	314
Viruses	*Group*	*Type*
Animal	65	2,259
Bacterial	62	500
Insect	8	68
Plant	30	307
Total	4,058	51,453

Table 7. Number of holdings in various collections in different regions of the world

Region	Algae	Bacteria	Fungi	Protozoa	Viruses				Total for Region
					Animal	*Bacterial*	*Insect*	*Plant*	
Africa	450	2,283	2,530		267				5,530
Asia	143	23,133	28,483	16	555	102	28	10	52,472
Australia – New Zealand – Papua New Guinea	156	32,685	31,525	21	420	484	34	38	34,829
Europe (Eastern) & USSR	1,252	36,550	19,266	68	900	254		27	58,317
Europe (Western)	8,098	127,369	72,370	333	2,362	380	816		212,728
Latin America	13	16,131	3,517	2	29	34			19,726
Middle East		851	94			186			1,131
North America	1,688	158,134	96,382	1,004	2,406	1,993	18	200	261,825
Total	11,800	397,136	254,169	1,444	6,939	3,433	80	1,091	646,558

some strains commonly used as standard references for comparative purposes are duplicated in a number of collections, all of them contain an important gene pool of microbial germplasm in the world. Agriculture, medicine, and industry all have benefited significantly from products derived from these cultures. Any of these cultures also offer new sources of biosynthetic products, new sources of genes for hybridization, or even sources of characteristics for genetic engineering. They are recognized as being a vital but often untapped resource for modern biotechnology.

Services provided by culture collections

Microbial culture collections may maintain a large number of strains of a few species, or a few strains of each species, or a large number of strains of all known species in culture. They may also specialize in material relevant to particular fields, such as agriculture, medicine, industry, or forestry, or serve educational needs as well as research and development needs and may be national or international in operation. They may be housed in an educational institution, a research organization, a taxonomic institute, an independent laboratory, or a commercial firm. Any of these collections can be interpreted as a germplasm resource equivalent to seed banks in that they conserve the diversity of microbes that are of past, present, or future interest to scientists in education, research, and industry [3, 4, 5, 72, 73, 77].

In general, there are three major types of culture collections, namely, service collections, institutional collections, and private collections. Service collections are usually well established as permanent resource centers, which are prime sources of reference and type cultures. These collections publish catalogues and charge a fee for cultures available on demand to the scientific community without discrimination. The size of the service collection may vary from a comprehensive range of all types of microbes, such as the American Type Culture Collection, to a limited range of a specific type, such as the Fungal Genetic Stock Center, which maintains primarily genetic stocks of *Neurospora crassa*.

Institutional collections are usually mission-oriented and are established primarily to facilitate ongoing research or teaching programs of the organizations of which they form a part, such as university departments, research institutes, government or industrial laboratories, or hospitals. These collections range in size from being even more diverse than some service collections to those specialized in medical, veterinary, agricultural, or industrial microbes. Usually no cataloges are available, and the use of such collections by others outside the immediate working unit is somewhat limited. The USDA-Agriculture Research Service's Culture Collection

(NRRL) at the Northern Regional Research Center in Peoria, IL, is an example of an institutional collection. It contains about 77,000 cultures of agricultural and industrial significance and includes fungi (both filamentous and yeasts), actinomycetes, and bacteria. The NRRL is internationally known for its research on penicillin and other antibiotics, riboflavin, organic acids, and more recently on food fermentations, mycotoxins, and biomass conversions [79, 80].

Private collections are usually built up by individual research scientists from their research work in a university or research institute. These collections are often highly specialized, containing strains with unusual attributes either cited in publications or kept under strict security as a proprietary collection. No cataloge of holdings is published. However, some researchers do make cultures available from time to time to other scientists free of charge or on an exchange basis when this will not jeopardize their research confidentiality. The continuity of these collections is consequently vulnerable and often at risk of loss due to inadequate storage methods or the termination of research projects or a scientist's career. It is always prudent for research scientists regularly to deposit representative strains with institutional or service collections as an insurance against the loss of this important germplasm.

It is obvious that the functions of the collections outlined above are not mutually exclusive, as many institutional and private collections provide a service function as well by supplying cultures on demand to other institutions and research scientists [77]. In addition to the basic functions of acquistion, authentication, preservation, and distribution of cultures and associated information, other services provided by culture collections may include:

1. *Identification and taxonomic analyses of cultures*, including extensive morphological and biochemical testing, and use of computerized identification systems,
2. *Safe deposits of proprietary cultures* of which the depositor retains ownership and confidentiality,
3. *Contract services* for microbiological testing of materials and components (e.g., toxicity, fungal resistance, sterility, preservative effectiveness, and biocompatibility), custom preparations (e.g., microbial, pharmaceutical and immunological products, and other research reagents and cultures), screening of strains for desirable properties, studies of growth requirements of cultures, development of educational and diagnostic kits, formulation and maintenance of computerized data resources, and any funded contract research and investigative work,
4. *Consulting services* on strain selection of specific uses and purposes; techniques on preservation, packaging and shipping regulations; good

laboratory practice requirements regarding the accuracy and traceability of measuring instruments; quality control procedures, record keeping, management, and administration of non-clinical experiments; and provision of specialist bibliographies and contract studies,

5. *Information services* on availability of cultures, specific facilities, and technical information through strain data management systems, laboratory information management systems, online searches, and regional and international strain data networks, and on exchange of a wide range of information resources by means of regular global and regional newsletters, electronic communication systems, and computer conferences,
6. *Patent depository* for strains referred to in patentable processes, including testing culture viability and providing information and assistance in biotechnology patent procedures,
7. *Publications* of cataloges, newsletters, and technique manuals on culture collection practices and management and on special applications and teaching of specific cultures as well as scientific articles and books,
8. *Sponsoring of workshops and training courses* related to the collection activities, such as preservation and maintenance techniques, biotechnology patent conferences, strain data management, identification and quality control of microbial cultures, recombinant DNA technology, clinical anaerobic bacteriology, and fermentation biotechnology, as well as on a variety of industrial and scientific topics as needs are identified, and
9. *Accommodation of visiting scientists* to pursue their interests relevant to all culture collection activities while at the same time assisting in the improvement of the collection.

Coordination of collection activities

Adequate mechanisms for interactions among existing culture collections could enhance knowledge, national and international cooperation, and security of funding for all valuable and indispensable collections of microbial germplasm. In addition to the World Network of MIRCENS described earlier, many other organizations promote administrative and scientific coordination and information exchange between different collections or between collections and appropriate scientific communities at the international, regional, and national levels [72, 73, 77].

Several international organizations have current and potential initiatives for promoting and enhancing conservation of microbial germplasm and exchange of culture collection information resources. These include: the International Council of Scientific Unions (ICSU), the UNESCO, the

UNEP, the International Cell Research Organization (ICRO), the WFCC, the International Union of Immunological Societies (IUIS), the International Union of Biological Societies (IUBS), the International Union of Microbiological Societies (IUMS), the Committee on Data for Science and Technology (CODATA), the World Intellectual Property Organization (WIPO), Food and Agricultural Organization (FAO), and the World Health Organization (WHO) [73].

World Federation for Culture Collections (WFCC)
At the recommendation of the Conference on Culture Collections held in 1962 in Canada, the International Association of Microbiological Societies (IAMS) set up a 'Section on Culture Collections'. In 1972 the Section was reorganized as the World Federation for Culture Collections (WFCC). The WFCC is now an Interdisciplinary Commission of the IUBS and a Federation within the IUMS. The WFCC was established to create and coordinate a worldwide network of culture collections. All of its activities are conducted through the General Assembly, composed of all members, the Executive Board, and various committees on specific subject matters such as patent depositories, postal and quarantine regulations, data coding, publicity, education, and the safeguarding of endangered collections. The Committees serve as a bridge between the Executive Board and the global community of culture collection. The WFCC engages in collaborative programs with organizations with similar aims and objectives such as UNESO, UNEP, ICRO, IUMS, ICSU, WIPO, FAO, and WHO. For example, the WHO has established international, regional, and national reference centers for various pathogens to provide microbial cultures for rapid and accurate identifications [81].

As a component of the WFCC, the World Data Center for Microorganisms is an information center responsible for the production of a species-oriented directory, including a comprehensive description of each culture collection registered. To facilitate communication between WDC and other organizations [such as WFCC, culture collections, federations, regional and/or national data centers, the MIRCENS and international data activities such as the Microbial Strain Data Network (MSDN)], WDC is using a variety of communication media including mail, telex, cable, facsimile, floppy disks, magnetic tapes, international information network services like electronic mail, and publications. Participating organizations include UNEP, UNESCO, and the Institute of Physical and Chemical Research (RIKEN) in Japan.

The WFCC is also coordinating the development of the Microbial Strain Data Network (MSDN), which is an international information network to refer scientists to sources of strain data on the specific properties of microbes and cell lines [73, 82]. Its major function is to facilitate information exchange rather than to collect strain data in a centralized system. The MSDN Secretariat keeps a central index of strain data held by culture collections or data centers. Nodes of the Network refer scientists to the centers holding the desired information. In addition to WFCC, the MSDN is also sponsored by CODATA and IUMS. Other participating organizations include UNEP, the U.S. National Institute of Dental Research (NIDR), the U.S. National Science Foundation (NSF), the Commission of the European Communities (CEC), and the Belgian Government. MSDN members include personal collections, research collections, national service collections, regional collections, microbial strain data repositories, and international information networks.

European Culture Collections' Organization (ECCO)
The European Culture Collections' Organization (ECCO) was set up in 1982 for the curators of European service culture collections in countries that have microbiological societies affiliated with the Federation of European Microbiological Societies (FEMS). ECCO exists to collaborate and exchange ideas on all aspects of culture collection activities [83]. Thirty-one of the members of ECCO are curators of culture collections in Belgium, Czechoslovakia, East Germany, Finland, France, Hungary, the Netherlands, Norway, Poland, Spain, Sweden, Switzerland, Turkey, the UK, U.S.S.R, and West Germany. Information about the holdings and services of ECCO collections is available from ECCO officers or the Secretary of FEMS.

Within the European Economic Community (EEC), the Commission of the European Community (CEC) supports the establishment of the Microbial Information Network Europe (MINE) which acts as an information exchange medium. MINE will consist of a series of nodes located in regional/national and local centers. Each center will provide dynamic databases containing detailed information on microbial strains with specific properties. Nodes are being established in Belgium, West Germany, the Netherlands, Portugal, and the United Kingdom.

In the UK, the Microbial Culture Information Service (MiCIS) was established by a cooperative effort between the UK Department of Trade and Industry (DTI) and the UK Federation for Culture Collections (UKFCC). MiCIS is a computer database that contains microbial strain data from all UK national culture collections. Eventually it will include data from private and other European collections. The data will include cataloge information, hazards, morphology, enzymes, culture conditions, maintenance requirements, industrial properties, metabolites, sensitivities, and tolerances.

In 1984, the Nordic Council of Ministers formed a Policy Group for Nordic Cooperation on Microbiology, with representatives from all the Nordic countries, which encompasses Denmark, Finland, Iceland, Norway, and Sweden. This group now supervises the Nordic Register of Microbiological Culture Collections, which was established in 1982. For practical reasons the first stage of the register project has been restricted to strains relevant to agriculture, forestry, and horticulture. The register will only facilitate exchange of strain data. Each collection will have the full responsibility for the maintenance and delivery of their own strains [73].

National Federations and Committees

To facilitate communication and to promote coordination, a number of countries throughout the world have established national federations for culture collections or committees of culture collections. The countries are Australia, Canada, China, Czechoslovakia, Japan, Korea, New Zealand, Turkey, the UK and the U.S. (Table 8). Primarily the federation promotes communication between the persons and organizations concerned with culture collections and between them and the users of the collections. The committee coordinates culture collection activities within the country by the persons working in the culture collections. For instance, China maintains a decentralized network of national centers for microbial culture collections, which are housed in seven research establishments with related interests. In 1979, the China Committee of Culture Collections of Microorganisms (CCCCM) was founded to coordinate these national centers for specialized culture collections [84]. On the other hand, the U.S. Federation for Culture Collections (USFCC) maintains liaison between persons interested in culture collections and related activities. USFCC members include any indi-

Table 8. National and regional federations and committees

1. Australian Collections of Microorganisms Committee
2. Australian Federation of Culture Collections
3. Canadian Permanent Committee on Culture Collections
4. China Committee for Culture Collections of Microorganisms
5. Czechoslovakian Federation of Culture Collections
6. European Culture Collections' Organization
7. Japan Federation for Culture Collections
8. Korean Federation of Culture Collections of Microorganisms
9. KUKEM Society for Culture Collections and Industrial Microbiology
10. Latin American Federation of Culture Collections
11. New Zealand Committee on Culture Collections of Microorganisms
12. United Kingdom Federation for Culture Collections
13. United States Federation for Culture Collections

vidual scientist with concern for culture collections and the users of the collections. Coordination is primarily through conferences, symposia, round table discussions, seminars, workshops, and training courses.

However, the prime functions of these national federations and committees, and the ways in which they collaborate with each other, are similar to those of the WFCC [73]. The general objectives of the WFCC, as presently defined, are as follows:

1. To establish effective liaison between persons and organizations concerned with culture collections and between them and the users of cultures,
2. To encourage the study of procedures for the isolation, characterization, conservation, and distribution of microorganisms and to publicize those procedures,
3. To promote the training of personnel in the operation of culture collections,
4. To promote the establishment of a world network of the data services concerned with the location of, and information about, microorganisms maintained in culture collections and to publish derivative information in the form of a world directory of collections, lists of species maintained therein, and related documents,
5. To promote and aid in the establishment of culture collections,
6. To promote the establishment of special reference collections and identification services and to aid existing identification services,
7. To establish official means of communication among culture collections, their personnel, their users, and related organizations,
8. To organize conferences and symposia on topics and problems of common interest,
9. To attempt solution of problems of distribution of cultures of microorganisms that may arise through postal regulations, quarantine rules, patent laws, public health problems, and other factors of international importance,
10. To attempt to provide for the perpetuation of important collections and cultures, and
11. To make recommendations through sponsoring organizations to national and international organizations on any question within these terms of reference when, in the judgement of the Federation, this is desirable or necessary.

References

1. U.S. Congress, Office of Technology Assessment. 1987. Technologies to Maintain Biological Diversity. U.S. Government Printing Office, Washington, DC. 344 pp.
2. U.S. Department of State. 1982. Proceedings of the U.S. Strategy Conference on Biological Diversity. Department of State Publ. 9262. U.S. Department of State, Washington, DC. 125 pp.
3. American Phytopathology Society. 1981. National Work Conference on Microbial Collections of Major Importance to Agriculture. Am. Phytopathol. Soc., St. Paul, MN. 52 pp.
4. U.S. Congress, Office of Technology Assessment. 1984. Commercial Biotechnology, an International Analysis. U.S. Government Printing Office, Washington, DC. 612 pp.
5. National Academy of Sciences. 1979. Microbial Processes: Promising Technologies for Developing Countries. Nat. Res. Council, Washington, DC. 198 pp.
6. Norris, J.R. and D.W. Ribbons (eds.). 1970. Methods in Microbiology. Vols. 3A and 3B. Academic Press, New York. 505 pp.
7. Booth, C. (ed.). 1971. Methods in Microbiology. Vol. 4. Academic Press, New York. 795 pp.
8. Dhingra, O.D. and J.B. Sinclair. 1985. Basic Plant Pathology Methods. CRC Press, Boca Raton, Florida. 355 pp.
9. Jong, S.C. 1981. Isolation, cultivation, and maintenance of conidial fungi. Pp. 551–575 *in* G.T. Cole and B.K. Kendrick (eds.), Biology of Conidial Fungi. Vol. 2. Academic Press, New York. 660 pp.
10. Jong, S.C. and W.F. Atkins. 1985. Conservation, collection and distribution of cultures. Pp. 153–194 *in* D.H. Howard (ed.), Fungi Pathogenic for Humans and Animals. Marcel Dekker, New York. 381 pp.
11. Jong, S.C. 1978. Conservation of the cultures. Pp. 119–135 *in* S.T. Chang and W.A. Hayes (eds.), The Biology and Cultivation of Edible Mushrooms. Academic Press, New York. 819 pp.
12. Jong, S.C. and E.E. Davis. 1987. Germplasm preservation of edible fungi in culture through cryogenic storage. Pp. 213–225 *in* P.J. Wuest, D.J. Royse and R.B. Beelman (eds.), Cultivating Edible Fungi. Elsevier, Amsterdam. 677 pp.
13. Kirsop, B.E. and J.J.S. Snell (eds.). 1984. Maintenance of Microorganisms. A Manual of Laboratory Methods. Academic Press, New York. 207 pp.
14. Smith, D. and A.H.S. Onions. 1983. The Preservation and Maintenance of Living Fungi. Commonwealth Mycological Institute, Kew. 51 pp.
15. Haynes, W.C., L.J. Wickerham, and C.W. Hesseltine. 1955. Maintenance of cultures of industrially important micro-organisms. Appl. Microbiol. 3: 361–368.
16. Carmichael J.W. 1962. Viability of mold cultures stored at $-20°C$. Mycologia 54: 432–436.
17. Sherf, A.F. 1943. A method for maintaining *Phytomonas sepedonia* in culture for long periods without transfer. Phytopathology 33: 330–332.
18. Buell, C.B. and W.H. Weston. 1947. Application of the mineral oil conservation method to maintaining collections of fungous cultures. Am. J. Bot. 34: 555–561.
19. Castellani, A. 1939. Viability of some pathogenic fungi in distilled water. J. Trop. Med. Hyg. 42: 225.
20. Castellani, A. 1960. A brief note on the viability of some pathogenic fungi in sterile distilled water. Imprensa Medica 24: 1–3.
21. Castellani, A. 1962. Long viability (more than 12 months) of many pathogenic fungi in sterile distilled water, suggesting a single method of maintaining fungal strains in mycological collections while apparently preventing pleomorphism of dermotophytes. Ann. N.Y. Acad. Sci. 93: 152–155.

22. Castellani, A. 1963. Further researches on the long viability and growth of many pathogenic fungi and some bacteria in sterile distilled water. Mycopathol. Mycol. Appl. 20: 1–6.
23. Castellani, A. 1967. Maintenance and cultivation of the common pathogenic fungi of man in sterile distilled water. Further researches. J. Trop. Med. Hyg. 70: 181–184.
24. Benedek, T. 1962. Fragmenta mycologica. II. On Oatellani's 'water cultures' and Benedek's 'mycotheca' in chlorallactophenol. Mycopathol. Mycol. Appl. 17: 255–260.
25. McGinnis, M.R., A.A. Padhye, and L. Ajello. 1974. Storage of stock cultures of filamentous fungi, yeasts and some aerobic actinomycetes in sterile distilled water. Appl. Microbiol. 28: 218–222.
26. Kelman, A. 1956. Factors influencing viability and variation in cultures of *Pseudomonas solanacearum*. Phytopathol. 46: 16–17.
27. Boesewinkel, H.J. 1976. Storage of fungal cultures in water. Trans. Brit. Mycol. Soc. 66: 183–185.
28. Marx, D.H. and W.J. Daniel. 1976. Maintaining cultures of ectomycorrhizal and plant pathogenic fungi in sterile water cold storage. Can. J. Microbiol. 22: 338–341.
29. Bakerspigel, A. 1953. Soil as a storage medium for fungi. Mycologia 45: 596–604.
30. Greene, H.C. and E.B. Fred. 1934. Maintenance of vigorous mold stock cultures. Ind. Eng. Chem. 26: 1297–1298.
31. Fennell, D.I. 1960. Conservation of fungous cultures. Bot. Rev. 26: 79–141.
32. Atkinson, R.G. 1954. Quantitative studies on the survival of fungi in five-year old dried soil cultures. Can. J. Bot. 32: 673–768.
33. Weiss, F.A. 1957. Maintenance and preservation of cultures. Pp. 99–119 *in* Manual of Microbiological Methods. Edited by the Society of American Bacteriologists, McGraw-Hill, New York. 315 pp.
34. Perkins, D.D. 1962. Preservation of *Neurospora* stock cultures with anhydrous silica gel. Can. J. Microbiol. 8: 591–594.
35. Perkins, D.D. 1977. Details for preparing silica gel stocks. Neurospora Newsl. 24: 16–17.
36. Barratt, R.W., G.B. Johnson, and W.N. Ogata. 1965. Wild-type and mutant stocks of *Aspergillus nidulans*. Genetics 52: 233–246.
37. Bell, J.V. and R.J. Hamalle. 1974. Viability and pathogenicity of entomogenous fungi after prolonged storage on silica gel at −20°C. Can. J. Microbiol. 20: 639–642.
38. Elliott, R.F. 1975. Viability of fungous cultures dried and stored over silica gel. N.Z. J. Sci. 18: 577–583.
39. Sleesman, J.P., P.O. Larsen, and J. Safford. 1974. Maintenance of stock cultures of *Helminthosporium maydis* (races T and O). Plant Dis. Rep. 58: 334–336.
40. Parina, O.V., V.V. Patrikeev, and S.V. Lysenko. 1972. Investigation of the survival and physiological activity of some yeast strains after long storage in silica gel. Microbiologia 41: 164.
41. Reinhardt, D.J. 1966. Silica gel as a preserving agent for the cellular slime mould *Acrasis rosea*. J. Protozool. 13: 225.
42. Watts, D.J. and J.M. Ashworth. 1970. Growth of myxamoebae of the cellular slime mould *Dictyostelium discoideum* in axenic culture. Biochem. J. 119: 171.
43. Mayne, R.Y., J.W. Bennett, and J. Tallant. 1971. Instability of an aflatoxin-producing strain of *Aspergillus parasiticus*. Mycologia 63: 644–648.
44. Norris, D. 1944. Preservation of tube cultures of fungi and bacteria with liquid paraffin. J. Aust. Inst. Agric. Sci. 10: 77.
45. Grivell, A.R. and J.F. Jackson. 1969. Microbial culture preservation with silica gel. J. Gen. Microbiol. 58: 423–425.
46. Goldie-Smith, E.K. 1956. Maintenance of stock cultures of aquatic fungi. J. Elisha Mitchell Sci. Soc. 72: 158–166.

47. Gherna, R.L. 1981. Preservation. Pp. 208–217 *in* P. Gerhardt (ed.), Manual of Methods for General Bacteriology. Am. Soc. for Microbiol., Washington, DC. 524 pp.
48. Stamp, L. 1947. The preservation of bacteria by drying. J. Gen. Microbiol. 1: 251–265.
49. Hunt, G.A., A. Gourevitch, and J. Lein. 1958. Preservation of cultures by drying on porcelain beads. J. Bacteriol. 76: 453–454.
50. Lang, B.J. and W.J.R. Boyd. 1968. Preservation of fungal spores by drying on porcelain beads. Phytopathol. 58: 1711.
51. Kirsop, B. 1955. Maintenance of yeasts by freeze-drying. J. Inst. Brew. 61: 466–471.
52. Wynants, J. 1962. Preservation of yeast cultures by lyophilization. J. Inst. Brew. 68: 350–354.
53. Mehrotra, B.S. and C.W. Hesseltine. 1958. Further evaluation of the lyophil process for the preservation of Aspergilli and Penicillia. Appl. Microbiol. 6: 179–183.
54. Jong, S.C. and E.E. Davis. 1976. Reproducibility of vitamin assays with mutants of *Neurospora crassa* maintained by freeze-drying. Can. J. Microbiol. 22: 1062–1064.
55. Hwang, S.W. 1966. Long-term preservation of fungus cultures with liquid nitrogen refrigeration. Appl. Microbiol. 14: 784–788.
56. Haskins, R.H. 1957. Factors affecting survival of lyophilized fungal spores and cells. Can. J. Microbiol. 3: 477–485.
57. Onions, A.H.S. 1971. Preservation of fungi. Pp. 113–151 *in* C. Booth (ed.), Methods in Microbiology, Vol. 4, Academic Press, New York. 795 pp.
58. Raper, K.B. and D.F. Alexander. 1945. Preservation of molds by the lyophil process. Mycologia 37: 499–525.
59. Haskins, R.H. and J. Anastasiou. 1953. Comparisons of the survivals of *Aspergillus niger* spores lyophilized by various methods. Mycologia 45: 523–532.
60. Annear, D.I. 1954. Preservation of bacteria. Nature 174: 359.
61. Annear, D.I. 1956. The preservation of bacteria by drying in peptone plug. J. Hyg. 54: 487.
62. Annear, D.I. 1962. Recoveries of bacteria after drying on cellulose fibers (A method for the routine preservation of bacteria). Aust. J. Exp. Biol. Med. Sci. 40: 1–8.
63. Iijima, T. and T. Sakane. 1973. A method for preservation of bacteria and bacteriophages by drying *in vacuo*. Cryo-biol. 10: 379–385.
64. Iijima, T. and T. Sakane. 1973. Method for preservation of bacteria and bacteriophages by drying *in vacuo*. IFO Res. Comm. 6: 4–17.
65. Banno, I., K. Mikata, and T. Sakane. 1979. Viability of various yeasts after L-drying. IFO Res. Comm. 9: 27–34.
66. Banno, I. and T. Sakane. 1979. Viability of various bacteria after L-drying. IFO Res. Comm. 9: 35–45.
67. Mikata, K., S. Yamauch, and I. Banno. 1983. Preservation of yeast cultures by L-drying: Variabilities of 1710 yeasts after drying and storage. IFO Res. Comm. 11: 25–46.
68. Yokoyama, T. and I. Asano. 1983. Preservation of ISP strains of Actinomycetes by L-drying. IFO Res. Comm. 11: 47–59.
69. Kusaka, T., K. Sato, and I. Asano. 1985. Preservation of antimicrobial activity in L-dried and agar-transplanted Actinomycetes. IFO Res. Comm. 12: 90–100.
70. Kaneko, Y., K. Mikata, and I. Banno. 1985. Maintenance of recombinant plasmids in *Saccharomyces cerevisiae* after L-drying. IFO Res. Comm. 12: 78–82.
71. Butterfield, W., S.C. Jong, and M.T. Alexander. 1978. Polypropylene vials for preserving fungi in liquid nitrogen. Mycologia 70: 1122–1124.
72. Porter, J.R. 1976. The world view of culture collections. Pp. 62–72 *in* R.R. Colwell (ed.), The Role of Culture Collections in the Era of Molecular Biology. Am. Soc. Microbiol., Washington, DC. 76 pp.

73. United Nations Educational, Scientific and Cultural Organization. 1984, 1985, 1986 & 1987. MIRCEN News. 6, 7, 8 & 9. UNESCO, Paris.
74. McGowan, V.F. and F.B.D. Skerman. 1986. World Catalogue of *Rhizobium* Collections. 3rd edition. World Data Center for Microorganisms, Queensland, Australia. 325 pp.
75. Holliday, J. and P. Somasegaran. 1984. The *Rhizobium* germplasm Resource of NifTAL Catalogue of strains. Univ. of Haw. NifTAL. Project and MIRCEN. Hawaii. 76 pp.
76. Rolz, C. and R. de Leon. 1986. Summary of current research activities in the Central American Biotechnology MIRCEN. MIRCEN J. 2: 221–230.
77. Sly, L.I. 1986. Culture collection technologies and the conservation of our microbial heritage. Pp. 1–37 *in* H.W. Doelle and C.G. Heden (eds.), Trends in Scientific Research 2. Reidel, Dordrecht, and UNESCO, Paris.
78. Staines, J.E., V.F. McGowan, and V.B.D. Skerman. 1986. World Directory of Collections of Cultures of Microorganims. World Data Center, Univ. of Queensl., Brisbane, Australia. 678 pp.
79. Kurtzman, C.P. 1986. The ARC culture collection: present status and new directions. Enzyme Microb. Technol. 8: 328–333.
80. Rossman, A.Y. 1987. USDA maintains microbial germplasm collections. USFCC Newsl. 17: (4): 3.
81. World Health Organization. 1968. The Work of WHO Virus Reference Centers and the Services They Provide. World Health Organization, Geneva. 120 pp.
82. Hill, L.R. and M.I. Krichevsky. 1985. Needs and Specifications for an International Microbial Strain Data Network. United Nations Environment Programme. Nairobi, Kenya. 58 pp.
83. Hawksworth, D.L. 1985. Fungus culture collections as a biotechnological resource. Biotechnol. Genet. Eng. Rev. 3: 417–453.
84. China Committee for Culture Collections of Microorganisms. 1983. CCCCM Catalogue of Cultures. Light Industry Publishing House, Beijing. 422 pp.

Gene transfer in plants

BENJAMIN F. MATTHEWS[1] and JAMES A. SAUNDERS[2]
[1]*Plant Molecular Biology Laboratory, Plant Sciences Institute, Agricultural Research Service – USDA, Beltsville, Maryland, 20705, U.S.A.*
[2]*Germplasm Quality and Enhancement Laboratory, Plant Sciences Institute, Agricultural Research Service – USDA, Beltsville, Maryland, 20705, U.S.A.*

Abstract With the specter of dwindling biotic diversity, need is increased for preservation of available germplasm. Plant germplasm may be preserved in a variety of ways including seed storage, cryopreservation, and tissue culture. A new method of germplasm preservation may be through the storage of DNA libraries, which maintain in test tubes the genes that encode life's diversity. This new area of plant research germinated more than a decade ago and is beginning to bear fruit today. To be a viable alternative to traditional methods, however, ways must be devised to return cryopreserved genes to living cells, where they can be expressed and utilized to man's benefit. DNA libraries also can be used to increase genetic diversity greatly through methods involving molecular biology and gene transfer.

Numerous strategies have been developed to transfer genes into plants. Some of these techniques can be used to combine (entire) whole genomes by fusing two plant protoplasts. These somatic hybridization procedures have been refined to allow the combination of nuclear and organelle genomes from different plant species. Transfer procedures also have been developed for moving genes from chloroplasts, mitochondria, or nuclei into alien cytoplasms. Still other techniques can shuttle single genes from one plant species to another. These gene transfer techniques are being developed at a rapid pace to complement equally rapid developments in gene isolation, cloning, and recombination techniques.

A multitude of barriers inhibiting the transfer of genes to plants plagued the agricultural scientists of the 1970's, and many problems face these scientists today. But significant progress is being made, and further refinements in current plant gene transfer techniques will provide a strong base for future crop improvement using biotechnology.

Key words: protoplast fusion, electrofusion, electroporation, Ti-plasmid, pollen, transformation

Introduction

Gene transfer is no longer a vision of the future. Gene transfer is a technology of today. Scientists now can transfer genes from one plant to another. Genes isolated from animals and bacteria can even be transferred into plants. We are now learning how to control the regulation of genes after transfer, so that those genes are expressed at the proper time of plant development, in the correct tissues, and at the optimum levels. Through gene transfer, gene diversity of plants can be increased by inserting new genes and devising new

L. Knutson and A.K. Stoner (eds.). Biotic diversity and germplasm preservation, global imperatives.
 ISBN 0-7923-0178-1

gene combinations, while bypassing sexual breeding barriers. Gene transfer may make it feasible to store germplasm and, at a future date, isolate desired genes from the stored germplasm and transfer it into commercial varieties. Perhaps someday rare germplasm may even be stored simply as DNA.

The new gene transfer techniques allow the possibility for the development of more productive crop plants at a pace much faster than before. Gene transfer and selection of the best plants has been the objective of breeding programs for years. Powerful new techniques for gene transfer recently have been developed for moving single genes and whole blocks of genes from one plant to another and even for moving genes from non-plants into plants. These new techniques can help mankind preserve the genetic diversity of plants by actually creating greater diversity through gene recombination. This is very important for future generations because, as human civilization advances and uninhabited areas are destroyed, natural genetic diversity dwindles. To preserve nature's rich legacy and to ensure mankinds's continuing progress and well-being, scientists are challenged to find not only means to preserve existing germplasm but procedures to create new germplasm. Newly developed techniques for gene transfer can help mankind in this struggle. These new gene transfer techniques, when blended with conventional breeding methods, promise to revolutionize agriculture in the 21st century.

Rationale for gene transfer

The development of new plant varieties possessing superior agronomic traits is limited by conventional breeding because of sexual breeding barriers. When attempts are made to sexually breed two widely divergent plant species, often no seeds are formed or low seed numbers are produced. Sometimes these seeds are sterile or produce plants that are of little commercial value unless backcrossed for numerous generations to remove unwanted traits introduced during the initial cross. This process is very time-consuming, costly, and does not assure success. In some instances, detrimental genes are closely associated with agronomically valuable genes and are difficult or impossible to separate by breeding. In other instances, no viable seeds are produced from a cross between genetically dissimilar breeds. The breeder may resort to embryo rescue and artifically culture the zygote or may try a bridge hybridization, using a plant that is sexually compatible with both plants. This technique is even more difficult and time consuming. Finally, some crosses simply cannot be made through use of any technique known.

However, with the new tools of gene transfer we can combine single genes, multiple genes, organelles, and even whole cells to produce new genetic combinations. These new gene transfer tools include protoplast fusion, electroporation, polyethylene glycol, and the use of *Agrobacterium tumefaciens* to introduce DNA into plants. These and other gene transfer tools will help the new plant breeder meet the agricultural demands of the 21st century.

Protoplast fusion

Many genes could contribute importantly to crop productivity if they could be incorporated into the desired crop plant. For example, genes encoding disease resistance, pest resistance, and tolerance to heat, salt, and drought could all improve crop yields. However, these genes often reside in diverse plant species, which make crosses difficult due to sexual incompatibility with the crop species susceptible to that biotic or abiotic stress. Protoplast fusion is one method that has been developed for the transfer of large numbers of genes, polygenic traits, and traits encoded by uncloned genes among sexually incompatible plants. Protoplast fusion has been the subject of numerous reviews [1–4]. In general, both the nuclear and mitochondrial genomes are combined, while the chloroplast genome oftentimes remains identical to that of one parent [5–8]. However, techniques recently have been developed that promote chloroplast genome recombination [9].

Protoplast fusion is now a routine laboratory technique, although it still requires considerable skill and attention. The walls of the cells to be combined are enzymatically removed, and the resultant protoplasts are merged by fusion (Fig. 1). The resultant fusion products must be separated from

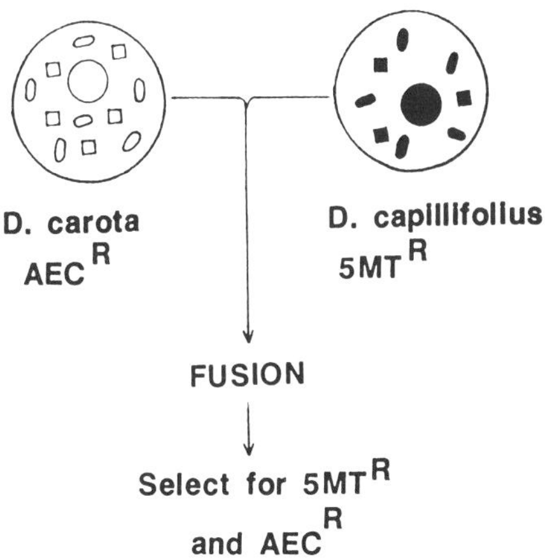

Figure 1. Selection scheme for somatic hybrids. Protoplasts from aminoethyl cysteine (AEC) resistant cells of *Daucus carota* and from 5-methyltryptophan (5MT) resistant cells of *D. capillifolius* are fused. Somatic hybrids are identified by their growth in the presence of both AEC and 5 MT.

unfused protoplasts and from products of self-fusion. This separation can be achieved by tagging the parental protoplasts with different selectable markers (for reviews see 1, 2). One of the simplest methods used to identify somatic hybrids employs visual markers. One parental type of protoplast may be isolated from green mesophyll cells, while the other parental type may be derived from cell suspension cultures and would be translucent. Another system for the identification of fusion progeny uses different fluorescent dyes to identify each parent [10, 11]. In each of these examples, the fusion products can be selected and removed from background protoplasts with a micropipette or cell sorter and the fused single cell then can be cloned.

Other schemes for the selection of somatic hybrids include the use of compounds that selectively inhibit the growth of one parental type [12–14]. Thus, only somatic hybrids that express a double resistance will grow when in the presence of both inhibitory compounds. Numerous other methods also have been employed.

Similarly, several methods of protoplast fusion also have been developed. The most extensively used techniques employ chemical fusigens such as polyethylene glycol or changes in pH and calcium levels. A more recently developed technique is electrofusion, which is the fusion of two cells by an electric charge [3].

Protoplast fusion can be used to combine the nuclear genomes of two plants, exchange chloroplasts, and recombine mitochondrial DNA. The presence of portions of both nuclear genomes often can be demonstrated by examination of the karyotype of the somatic hybrid, which may contain all or most of the chromosomes of both parents. Our laboratory has succesfully used isoenzymic patterns on electrophoretic gels to confirm the identify of *Daucus* somatic hybrids [15]. When an isoenzyme from one parent migrates on a polyacrylamide gel at a different rate than that of the other parent, the isoenzymic pattern of the somatic hybrid often shows the presence of both isoenzymes. In fact, we have shown new isoenzymic forms present in the hybrid as a result of mixing of subunits of the enzyme in the somatic hybrid. We also have examined mitochondrial DNA recombination in *Daucus* somatic hybrids by analyzing patterns of mitochondrial DNA endonuclease restriction fragments on agarose gels (Fig. 2). The somatic hybrid contained some of the mitochondrial DNA fragments specific to each of the parents and also contained new DNA fragments not present in either parent. Thus, protoplast fusion can be used to transfer large blocks of nuclear and cytoplasmic genes.

Electrofusion

The fusion of somatic cells by use of chemical fusogenic agents, initiated in the 1960's by Cocking [16], represented an important beginning in the

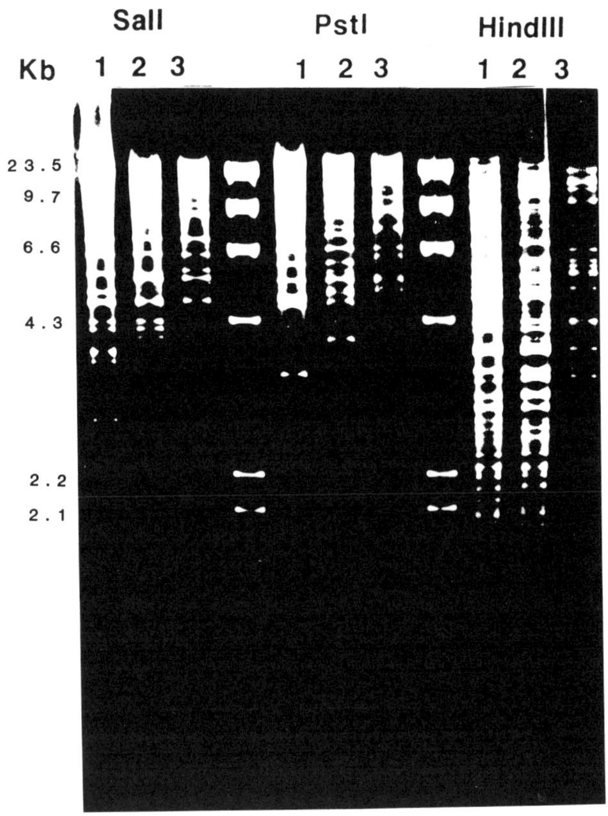

Figure 2. Pattern of mitochondrial DNA restriction fragments of *D. carota* (1), a somatic hybrid (2), and *D. capillifolius* (3). The somatic hybrid has some of the mitochondrial DNA fragments present in each parent as well as unique fragments, indication of mitochondrial DNA recombination. Restriction enzymes are listed above the lanes.

development of gene transfer techniques in plants. This was the first time that non-gametic cells from higher plants had been used to combine nuclear material from two distinct cells without the use of sexual reproduction. These resultant cell lines and later regenerated hybrid plants could be produced without the necessity of using gametic cells and confrontation with sexual incompatibility barriers. Although numerous research studies were to result from this important discovery, commercial application of this technique in major agronomic crops was scant over the next twenty years. In the early 1980's, use of these chemical fusogens was replaced in many research laboratories by the simple yet elegant technique of electrofusion of protoplasts [3, 17]. This reversible procedure breaks down the cell membrane of adjacent cells by use of brief high voltage discharges. Upon resealing, many adjacent cells reanneal at the point of cell membrane disruption and fuse, creating a single cell with two nuclei. In plants, with the subsequent nuclear fusion of these two nuclei, a hybrid cell line would be formed. In some cases, this cell line could be induced through tissue culture techniques to grow and regenerate a hybrid plant.

In electrofusion experiments with plants, it has been necessary first to isolate protoplasts by mechanically or enzymatically stripping off the cell wall, which represents a substantial physical barrier for the fusion of the internal cell membranes. This, of course, is much less of a problem in fusion of animal cells than with plant cells because of the lack of the cell wall in animal cells; however, animal systems lack the potential to regenerate into mature organisms and therefore only can be induced to form hybrid cell lines by use of this technique. Whether plant or animal cells are used for electrofusion, the cells must be brought into close proximity so that adjacent cell membranes can fuse during the reannealing phase. Cells have been placed in juxtaposition by a number of procedures, including cell-surface binding agents [18] and packed cell densities [19]. In addition to these techniques, one of the simplest procedures used to move nearby cells into close proximity has been the process of dielectrophoresis [20]. This procedure exposes the cells in a low ionic media to an alternating sine wave (A.C.) low-voltage current. The dielectrophoretic current creates charged dipoles in the cells in this field because of the inherent ionic salts present in individual cells. In a non-homogenous field, the cells migrate to areas of highest field strength. The induced attractive forces of the dielectrophoretic current are sufficient to overcome the weaker repulsive forces on the surface of the membrane. As the field strength of areas rich in cells is increased, due to conductivity of the cells, additional cells are drawn into the region, and this further enhances the probability of successful fusion events.

Although electrofusion techniques now are used commonly in many laboratories, very few reports have appeared in the literature citing recovery of stable regenerated hybrids from electrofused plant protoplasts. This may, in part, be a result of the dynamic nature of this rapidly expanding field. Alternatively, it may result from difficulties encountered in either the stability of the somatic hybrid during subsequent tissue culture procedures or unexplained modifications to the cell lines that present problems in the regeneration of mature plants. In cooperation with Dr. Steven Sinden of the Vegetable Laboratory, USDA, Beltsville, Maryland, our laboratory has successfully fused and regenerated protoplasts from two different potato species by use of electrofusion techniques [Table 1.] In these experiments, mesophyll protoplasts isolated from *Solanum tuberosum* were electrofused with leaf mesophyll protoplasts isolated from *Solanum chacoense* by a single 40 μsec D.C. pulse of 1.2 kvolts/cm. These fused cell lines, cultured and regenerated on modified MS media, show unique biochemical characteristics

Table 1. List of characteristics of hybrid potato

Solanum tuberosum	*Solanum chacoense*	*Hybrid*
(USW)	(55–1)	
Regenerates in tissue culture	Never regenerates in tissue culture	Regenerates in tissue culture
Grows slowly in culture	Grows slowly in culture	Grows rapidly in tissue culture
Contains tomatidinol	Contains acetylated solanidine	Contains both tomatidinol and acetylated solanidine
Pubescent leaves	Glabrous leaves	Pubescent leaves
No anthocyanidin	Anthocyanidins present in leaf and stem	Anthocyanidins very apparent in leaf and stem
Flowers white	Flowers blue	Flowers bluish

of both parental cell lines (Fig. 3). This particular hybrid was partially selected by the inability of *chacoense* to regenerate in tissue culture and the inability of *tuberosum* to produce the readily apparent red anthrocyanin pigment on the leaves and stem of the regenerated plant.

Figure 3. Comparison of *Solanum chacoense* (A), fusion hybrid (B), and *S. tuberosum* (C). The fusion hybrid, with characteristics of each parental type, was formed through electrofusion of mesophyll protoplasts of the parent potato species shown.

The increase in chromosomes or ploidy number of the resultant hybrid cell line is one inherent difficulty that must be addressed in somatic fusion procedures, whether they are produced through chemical fusogens or through electrofusion. In plant species known to be tolerant of variation in ploidy numbers (e.g., tobacco and potato), successful somatic hybrids can be obtained much more frequently than in other species less tolerant of modifications in chromosome numbers. This difficulty has been addressed partially by use of enucleated cells as one of the fusion parents. In this way, cytoplasmically encoded genes can be transferred to fully developed somatic cells through electrofusion procedures without an increase in the chromosome number. This procedure, however, is only useful for those traits that are expressed by chloroplast or mitochondrial genomes, which creates a severe limitation on the number of genes that can be transferred by cell enucleation.

Organelle transfer

In many cases, it is desirable to transfer only organellar traits. Several important genes are encoded within the organelles. Genes encoding resistance to the triazine herbicides [21, 22], such as atrazine, are good examples of chloroplast encoded traits. In addition, the large subunit of ribulose bisphosphate carboxylase, which fixes carbon in plants, also is located in the chloroplast. An important trait encoded within the mitochondrion is cytoplastic male sterility [23, 24]. Cytoplastic male sterility is quite important in breeding programs throughout the world and makes hybrid seed production substantially easier. The transfer of organelles containing these genes from one plant species to another has great commercial implications. The transfer of chloroplasts and mitochondria from a donor protoplast to a recipient protoplast without the transfer of nuclear DNA into the recipient protoplast allows much more specific transfer of only a subset of the total genetic material of a specific cell. It is desirable in some cases not to move nuclear DNA into the recipient protoplast, if possible, because the donor crop species may have nuclear traits that are totally undesirable. These traits could have a devastating effect on a breeding program. If inserted into a commercial crop species, it could decrease the commercial value of the newly developed hybrids. Some techniques recently have been refined so that organelles can be transferred without the nucleus. Several methods of organelle transfer have been shown to work. One such method (Fig. 4) involves the enucleation of the donor protoplasts, which produces a cytoplast containing only organelles [2–28]. The cytoplasts then are fused with the recipient protoplasts, which may be treated with iodoacetate to inactivate those organelles. In another method developed in Galun's laboratory [29–32], the donor nucleus is disrupted by X-ray irradiation so that it is no

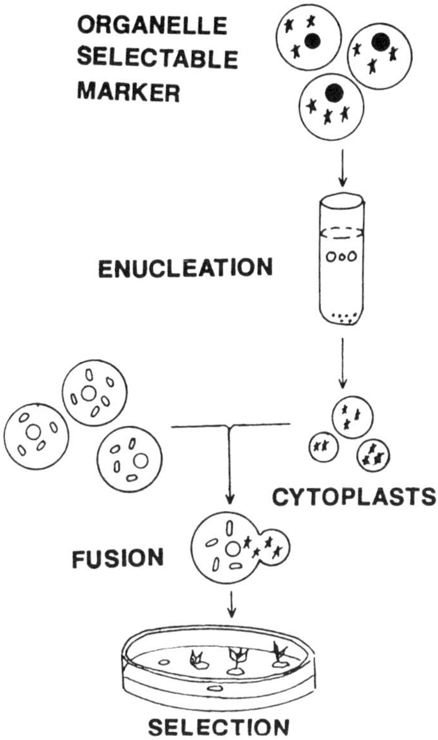

Figure 4. Enucleation of protoplasts by centrifugation yields cytoplasts containing organelles with selectable markers. These cytoplasts can be fused with protoplasts to transfer the organelles. Hybrids are recovered on selection medium.

longer viable (Fig. 5). The recipient protoplast is treated with iodoacetate to prevent these organelles from replicating. The protoplasts are fused and hybrids selected. These methods more precisely incorporate organelle-encoded traits than do techniques that fuse protoplasts with intact, viable nuclei.

Single gene transfer

As genes are identified and better understood, it has become apparent that the method of choice for most gene transfers is one that transfers only the specific DNA sequences desired and which inserts these sequences in the desired position on the chromosome or organelle genome. Thus only defined sequences of DNA would be inserted into the recipient genome. There would be no transfer of unknown genes or sequences and no transfer of undesirable genes or sequences. The appropriate promoters and regulators

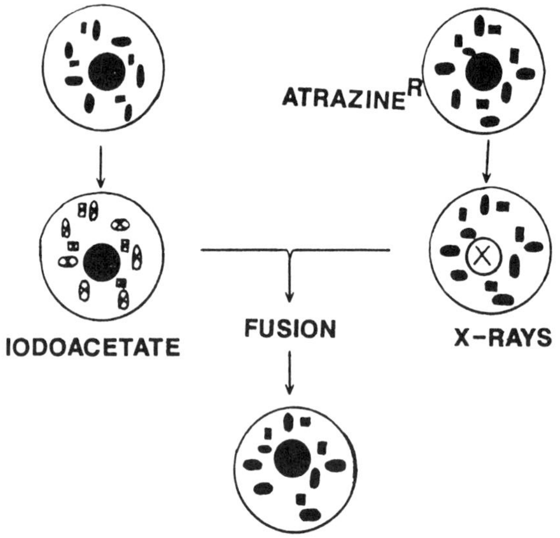

Figure 5. Fusion of donor-recipient protoplasts is used to transfer organelles from one plant species to another without transferring and combining nuclei. The organelle donor is treated with X-rays to disrupt the nucleus, and the recipient is treated with iodoacetate to disrupt the organelles. After fusion, the hybrids are selected by means of the proper functioning of both nuclear and organellar genomes.

would be incorporated into the gene construct so the gene is translated at the proper time in the desired cells and in amounts suitable to the plant.

Ti plasmid-mediated transfer

Numerous investigators are making remarkable strides towards gene transfer mediated through *Agrobacterium tumefaciens*. The tumor inducing (T)-DNA of *A. tumefaciens* has been extensively studied and developed as a vector for the stable integration of foreign DNA into plant cells.

The transfer of T-DNA into plant cells relies upon the presence of two border sequences, which are 24-bp imperfect direct repeats flanking the T-DNA. The left-hand border repeat can be deleted without significantly decreasing pathogenicity; however, if the right-hand border is deleted, pathogenicity is totally eliminated [33, 34]. This border sequence is also polar, because alteration of the orientation of the right-hand sequence with respect to the Ti plasmid attenuates the effectiveness of transfer and integration. The T-DNA is processed by nicking at unique, identical positions of the border sequences [34]. The major intermediate of T-DNA transfer is believed to be a linear, single-stranded T-DNA molecule [34, 35]. In binary vectors, only one copy of the 25-nucleotide T-DNA border sequence is necessary to direct and allow stable integration of T-DNA in plants [33]. In some cases the target DNA of the plant is rearranged upon the integration of the T-DNA [36]. The variety of insertion, deletion, and duplication of the target DNA indicates that T-DNA integration is a multistep process that includes replication and repair processes conducted by enzymes of the recipient plant cell.

Regenerated plants transformed by Ti plasmid T-DNA transmit the T-DNA in a Mendelian manner. In most cases, the trait is inherited as a single locus [37, 38]; however, some do segregate as two independent loci [37]. Although molecular and genetic analysis of transgenic plants is still limited, it appears that the T-DNA integrates at different sites. Wallroth *et al.* [39] demonstrated that six transgenic *Petunia hybrida* plants were independently transformed at six different chromosome loci. Similarly, Ambros *et al.* [40] showed that the T-DNA integrated into the chromosomes of transgenic *Crepis capillars* at different sites. Two tranformed lines of *C. capillars* had one T-DNA insertion, while two other lines had two copies of T-DNA. In one case the T-DNAs were on separate chromosomes, whereas in the other case the T-DNAs were integrated into the same chromosome close together, although not in tandem.

Much new knowledge is being gained about the sequences controlling the developmental regulation of specific genes through the use of gene transfer and stable insertion of foreign genes into plant cells via the Ti plasmid system. The gene encoding the small subunit of ribulose-1,5-bisphosphate carboxylase (Rubisco) has been examined extensively [41, 42]. Mutations in the putative regulatory regions of the Rubisco gene have been constructed, and the effects of these alterations on gene expression have been examined. Broglie *et al.* [43] have inserted different constructs of the gene for the small subunit of pea Rubisco into petunia protoplasts. The resulting tranformed *Petunia* calli synthesized pea Rubisco small subunit protein. This synthesis was light-regulated, and there was a 50-100 × increase in the transcript level of the small subunit of pea Rubisco.

Two different pea genes encoding the small subunit of Rubisco were examined by Fluhr and Chua [44] to determine the effects of light quality on the light-induced expression of the two genes. Different responses were obtained from the two genes depending upon the developmental stages of the leaves examined. Phytochrome mediated the response of the two genes in etiolated leaves, while phytochrome and a blue-light receptor mediated the responses of the two genes in mature leaves. Through gene construction and transfer experiments such as these, it now is possible to dissect the gene to determine which sequences are important in the regulation of transcription of that gene.

Because of the limited host range of *A. tumefaciens*, it has been much more difficult to transfer genes to monocots than to dicots by use of the Ti plasmid. Several new approaches may allow genes to be transferred readily to monocots as well as to dicots. One general method, which holds great promise, electrically induces the uptake of DNA into plant protoplasts by a process very similar to electrofusion. This method, called electroporation, avoids the use of *A. tumefaciens* and potentially can be used on a wider range of crop plants.

Electroporation

In the process of electrofusion research, it was noted that the reversible breakdown of the cell membrane occurred at localized regions on the cell so that the disruption caused a small hole or pore in the membrane [45]. It was in these regions that adjacent membranes would reaneal to form a common cell or hybrid. In 1982, Neuman [46] reported that cloned DNA could be taken up through these pores if the cloned DNA in the cell suspension media was at a relatively high concentration. This process, termed electroporation, represents an effective modification of the electrofusion procedures by eliminating some of the drawbacks associated with the increases in chromosome numbers during electrofusion. The electroporation process is somewhat less complex than electrofusion because cells or protoplasts need not be brought into close contact to effect gene transfer. Thus the need for the dielectrophoresis step of the fusion process is eliminated. This modification of the electrofusion process also enables the researcher to employ a fundamentally simpler and less expensive electroporation device. Two different electroporation machines currently are in use. The first uses a square-wave D.C. pulse to provide a specified pulse of known voltage over a controlled time. The second approximates this square-wave pulse by using an exponentially decaying capacitor-discharge wave. The capacitor pulse decays over a much longer time period than the duration of the square-wave pulse. A great deal of confusion has been introduced into the literature because of the use of these two different types of wave forms for the electroporation of cells. Using *Nicotiana tabacum* protoplasts, we found that viral RNA was optimally incorporated with a square-wave pulse of 60–80 uSec and a field strength of approximately 2 kvolts/cm. These conditions maintain viability of the protoplasts at greater than 95% and yield incorporation rates for the uptake and expression of cloned RNA at 50%. Using a capacitor-discharge D.C. pulse generator, incorporation of RNA into an aliquot of the same protoplasts was 25% at an optimal field strength of 1 kvolts/cm (maximum peak). By use of a minimum capacitance of 50 uF, the viability of these protoplasts was decreased to a far greater extent with the exponential pulse than the square-wave pulse (36% of the controls) [Saunders *et al.* in press].

One disadvantage inherent in both protoplast fusion and protoplast electroporation is the necessity to culture and regenerate the genetically modified germplasm into mature plants. Although the regeneration of many plants from protoplast culture has been reported, important groups of economic crops are recalcitrant to regeneration. At best the procedure is an arduous, time-consuming task that may be prone to difficulties associated with contamination of the tissue cultures or to unstable germplasm lines.

Other methods of direct gene transfer

Although the use of the Ti plasmid is the most common method used to transfer single genes into plants, electroporation and other methods are being developed as alternatives. These methods may be most useful with certain plants, such as monocots, which are not readily infected by *A. tumefaciens*. These methods also may prove to be simpler and easier, but they also must be refined into highly efficient and reliable methods.

One new method for direct gene transfer involves the incubation of naked DNA with plant protoplasts. Polyethylene glycol is added as a fusigen to yield transformants [47, 48]. Protoplasts of several plant species, such as tobacco [47, 49], soybean [50], and petunia [51], have been successfully transformed by direct gene transfer. This method may be useful in transferring genes into plants that are not susceptible to *A. tumefaciens* infection, but the regeneration of intact plants from protoplasts does pose a major problem with some species.

Our laboratory is currently developing a possible alternative to the elec-

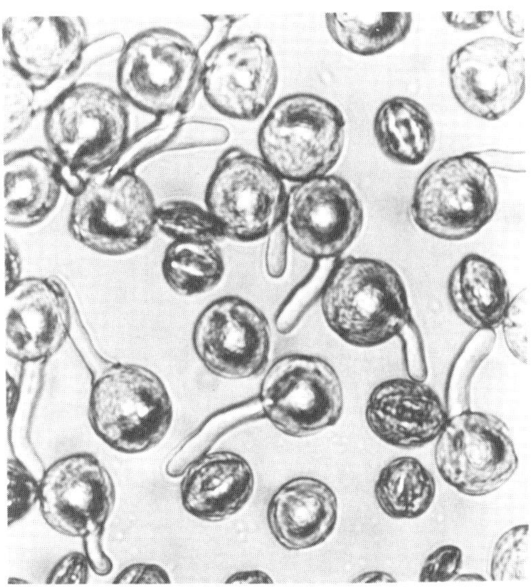

Figure 6. Germinating tobacco pollen for transformation studies. The pollen is incubated in a germination medium to promote the initiation of the pollen tube, which has a thin cell wall.

troporation of protoplasts and subsequent tissue culture steps to introduce foreign genes into plants. We are exploring the possibility of electroporating cloned DNA directly into germinating pollen grains and then fertilizing developing ovules with the genetically modified pollen. This technique may completely circumvent tissue culture procedures required for current procedures. We have used germinating pollen with an extremely thin cell wall as the recipient of the electroporated DNA (Fig. 6). Naked DNA is then added to the pollen, which is treated with polyethylene glycol or by electroporation to insert the DNA into the germinating pollen tube containing the generative nucleus. After the DNA has been incorporated, the pollen is placed on the stigma of a receptive plant species. After seed set, the seeds are collected and germinated on selective media allowing the identification of transformants. The transformants are rescued and grown to full plants, selfed, and the resulting seed is examined for the presence of the marker gene and other incorporated genes. If this method proves successful and reliable, regeneration of plants from protoplasts normally will not be a problem in direct gene transfer experiments.

Summary

Several reproducible methods are now available for transferring genes into plants. Large blocks of genes may be moved among closely related plants species, and whole plants may be obtained from this process. As the parents of choice are more distantly related, the problem of obtaining regenerated plants becomes dominant.

The transfer of single or multiple traits encoded on cloned, defined DNA is also possible and will become the method of choice as more and more genes are identified and characterized. Several methods are now available for the transfer of this DNA and may include the use of the Ti-plasmid or the use of electroporation or polyethylene glycol. Some techniques being developed may eliminate host range barriers of *A. tumefaciens* and may eliminate the need to regenerate plants by use of tissue culture. The real challenge of the future will be to clone genes of high agronomic value and insert them into crop plants, where they function properly at the correct time and place, without compromising other traits of the plant. However, we must have the genes available to transfer. Germplasm preservation is important for the ultimate success of gene transfer. The loss of one species is the loss of tens of thousands of genes, many of which are unique, encoding that species. Can we afford that loss?

References

1. Evans, D.A. 1983. Protoplast fusion, pp. 291–321. In Evans, D.A., Sharpe, W.R., Ammirato, P.V. and Yamada, Y. (eds.), Handbook of Plant Cell Culture. Macmillan, New York.
2. Gleba, Y.Y. and Evans D.A. 1983. Genetic analysis of somatic hybrid plants, pp. 322–357. In Evans, D.A., Sharpe, W.R., Ammirato, P.V. and Yamada, Y. (eds.), Handbook of Plant Cell Culture. Macmillan, New York.
3. Bates, G.W., Nea, L.J. and Hassenkampf, C.A. 1987. Electrofusion and plant somatic hybridization, pp. 479–496. In Sowers, A.E. (ed.), Cell Fusion. Plenum Press, New York.
4. Saunders, J.A. and Bates, G.W. 1987. Chemically induced fusion of plant protoplasts, pp. 497–520. In Sowers, A.E. (ed.), Cell Fusion. Plenum Press, New York.
5. Scowcroft, W.R. and Larkin, P.J. 1981. Chloroplast DNA assorts randomly in intraspecific somatic hybrids of *Nicotiana debneyi*. Theor. Appl. Genet. 60: 179–184.
6. Schiller, B., Herrmann, R.G. and Melchers, G. 1982. Restriction endonuclease analysis of plastid DNA from tomato, potato and some of their somatic hybrids. Molec. Gen. Genet. 186: 453–459.
7. Cseplo, A., Nagy, F. and Maliga, P. 1984. Interspecific protoplast fusion to rescue a cytoplasmic lincomycin resistance mutation into fertile *Nicotiana plumbaginifolia* plants. Molec. Gen. Genet. 198: 7–11.
8. Pental, D., Hamill, J.D. and Cocking, E.C. 1984. Somatic hybridisation using a double mutant of *Nicotiana tabacum*. Heredity 53(1): 79–83.
9. Medgyesy, P., Erzsebet, F. and Maliga, P. 1985. Interspecific chloroplast recombination in a *Nicotiana somatic* hybrid. Proc. Natl. Acad. Sci. U.S.A. 82: 6960–6964.
10. Galbraith, D.W. and Mauch, T.J. 1980. Identification of fusion of plant protoplasts: II. Conditions for the reproducible fluorescence labelling of protoplasts derived from mesophyll tissue. Z. Pflanz. 98: 129–140.
11. Afonso, C.L., Harkins, K.R., Thomas-Compton, M.A., Krejci, A.E. and Galbraith, D.W. 1985. Selection of somatic hybrid plants in *Nicotiana* through fluorescence-activated sorting of protoplasts. Bio/Technology. 3: 811–821.
12. Power, J.B., Berry, S.F., Frearson, E.M. and Cocking, E.C. 1977. Selection procedures for the production of inter-species somatic hybrids of *Petunia hybrida* and *Petunia parodii* I Nutrient media and drug sensitivity complementation selection. Pl. Sci. Ltrs. 10: 1–6.
13. Horn, M.E., Kameya, T., Brotherton, J.E. and Widholm, J.M. 1983. The use of amino acid analog resistance and plant regeneration ability to select somatic hybrids between *Nicotiana tabacum* and *N. glutinosa*. Molec. Gen. Genet. 192: 235–240.
14. De Vries, S.E., Jacobsen, E., Jones, M.G.K., Loonen, A.E.H.M., Tempelaar, M.J., Wijbrandi, J. and Feenstra, W.J. 1987. Somatic hybridization of amino acid analogue-resistant cell lines of potato (*Solanum tuberosum* L.) by electrofusion. Theor. Appl. Genet. 73: 451–458.
15. Matthews, B.F. and Widholm, J.M. 1985. Organelle DNA compositions and isoenzyme expression in an interspecific somatic hybrid of *Daucus*. Molec. Gen. Genet. 198: 371–376.
16. Cocking, E.C. 1960. A method for the isolation of plant protoplasts and vacuoles. Nature 187: 962–963.
17. Fromm, M., Taylor, L.P. and Walbot, V. 1985. Expression of genes transferred into monocot and dicot plant cells by electroporation. Proc. Natl. Acad. Sci. U.S.A. 82: 5824–5828.
18. Lo, M.M.S., Tsong, T.Y. and Conrad, M.K. 1984. Monoclonal antibody production by receptor-mediated electrically induced cell fusion. Nature 310: 792–794.

19. Chapel, M., Montane, M.H., Benoit, R., Teissie, J. and Alibert, G. 1986. Viable somatic hybrids are obtained bv diect current electrofusion of chemically aggregated plant protoplasts. FEBS 196(1): 79–86.
20. Pohl, H.A. 1978. Dielectrophoresis. Cambridge University Press, London.
21. Goloubinoof, P. and Edelman, M. 1984. Chloroplast-coded atrazine resistance in *Solanum nigrum*: *psbA* loci from susceptible and resistant biotypes are isogenic except for a single codon change. Nuc. Acids Res; 12: 9489–9496.
22. Steinback, K.E., McIntosh, L., Bogorad, L. and Arntzen, C.J. 1981. Identification of the triazine receptor protein as a chloroplast gene product. Proc. Natl. Acad. Sci. U.S.A. 78: 7463–7467;
23. Pring, D.R. and Levings, C.S., III, 1978. Heterogeneity of maize cytoplasmic genomes among male-sterile cytoplasms. Genetics 89: 121–136.
24. Laughnan, J.R. and Gabay-Laughnan, S. 1983. Cytoplasmic male sterility in maize. Annu. Rev. Genet. 17: 27–48.
25. Bradley, P.M. 1978. Production of enucleated plant protoplasts of *Alium cepa*. Pl. Sci. Ltrs. 13: 287–290.
26. Wallin, A., Glimelius, K. and Eriksson, T. 1978. Enucleation of plant protoplasts by cytochalasin B. Z. Pflanz. 87 S: 333–340.
27. Archer, E.K., Landgren, C.R. and Bonnett, H.T. 1982. Cytoplast formation and enrichment from mesophyll tissues of *Nicotiana* spp. Pl. Sci. Ltrs. 25: 175–185.
28. Lotz, H., Paszkowski, J., Dierks-Ventling, C. and Potrykus, I. 1981. Isolation and characterization of cytoplasts and miniprotoplasts derived from protoplasts of cultured cells. Physiol. Plant. 55: 385–391.
29. Zelcer, A., Dvora, A. and Galun, E. 1978. Interspecific transfer of cytoplasmic male sterility by fusion between protoplasts of normal *Nicotiana sylvestris* and X-ray irradiated protoplasts of male-sterile *N. tabacum*. Z. Pflanz. 90: 3997–4007.
30. Aviv, D., Arzee-Gonen, P., Bleichman, S. and Galun, E. 1984. Novel alloplasmic *Nicotiana* plants by 'donor-recipient' protoplast fusion: cybrids having *N. tabacum* or *N. sylvestris* nuclear genomes and either or both plastomes and chondriomes from alien species. Molec. Gen. Genet. 196: 244–253.
31. Fluhr, R., Aviv, D., Galun, E. and Edelman, M. 1984. Generation of heteroplastidic *Nicotiana* cybrids by protoplast fusion: analysis for plastid recombinant types. Theor. Appl. Genet. 67: 491–497.
32. Aviv, D., Bleichman, S., Arzee-Gonen, P. and Galun, E. 1984. Intersectional cytoplasmic hybrids in *Nicotiana*. Theor. Appl. Genet. 67: 499–504.
33. Horsch, R.B. and Klee, H.J. 1986. Rapid assay of foreign gene expression in leaf discs transformed by *Agrobacterium tumefaciens*: Role of T-DNA borders in the transfer process. Proc. Natl. Acad. Sci. U.S.A. 83: 4428–4432.
34. Albright, L.M., Yanofsky, M.F., Leroux, B., Ma, D. and Nester, E.W. 1987. Processing of the T-DNA of *Agrobacterium tumefaciens* generates border nicks and linear, single-stranded T-DNA. J. Bacteriol. 169(3): 1046–1055.
35. Stachel, S.E., Timmerman, B. and Zambryski, P. 1986. Generation of single-stranded T-DNA molecules during the initial stages of T-DNA transfer from *Agrobacterium tumefaciens* to plant cells. Nature 322: 706–712.
36. Gheysen, G., van Montagu, M. and Zambryski, P. 1987. Integration of *Agrobacterium tumefaciens* transfer DNA (T-DNA) involves rearrangements of target plant DNA sequences. Proc. Natl. Acad. Sci. U.S.A. 84: 6169–6173.
37. Budar, F., Toong, T., Van Montagu M. and Hernalsteens J.P. 1986. *Agrobacterium*-mediated gene transfer results mainly in transgenic plants transmitting T-DNA as a single Mendelian factor. Genetics 114: 303–313.

38. Horsch, R.B., Fraley, R.T., Rogers, S.G., Sanders, P.R., Lloyd, A. and Hoffmann, N. 1984. Inheritance of functional foreign genes in plants. Science 223: 496–498.
39. Wallroth, M., Geraats, A.G.M., Rogers, S.G., Fraley, R.T. and Horsch, R.B. 1986. Chromosomal localization of foreign genes in *Petunia hybrida*. Molec. Gen. Genet. 202: 6–15.
40. Ambros, P.F., Matzke, M.A. and Matzke, A.J.M. 1986. Detection of a 17 kb unique sequence (T-DNA) in plant chromosomes by *in situ* hybridization. Chromosoma 94: 11–18.
41. Coruzzi, G., Broglie, R., Edwards, C. and Chua, N.H. 1984. Tissue-specific and light-regulated expression of a pea nuclear gene encoding the small subunit of ribulose-1,5-bisphosphate carboxylase. Eur. Molec. Boil. J. 3: 1671–1679.
42. Poulsen, C., Fluhr, R., Kauffman, J.M., Boutry, M. and Chua, N.H. 1986. Characterization of an rbcS gene from *Nicotinia plumbaginifolia* and expression of an rbcS-CAT chimeric gene in homologous and heterologous nuclear background. Molec. Gen. Genet. 205: 193–200.
43. Broglie, R., Coruzzi, G., Fraley, R.T., Rogers, S.G., Horsch, R.B., Niedermeyer, J.G., Fink, C.L., Flick, J.S. and Chua, N.H. 1984. Light-regulated expression of a pea ribulose-1,5-Bisphosphate carboxylase small subunit gene in transformed plant cells. Science 224: 838–843.
44. Fluhr, R. and Chua, N.H. 1986. Developmental regulation of two genes encoding ribulose-bisphosphate carboxylase small subunit in pea and transgenic petunia plants: Phytochrome response and blue-light induction. Proc. Natl. Acad. Sci. U.S.A. 83: 2358–2362.
45. Bates, G.W., Saunders, J.A. and Sowers A.E. 1987. Electrofusion. Principles and Applications, pp. 367–395. In A.E. Sowers (ed.) Cell Fusion. Plenum Press, New York.
46. Neumann, E., Schaefer-Ridder, M., Wang, Y. and Hofschneider, P.H. 1982. Gene transfer into mouse lyoma cells by electroporation in high electric fields. Eur. Molec. Boil. Org. J. 1: 841–845.
47. Waldron, C., Malcolm, S.K., Murphy, E.B. and Roberts, J.L. 1985. A method for high-frequency DNA-mediated transformation of plant protoplasts. Pl. Molec. Biol. Rep. 3: 169–175.
48. Kruger-lebus, S. and Potrykus, I. 1987. A simple and efficient method for direct gene transfer to *Petunia hybrida* without electroporation. Pl. Molec. Biol. Rep. 5: 289–294.
49. Shillito, R.D., Saul, M.W., Paszkowski, J., Muller, M. and Potrykus, I. 1985. High efficiency direct gene transfer to plants. Bio/Technology 3: 1099–1103.
50. Lin, W., Odell, J.T. and Schreiner, R.M. 1987. Soybean protoplasts culture and direct gene uptake and expression by cultured soybean protoplasts. Pl. Physiol. 84: 856–861.
51. Meyer, P., Heidmann, I., Forkmann, G. and Saedler, H. 1987. A new petunia flower colour generated by transformation of a mutant with a maize gene. Nature 330: 67–678.

Alteration of the genome as a method to increase diversity in animal germplasm resources

ROBERT J. WALL and CAIRD E. REXROAD, Jr.
Reproduction Laboratory, Beltsville Agricultural Research Center, Agricultural Research Service, U.S. Department of Agriculture, Beltsville Maryland, 20705, U.S.A.

Abstract During this decade, methods have been developed to introduce new and sometimes unique genetic information into animals as diverse as *Drosophila* and swine. These techniques theoretically could be applied to any animal. The methods currently practiced utilize recombinant DNA technology to modify existing genetic coding sequences before introduction into embryos. If the foreign gene (transgene) is incorporated into the genome of the one-cell embryo, all of the cells of the mature transgenic animal, including its germ cells, will contain the transgene. Transgenic animals have been used successfully to study developmental and tissue-specific gene expressions, genetic regulation of the immune system, oncogenesis, globin regulation, and pancreatic function. Practical applications of the transgenic model are currently focused on enhancing disease resistance, improving growth characteristics, producing pharmaceuticals, and creating models for the study of human disease. Low efficiency of gene insertion has restricted widespread use of the transgenic model to *Drosophila* and mice, though transgenic farm animals have been produced. Improved efficiency of gene insertion and improved regulation of gene function will provide a means of specifically introducing genetic alterations into the gene pool and thereby increasing genetic diversity in the animal kingdom.

Key words: transgenic animals, gene transfer transgene, recombinant DNA

Introduction

Scientists have successfully introduced 'foreign' genes (recombinant DNA sequences: transgenes) into both plants and animals. In many instances, the physiology of the transgenic organism is dramatically altered, reflecting the consequences of carrying the newly introduced gene. Production of transgenic plants and animals is a strategy for increasing diversity. Each new transgenic animal represents an increase in the genomic diversity of that species even if the inserted gene was derived from that species. Even when embryos are used from highly inbred strains, the site at which the transgene integrates into the genome varies among individuals, as does the number of copies integrated. It is thought that the site of integration influences the expression of the transgene and thus the amount of the gene product produced. Therefore, two transgenic animals with the same number of

L. Knutson and A.K. Stoner (eds.). Biotic diversity and germplasm preservation, global imperatives.
 ISBN 0-7923-0178-1

copies of a particular gene, located at different sites in the genome, have, in essence, different genotypes and phenotypic traits. For the practitioners of transgenic animal research, the inability to control the site of integration and number of copies inserted is considered to be a problem to be overcome. For those interested in increasing diversity in the gene pool, these perceived inadequacies actually may be regarded as beneficial tools.

The ability to introduce a new gene in a single generation is an enormous accomplishment, but it is easily overshadowed by the potential capabilities that are likely to evolve in the future. At present, the usual practice is to introduce a single gene construction into transgenic animals. Controlled modification of production traits may be achieved only when sets of genes are introduced simultaneously. The limiting step at the moment is identification of those sets of genes responsible for quantitative variation in production traits. The logical progression from introducing families of genes leads to introduction of synthetic chromosomes containing a number of traits that modify gene families. It is unrealistic to think in terms of transferring synthetic nuclei (composed of a set of synthetic chromosomes) at this time, but it may not be impossible in the next century. As the scientific community's ability to identify and transfer genes becomes more sophisticated, radically new methods of storing genetic information contained in germplasm also will be possible. Genetic information is now stored in the form of gametes or as DNA molecules in freezers and as sequences in computer-readable storage devices. When genomes are mapped and sequenced, as has been proposed for the human genome, the storage device of choice will be the computer. Given this, one can propose a scenario that involves retrieving desired genetic characteristics from a computer, feeding that information into a DNA synthesizer, and then transferring the fabricated molecule into a developing organism. Suffice it to say, we do not know enough at this time to recognize the flaws in this scheme.

The remainder of this paper will deal with reality. We will discuss methods employed for making transgenic animals and what is known about incorporation and expression of transgenes, then present a brief overview of the genes currently being used to produce transgenic livestock.

Transgenic animal model system

A transgenic animal is an animal with a recombinant DNA molecule integrated into its genome. The recombinant DNA molecule may or may not be a clone (copy) of a naturally occurring gene, and its origin may or may not be the species into which the recombinant gene has been introduced.

Historical landmarks

In 1980, Jon Gordon, working with colleagues in Frank Ruddle's laboratory at Yale, demonstrated that it was possible to 'transform' mice by injecting recombinant DNA into the pronuclei of recently fertilized eggs [1]. The following year, Gordon reported that the microinjected DNA sequence was apparently stably incorporated into the genome of the mice and that the foreign gene could be transmitted to progeny [2]. These results were quickly substantiated by publications from several other laboratories [3–5]. It was immediately apparent to the scientist in the field that the transgenic animal model system could serve as a powerful tool for studying genetic control of development, cell differentiation, and tissue-specific gene expression. A number of laboratories became involved in these studies. One of the most outstanding efforts has been made by collaborators Ralph Brinster (University of Pennsylvania) and Richard Palmiter (University of Washington). They pioneered the fusion gene concept that has become the standard approach to exploiting the transgenic animal model [6]. By linking the promoter region (a regulatory element) of one gene with the coding region (structural sequences) of another gene, they were able to begin to functionally dissect gene regulation in whole animals. They found that fusion genes could be built such that the structural sequences are transcribed in tissues other than their normal site of expression. Theoretically at least, the regulatory element of the fusion gene could control when, during development, a gene-coding region is expressed or 'turned on'. This technique also afforded the opportunity to induce expression of a foreign gene in response to an easily manipulated environmental cue. It was the latter approach that was taken when the first transgenic farm animals were produced [7, 8]. To date, transgenic fish, mice, rabbits, sheep, and pigs have been born, and it is likely that the first successes in producing transgenic cattle will soon be reported.

Methods for producing transgenic animals

Three methods have been utilized to produce transgenic animals: pronuclear microinjection, virus-mediated gene transfer, and chimera formation using transformed cells. Other methods, such as electroporation, have been proposed or attempted with limited success. Each of the three most successful methods will be described briefly, and their advantages and disadvantages will be mentioned.

Gene transfer by pronuclear microinjection. Direct injection of a solution containing recombinant DNA into the male pronucleus of a recently fertilized egg is the most straightforward and most popular method of gene transfer for mammalian eggs. The eggs are placed in a petri dish or a depression slide and examined, usually with an inverted microscope fitted with interference contrast optics. A micropipet is used to hold an egg, and then a needle is introduced into the male pronucleus and a small aliquot of the solution containing the genes is deposited [9]. With cattle and swine, the eggs are first centrifuged to displace the optically dense material in the cytoplasm that normally obscures the pronuclei [10].

The advantages of this approach are many:

1. The procedure is simple – it has been applied successfully to most mammalian species so far tested,
2. The procedure is relatively fast, eliminating the need to culture the eggs for extended periods of time,
3. The length of the gene being injected is unlimited, and
4. No additional manipulation of the DNA is required after it is cloned and purified.

Although it is the most frequently used method because of these advantages, microinjection also has numerous disadvantages:

1. The integration rate (number of transgenics/number of eggs injected) is low,
2. Equipment is specialized and relatively expensive (a good microscope, micromanipulators, and ancillary equipment to make the microtools is required),
3. The person performing the injections must possess a high degree of dexterity; and
4. A significant training period is required.

Viral-mediated gene transfer. Viral-mediated gene transfer takes advantage of the inherent characteristic of a retrovirus to introduce its own genetic material into the genome of eukaryotic cells. The retroviral genome, which is stored in the virion as RNA, is reverse transcribed when it enters a eukaryotic cell and the proviral DNA molecule so produced becomes incorporated into the cell's genome. Under appropriate conditions, the virus replicates in the host and eventually can destroy the cell or organism in which it resides. To prevent viraemia, several techniques have been developed to render the virus defective [11–15]. In general, defective viruses are produced by removing one or more of the viral genes associated with packaging the viral particle or by mutating or deleting DNA sequences that regulate viral expression. As an example, genes coding for the viral coat proteins could be excised from proviral DNA. The deleted genes would then be introduced into eukaryotic cells to form a helper cell line. When the defective viral DNA is introduced into the helper cells, the virus can replicate (because removal of the coat protein genes does not affect replication) and be packaged properly with the aid of the products of the genes resident in the helper cells. The newly formed viral particles emerging from the helper cells are capable of infecting another cell. However, once the defective virus inserts its genome into the new host cell's genome, the virus becomes trapped because it lacks critical packaging information.

Because it is difficult to obtain high titers of viral particles, eggs are usually cultured on a feeder layer of cells that are actively releasing virus particles. Prior to co-culturing the eggs on the feeder layer, the egg's zona pellucida, which is an extremely effective viral barrier, must be removed. After 24 hr of co-culture and further incubation in virus-free medium, the eggs are transferred to the oviducts of a foster mother [14, 16–18]. Because the culturing of livestock embryos has met with only limited success, Mettus and Petters have proposed a novel alternative approach [19, 20]. In their scenario, a helper cell line is transformed with the defective virus genome (the genome carrying the gene to be transferred). Those cells are then injected into the blastocoel cavity of the developing embryo. Helper cells survive only a few days in the blastocoel cavity, during which time they produce defective viral particles to infect the embryo (from the 'inside').

To date, there are few large-animal practitioners of viral-mediated transfer. However, when the operational difficulties of this approach are resolved, it may become the gene transfer method of choice because of its potential for producing transgenic animals at high efficiency and because it could limit integrations to a single copy per cell of the transferred gene. Retroviral vectors have several disadvantages:

1. One has to 'engineer' the helper cell line and the virus before it can be used as a gene carrier,
2. The procedure is limited to species in which embryo culture is successful (virtually eliminating all species of farm animals, although this problem may be overcome by the approach proposed by Petters),
3. The size of the gene carried by retroviruses is limited to 8 kb (not a serious limitation at this time, but it could become so as multiple genes are transferred simultaneously),
4. Embryos may possess a mechanism by which they render permanently inactive the expression of the gene contained in the virus [21]; and
5. Components of the viral genome (namely, long terminal repeats) may alter the desired level and specificity of expression of the introduced gene [22].

The processing of the RNA intermediate during the retroviral life cycle may introduce complications in transcription of transgenes containing introns and/or polyadenylation signals. Furthermore, many viruses have limited host ranges, possibly requiring preparation of species-specific viral vectors.

Chimera formation. A number of approaches have been proposed in which cells containing the desired foreign genes would be used to form chimeras with a recipient embryo. In one approach that has been successful in mice, pluripotent embryonal stem cells are simultaneously transfected with two genes, the desired foreign gene and an antibiotic-resistance marker gene. Stem cells are grown in antibiotic-containing medium to select those that have been transformed (contain the antibiotic-resistance gene and presumably also the gene of interest). Once colonies of stem cells that survive the selection pressure have been established, their DNA can be screened to verify that the gene of interest has integrated properly. After screening, the transformed stem cells are either injected into the blastocoel cavity of a developing embryo or used to form aggregate chimeras with cleavage stage embryos [13, 23, 24]. A relatively high proportion of blastocysts incorporate injected stem cells, which contribute to various tissues of the developing embryo. Production of transgenic animals by this approach has a number of important advantages in addition to its relatively high efficiency. The opportunity to analyze the transformed stem cells (or other pluripotent cell lines that might be used) before they are used to produce transgenic animals gives the scientist a chance to choose among inserted construct configurations, such as number of copies desired and site of insertion. This selection process may enable site-specific gene integration. No other gene transfer method provides this flexibility. However, this approach has a serious drawback; all founder transgenic animals produced will be chimeras. That is, only some of the animal's tissues will arise from the transformed stem cells. If no gonadal germ cells are derived from the stem cells, the transgenic animal produced will be of limited value. For this technique to be successful, it is likely that species-specific stem cell lines will have to be established.

Efficiency of producing transgenic animals

The primary limitation in producing transgenic animals is the number of eggs available. Therefore, no matter which gene transfer technique is used, donor females are usually superovulated (administered hormones to stimulate follicle development and ovulation) to maximize the number of eggs harvested. The eggs are either collected at necropsy or surgically flushed from the donor's oviducts. After the genes have been transferred, eggs are returned to the oviducts of a foster mother that has been selected (or hormonally induced) to be in the appropriate reproductive state. One exception to this general approach is with cattle, with which the manipulated eggs usually are surgically transferred to the oviducts of a temporary recipient, such as a rabbit [25] or sheep [26] until they reach the morula or blastocyst stage of development. The embryos then are recovered and retransferred, non-surgically, to the uterus of a recipient cow.

Because of the paucity of data for alternative methods, the remainder of this discussion will be restricted to results obtained by the pronuclear-microinjection approach for producing transgenic animals. Of the six mammalian species studied (mouse, rabbit, sheep, goat, pig, and cattle), the vast majority of the data has been generated from studies conducted with mice. Hundreds of papers have been published in the last five years documenting the successful use of the transgenic mouse model for studies ranging from disease to growth. However, since relatively high levels of inefficiency can be tolerated when working with the mouse, investigators have not felt the need to focus their attention on issues associated with the transgenic model system itself. On the other hand, those who work with large animals are very much concerned with the efficiencies associated with producing transgenic livestock. For the mouse, approximately 20% of the injected eggs develop to term and are born [27], whereas the percentage of eggs that survive in pigs and sheep is closer to 8% (range from 3 to 25%) [28, 29]. Likewise, the percentage of mouse pups born (or taken as fetuses just before birth) that retain the injected DNA is significantly higher for mice (20–30%) than for sheep and pigs (about 6.5%). The overall efficiency of producing transgenic mice ranges from 3 to 6% of eggs injected. Overall efficiency for all other mammals tested is usually an order of magnitude lower. Given these present constraints, employment of the transgenic animal system to increase germplasm diversity in mammals must be restricted to those species from which a large number of eggs can be harvested.

Although still sparse, accumulating evidence suggests that transgenic rabbits may be produced at higher efficiencies than any other mammalian species [7, 30]. If further experimentation confirms these initial observations, it would be interesting to determine the causative factor. One might speculate that the large, clearly visible, firm pronuclei of rabbit zygotes are more easily injected than pronuclei of zygotes from other species and that therefore injections are more frequently successful when rabbit eggs are used. The construct injected and the species of eggs being injected may also interact. In several instances, gene constructs that successfully produced transgenic pigs failed to produce transgenic sheep, and in at least one case the reverse was true [28]. Furthermore, an apparent interaction exists between the stage of the egg at the time of injection and the species. Two-cell mouse eggs do not survive injections as well as 1-cell mouse eggs [27], whereas pig eggs injected at the 2-cell stage survive better than those injected at the 1-cell stage [31]. Site of injection also plays a role in overall efficiency of producing transgenic animals. Injection of genes into pronuclei results in a higher proportion of transgenic mice than injection into cytoplasm [27]. However, choice of pronucleus injected (male versus female) has no detectable effect on efficiency.

Characteristics of integrated transgenes

The molecular mechanism by which the injected genes integrate into the genome of the host is unknown, but a number of studies have documented what appears to be characteristic features of the event. The sites of integration of foreign genes into genomes vary and are thought to be random. It should be pointed out that only a few studies have addressed this issue and that when more exhaustive investigations have been reported, some common characteristics of sites of integration may emerge. However, at this time it is thought that the only prerequisite for integration is a break in the genome [27]. If integration occurs at random sites, one would expect a proportion of transgenic animals to exhibit mutations caused by the integrating gene disrupting a resident gene. Palmiter and Brinster [32] estimate that 7% of transgenic mice, when bred to homozygosity, would display mutations [33]. This relatively low incidence of insertional mutagenesis may be a consequence of the organization of the genome. Most (about 90%) of DNA in eukaryotic cells appear to be configured as transcriptionally inactive heterochromatin [34]. Therefore, on the basis of the random integration hypothesis, transgenes have only a 10% chance of disrupting an actively transcribed region of the genome.

The number of copies of the injected gene that become integrated varies among individuals from less than one to several hundred copies. In almost all cases, the multiple copies integrate at a single location and are arranged in a head-to-tail tandem array. If two genes are co-injected, they also integrate at a single site [27], which may suggest that the site at which genes integrate is uniquely receptive (a chromosomal break?). Brinster and colleagues [27] have proposed a model to explain the observations of multi-copy integrations. They suggest that a single copy of the gene integrates at a chromosomal break point and that other copies become circularized with each other, then by homologous recombination combine with the integrated copy. This hypothesis remains to be tested. At this time there is no means known by which to control the number of copies of the gene that integrate. In experiments designed to study gene regulation, multi-copy integrations can confound interpretations because it is not known what, if any, influence adjacent gene copies have on gene expression.

Not all transgenic animals express their transgene. On average, it appears that transcription and translation of the foreign gene occurs in 60% of transgenic animals [32]. The reason that 40% do not express their foreign gene is not known, but a widely held theory suggests that the transgenes of non-expressing animals may have integrated into regions of heterochromatin (where transcription does not normally occur). This 'local environment influence' argument has not been tested, but, to add credence to the hypothesis of local DNA effects, it has also been observed that prokaryotic

DNA attached to a transferred gene reduces or completely prevents expression [35, 36]. A non-expressing animal may have integrated one or multiple copies of the transgene.

Somewhat surprisingly, copy number does not appear to be related to the level of expression [7]. The literature is replete with examples of both low and high copy number transgenic animals expressing either low or high levels of the transgene product. A modification of the 'local environment influence' argument has been suggested to explain this observation. Gene copies close to the junction with the genome (*i.e.* ends of the tandem array) may be the only copies that are expressed correctly; therefore, the copies internal to the array may not produce useful transcription products (R. L. Brinster, personal communication). No definitive experiment has been performed to test this hypothesis, but the results of a recent experiment may provide interesting new information regarding this phenomena. Grosveld and co-workers [37] have identified control regions for the human β-globin locus that, when introduced with the β-globin gene into transgenic mice, confer correct tissue specificity and result in levels of β-globin expression that are proportional to the number of gene copies integrated. It remains to be seen whether these dominant control regions exist for other genes, as Grosveld predicts they might.

Transmission of transgenes to offspring

One of the most powerful aspects of the transgenic animal model is the potential for stable integration of the transgene into the genome and consequently for permanent genetic modification of the germ line. Indeed, the technique of introducing the foreign DNA at the pronuclear stage of development is used to achieve germline modification. The practical usefulness of transgenic animals would be greatly diminished if a proportion of their offspring did not carry the transgene. In an exhaustive study of pedigrees from more than 250 transgenic mice, 70% of transgenic founder animals were found to transmit their transgene to their progeny as a single heterozygous allele [38]. The remaining 30% of the population studied either transmitted their transgene to less than 50% of their progeny or did not transmit the gene at all. It is most likely that those animals which fail to transmit the transgene in a Mendelian fashion are either mosaic for the transgene in their gametes or the presumptive germ cells did not receive a copy of the transgene. Such a condition would be observed if the foreign DNA was integrated sometime after the 1-cell stage of development. In that case, some tissues of the developing embryo might not receive a copy of the gene. Alternatively, for some reason, the integrated gene may have been excised from the genome of the developing germ cell tissues. These hypotheses have not yet been tested.

In most instances, the progeny of founder transgenic animals that transmit the transgenes in a Mendelian fashion exhibit levels of expression similar to that of their founder parent. There are, however, examples of progeny whose level of expression differ from that of their founder, presumably resulting from the influence of the genetic environment contributed by the non-transgenic parent [6, 39]. Nevertheless, if the founder transgenic parent does not express the transgene, neither do its offspring. Whether or not the transgenic animal expresses the transgene, progeny and subsequent generations transmit the foreign gene in a predictable manner.

Applications of the transgenic animal model to livestock

Added to the above-mentioned constraints imposed by the extremely low efficiency of producing transgenic farm animals and the attendant high cost is the general scarcity of adequately characterized genes that have the potential for beneficial modification of animals. As quantitative geneticists join forces with molecular biologists, a better understanding of the genetic control of various physiological systems will lead to the increased availability of useful gene constructs. This knowledge will take time to accumulate, giving those engaged in improving the efficiency of transgenesis in farm animals the time to make the procedure an economically viable tool for animal production. Nevertheless, preliminary steps have been taken to demonstrate the feasibility of producing transgenic farm animals with desirable characteristics.

Proposed strategies

Scientists currently producing transgenic livestock are pursuing two divergent objectives. Toward one objective, experiments have been proposed or are currently in progress in which researchers are attempting to modify production traits of animals by introducing genes designed to alter growth, disease resistance, wool production, and reproductive performance. The other objective has been to design and introduce gene constructs intended to produce biologically important molecules that can be harvested easily from transgenic animals. Both objectives rely on the use of fusion genes (regulatory sequence from one gene linked to the structural sequence of another). When the intention is to modify the function of a particular physiological system, it is desirable to remove the regulation of the transgene from the control system that usually maintains homeostatis of that system. For example, if one intends to modify growth, one obvious approach would be to introduce an additional gene that codes for growth hormone. However, if the entire growth hormone gene, with its endogenous regulatory sequences, were used, little or no benefit would be realized because the new growth-hormone gene would be subject to the same regulation as the animal's native growth-hormone gene [40]. This problem could be overcome by use of a heterologous promoter linked to the growth-hormone coding sequences [41–43].

Those researchers interested in producing biologically active proteins in transgenic animals have somewhat different problems. Ideally, they would like to produce the product in high concentrations, but they must insure that the animal is not adversely affected by the presence of the new biologically active compounds. Such researchers basically have two choices: They either can engineer the gene to make biologically inactive molecules that would be 'activated' after harvest, or they can restrict the expression of the transgene to a compartment within the animal that precludes the gene product from achieving its normal biological effect [44]. This would be accomplished by use of an appropriate tissue-specific promoter. Directing the expression exclusively to the mammary gland is an approach currently being attempted. There is also discussion of directing expression to the kidney which would secrete the product into urine (Bishop, personal communication).

Areas of active research – a brief review

The production of transgenic farm animals was first reported in 1985 [7, 8]. Those two research groups demonstrated that it was possible to introduce foreign genes into rabbits, sheep, and pigs. Independently, each group pursued the same strategy of introducing fusion genes consisting of the metallothionein (MT) promoter fused to the structural gene encoding human growth hormone (hGH). Since that time, the MT promoter has been used to drive the coding sequence of bovine growth hormone (bGH), ovine growth hormone (oGH) and porcine growth hormone (pGH) in the production of transgenic sheep [42, 45] and pigs [44, 46–49]. Use of the promoter from the MT gene, which is expressed predominantly in the liver, kidney, and gut, was thought to be a good choice because:

1. It was reasonably well characterized,
2. It would not be subject to the normal regulatory signal controlling growth hormone,
3. It could be induced (turned on) by feeding the transgenic animals zinc, and
4. When linked to GH it dramatically increased growth in transgenic mice [50].

The choice of the MT promoter has turned out to be a mixed blessing. Many of the desired results have been observed, such as increased growth rate [46, 47], reduced body fat [45], and improved feed efficiency [47]. However, attendant with the extremely high levels of circulating growth hormone in some animals are some of the clinical symptoms of acromegalia. These unforeseen problems have been attributed to imperfect control of the growth hormone genes by the MT promoter [28]. Pursel and colleagues [45] reported that expressing transgenic piglets weigh less than non-expressing transgenic or non-transgenic litter mates at birth, demonstrating that the transgene has a negative impact on the piglets *in utero*. MT-promoter-driven expression of the growth hormone genes is observed at birth, and the MT promoter continues to cause expression in the absence of stimulation (i.e. zinc feeding). The constitutive action of the MT promoter, combined with its ability to cause expression in tissues other than the pituitary gland, makes the choice of the MT promoter less attractive than was first thought. One laboratory, apparently not convinced that constitutive expression of growth hormone is detrimental, has reported the birth of a pig containing the fusion gene MLVrGH [Moloney murine leukemia virus promoter (a 'strong' promoter that is always turned on) fused to the coding sequences for rat growth hormone]. At nine months of age, when the pig was sacrificed, it was 26% heavier than was its litter mate [51]. At necropsy, joint pathology was similar to that observed for MT-xGH transgenic pigs.

To extend the growth studies, an MThGRF fusion-gene construct (MT promoter linked to the coding sequence for growth hormone-releasing factor) has been introduced into pigs and sheep [52]. The rationale for use of this particular construct was based on the desire to stimulate production of the animal's endogenous growth hormone and to restrict the expression of growth hormone to its normal site, the pituitary. It was thought that if sufficient GRF was produced in the liver, kidney, and gut, circulating levels would be great enough to stimulate the pituitary to produce excess amounts of growth hormone. High levels of GRF were observed in an expressing lamb; however, no concomitant increase was observed in GH levels [28]. Results for the transgenic pigs remain to be reported.

As mentioned above, some laboratories are pursuing an altogether different objective to take advantage of the transgenic farm animal model. Their goal is to produce pharmaceuticals or other normally occurring complex biologically active molecules, and their strategy is to restrict the production of those compounds to the mammary gland. Production of these compounds in the mammary gland is advantageous for two reasons. Obviously, if the compounds are secreted into the milk, they will be easy to harvest. In addition, the normal direction of hydrodynamic flow is towards the mammary gland. Thus the chance of these biologically active compounds (which could be toxic when produced at high concentrations) entering the general circulation is reduced.

So far, two research groups have reported successful attempts to produce biomedically important compounds in milk. Transgenic mice have been produced carrying the fusion gene WAP-tPA (mouse whey acidic protein promoter ligated to the tissue plasminogen activator coding sequence) [53]. In the normal lactating murine mammary gland, WAP accounts for 15% of the total mRNA. Thus there is a potential for producing large amounts of tPA in the mouse mammary gland. The tPA concentrations were estimated to be between 200 and 400 ng/ml in the milk of the expressing WAP-tPA mice. These concentrations were lower than expected. The authors suggest that the problem may have to do with tPA mRNA instability rather than level of expression of the fusion gene. The other group of scientists (located in Edinburgh, Scotland) have produced transgenic mice carrying the β-lactoglobulin (BLG) gene, a major whey protein in ruminant milk [54], and transgenic sheep with a fusion gene containing the β-lactoglobulin promoter fused to either the coding sequences for blood coagulation factor IX (FIX) or α1-antitrypsin (α1AT) [55]. No data are yet available regarding the levels of expression in the transgenic sheep. Nevertheless, the production of those transgenic sheep is a major first step. The BLG promoter should function properly in both sheep and cows.

The first report heralding the eventual production of transgenic cattle was published recently [26]. The marker fusion gene RSV-CAT (Rous sarcoma virus promoter fused to the bacterial chloramphenicol acetyl transferase gene) was injected into over 1,000 cattle zygotes. Zygotes were transferred to sheep oviducts and recovered six days later, and the viable embryos were retransferred to recipient cattle. At 60 days of gestation, fetuses were recovered and analyzed for the presence of RSV-CAT. Four transgenic fetuses were identified, and CAT expression was detected in the placenta of one of those fetuses. This work represents a small but significant step in development of the technology necessary to apply the transgenic model to another economically important species.

Finally, although the focus of this paper is on transgenic mammals, and more specifically on transgenic farm animals, we would be remiss not to mention the work underway with transgenic chickens. The project, being conducted at the USDA Agricultural Research Service's Regional Poultry Research Laboratory in East Lansing, Michigan, has focused on employing retroviral genes to 'genetically immunize' chickens. A strain of endogenous avian leukosis virus (ALV) was engineered to contain the sequence coding for one of the envelope protein genes of a highly pathogenic ALV strain. This newly engineered virus, RAV-0-A_1, then was injected into freshly laid eggs. RAV-0-A_1 became stably incorporated in the genome of a number of the chickens, and some of the males have transmitted the proviral DNA to their offspring. RAV-0-A_1 has now been successfully passed through several generations [56]. These workers hypothesize that the protein coded for by the envelope gene sequence will protect the transgenic chickens from infection by the highly pathogenic ALV strain from which it was derived. Work is now in progress to test this hypothesis [57; L. B. Crittenden, personal communication].

Conclusions

Scientists now have the tools, inefficient and unrefined though they may be, to introduce new genetic information into animals as diverse as *Drosophila* and cattle. The foreign DNA can become stably incorporated into the animal's genome and can be passed on to progeny. Furthermore, the introduced genes are transcribed and the translation products are biologically active. Refinement of the methods in the next decade will allow these procedures to be performed routinely. With efficient methods in hand and precise control over the introduced genes and the range of animal species extended, the pressing questions become:

1. Are there consequences of producing transgenic animals that are not yet appreciated, and
2. What genes should be transferred and to what ends?

One only can speculate about the unforeseen impact of the transgenic animals on the gene pool. Will the transgene 'behave itself' when it is bred into the general population of a given species, or will its behavior be altered by introducing it into a variety of genetic backgrounds? Will gene transfer become the ultimate solution for increasing diversity in the germplasm? Some of the answers to the last question are obvious now. Transgenic animals will serve the biomedical community as models for disease studies (as the first patented animal does), for genetic disorders, and for basic research questions related to genetic regulation of development and homeostasis. The agricultural community will be able to alter production characteristics in much more dramatic ways than now possible, making animal production more efficient, producing breeds of animals better suited to their production environment, creating new uses for animals and their products, and producing more healthful food products.

It is clear that the road to advancing the transgenic animal model from a laboratory curiosity to an applied tool will be strewn with scientific surprises. What is less clear is society's acceptance of this approach. A full debate may have to await the emergence of a clearer vision of the benefits and risks of the transgenic animal model.

References

1. Gordon, J. W., G. A. Scangos, D. J. Plotkin, J. A. Barbosa, and F. H. Ruddle. 1980. Genetic transformation of mouse embryos by microinjection of purified DNA. Proc. Natl. Acad. Sci. USA 77: 7380.
2. Gordon, J. W. and F. H. Ruddle. 1981. Integration and stable germ line transmission of genes injected into mouse pronuclei. Science 214: 1244–1246.
3. Constantini, F. and E. Lacy. 1981. Introduction of a rabbit β-globin gene into the mouse germ line. Nature 294: 92–94.
4. Brinster, R. L., H. Y. Chen, M. E. Trumbauer, A. W. Senear, R. Warren, and R. D. Palmiter. 1981. Somatic expression of herpes thymidine kinase in mice following injection of a fusion gene into eggs. Cell 27: 223–231.
5. Wagner, T. E., P. C. Hoppe, J. D. Jollick, D. R. School, R. L. Hodinka, and J. B. Gault. 1981. Microinjection of a rabbit β-globin gene in zygotes and its subsequent expression in adult mice and their offspring. Proc. Natl. Acad. Sci. USA 78: 6376–6380.
6. Palmiter, R. D., H. Y. Chen, and R. L. Brinster. 1982. Differential regulation of metallothionein-thymidine kinase fusion genes in transgenic mice and their offspring. Cell 29: 701–710.
7. Hammer, R. E., V. G. Pursel, C. E. Rexroad, Jr., R. J. Wall, D. J. Bolt, K. M. Ebert, R. D. Palmiter, and R. L. Brinster. 1985. Production of transgenic rabbits, sheep and pigs by microinjection. Nature 315: 680–683.
8. Brem, G., B. Brenig, H. M. Goodman, R. C. Selden, F. Graf, B. Kruff, K. Springman, J. Hondele, J. Meyer, E.-L. Winnaker, and Z. F. Krausslich. 1985. Production of transgenic mice, rabbits and pigs by microinjection into pronuclei. Zuchthygiene 23: 251–252.
9. Hogan, B., F. Constantini, and E. Lacy. 1986. Manipulating the mouse embryo: A laboratory manual. Cold Spring Harbor Laboratory, Cold Spring Harbor, New York.
10. Wall, R. J., V. G. Pursel, R. E. Hammer, and R. L. Brinster. 1985. Development of porcine ova that were centrifuged to permit visualization of pronuclei and nuclei. Biol. Reprod. 32: 645–651.
11. Mann, R., R. C. Mulligan, and D. Baltimore. 1983. Construction of a retrovirus packaging mutant and its use to produce helper-free defective retrovirus. Cell 33: 153–159.
12. Watanabe, S. and H. Temin. 1983. Construction of a helper cell line for avian reticuloendotheliosis virus cloning vectors. Molec. Cell. Biol. 3: 2241–2249.
13. Wagner, E. F., G. Keller, E. Gilboa, U. Ruther, and C. Stewart. 1985. Gene transfer into murine stem cells and mice using retroviral vectors. Cold Spring Harb. Symp. Quant. Biol. 50: 691–700.
14. Rubinstein, J.L.R., J. F. Nicolas, and F. Jacob. 1986. Introduction of genes into preimplantation mouse embryos by use of a defective recombinant retrovirus. Proc. Natl. Acad. Sci. USA 83: 366–368.
15. Yu, Sheau-Fung, T. von Ruden, P. W. Kantoff, C. Garber, M. Seiberg, U. Ruther, W. F. Anderson, E. F. Wagner, and E. Gilboa. 1986. Self-inactivating retroviral vectors designed for transfer of whole genes into mammalian cells. Proc. Natl. Acad. Sci. USA 83: 3194–3198.
16. Jaenisch, R., D. Jahner, P. Nobis, I. Simon, J. Lohler, K. Harbers, and D. Grotkopp. 1981. Chromosomal position and activation of retroviral genomes inserted into the germ line of mice. Cell 24: 519–529.
17. van der Putten, H., F. M. Botteri, A. D. Miller, M. G. Rosenfeld, H. Fan, R. M. Evans, and I. M. Verma. 1985. Efficient insertion of genes into the mouse germ line via retroviral vectors. Proc. Natl. Acad. Sci. USA 82: 6148–6152.
18. Soriano, P., R. D. Cone, R. C. Mulligan, and R. Jaenisch. 1986. Tissue-specific and

ectopic expression of genes introduced into transgenic mice by retroviruses. Science 234: 1410–1413.

19. Mettus, R. V., R. M. Petters, B. H. Johnson, and R. M. Shuman. 1986. In vitro development of porcine blastocysts injected with a retrovirus helper cell line. J. Anim. Sci. 63 (Suppl. 1): 236. (Abstr.).
20. Petters, R. M., R. M. Shuman, B. H. Johnson, and R. V. Mettus. 1987. Gene transfer in swine embryos by injection of cells infected with retrovirus vectors. J. Exp. Zool. 242: 85–88.
21. Jahner, D., K. Haase, R. Mulligan, and R. Jaenisch. 1985. Insertion of the bacterial gpt gene into the germ line of mice. Proc. Natl. Acad. Sci. USA 82: 6927–6931.
22. Stewart, C. L., H. Stuhlmann, D. Jahner, and R. Jaenisch. 1982. De novo methylation, expression, and infectivity of retroviral genomes introduced into embryonal carcinoma cells. Proc. Natl. Acad. Sci. USA 79: 4098–4102.
23. Stewart, C. L., M. Vanek, and E. F. Wagner. 1985. Expression of foreign genes from retroviral vectors in mouse teratocarcinoma chimeras. EMBO J. 4: 3701–3709.
24. Wagner, E. F. and B. Mintz. 1982. Transfer of nonselectable genes into mouse teratocarcinoma cells and transcription of the transferred human β-globin gene. Molec. Cell. Biol. 2: 190–198.
25. Wall, R. J. and H. W. Hawk. 1988. Development of centrifuged cow zygotes cultured in rabbits oviducts. J. Reprod. Fert. 82: 673–680.
26. Biery, K. A., K. R. Bondioli, and F. J. De Mayo. 1988. Gene transfer by pronuclear injection in the bovine. Theriogenology 29: 224. (Abstr.).
27. Brinster, R. L., H. Y. Chen, M. E. Trumbauer, M. K. Yagle, and R. D. Palmiter. 1985. Factors affecting the efficiency of introducing foreign DNA into mice by microinjecting eggs. Proc. Natl. Acad. Sci. USA. 82: 4438–4442.
28. Rexroad, C. E. Jr. and V. G. Pursel. 1989. Status of gene transfer in domestic animals. In: Proc. 11th Internat. Congr. Anim. Reprod. Artif. Insem., Dublin. In press.
29. Rexroad, C. E. Jr. and R. J. Wall. 1987. Development of one-cell fertilized sheep ova following microinjection into pronuclei. Theriogenology 27: 611–619.
30. Strojek, R. M., A. K. Voss, and J. Hahn. 1987. Production of transgenic rabbits by injection of an uteroglobin-promoter-CAT fusion gene into fertilized rabbit oocytes. Dtsch. Tierarztl. Wschr. 94: 476–478.
31. Pursel, V. G., K. F. Miller, C. A. Pinkert, R. D. Palmiter, and R. L. Brinster. 1989. Effect of ovum cleavage stage at microinjection on embryonic survival and gene integration in pigs. In: Proc. 11th Internat. Congr. Anim. Reprod. Artif. Insem., Dublin. In press.
32. Palmiter, R. D. and R. L. Brinster. 1986. Germ-line transformation of mice. Annu. Rev. Genet. 20: 465–499.
33. Wagner, E. F., L. Covarrubias, T. A. Stewart, and B. Mintz. 1983. Prenatal lethalities in mice homozygous for human growth hormone gene sequences integrated in the germ line. Cell 35: 647–655.
34. Bruce, A., D. Bray, J. Lewis, M. Raff, K. Roberts, and J. D. Watson. 1983. Molecular Biology of the Cell. Garland Publishing, New York.
35. Townes, T. M., J. B. Lingrel, H. Y. Chen, R. L. Brinster, and R. D. Palmiter. 1985. Erythroid-specific expression of human β-globin genes in transgenic mice. EMBO J. 4: 1715–1723.
36. Krumlauf, R., R. E. Hammer, S. M. Tilghaman, and R. L. Brinster. 1985. Developmental regulation of α-fetoprotein genes in transgenic mice. Molec. Cell. Biol. 5: 1639–1648.
37. Grosveld, F., G. B. van Assendelft, D. R. Greaves, and G. Kollias. 1987. Position-independent, high-level expression of the human β-globin gene in transgenic mice. Cell 51: 975–985.

38. Wilkie, T. M., R. L. Brinster, and R. D. Palmiter. 1986. Germline and somatic mosaicism in transgenic mice. Dev. Biol. 118: 9–18.
39. Renard, Jean-Paul and C. Babinet. 1987. Genetic engineering in farm animals: the lessons from the genetic mouse model. Theriogenology 27: 181–200.
40. Hammer, R. E., R. D. Palmiter, and R. L. Brinster. 1984. Partial correction of murine hereditary growth disorder by germ-line incorporation of a new gene. Nature 311: 65–67.
41. Palmiter, R. D., G. Norstedt, R. E. Gelinas, R. E. Hammer, and R. L. Brinster. 1983. Metallothionein-human growth hormone fusion genes stimulate growth of mice. Science 222: 809–814.
42. Nancarrow, C., J. Marshall, J. Murray, I. Hazelton, and K. Ward. 1987. Production of a sheep transgenic with the ovine growth hormone gene. Theriogenology 27: 263. (Abstr.).
43. Brinster, R. L. and R. D. Palmiter. 1982. Introduction of foreign genes in animals. Trends Biochem. Sci. 7: 438–440.
44. Clark, A. J., P. Simons, I. Wilmut, and R. Lathe. 1987. Pharmaceuticals from transgenic livestock. Trends Biotechnol. 5: 20–24.
45. Pursel, V. G., C. E. Rexroad, Jr., D. J. Bolt, K. F. Miller, R. J. Wall, R. E. Hammer, C. A. Pinkert, R. D. Palmiter, and R. L. Brinster. 1987. Progress on gene transfer in farm animals. Vet. Immunol. Immunopathol. 17: 303–312.
46. Seamark, R. F. 1987. The potential of transgenic pigs and related technology for the pig industry. Proc. Australian Pig Science Assoc.
47. Pursel, V. G., R. G. Campbell, K. F. Miller, R. R. Behringer, R. D. Palmiter, and R. L. Brinster. 1988. Growth potential of transgenic pigs expressing a bovine growth hormone gene. J. Anim. Sci. (Suppl. 1). In press.
48. Vize, P. D. 1987. Expression of porcine growth hormone in bacteria and transgenic animals. Ph.D. Thesis, University of Adelaide, Adelaide, South Australia. 129 pp.
49. Ward, K. A., J. D. Murray, C. D. Nancarrow, M. P. Boland, and R. Sutton. 1984. The role of embryo gene transfer in sheep breeding programs. pp. 279–284. In D. R. Lindsay and D. T. Pierce (eds.), Reproduction in Sheep. Aust. Acad. Sci., Canberra.
50. Palmiter, R. D., R. L. Brinster, R. E. Hammer, M. E. Trumbauer, M. G. Rosenfeld, N. C. Birnberg, and R. N. Evans. 1982. Dramatic growth of mice that develop from eggs microinjected with metallothionein-growth hormone fusion genes. Nature 300: 611–615.
51. Ebert, K., M. Low, E. Overstrom, F. Buonomo, C. F. Baile, T. M. Roberts, A. Lee, G. Mandel, and R. Goodman. 1989. A Moloney MLV-RAT somatotropin fusion gene produces biologically active somatotropin in a transgenic pig. Molec. Endocrinol. In press.
52. Pinkert, C. A., V. G. Pursel, K. F. Miller, R. D. Palmiter, and R. L. Brinster. 1987. Development of 1-cell and 2-cell pig ova after microinjection of genes. J. Anim. Sci. 65 (Suppl. 1): 260. (Abstr.).
53. Gordon, K., E. Lee, J. A. Vitale, A. E. Smith, H. Westphal, and L. Hennighausen. 1987. Production of human tissue plasminogen activator in transgenic mouse milk. Biotechnol. 5: 1183–1187.
54. Simons, J. P., M. McClenaghan, and A. J. Clark. 1987. Alteration of the quality of milk by expression of sheep β-lactoglobulin in transgenic mice. Nature 328: 530–532.
55. Simons, J. P., I. Wilmut, A. J. Clark, A. L. Archibald, J. O. Bishop, and R. Lathe. 1988. Gene transfer into sheep. Biotechnol. 6: 179–183.
56. Salter, D. W., E. J. Smith, S. H. Hughes, S. E. Wright, A. M. Fadly, R. L. Witter, and L. B. Crittenden. 1986. Gene insertion into the chicken germ line by retroviruses. Poultry Sci. 65: 1445–1458.
57. Crittenden, L. B. and D. W. Salter. 1989. Insertion of retroviruses into avian germ line. In B. S. Weir, E. J. Eisen, M. M. Goodman, and G. Namkoong (eds.), Proc. 2nd Internat. Conf. Quantitative Genetics. In press.

Enhancement of plant germplasm

JAMES H. ELGIN, Jr. and PHILIP A. MILLER
U.S. Dept. Agric., Agricultural Research Center, Beltsville, Maryland, 20705, U.S.A.

Abstract Enhancement is the process of transferring genes that control desirable traits of wild, unadapted germplasm to adapted germplasm that can be used in advanced breeding programs. The wild germplasm we collect, store, and evaluate is of little value unless it will be used by plant breeders. The emphasis in germplasm enhancement is to bring wild plant germplasm to a level at which the plant breeder can use it directly in cultivar development.

If we were to examine the source of the attributes of our present-day crop cultivars, we would discover that wild-growing germplasm provided the original genes for such traits as adaptability to hostile environments, resistance to diseases and insects, and superior quality characteristics. There are numerous examples: tomatoes, alfalfa, cotton, soybeans, wheat, strawberries, oats, crested wheatgrass, pearl millet, bermudagrass, etc. For the future, wild germplasm is expected to provide additional genes for our improved seed, fiber, and forage crop cultivars.

Since commercial seedsmen rarely have the incentive to invest long-term efforts in enhancement activities, germplasm enhancement using raw, wild germplasm will be primarily a responsibility of governmental agencies. Therefore, public policies will greatly influence future germplasm enhancement activities. These activities will be strengthened with the continued development and maintenance of a strong National Plant Germplasm System, including the continued promotion of free exchange of germplasm and genetic information among plant scientists worldwide.

Key words: pre-breeding, plant breeding, cultivar development, disease resistance, insect resistance

Introduction

Enhancement is the process of transferring genes that control desirable traits of wild, unadapted germplasm that can be used in advanced breeding programs. The wild germplasm we collect, store, and evaluate is of little value to us unless we can use it in our plant breeding and improvement programs. The emphasis in germplasm enhancement is to bring the wild plant germplasm to a level at which plant breeders can use it directly in variety development programs. One might think of the process as 'pre-breeding'. The methods used in these efforts might range from conventional back-crossing procedures to gene transfer at the cellular level.

The commercial plant breeders generally must provide a continuing flow of new and improved varieties from their breeding programs to maintain seed sales for their companies. This generally means crossing

L. Knutson and A.K. Stoner (eds.). Biotic diversity and germplasm preservation, global imperatives.
 ISBN 0-7923-0178-1

'good' × 'good' parents with the expectation of recovering an improved segregate within a few (4–6) generations. Commercial breeders thus usually use as parents in their breeding programs only germplasm that:

a. Is fully fertile with commercial varieties,
b. Will flower and produce mature seed in the breeding nursery, and
c. Has at least minimal qualities for plant type, e.g., non-shattering, lodging resistant, acceptable grain color, not viney, etc.

The great majority of the accessions in our germplasm collections do not meet these criteria. Yet these accessions possess many individual genes of interest. Enhancement to meet these criteria thus is essential if the material is to be used as a source of new and needed genes in commercial cultivars. Vogel *et al.* [1] explain the process this way:

> 'Plant breeders take a raw product, plant germplasm, and by genetic manipulations produce products such as cultivars that are of increased value to mankind. Hence, plant breeding can be described as man-directed evolution. The process is analogous to manufacturing in which raw materials are transformed into items that are of increased value by the manufacturing process. In each instance, value is added to the raw material by the breeding or manufacturing process and this improved value fulfills specific needs of mankind. Cultivars and other products of plant breeding must fulfill specific needs of mankind if they are to have any value.
>
> The value that is added to germplasm in the breeding process occurs in incremental steps. A superior grass genotype or plant growing on the plains of Africa has value only to the owner of the cow that may graze it once or twice a year or to a wild herbivore. Although it has intrinsic value, its economic value is limited. If this plant is collected, increased, evaluated, used as a source of desirable genes in a breeding program, production tested, increased and released as a cultivar that is planted on millions of acres, its value is greatly multiplied (Fig. 1). This procedure was basically used in the development of 'Coastal' bermudagrass (*Cynodon dactylon* (L.) Pers.) that is planted on millions of acres in the southern United States and that has added billions of dollars to the economy of this and other countries (Burton [2]). Similar examples could be given for many other grasses. Each step in this breeding procedure (Figure 1) adds value to the germplasm. As in the example above, uncollected germplasm has little economic value; it has value to mankind only after it has been collected and its attributes are known'.

Although we seldom stop to think about the source of the attributes of our present-day crop varieties, if we were to do so we would discover that they derive from a continuing effort by plant breeders over the centuries to

GERMPLASM

↓

COLLECTION

↓

EVALUATION

↓

ENHANCEMENT

↓

ADVANCED BREEDING

↓

TESTING

↓

CULTIVAR RELEASE

↓

UTILIZATION

Figure 1. The process of development and release of a cultivar (Adapted from Vogel *et al.* [1]).

select plants with the most valuable characteristics for seed, forage or fiber production. Many of these efforts have made use of wild-growing germplasm that produces more seed or withstands cold better or resists disease. Superior plants were identified, selected, and crossed to other plants with superior traits. Examples of contributions from wild germplasm to our modern-day varieties are numerous.

Histories of crops

To allow you to appreciate the impact of wild germplasm on some of our commercial crops, let us share as examples with you the history of cotton, soybeans, small fruits, wheat, oats, rye, and several other grasses.

Cotton

Until 1978 all commercial species and varieties of cotton grown in the U.S. contained nectaries, structures on the underside of the leaf on the mid-rib, at the base of the involucral bracts, and between the calyx and the bracts. However, a wild species of cotton, *Gossypium tomentosum* Nuttall et Seemann, endemic to the Hawaiian Archipelago, contains no nectaries. Meyer and Meyer [3] determined that duplicate recessives, ne_1 ne_2, control the expression.

Meredith *et al.* [4] transferred this trait from the wild species into cultivated cotton, *G. hirsutum* L., and reported that the use of nectariless cottons reduced tarnished plant bug populations in the fields by about 50%. Tarnished plant bugs are a major pest on cotton in the Mississippi Delta. It later was demonstrated (Meredith [5]) that differences were insignificant under insect-free regimes in either yield or fiber quality between normal nectaried and nectariless near-isolines. However, when early season insects, especially tarnished plant bugs, were present in large numbers, nectariless isolines produced about 34% more lint (215 kg/ha) than did their nectaried recurrent parents. Similar results have been reported by other researchers in the southern U.S. Besides conferring resistance to tarnished plant bugs, the nectariless trait confers resistance or tolerance to at least six other insect pests of cotton. The exact mechanism of resistance is unknown but is probably related to the absence of sugars secreted by nectaries and used as food by insects.

Enhanced cotton germplasm with the nectariless trait was released to the public in 1972. These genetic stocks were used by public and private cotton breeders to develop and distribute the seed of nectariless cultivars to growers. By 1985, about 12% of the U.S. hectares and 30% of the Mississipi hectares were planted to these resistant types. As agriculture moves towards more limited use of pesticides to reduce production costs and protect groundwater quality, germplasm with host-plant resistance to pests will be increasingly important.

Soybeans

The soybean cyst nematode (SCN), *Heterodera glycine* Ichinohe, is a major soybean pest in the U.S. Losses caused by SCN range from slight to as high as 90%, depending upon the level of infestation, environmental conditions, and other factors. A search for host-plant resistance was initiated soon after the nematode first was identified in the U.S. A field screening of more than 4,000 soybean germplasm lines in an SCN-infested field in North Carolina in 1957 revealed apparent resistance in about 0.3% of the lines (Hartwig [6]). All lines identified as resistant were black-seeded and originated from northeastern China.

Matson and Williams [7] concluded that resistance in these lines was determined by three recessive genes plus one dominant gene. Furthermore, the dominant gene was closely linked with the gene responsible for the black seed color. Breeders thus had to cross the resistant germplasm with locally adapted lines and select for commercially acceptable yellow-seeded types that also were homozygous for the resistance trait. In addition, it was essential to combine this resistance into non-shattering, phytophthora rot-resistant types with the desired maturity adaptation. This process required extensive and intensive germplasm enhancement programs to provide growers with commercially acceptable, resistant cultivars.

Brim and Ross [8] distributed foundation seed of the 'Pickett' cultivar for increase in North Carolina, Tennessee, Missouri, and Arkansas in 1966. An additional cultivar, 'Dyer', developed by Hartwig and Epps [9], was released in 1967. Since then, four races of SCN have been recognized. Sources of resistance to each of the races have been identified in the USDA Soybean Germplasm Collection and used in enhancement programs to provide farmers with regionally resistant cultivars (Hartwig [6]). Enhancement research has been particularly effective in developing resistance to multiple SCN races and in combining the resistance with the array of other essential agronomic traits required for commercial production.

A special study by the USDA Economic Research Service (Bradley and Duffy [10]) documented that use of the resistant variety 'Forrest' accounted for a savings to soybean farmers of approximately $401 million (constant 1980 dollars) in the 6-year period following its release. 'Forrest' is only one of 20 different SCN-resistant varieties that have been released for growers in both the southern and northern soybean production regions of the U.S.

Small fruits

Unadapted wild-growing germplasm has played an important role in the enhancement of small fruits – strawberry, blackberry, and blueberry. From such germplasm have come the high yielding, high quality, everbearing strawberry cultivars 'Tribute' and 'Tristar' (Draper *et al.* [11]); the blackberry cultivars 'Hull Thornless' (Galleta *et al.* [12]) and 'Chester'; and the blueberry cultivars 'Gulfcoast', 'Cooper', 'Georgiagem' (Austin and Draper [12]), 'Oneal', 'Blueridge', and 'Capefear'.

Short stature in wheat

In 1874, the U.S. agricultural attache to Japan described in routine reports the 'Daruma' type of short straw used in that country, but no one in the USDA paid attention to the information. Not until 1946 (72 years later, when Dr. S.C. Salmon, USDA agronomist with Gen. Douglas McArthur's occupation program, observed the same types) were they recognized as useful germplasm. That year Dr. Salmon brought back to the U.S. several short types, including 'Norin 10'. 'Norin 10' was assigned the Plant Inventory (PI) number 156641. Salmon had noted the tremendous productive capacity of 'Norin 10' when grown in Japan on irrigated land that had been heavily fertilized. The plants had many tillers (stems); its normal size spikes (heads), were only about 61 cm high, and had not lodged.

Following its introduction in 1946, 'Norin 10' was grown in a detention nursery to assure that no new diseases or insect pests had been introduced with it. Along with 23 other Japanese short-stature introductions, 'Norin 10' was grown at seven locations in the U.S. in 1948. It was described that year as a very short winter wheat that was highly susceptible to leaf rust and powdery mildew. Seed was then made available to interested U.S. wheat breeders.

The first person to realize the real potential of the Japanese Norin wheats was Dr. O.A. Vogel of the USDA at Washington State University, Pullman. In 1949 'Norin 10' was crossed with 'Brevor', an adapted cultivar then grown in the Pacific Northwest.

Dr. Vogel was doggedly persistent, and after intensive selection for self-fertility, he produced excellent selections of 'Norin 10' × 'Brevor'. Those selections became the foundation for the major wheat-breeding effort in the Pacific Northwest (Washington, Oregon, Idaho) and were shared with many other wheat breeders, including Dr. Norman Borlaug, who at that time directed the Rockefeller-Ford Foundation Wheat Improvement Program in Mexico.

The first semidwarf wheat variety to be released for production was 'Gaines', which was distributed in 1961 and which gained worldwide attention because of its remarkably high level of productivity. A new era in wheat improvement began with 'Gaines' – it had a tremendous impact on wheat research in the U.S. and later in wheat-growing areas worldwide. The 'Norin 10' × 'Brevor' selections developed in Dr. Vogel's program were used by Dr. Borlaug in 1954 as base material in his outstanding development of high-yielding, disease-resistant spring wheats, which sparked the Green Revolution and for which he was awarded the Nobel Peace Prize in 1970 (Reitz [14], Briggle [15]).

Resistance to oat crown rust

Since the mid-1960's a large number of wild oat (*Avena sterilis* L.) collections have been obtained from countries bordering the Mediterranean Sea. Most germplasm has come from Israel through a series of PL-480 and other Foreign Currency programs that began in the early 1960's.

The greatest resistance to oat crown rust (*Puccinia coronata* Cda.) in oat cultivars grown in the U.S. today traces back to those accessions, today stored in the National Small Grain Collection (NSGC). That particular species has proven to be a veritable gold mine as a source of resistance not only to crown rust but to other oat diseases as well (Frey *et al.* [16, 17]; Frey and Browning [18]; McDaniel [19, 20]).

Resistance to cereal leaf beetle

Long before the cereal leaf beetle (*Oulema melanopus* L.) reached Michigan (about 1962), resistant strains of wheat had already been accessioned in the

NSGC. An introduction had been obtained from Saratov, USSR, in 1927, and had been assigned the CI (Cereal Investigations) number 9321. No one recognized its value for 40 years, and no one knew why such an agronomically miserable wheat had been kept. However, it provided the genetic source for pubescent (hairy) leaves, which is the current mechanism of resistance to the cereal leaf beetle. The first wheat variety, 'Vel', resistant to the cereal leaf beetle was released in 1976 (Reitz [14]; Briggle and Craddock [21]).

Male sterility in wheat
A collection of *Triticum timopheevii* (Zhuk.) Zhuk., a wild relative of wheat, became a part of what is now the NSGC in 1930. It came from the USSR and was assigned the PI number 94760. Cytoplasmic male sterility and some fertility-restoring genes were discovered in this accession around 1960, some 30 years later. This source of male sterility became the principal genetic tool in the development of hybrid wheat in the U.S. during the early 1960's and later. Although hybrid wheat has not become a major part of commercial wheat production, minor acreage is being grown (Reitz [14]).

Russian wildrye
'Bozoisky-Select' Russian wildrye was released cooperatively with the Utah Agricultural Experiment Station and the USDA Soil Conservation Service in 1984. The parental materials were obtained from PI 440627 from the USSR. The breeding population was screened in the field and laboratory largely on the basis of characters related to stand establishment under environmental stress. 'Bozoisky-Select' has been significantly more vigorous and productive than the check cultivar 'Vinall' in range seedings. At eight semi-arid range locations in Utah, Idaho, and Wyoming, 'Bozoisky-Select' yielded 23% more forage than did 'Vinall' during the first two years of production (Asay *et al.* [22]).

Crested wheatgrass
Crested wheatgrass used in North America has been either diploid (2n = 14) or tetraploid (2n = 28). Hexaploids have had little or no impact, but selections from hexaploid introductions obtained from Turkey, Iran, and the USSR have been extremely promising. PI 406442 (from the USSR) is particularly noteworthy. This accession has leaves that are 2 to 3 times wider than those of other crested wheatgrasses, and its seedlings have excellent vigor. Attempts are presently being made to introgress the leafiness and quality attributes of hexaploid germplasm into tetraploid breeding populations.

Crested wheatgrass is typically a bunchgrass and is used primarily as a forage grass. Accessions recently introduced into the U.S. from the People's Republic of China, Iran, and Turkey are rhizomatous and show other adaptations for turf. They have finer leaves and are more procumbent than are the crested wheatgrasses used for forage (Asay *et al.* [22]).

Weeping lovegrass

'Morpa' weeping lovegrass is an increase from PI 299924 (Voigt [23]). 'Morpa' selected for increased palatability over that of other PIs and common 'A-67', (also an increase from a PI) produced 12% greater gain per head than common weeping lovegrass in grazing studies.

Old world bluestem

Old world bluestem cultivars 'WW-Spar" and the recent release 'WW-Iron Master' are direct selections from apomictic PIs (Shoop *et al.* [24]).

Kleingrass

A major limitation to seed production and establishment of kleingrass, *Panicum coloratum* L., is seed shattering. Although there may be minor variation in seed retention within existing populations of kleingrass, no progress was made toward the development of increased resistance to seed shattering until recently, when a new source of resistance was identified in PI 410177 (Young [25]). This germplasm is now being used in a recurrent selection program that is increasing seed retention in kleingrass.

Buffelgrass

A series of buffelgrass PIs collected from southern Africa were more winterhardy than any other buffelgrass cultivars. These PIs contained an extra set of chromosomes compared to normal tetraploid buffelgrass. The genes for increased winterhardiness apparently are located on the extra genome. Apomictic breeding through hybridization with sexual germplasm is being used to transfer the winterhardiness of these apomictic PIs to less hardy, rhizomatous buffelgrass germplasm. Common buffelgrass was selected from the apomictic PI 153671. It is estimated that over 90% of the revegetated land in south Texas has been seeded to buffelgrass (Bashaw and Johns [26]).

Digitgrass

'Pangola' digitgrass was released by the Florida Agricultural Experiment Station in 1943. It traces to an introduction received by the USDA as vegetative material (PI 111110) from Pretoria, South Africa, in 1935. Today it occupies approximately 200,000 hectares in Florida and millions of hectares in other tropical areas (Schank *et al.* [27]).

Pearl millet

In April 1979, three accessions of a species related to pearl millet were sent to the USDA, ARS laboratory at Tifton, Georgia, by a French plant breeder in Senegal for determinations of chromosome numbers. By 1980 these accessions had been observed to be highly resistant to rust and were being used in crosses with rust-susceptible pearl millet lines. The resistance imparted by these accessions provided the basis for the recently released 'Tifleaf 2' pearl millet that is virtually immune to rust and which yields as much as 22% more forage than the previous 'best' cultivar (Hanna *et al.* [28]).

Bermudagrass

'Tifton 44' bermudagrass was the best of several thousand F_1 hybrids screened for winterhardiness in northern Georgia. 'Tifton 44' is a cross between Coastal bermuda and a bermudagrass that had survived in Berlin, Germany, for 15 years before it was collected in 1966. The winterhardiness introduced by this accession has greatly extended the range of bermudagrass in the southern U.S. (Burton [29. 2]).

Alfalfa

Alfalfa is a crop recently introduced to North America. Spanish-derived alfalfa was introduced into the southwestern U.S. about 1850. Three winter-hardy germplasm sources also were introduced from Europe and Russia into the upper Midwest and Canada between 1858 and 1919. Three nonwinter-hardy germplasms were introduced from Peru (1899), India (1913 and 1956), and Africa (1924). Two intermediate winterhardy types were introduced from France (1947) and from the general area of southern Russia, Iran, Afghanistan, and Turkey between 1898 and about 1925. These nine sources (through extensive enhancement efforts) represent most of the basic alfalfa germplasm in current U.S. cultivars (Barnes *et al.* [30]).

Clover

From 1910 to 1950, numerous introductions of ladino clover were made from Italy. The cultivars 'Merit', 'Regal', and 'Tillman' trace to these introductions. The 1956 crimson clover introduction PI 233812 from the Mediterranean region was the base for the cultivars 'Frontier' and 'Tibbee'. The arrowleaf clover cultivars 'Amclo', 'Meechee', and 'Yuchi', grown in many parts of the South, originate from three direct introductions from Italy, also in 1956. Likewise, the alsike and red clovers grown in the U.S. trace to introductions from Europe and the Mediterranean region (Taylor [31]).

Other crops

Numerous other crops also have had their origin by way of enhancement of wild germplasm. These examples include the development of high solids in tomatoes; long-day adaptation in corn and sorghum; and disease resistance and quality components in snap beans, potatoes, and lettuce. We seldom stop to realize how indebted we are to our forefathers for their foresight in their germplasm enhancement efforts that led to the many crop cultivars we have today.

Continuing need for enhanced germplasm

It is worth noting that, in all of the preceding examples, few of the commercially usefull varieties were direct introductions from the NPGS. A significant pre-breeding enhancement effort was required to transfer the desired genes into genetic material that could then be used directly in applied breeding programs. Germplasm enhancement thus is an essential link between collection/evaluation of germplasm and its utilization to benefit agriculture.

After all that has been done in the past, one may ask, 'Why do we need to have more enhancement – more moving genes from wild, unadapted germplasm of little use in this era of today's high yielding crops'? The example of the corn blight epidemic of the early 1970's provides an excellent reason. More than 90% of the U.S. hybrid corn varieties of those years carried the Texas 'T' cytoplasm that rendered them highly susceptible to *Helminthosporium maydis* Nisik. & Miyake, the corn blight organism. Even today's varieties are often built on narrow or limited genetic bases rendering them potentially vulnerable to new races and strains of pests.

Emphasis on new breeding objectives, such as increased stress tolerance, biological efficiency with reduced inputs, etc., also are causing plant breeders to look once again at wild germplasm for sources of genes regulating these traits. To keep up with the ever-increasing needs of our modern agricultural system, the development of broad-based germplasm pools carrying genes for resistance to diseases, insects, and nematodes, tolerance to hostile environmental stresses, and potential for high productivity of superior quality must be available to the plant breeders if they are to be successful in developing competitive new varieties.

Enhancement technology

For the most efficient effort at germplasm enhancement in today's research environment, we need to know much more about the plant than we did

20–30 years ago. If we are striving to improve the biological efficiency (photosynthesis, nitrogen fixation, etc.) of a crop, it is necessary to understand the physiological/biochemical pathways involved as well as the genetic regulation of these pathways. With such understanding, today's breeder is better equipped than before to make significant progress in moving genes regulating these pathways by use of biotechnological techniques in conjunction with conventional plant-breeding methods. Germplasm enhancement of the future thus should revolve around the transfer of genes regulating basic biological processes in wild germplasm to improved populations from which tomorrow's cultivars will be developed.

Public role in germplasm enhancement

Since commercial seedsmen rarely have the incentive to invest long-term efforts in enhancement activities, germplasm enhancement using raw, wild germplasm will be a responsibility primarily of public agencies, i.e., the NPGS. Public policies will greatly influence germplasm enhancement activities in the future. These activities will be strengthened with the continued development and maintenance of a strong NPGS consisting of vigorous germplasm acquisition, maintenance, and evaluation programs, and an easily accessible and comprehensive information base.

Continued promotion of the free exchange of germplasm and genetic information among all plant scientists throughout the world – public and private – also will significantly increase the successes associated with future germplasm enhancement. In addition, support of interdisciplinary approaches involving molecular geneticists, plant breeders, physiologists, biochemists, pathologists, entomologists, etc., with appropriate incentives for team research, will greatly encourage germplasm enhancement activities. Finally, we must expand the training of scientists with expertise in germplasm enhancement to keep pace with the ever-increasing demands for the new cultivars of tomorrow.

References

1. Vogel, K.P., H.J. Gorz and F.A. Haskins. 1989. Breeding grasses for the future. In D.A. Sleper and K.A. Asay (eds.), Contributions from Breeding Forage and Turf Grasses. Spec. Publ. Crop Sci. Soc. Am., Madison, Wisconsin. In press.
2. Burton, G.W. 1989. Progress and benefits to mankind from breeding warm-season forage grasses. In D.A. Sleper and K.A. Asay (eds.). Contributions from Breeding Forage and Turf Grasses. Spec. Publ. Crop. Sci. Soc. Am., Madison, Wisconsin. In press.
3. Meyer, J.R. and V.G. Meyer. 1961. Origin and inheritance of nectariless cotton. Crop Sci. 1: 167–169.
4. Meredith, W.R., Jr., M.L. Laster, C.D. Ranney and R.R. Bridge. 1973. Agronomic potential of nectariless cotton (*Gossypium hirsutum* L.). J. Environ. Qual. 2: 141–144.

5. Meredith, W.R., Jr. 1978. Nectar glands role in breeding for plant resistance. Proc. 3rd Biennial Nat. Plant Resist. Workshop, Gainesville, Florida: 24–35.
6. Hartwig, E. E. 1985. Breeding productive soybeans with resistance to the soybean cyst nematode. In R. Shibles (ed.) World Soybean Research Conference III: 394–399. Westview Press, Boulder and London.
7. Matson, A.L. and L.F. Williams. 1965. Evidence of four genes for resistance to the soybean cyst nematode. Crop Sci. 5: 477.
8. Brim, C.A. and J.P. Ross. 1966. Registration of Pickett soybeans. Crop Sci. 6: 305.
9. Hartwig, E.E. and J.M. Epps. 1968. Dyer soybean. Crop Sci. 8: 402.
10. Bradley, E.B. and M. Duffy. 1982. The value of plant resistance to soybean cyst nematodes: A case study of Forrest soybeans. Nat. Res. Econ. Div., Econ. Res. Serv., U.S. Dept. Agric., Washington DC.
11. Draper, A.D., G.J. Galletta and H.J. Swartz. 1981. 'Tribute' and 'Tristar' everbearing strawberries. Hort. Sci. 16: 794–795.
12. Galletta, G.J., A.D. Draper, R.G. Hill, Jr., R.C. Blake and R.M. Skirvin. 1981. 'Hull Thornless' blackberry. Hort. Sci. 16: 796–797.
13. Austin, M.E. and A.D. Draper. 1987. 'Georgiagem' blueberry. Hort. Sci. 22: 682–683.
14. Reitz, L.P. 1976. Improving germplasm resources. Am. Soc. Agron. Publ. 16.
15. Briggle, L.W. 1968. Breeding short stature, disease resistant wheats in the United States. Euphytica. Suppl. 1.
16. Frey, K.J., J.A. Browning and R.L. Grindeland. 1971a. Registration of Multiline E68, Multiline E69, and Multiline E70 Oat Cultivars. Crop Sci. 11: 939.
17. Frey, K.J., J.A. Browning and R.L. Grindeland. 1971b. Registration of Multiline M68, Multiline M69, and Multiline M70 Oat Cultivars. Crop Sci. 11: 940.
18. Frey, K.J. and J.A. Browning. 1976. Registration of Multiline E72, Multiline E73, and Multiline E74 Oat Cultivars. Crop Sci. 16: 312–312.
19. McDaniel, M.E. 1974a. Registration of TAM 0-301 Oats. Crop Sci. 14: 127–128.
20. McDaniel, M.E. 1974b. Registration of TAM 0-312 Oats. Crop Sci. 14: 128.
21. Briggle, L.W. and J.C. Craddock. 1978. Maintaining world cereal grain genetics seed stock vital to plant breeding. Wheat Grower 1, October 1978.
22. Asay, K.H., W.H. Horton and W.T. Hansen II. 1985. New grasses for intermountain rangelands. Utah Sci. 46(4): 119–123.
23. Voigt, P.W. 1971. Registration of Morpa weeping lovegrass. Crop Sci. 11: 312–313.
24. Shoop, M.C., E.H. McIlvain and P.W. Voigt. 1976. Morpa weeping lovegrass produces more beef. J. Range Manage 29: 101–103.
25. Young, B.A. 1986. A source of resistance to seed shattering in Kleingrass, *Panicum coloratum* (L.). Euphytica 35: 687–694.
26. Bashaw, E.C. and C.W. Johns. 1983. Buffelgrass germplasm research for the southern great plains. Proc. 39th Southern Pasture and Forage Crop Improv. Conf.: 31–34.
27. Schank, S.C., E.M. Hodges, B.C. Killinger and D.E. McCloud. 1972. Registration of *Pangola* digitgrass. Crop Sci. 12: 715.
28. Hanna, W.W., H.D. Wells and G.W. Burton. 1985. Dominant gene for rust resistance in pearl millet. J. Hered. 76: 134.
29. Burton, G.W. 1986. Developing better forages for the South. J. Anim. Sci. 63: 955–961.
30. Barnes, D.K., E.T. Bingham, R.P. Murphy, O.J. Hunt, D.F. Beard, W.H. Skrdla and L.R. Teuber. 1977. Alfalfa germplasm in the United States: Genetic Vulnerability, Use, Improvement, and Maintenance. U.S. Dept. Agric. Agric Res. Serv. Tech. Bull. 1571. U.S. Government Printing Office, Washington, DC. 21 pp.
31. Taylor, N.L. (ed.). 1985. Clover science and technology. Am. Soc. Agron. Spec. Publ. 25, Madison, Wisconsin. 616 pp.

The impact of biotechnology on conventional germplasm conservation and use

PETER R. DAY

Center for Agricultural Molecular Biology, Cook College, Rutgers University, New Brunswick, New Jersey 08903, U.S.A.

Abstract Genetic information exists in a variety of forms. Molecular biologists deal with DNA sequences that can be stored in a computer database, transmitted electronically, and used to synthesize molecules of 100 or so units in length. This technology has had little impact so far on conventional germplasm conservation. The extent to which it might replace or supplement conventional germplasm conservation is far from clear. Comparison of DNA sequences of genes from different sources may well prove to be valuable in evaluating and classifying accessions. For those species that cannot be conserved as seed or which are clonally propagated, conservation by micropropagation techniques has had a major impact. The expense and inconvenience of cryopreservation, however, is still an obstacle to long-term conservation of tissue-culture materials. New tools for more efficient cataloging and description of accessions will provide valuable adjuncts to conventional methods.

Key words: DNA sequences, disease resistance, quality components, *in vitro* culture, plant molecular biology, transformation, RFLPs

In 1972 the National Research Council's (NRC) report on Genetic Vulnerability [1] addressed the problem of preparing to meet future challenges to the adequacy of crop plant genetic resources. A committee of the NRC Board on Agriculture had been established to consider the implications of the southern corn leaf blight epidemic. In the concluding chapter, in a section dealing with germplasm collections, the report stated:

> It would be foolish to think of the nation's gene pools as warehouses stocked with finished products. They are much more like stores of outdated military surplus equipment awaiting disassembly and reassembly into new, more useful, forms. The items in a gene pool cannot be stored as disassembled parts but only within a living organism whose constituent properties are catalogued.

At the time this was written, the first recombinant DNA molecules had been generated at Stanford University [2]. The following year foreign DNA fragments were inserted deliberately into bacterial plasmid DNA and reinserted into *E. coli* demonstrating the potential for cloning genes. The rapid advancement of recombinant DNA technology and plant molecular

L. Knutson and A.K. Stoner (eds.). Biotic diversity and germplasm preservation, global imperatives.
 ISBN 0-7923-0178-1

biology during the intervening 16 years has led to the successful isolation and cloning of probably more than 100 different plant genes that are indeed stored as disassembled parts in the form of DNA-ethanol precipitates in laboratory freezers. Many of these genes have been sequenced and can also be 'stored' in the form of long strings of symbols in laboratory notebooks, publications, and, even more usefully, in computer databases. Although in theory it is possible to resynthesize large DNA molecules by assembling oligonucleotides of about 100 bases each, presently it is more convenient to use them as probes to identify and isolate molecules of thousands of bases from DNA libraries prepared from the organisms themselves.

Looking to the future we may ask if this technology will eventually replace conventional germplasm collection and plant exploration by providing a database that is sufficiently complete so that not only existing DNA sequences can be stored and resynthesized at will, but also new ones can be synthesized with the properties that breeders and end-users desire. In my view, this is still remote. For the foreseeable future we will continue to rely on existing systems. It is necessary, therefore, to ensure that these systems are as effective and as efficient as we can make them. Increasing use will be made of new methods to expand the range of germplasm uses and to simplify presently laborious procedures. Germplasm resources are, in any case, an important resource for molecular biology [3]. In this paper, I review some current impacts of biotechnology on plant germplasm conservation and use and speculate on likely impacts in the future.

First it is necessary to define biotechnology in this context. In the broadest sense, the word means 'the manipulation of living forms for economic benefit.' Two broad areas of biotechnology are relevant to germplasm conservation. The first is cryopreservation, a method of maintaining living material *in vitro*. The second is molecular biology, which provides the means to manipulate directly as well as to evaluate the genomes of organisms.

Cryopreservation

Animal breeders and geneticists have used cryopreservation extensively, principally in the form of frozen semen of livestock. Frozen embryos also now are used for shipping valuable germplasm from one country to another. Cryopreservation of plants is largely based on *in vitro* cultures. This method ensures the preservation of clonal material that is either very valuable or which cannot be preserved easily in other ways. Since other authors deal with these subjects in these proceedings, I will say little more about them now except to comment on the growing interest in using tissue culture as a means of expanding field-collecting activities for certain crop plants. Withers [4] has pointed out that collection of plant tissue in the field with aseptic excision techniques can be a useful supplement to laboratory-based procedures. For example, the use of sterile glove boxes, hypochlorite solution as a sterilizing agent, and unspillable liquid culture vessels all facilitate field collection. Even though the cultures are subject to contamination with microorganisms, this can be kept to acceptable levels. Even so, the techniques of tissue culture and cryobiology applied to preserving plant material remain cumbersome, elaborate, and expensive and, for the foreseeable future, are likely to be limited to special cases.

Molecular biology

Since molecular biology is having so widespread an influence on the whole of biology, I will first consider its impact on the individual components of a plant germplasm conservation operation. I choose to use plants to illustrate my points only because I am more familiar with them than with animals. Hawkes [5] has described the chain of genetic resource activities (Fig. 1). Conservation normally begins with a survey, for each crop, of materials that already exist in gene banks and for germplasm that is likely still to be available in the field. This survey, an important part of the establishment of programs and objectives, is followed by exploration and collection, conservation, evaluation, enhancement, breeding and trials, and the establishment of databases and research programs. Other aspects of the chain, such as training, need not be discussed here. At first sight, molecular biology is

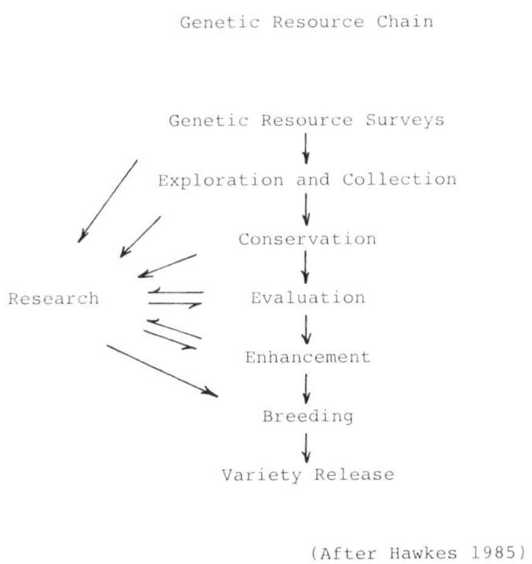

Figure 1. Sequence of activities involved in genetic resource conservation and use.

relevant principally to the subject of germplasm enhancement. However, it already is obvious that this is but one aspect.

Genetic resources surveys

As Plucknett *et al.* [3] pointed out, there is much interest among plant molecular biologists in exploring variation in the structure and expression of major genes. Although *Arabidopsis thaliana* is attractive as a model system because of its low DNA content, small chromosome number, and convenience as a laboratory organism, many plant molecular biologists work with other plants, including crop plants. This is illustrated by the fact that 53 different genera of higher plants were the subjects of some 275 abstracts of talks and posters at the 1988 UCLA Symposium on the molecular basis of plant development [6]. The most widely studied species were maize (12%), tobacco (10%), *Arabidopsis* (9%), soybean (6%), and wheat (5.5%). Also described was research on trees (larch, plane tree, Brazil nut), aquatics (*Isoetes* or quillwort, *Elodea*, *Vallisneria*, *Lemna*), ornamentals, weeds, and other crop species. Twelve of the abstracts did not identify the plant source of the systems they described! Maize workers often need to supplement studies with standard inbreds by using variants with both demonstrated and putative alterations in the pathways or structures they are concerned with. Surveys could well address these needs, particularly where the variant characters are economically significant. Although such forms are maintained in genetic stock collections, the latter consist principally of experimental materials described in earlier genetic studies. They are, of course, very important, but they are limited in terms of the variation they encompass in comparison to unexplored materials. As the partnership between molecular biology and plant breeding develops, the needs of the former will begin to influence surveys.

Exploration and collection

The nature and quality of material conserved in gene banks depends ultimately on what is collected. For some major crops, such as rice, the existing collections would appear to be very thorough and, at least to the outsider, almost complete. For many other crops and minor species, resources are limited and compromises are necessary. In the section on evaluation, I describe the usefulness of restriction fragment length polymorphisms (RFLPs) for linkage studies and for discriminating among different genetic forms as well as how these techniques could be useful in classifying and searching collections. It seems likely that RFLPs also may be useful in testing sampling and collection methods from natural populations and could have an impact on how these procedures are designed, managed, and executed.

I am not aware of any plant germplasm collections that were originally set up as, and are now maintained in, the form of DNA samples. However, this is clearly possible and is, after all, only a logical extension of cryopreservation. Needless to say, no germplasm collection can ever be complete. There is, therefore, great interest in deciding how far those who maintain collections should go in striving for completeness. Molecular biology currently has little to contribute to this debate except to state its needs for materials.

DNA collections are clearly useful and extremely important as reference sources for research, and it is necessary to plan for their growth and maintenance in the future. However, their ultimate utility as sources of raw material for crop plant improvement depends on the extent of the characterization and evaluation of accessions and on the availability of transformation methods for their reintroduction into whole plants for germplasm enhancement.

Not surprisingly, characterization and evaluation introduce a problem several orders of magnitude more complicated than the parallel problem of evaluating the germplasm accessions themselves. For example, the International Rice Research Institute (IRRI) germplasm bank has more than 80,000 rice accessions, yet each one has a comparable number of gene-sized DNA sequences with comparable internal repetitiveness and redundancy. Like the accessions in the collection itself, only one or a few of the DNA sequences from an accession are ever likely to be of interest to a gene bank user. Fortunately, DNA probes greatly simplify the problem of recovering sequences that are identical or related to the probe sequence. But the question still remains, would it not be better to have viable material available for repeated DNA isolation long into the future rather than either a stored DNA sample or dried tissue from which it may be prepared, both of which sources eventually will be exhausted? Storing seeds or live plant tissue clearly is preferable, but the lower costs of collecting, preparing, and storing dried tissue to be used as a source of DNA still may be advantageous for some materials.

Another view suggests that eventually very little single-gene DNA need be stored as base sequences ready to be spliced into transformation vectors. The genes required would be synthesized on demand from computerized databases. As the technology evolves, the search for new genes would be replaced by directed synthesis of sequences with predicted, desired effects on phenotype. This brave new world of germplasm conservation and enhancement is, of course, some time off in the future. We are still only beginning to grapple with the problems of understanding the relationship between genotype and phenotype in plant and animal development.

The availability of transformation methods for making use of genes in the form of DNA sequences presents less of a problem for crop plants such as potato, tomato, and tobacco in the family Solanaceae than it does for some other economically important species, notably cereals. In the former, the use of vectors based on disarmed tumor-inducing plasmids of the crown gall bacterium *Agrobacterium tumefaciens* now enable DNA constructs to be introduced routinely into tissue explants that can be regenerated to form fertile seed-bearing plants. Unfortunately, there are still no comparable, widely used methods for transforming maize, wheat, barley, soybean, and rice. This is chiefly because these plants are not hosts of *A. tumefaciens*, and it is difficult to regenerate plants from their protoplasts. There are intensive efforts in many laboratories to develop transformation technology by refining existing methods, developing new ones, and extending them to other species and genera.

Setting up DNA germplasm banks as an alternative rather than as a supplement to existing banks has overtones of the bizarre practice of preserving human cadavers in liquid nitrogen against the day medical science advances to the point when they might be revived and their terminal disorders rectified. There are simply too many problems to solve before this will be feasible.

Most plant breeders now recognize the potential for transformation and its implications for germplasm use. Their search for new, useful genetic information is, in theory if not yet in practice, no longer limited by their ability to transfer genes by sexual means. However, many of their requirements presently are beyond the capability of plant molecular biology. Simmonds [7] has pointed out that most plant breeding proceeds by the manipulation of many genes of individually small effect (polygenes). Polygenes do not lend themselves to identification, isolation, and cloning to effect directed change. It is, therefore, likely that biotechnology, for the time being, will contribute specific major genes. These now include genes for resistance to pests, viruses, and herbicides and will soon extend to genes that control aspects of quality and resistance to fungal and bacterial pathogens.

Directed genetic change requires a relatively complete understanding of the nature of the character, its genotypic structure, and how this is translated through biochemistry and developmental controls to produce the phenotype. Characters such as yield, plant form, daylength response, and tolerance of environmental extremes (heat, cold, drought, flood, salinity, etc.) are sufficiently complicated that only much more research will enable us to understand how these characters are expressed and which pathways are amenable to genetic manipulation. Isolation and cloning of such genes for effective germplasm enhancement must wait on the development of this knowledge. Fortunately, molecular biology itself provides some of the most powerful methods for this research.

Conservation

The literature on *ex situ* and *in situ* conservation is large, and I can do no more than make some observations and suggestions. *Ex situ* conservation is principally the preserve of seed physiologists, who determine the optimal developmental stage of the seed for storage and the best combinations of low temperature, reduced relative humidity, and oxygen concentration to maximize longevity and minimize genetic change during storage. The ultimate test is seed viability. Provided that viability is high and germination vigorous, most germplasm bank users assume that genetic change has been minimal since intact DNA repair systems will have maintained viability and, therefore, genetic integrity. RFLPs could provide a sensitive means to test these assumptions by observing the extent to which storage introduces additional polymorphisms as a result of breakage, base substitution, deletion, and translocation in specific regions of the DNA. For seeds, even those of inbred self-pollinating species, appropriately sized samples would be required, before and after storage, to establish whether RFLP differences were significant. Probably the simplest experimental material would be seed of an apomictic species (e.g., Kentucky bluegrass) in which there are few or no pre-existing RFLPs when the problems of sampling are analogous to those for members of a clone.

In contrast, *in situ* conservation may involve many disciplines: ecology, agronomy, resource management, soil science, genetics, entomology, and plant pathology, just to name the obvious ones. Here biotechnology offers the prospect of RFLP and allozyme markers to identify genotypes that are more detailed than descriptions of morphology.

In 1973 a fish called the Sonoran topminnow (*Poeciliops occidentalis*) was added to the Federal list of endangered species. It had flourished in the Gila River system of Arizona but groundwater pumping, stream diversion, and the introduction of a predator fish (*Gambusia affinis*) had drastically reduced its population density. The Sonoran topminnow breeds well in captivity and is the subject of intensive population genetics studies. A restocking program at first concentrated on a population from a Gila tributary that laboratory studies [8] had shown were inbred and less fertile than other populations from harsher environments such as the Rio Maya near the Mexican border. Vrijenhoek et al. [8] measured allelic frequencies at 12 loci showing protein polymorphism and have since extended their studies to include variation in mitochondrial DNA [9]. As a consequence of these studies, the restocking program is continuing but is now making use of contrasted populations to test the suggestion that the more variable populations will establish more successfully than the inbred populations in environments similar to that where the inbred population originally evolved.

In population genetics, one assumes that polymorphisms for observable markers allow valid generalizations to be made for genes that are directly concerned with fitness and survival. The information on population structure and genetic polymorphism is useful but must be interpreted with caution.

Evaluation

The traditional method of germplasm evaluation involves screening material for characters that are known to be of interest to plant breeders. This is most effective when carried out alongside a breeding program and when the two activities are coordinated properly. Most breeders soon can become overwhelmed with germplasm. They tend to concentrate most on elite materials that are similar to their own, which they use as sources of specific characters. The fewer non-adaptive genes or 'junk' they introduce, the better.

The great majority of plant breeders thus have little interest in working with uncharacterized, unimproved germplasm. This means that gene banks must spend a great deal of time in describing and recording the features of their accessions. Passport data are of prime importance since they indicate the nature of the collection site and are a guide to the environmental adaptation of the accessions. A large number of other morphological descriptors are commonly used, but most of these are unimportant to breeders seeking resistance to pests, diseases, and environmental extremes (heat, cold, drought, salinity, soil toxicity, etc.). For most of these physiological characters, there is no substitute for direct screening procedures. However, once the functional basis of the character is understood it becomes possible to screen for its components, employing more efficient methods. The following example, concerned with wheat grain quality, indicates how evaluation may make use of biotechnology.

Several review articles [10, 11] have summarized recent work on understanding the molecular basis of bread-making quality in winter wheat. The high molecular weight (HMW) glutenin and gliadin protein subunits found in the grain endosperm are largely responsible for the viscoelastic properties of bread doughs. To produce loaves with a satisfactory texture and crumb structure, a protein network must be formed when the dough is worked that is both sufficiently elastic to retain and trap the CO_2 produced by the yeast but viscous enough that the loaf volume does not increase excessively. Individual HMW glutenin subunits that confer these properties have been identified, which has in turn enabled breeders to assemble systematically the genes for complementary HMW subunits in new lines. The subunits are identified by SDS polyacrylamide gel-electrophoresis (SDS-PAGE) followed by protein staining. In the breeding program, PAGE can be conducted on a quarter of a single grain. The remainder, containing the embryo of the grain, can be planted to give rise to a healthy, vigorous seedling.

Progress in cloning and sequencing subunit genes [12–14] should lead eventually to characterization of the features that distinguish 'good' HMW glutenin subunits from those that have either no effect or a negative effect on dough characteristics. Current theory suggests that cysteine residues located near the ends of the subunits are crucial to establishing the sulfhydryl bonds between subunits to form the elastic network. PAGE of alien relatives of wheat reveals a broad range of HMW glutenin subunits that appear to increase the number already available among cultivars of *Triticum aestivum*. Many of these alien relatives (*Triticum* spp., *Aegilops* spp., etc.) can be crossed with cultivated wheat only with difficulty. If the alien HMW subunits could be compared at the DNA sequence level to identify those most likely to be valuable in a breeding program, it would be possible to concentrate on introducing only them by crossing. Of course, once a transformation method is available for wheat, new genes for new HMW submits would be still more readily accessible to breeders.

Enhancement and breeding

I have already referred several times to the role of DNA transformation as a tool for creating new forms of germplasm for breeders. The use of RFLP markers for the rapid construction of linkage maps now is receiving detailed attention. The objective is to saturate the genome with markers, spaced an average of 10 centimorgans (cM) apart, so that any economically important single gene will be no more than about 5 cM from a known marker. The nearest linked marker then can be used as an aid to the selection of genes that are not easy to score or assess, such as stress tolerance, insect resistance, grain quality, fiber length, oil quality, etc.

RFLPs also can help in assessing relatedness among accessions. For example, many vegetable varieties with different names are identical. RFLPs could help to eliminate unnecessary duplication in collections.

The procedure for using RFLPs is briefly as follows: A population known to be segregating for visible markers is sampled by taking a small piece of tissue (less than 1.0 g) from each of a number of characterized individuals. DNA prepared from the sample is digested with a restriction enzyme that cuts the DNA only at specific sites. The DNA fragments are separated according to size by one-dimensional electrophoresis on agarose gel. The separated fragments are transferred from the gel to a membrane by blotting. At this point a radio- or fluorescence-labeled probe DNA may be used to detect any DNA fragments on the membrane that, under suitable conditions, hybridize with the probe. These fragments are revealed as bands that form consistent patterns for each individual in the population. The patterns vary according to the DNA probe and the restriction enzyme used to make the digest. However, variations frequently are seen among individuals in a population treated with the same enzyme and probe. Such variations often result from point changes in the DNA of a fragment that hybridizes to the probe. Small deletions or additions of DNA bases alter the size of the fragment, causing it to move either faster or slower in the gel. The site of the point change is a genetic marker, which can be mapped by observing its segregation, as revealed by the difference in band patterns relative to other markers in the population.

The selection of probes is crucial. These are often cDNAs, or complementary DNAs synthesized from messenger RNAs, from the same or a related species. The cDNA probes will hybridize with the transcribed regions of DNA in the target genome. The band patterns are usually simple, e.g., a single band in a homozygous diploid, and thus are easy to interpret. By repeated washing to remove earlier probes, one membrane can be tested with 12 or more different probes in succession. RFLP maps of maize, rice, wheat, and tomato already are available, and mapping is proceeding with other species including humans, swine, and even plant pathogenic fungi.

Although few gene banks are likely to have laboratory facilities for RFLP mapping, several of the international agricultural research centers are starting to employ this technique in germplasm enhancement work associated with their gene banks. Methods for automating DNA extraction, digestion, electrophoresis, and probing will help in making the method more widely available.

Databases and research

I mentioned earlier that no germplasm bank can ever hope to be complete. Several authors have addressed the question of how best to make the compromise between collecting every new accession available or building a bank in which only the more important variation is adequately represented in a form that can be evaluated and maintained within the resources available. Frankel [16] and Brown [17] both advocate the concept of core collections in which representative samples, amounting to no more than about 5-10% of the much larger collections, receive priority for all gene bank activities. The remainder of the collection is not discarded but kept in reserve in the knowledge that unless regenerated it eventually will become useless. Because of the risk that some large collections will become repositories of poorly described, uncatalogued, inviable material, the core- collection concept likely will have many supporters, particularly for minor species that attract little funding or with breeding systems that make *in situ* conservation or maintenance as mixed populations impractical.

Just as biotechnology may be useful in judging the effectiveness of population sampling, it may also be used in helping to select the components of a core collection and for monitoring its progress through regeneration and enhancement. A core collection will be most useful when associated directly with the pulling effect of a breeding program that needs the products of germplasm enhancement.

Many benefits clearly can be gained from research in cryopreservation and molecular biology, the two aspects of biotechnology I have tried to relate to germplasm conservation and use. While it would be impractical for every gene bank to acquire the means to perform RFLP mapping or gene cloning, the major centers must adopt new methods that increase their efficiency and enhance the quality of their work.

References

1. National Research Council. 1972. Genetic Vulnerability of Major Crops. National Academy of Sciences, Washington, DC.
2. Watson, J.D. and J. Tooze. 1981. The DNA Story: A Documentary History of Gene Cloning. Freeman, San Francisco.
3. Plucknett, D.L., N.J.H. Smith, J.T. Williams, N.M. Amshetty. 1987. Gene Banks and the World's Food. Princeton University Press, Princeton.
4. Withers, L.A. 1987. *In vitro* methods for collecting germplasm in the field. Pl. Genet. Res. Newsl. 69: 2–6.
5. Hawkes, J.G. 1985. Plant genetic resources: The impact of the International Agricultural Research Centers. Consultative Gp. on Internat. Agric. Res. Study Paper 3, World Bank, Washington, DC.
6. Goldberg, R. 1988. UCLA symposium on the molecular basis of plant development (abstracts). J. Cell. Biochem.Suppl. 12C: 126–226.
7. Simmonds, N.W. 1983. Plant Breeding: The State of the Art. pp. 5–25 in T. Kosuge, C.P. Meredith and A. Hollaender (eds.), Genetic Engineering of Plants: An Agricultural Perspective. Plenum, New York.
8. Vrijenhoek, R.C., M.E. Douglas, and G.K. Meffee. 1985. Conservation genetics of endangered fish populations in Arizona. Science 229: 400–402.
9. Avise, J.C. and R.C. Vrijenhoek. 1987. Mode of inheritance and variation of mito-

chondrial DNA in hybridogenetic fishes of the genus *Poeciliopsis*, Molec. Bio. Evol. 4: 514–525.

10. Blackman, J.A. and P.I.Payne. 1987. Grain quality. pp. 455–485 in F.G.H. Lupton (ed.), Wheat Breeding: Its Scientific Basis, Chapman, Hall, London.
11. Day, P.R., J. Bingham, P.I. Payne, and R.D. Thompson. 1986. The way ahead: Breeding for quality improvement. pp. 251–261 in J.M.V. Blanshard, P.J. Frazier, and T. Galliard (eds.),Chemistry and Physics of Baking, Special Publ. 95, Royal Society of Chemists, London.
12. Thompson, R.D., D. Bartels, and N.P. Harberd. 1985. Nucleotide sequence of a gene from chromosome 1D of wheat encoding a HMW-glutenin subunit. Nucleic Acids Res. 13: 6833–6846.
13. Harberd, N.P., R.B. Flavell, and R.D. Thompson. 1987. Identification of a transposon-like insertion in a Glu-1 allele of wheat. Mol. Gen. Genet. 209: 326–332.
14. Colot, V., L.S. Robert, T.A. Kavanagh, M.W. Bevan, R.D. Thompson. 1987. Localization of sequences in wheat endosperm protein genes which confer tissue-specific expression in tobacco. Europ. Molec. Biol. Org. J. 6(12): 3559–3564.
15. Flavell, R.B., R.J. Kemble, R.E. Gunn, A. Abbott, and D. Baulcombe. 1983. Applications of molecular biology in plant breeding: the detection of genetic variation and viral pathogens. pp. 198–212 in Better Crops for Food. Ciba Foundation Symposium 97. Pitman, London.
16. Frankel, O.H. 1984. Genetic perspectives of germplasm conservation. pp. 161–170 in W. Arber, R. Illmensee, W.J. Peacock, P. Starlinger (eds.), Genetic Manipulation: Impact on Man and Society. Cambridge University Press, Cambridge.
17. Brown, A.H.D. 1989. The case for core collections. pp. 136–156 in A.H.D. Brown, O.H. Frankel, D.R. Marshall and J.T. Williams (eds.), The Use of Plant Genetic Resources. Cambridge University Press, Cambridge.

SECTION FOUR

Germplasm collection and data management

Cryopreservation of animal germplasm resources

V. G. PURSEL and L. A. JOHNSON
Reproduction Laboratory, Beltsville Agricultural Research Center, Agricultural Research Service, U.S. Department of Agriculture, Beltsville, Maryland, 20705, U.S.A.

Abstract Spermatozoa were first frozen and thawed successfully in 1949. Since then, semen-freezing procedures have been developed for almost all agriculturally important mammalian and avian species. In addition, semen has been frozen with varying degrees of success from more than 50 species of non-domestic mammals, from more than 25 species of fish, and from several endangered species of birds and mammals.

Mouse embryos were successfully frozen and thawed in 1972. Since then, living young have been produced from frozen embryos for most domestic and from several wild mammalian species. Cryopreservation of embryos holds particular interest because the risk of disease transmission is less for embryos than for semen and because the embryo is diploid.

Optimum cryopreservation procedures differ considerably among species. The most effective buffer solution, type and concentration of cryoprotective agent, and rate of cooling and thawing usually must be determined for each species. Although spermatozoa/embryo viability can be lost at any phase of the cryopreservation procedure, the most critical phase is passage through the temperature zone of 0 °C through − 50 °C during cooling and again during thawing. If spermatozoa or embryos are stored in liquid nitrogen (− 196 °C), they should remain viable for 3,000 to 10,000 years, since background ionizing radiation is thought to be the only source of damage at that temperature.

Cryopreservation of spermatozoa or embryos can prevent loss of genetic diversity of virtually all domestic species and many of the endangered species if storage banks are established that contain a broad sampling of the current base population.

Key words: spermatozoa, embryos, cryopreservation, freezing, germplasm

Introduction

Although efforts to freeze spermatozoa date back to Mantegazza (1866) [1], modern cryobiology began when Polge, Smith, and Parks [2] reported that glycerol provided spermatozoa with protection during freezing to − 79 °C and subsequent thawing. Since then, spermatozoa from nearly all species of domestic mammals and birds have been frozen and thawed successfully on at least a limited scale. Semen from dairy breeds of cattle is frozen on the largest scale of all species because as many as 300 insemination doses can be frozen per ejaculate, artificial insemination is compatible with other

L. Knutson and A.K. Stoner (eds.). Biotic diversity and germplasm preservation, global imperatives.
 ISBN 0-7923-0178-1

routine management requirements of dairy cows, and selection of bulls on the basis of milk production of their daughters is highly advantageous. These advantages do not prevail for most other species.

Cryopreservation and transplantation of mammalian embryos has received considerable attention since the first reports of successful freezing and thawing of mouse embryos [3, 4]. Cattle embryos were frozen successfully soon after those of the mouse. Survival of frozen and thawed cow embryos is sufficiently high to permit commercial application of this technology in several countries. Cryopreservation of embryos of a number of other mammalian species, including some wild animals, also has been reported.

Once spermatozoa or embryos are successfully stored in liquid nitrogen (−196 °C), they should retain their viability for several thousand years. Thus establishment of germplasm banks provides a means of conserving threatened breeds of farm animals and endangered species of non-domestic animals. Modern breeding practices that are dictated by economic pressures are continually narrowing the genetic base of farm animals. The economic expense of losing domestic breeds is unknown because the genetic value of these endangered breeds has not been determined. Extinction of animal species is similarly immeasurable on a monetary basis.

The objective of this paper is to review the technical aspects of cryopreservation of animal spermatozoa and embryos.

Basic principles in cryopreservation

Freezing

During the process of cryopreservation, gametes or embryos can be irreversibly damaged by inappropriate conditions at any time from collection until deposition of spermatozoa or embryos into the appropriate host female. However, for most living cells, the primary area of lethality is a critical temperature span from the freezing point of the cell suspension to about −50 °C. This critical temperature zone must be traversed twice, once during cooling and again during thawing.

Two main factors destroy living cells as they pass through the critical temperature zone. One factor is formation and growth of ice crystals within the cytoplasm; the other factor is exposure of cell membranes to increasing concentrations of solutes as ice forms in the suspending medium. Minimization of the effects of these two factors depends on the rate of cooling and warming, the particular cryoprotectant and its concentration, the composition of the suspending medium, and the physical characteristics of the cell.

The optimum rate of cooling is one that is slow enough for water to pass out of the cell, allowing it to shrink as ice forms in the medium, but fast enough to prevent the increasing salt concentrations from damaging the cell membranes. When the cooling rate is too rapid, water cannot move out of the cells fast enough, and intracellular ice crystals form. In contrast, when the rate of cooling is slower than optimum, cells are exposed to the concentrated solutes for a longer duration at a higher temperature.

Permeating cryoprotectants have low molecular weights, high solubility in water, and freeze at very low temperatures. Glycerol, dimethyl sulfoxide (DMSO), ethylene glycol, and acetamide are the most commonly used permeating cryoprotectants. They play an important role in preventing membrane damage by reducing the amount of ice formed at any particular temperature and by maintaining the salts in the liquid phase. In addition, the permeating cryoprotectants can diffuse into the cytoplasm as water effluxes from the cell as more and more water freezes.

Sugars, polymers, and lipoproteins function as non-permeating cryoprotectants by reducing the amount of ice formed and by reducing the concentration of extracellular salts.

The optimum rate of cooling can be calculated by use of mathematical formulas proposed by Mazur [5] if one knows the cell's surface area, molar water volume, and membrane permeability to water and cryoprotectant. Unfortunately, permeability constants have not been determined for most types of cells, and many factors may alter the permeability characteristics of cells.

Thawing

When frozen cells are warmed, the optimum warming rate depends upon the rate of cooling and the cryoprotectant used for freezing the cells. If cooling has been fast enough to induce intracellular ice crystals, then a slower than optimum warming rate permits ice crystal growth by recrystallization that may physically disrupt cell membranes. In this instance, a rapid rate of warming minimizes damage. Cells frozen with a slow rate of cooling can be damaged by excessive osmotic swelling if they are warmed too fast and the intracellular cryoprotectants cannot diffuse from the cells fast enough.

Duration of storage

Cells that are stored at temperatures below $-130\,^\circ C$, which is the glass transition temperature of water, should remain unchanged for centuries. Storage of frozen tissue in liquid nitrogen at $-196\,^\circ C$ is currently the method of choice. At $-196\,^\circ C$, no known thermally driven chemical reactions occur. However, background radiation and cosmic ray protons produce ionization that accumulates over time. Such radiation has been calculated to kill 60 to 90% of mammalian cells in 3,000 to 20,000 years [6, 7].

The effect of low temperature preservation on mutagenesis of germplasm has not been studied adequately [8]. However, mutagenesis was nil when microorganisms [7] and DNA [9, 10] were cryopreserved at low temperatures. Background radiation will contribute to mutations during low-temperature storage; however, background radiation contributes less than 1% of the spontaneous mutation rate at physiological temperatures [11]. Thus the mutation rate at $-196\,^{\circ}C$ should be considerably lower than the spontaneous rate [6, 7].

Cryopreservation of semen

Semen collection

Semen can be collected from domestic mammals by use of an artificial vagina or through manual stimulation, electroejaculation, or postmortal flush of the caudal epididymides. Under some circumstances, non-domestic mammals permit semen collection by manual massage or by use of an artificial vagina, but electroejaculation is usually employed under anesthesia [12]. Electroejaculation is used most successfully in species that ejaculate in a single burst (ram, bull). Electroejaculation is usually ineffective in species that ejaculate in a series of propulsive waves (boar, stallion, dog) [13]. Semen has been collected by electroejaculation from over 100 species of non-domestic animals [12, 14].

Semen usually is collected from birds by manual massage, with semen being expressed from the ejaculatory papillae by gentle pressure [15]. Semen can be collected from Falconiformes by imprinting the male birds to a handler, who then elicits the ejaculation [16]. Electroejaculation also has been used successfully for ducks and geese [17, 18].

Semen usually is collected from fish by hand stripping the milt from the urogenital papilla. Larger fish are usually anesthetized before semen is hand stripped.

Semen of rodents, primates, and macropod marsupials coagulates immediately after ejaculation and must be liquified before preservation. Part of the semen of the boar, rabbit, and stallion is composed of a gel that is readily removed by filtration at the time of collection. Goat semen contains an accessory fluid component that must be removed from the spermatozoa before preservation; otherwise, the spermatozoa undergo premature acrosome reaction, and fertilizing capacity is lost during cryopreservation [19]. Freeze-thaw survival is enhanced by the removal of seminal plasma before freezing of spermatozoa of boar, dog, stallion, and several non-domestic species.

Dilution

A variety of semen extenders are used to enhance spermatozoal survival of various animal species. In general, extenders provide osmolarity similar to the ejaculate, a source of energy, a buffer to maintain the pH at about 7.0, necessary ions, and antimicrobial compounds to control bacteria. Spermatozoa from some mammalian species benefit from the presence of chelating agents such as citrate, EDTA, or zwitterionic buffers (TRIS, TES, HEPES, etc.). Frequently, extenders used for short-term preservation differ considerably from extenders that are effective for cryopreservation.

Initial cooling

Species vary greatly in the susceptibility of their spermatozoa to irreversible damage during rapid cooling, a phenomenon referred to as cold shock. Human, avian, and fish spermatozoa are relatively unaffected by cold shock, whereas bull and ram spermatozoa are sensitive and boar spermatozoa are extremely sensitive to cold shock [20]. Egg yolk and milk are effective extender components for preventing cold shock damage of ram and bull spermatozoa, whereas only the low density lipoprotein fraction of egg yolk, casein, and BHT are effective in reducing effects of cold shock on boar spermatozoa [21]. Cold shock is avoided for spermatozoa of most species by a combination of the use of protective extenders with slow cooling of the spermatozoa from body temperature to refrigeration temperature (5 °C). Cryoprotectants such as glycerol and DMSO do not protect spermatozoa from cold shock.

Avian and fish spermatozoa are not susceptible to cold shock, and their survival diminishes rapidly if spermatozoa are held at body temperature after collection. Thus the semen can be cooled abruptly to 5 °C at collection.

Addition of cryoprotectants

Non-penetrating cryoprotectants such as sugars and lipoproteins are usually part of the semen extender and are added at the time of dilution. Penetrating cryoprotectants are usually added to the semen after it has been cooled to 5 °C, but this differs among species.

Glycerol is the most commonly used cryoprotectant for mammalian spermatozoa. For bull semen, 7% glycerol usually is used for egg yolk-containing extenders, and 10% glycerol is used for milk extenders. Swine spermatozoa do not tolerate glycerol as well as do bull spermatozoa, so usually only 1 or 2% glycerol is used. The optimum level of glycerol is 4 to 6% for ram spermatozoa [22, 23], 4 to 5% for stallion spermatozoa [24–26], and 7% for goat spermatozoa [27, 28]. Optimum levels of glycerol have not been determined for most other mammals.

The optimum level of glycerol for freezing of chicken semen is around 8%

[15], but the concentration must be reduced to less than 2% after thawing and before vaginal insemination or fertility is nil [29]. Thus 4% DMSO [30] and 8% dimethylacetamide [31] have been used as alternatives to glycerol for bird semen.

Glycerol, DMSO, ethylene glycol, propylene glycol, and methanol all have been used successfully for cryoprotecting semen of various species of fish [32]. Salmon spermatozoa have been frozen in 6.8 to 12.8% DMSO with little difference in post-thaw fertility [32], whereas glycerol is reported to be toxic to salmon spermatozoa [33]. A DMSO concentration of 10% was reported as optimum for sea bass [34], but optimum levels have not been determined for most other species.

Packaging

The most common storage vessels for semen are glass ampules, plastic straws, and pellets.

The glass ampule initially was used for freezing of bull semen in alcohol baths that were cooled by addition of granules of solid CO_2. The tip of the ampule is sealed by melting the glass. The main disadvantages of the use of ampules are inefficiency of storage space and difficulty in achieving high rates of cooling or warming because of the glass thickness and package geometry. In addition, thawing of ampules can be dangerous if liquid nitrogen enters the ampule as a result of an imperfect seal.

Straws were first introduced by Jondet [35] for use with bull semen and are usually frozen in the vapor phase of liquid nitrogen. Initially straws of 1.2 ml volume were introduced. Currently, straws of 0.5 ml or mini-straws of 0.2 to 0.25 ml are commonly used for bull, ram, and goat semen. Large straws of 4.0 to 6.0 ml have been used for boar and stallion semen. Straws are amenable to high cooling and warming rates and require minimal storage space. The straw also can be used as part of the insemination device, thereby eliminating the need to transfer the semen to an insemination pipet, which results in loss of spermatozoa.

The pellet-freezing method was introduced by Nagase and Niwa [36]. Pellets are formed by pipetting drops of extended semen directly into indentations on the surface of dry ice. Pellet volumes of 0.1 to 0.2 ml are most commonly used; however, pellet volumes up to 5.0 ml for boar semen have been investigated [37]. The pellets freeze in several minutes and then are transferred into liquid nitrogen for storage. If pellets are stored in bulk, they require a minimum of space. The main disadvantages in their use are 1) difficulty in maintaining identification of the animal on each pellet and 2) possible contamination of the pellets with microorganisms if the liquid nitrogen becomes contaminated.

Other freezing containers used on a limited scale include pipettes, gelatin capsules, aluminum packages or tubes, and plastic bags and vials.

Cooling rate

Spermatozoa from domesticated farm mammals appear to tolerate a rather wide range of cooling rates. Average cooling rates of 6.5 °C/min to 105 °C/min provided similar post-thaw survival rates for bull spermatozoa [38]. Likewise, Fiser and Fairfull [22] found that ram spermatozoa responded quite similarly to cooling rates covering the range of 10 °C/min to 100 °C/min but that cooling rates of 1, 2 and 5 °C/min resulted in lower post-thaw survival than did rates of 10 °C/min and faster. For stallion and pig spermatozoa [39, 40], cooling rates faster than 10 °C/min appear to be superior to slower cooling rates.

Cooling rates between 1 °C/min and 5 °C/min are the reported optimum as chicken spermatozoa are cooled from +5 °C to −20 °C or −35 °C; about 30 °C/min seems optimal for the final stage of the freezing process [30, 31, 41].

Limited systematic research on cooling rates of fish spermatozoa have been conducted. Stoss [32] suggests that spermatozoa of saltwater-spawning species may tolerate slow freezing rates better than spermatozoa of freshwater species. Spermatozoa from some saltwater-spawning species, such as *Clupea harengus* and *Gadus morhua*, tolerated 1 to 5 °C/min cooling rates [42, 43], while the optimum cooling rate for sea bass spermatozoa was 10 to 20 °C/min [34] and for rainbow trout spermatozoa, 30 °C/min [44]. Fast cooling rates obtained by immersion of spermatozoa samples directly into liquid nitrogen were destructive to *Ictalurus punctatus* [45], *Cyprinus carpro* [46], and *Salmo salar* [47]. However, use of fast thawing rates for *Salmo salar* and *Gadus morhua* spermatozoa frozen by liquid nitrogen submersion resulted in high levels of post-thaw fertility [48]. Milkfish spermatozoa also had high levels of post-thaw survival when a cooling rate of 100 °C/min was used for freezing [49]. The pellet-freezing method, which provides a variable cooling rate throughout the pellet [40, 50], has been used successfully for spermatozoa from salmonids [33, 51], whitefish [52], and pike [51, 53].

Warming rate

Semen that has been stored in ampules or straws usually is thawed by submersion in a water bath varying in temperature from 0 to 75 °C. In general, post-thaw recovery is improved by a fast rate of warming, but many precautions need to be observed. Several recent studies have shown that acrosomal membranes of bull, boar, and rabbit spermatozoa are damaged severely when spermatozoa are warmed too rapidly from 0 or +5 °C to 37 °C [54–58]. Post-thaw recovery of chicken and turkey spermatozoa is higher if straws are thawed in ice water instead of warm water. Whether rapid warming of avian spermatozoa after thawing is detrimental has not been determined.

Pelleted semen is thawed in heated tubes or in a heated semen extender. The composition of the thawing solution is important to subsequent fertility of boar spermatozoa [59]. Bamba and Cran [54] recently have shown that in the case of boar spermatozoa, the simultaneous and combined effects of dilution and rapid warming cause considerably more damage to the acrosome than either event by itself.

Fertility

Even under optimum freezing and thawing conditions, a variable percentage of spermatozoa lose their motility, have damaged acrosomal membranes, and have a shorter duration of survival after insemination-all of which lead to reduced fertilizing capacity.

In some species, this reduced capacity to fertilize an ovum can be partially overcome either by increasing the number of frozen-thawed spermatozoa per insemination or the frequency of insemination over that required for fresh semen. In the ewe, deposition of frozen-thawed semen in the uterus instead of in the cervix improves fertility. In the sow, insemination closer to the time of ovulation improves fertility with frozen-thawed spermatozoa. Differences in fertility between fresh and frozen semen are minimal and less variable among males for cattle than for other species.

As a general rule, for virtually all species, artificial insemination with frozen-thawed semen results in a lower rate of fertilization than does use of fresh semen. From the standpoint of preserving germplasm resources, the low fertilizing capacity of frozen spermatozoa is not an important issue so long as the unique genetic complement is passed on to progeny.

Cryopreservation of embryos

Mouse embryos were first successfully frozen and thawed in 1972 [3, 4]. Since then, live young have been born from frozen-thawed embryos of cattle [60], rabbit [61], sheep [62], rat [63], goat [64], horse [65], antelope [66], human [67], baboon [68], marmoset monkey [69], and cynomolgus monkey [70]. Ova or embryos of birds and fish have not been cryopreserved successfully. The large volume of yolk and lack of permeability of membranes to water prevents fish embryos from being sufficiently dehydrated during cooling to prevent the formation of intracellular ice [32, 71, 72].

Embryo recovery

Embryos can be recovered from cattle and horses by flushing the embryos from the uterus by non-surgical means. At 6 to 8 days after ovulation, a catheter is passed through the cervix into the uterus, and medium is flushed into and back out of the uterus into a collection vessel. This non-surgical method permits repeated recovery in the same animal with minimal effect on the ovaries and uterus.

Embryos from other mammalian species can be recovered under general anesthesia by use of the laparoscopy technique [73], by surgical laparotomy, or immediately after slaughter.

Techniques for superovulation have been developed for most domesticated mammalian species. Administration of follicle-stimulating hormone or pregnant mare's serum gonadotropin at the appropriate stage of the estrous cycle produces a variable but effective increase in the number of ovulations and embryos. If ovarian stimulation is excessive, fertility may be reduced and embryo quality and survival may decline.

Addition of cryoprotectant and cooling

Glycerol and DMSO are the most frequently used cryoprotectants for embryos, but a wide variety of penetrating cryoprotectants have been used, including methanol, erythritol, ethylene glycol, diethylene glycol, triethylene glycol, polyethylene glycol, propylene glycol, and acetamide. The concentrations of cryoprotectants used most frequently are around 1 to 1.5 M.

The cryoprotectant may be added at 37 °C, at room temperature, or at 0 to 5 °C, depending on the species, the cryoprotectant, and the particular freezing protocol. The temperature at the time of addition and the method of addition, i.e., single step versus multiple steps, are important factors that influence the cooling and warming rates required for maximum survival. Cryoprotectants permeate the embryos faster when the embryos are warm than when they are cooled to 0 °C.

When cryoprotectants are added to sheep or cattle embryos at room temperature or higher, the embryos have a higher freeze-thaw survival rate if they are cooled gradually (1 °C/min) to 0 °C or −7 °C than when they are cooled abruptly [74, 75]. However, other investigators have successfully cooled cattle embryos abruptly to 0 °C or below during the prefreeze-cooling period with no ill effects on embryo survival [76–78].

Seeding

Embryos usually are cooled to −5° to −7 °C, and then ice crystallization is induced by seeding. Seeding usually is achieved by touching the outside of the straw or ampule with metal forceps that have been cooled in liquid nitrogen. The phase change from water to ice results in production of heat, which must be dissipated before cooling can continue. Seeding prevents samples from supercooling to −10 to −15 °C. At that temperature the water would spontaneously seed and subject the embryos to the possibility of intracellular ice formation.

Cooling and warming rates

The optimum rates of cooling for embryos are much slower than for most other types of cells. Two cooling protocols have provided the most consistent embryo survival results across several mammalian species. The first protocol consists of cooling at a rate of 0.2 to 1.0 °C/min from seeding until −60 °C before transfer into liquid nitrogen. Embryos frozen by this protocol become dehydrated and require a slow warming rate (less than 20 °C/min) for maximum survival to avoid the osmotic effects produced by sudden rehydration [79].

The second protocol consists of cooling at a rate of 0.2 to 1.0 °C/min from seeding until between −20 °C and −48 °C before transfer into liquid nitrogen [75]. Embryos frozen by this protocol are less dehydrated than are embryos slowly cooled to −60 °C, so some intracellular ice probably forms. Maximum embryo survival is obtained by using a rapid warming rate (about 300 °C/min) during thawing. Rapid warming prevents growth of small ice crystals during thawing and recrystallization of vitrified water, which would occur with slow warming.

A rather different approach to cryopreservation of mouse embryos was described by Rall and Fahy [80]. This procedure requires use of high concentrations of several cryoprotectants that will permit embryos to become supercooled to very low temperatures and become vitrified without any crystallization of ice. Since no ice forms, exposure to concentrated solutes is avoided. Application of this vitrification method to other species hinges on determining which combinations of cryoprotectants are non-toxic in the concentrations required to permit vitrification. The advantage of the vitrification procedure is its simplicity of freezing and thawing–embryos can be plunged directly into liquid nitrogen and subsequently thawed rapidly.

Removal of cryoprotectant

If frozen-thawed embryos are transferred directly into a medium that contains no cryoprotectant, they are damaged by the osmotic stress created by the influx of water into the embryo during restoration of osmotic equilibrium.

A frequently used method of avoiding osmotic shock is to dilute the cryoprotectant gradually by transferring the embryo stepwise into progressively lower concentrations of cryoprotectant. Six-step dilutions have been used frequently for bovine embryos. Mouse embryos are more tolerant than are bovine embryos, so fewer steps are necessary for that species [81].

An alternative to the stepwise dilution method is to transfer the embryo into a 1-molar sucrose solution. Since sucrose does not permeate the embryo, water leaves the embryo faster than does the cryoprotectant. Thus the embryo shrinks until osmotic equilibrium is achieved. The embryo then can be transferred into the recipient female or into isotonic medium for incubation.

When embryos are frozen in 0.25 ml straws, a sucrose solution can be separated from the embryo-containing medium by an air bubble. After thawing, the two solutions can be mixed by shaking the air bubble to one end of the straw. This innovative procedure eliminates the need to locate the embryo under a stereoscopic microscope and manually transfer it into other media, thereby simplifying procedures that must be performed under field conditions [82].

Stage of embryonic development

Mouse and human embryos from the 2-cell stage to the blastocyst stage of development tolerate freezing and thawing quite well [83], but optimum freezing conditions may differ with each stage [84]. Variation in optimal freezing conditions probably results from changes in membrane permeability of the developing mouse embryos [85].

Several investigators found that 2-cell to 16-cell sheep and cattle embryos suffered a high rate of death when cooled to 0°C [74, 86, 87]. In contrast, late morula and early blastocysts readily survive cooling to 0°C and have been successfully frozen and thawed by use of a variety of protocols. Comparisons of freeze-thaw survival of different embryonic stages for other species have not been made.

Pregnancy rates

Live births have resulted from transfers of frozen-thawed embryos in 12 mammalian species. However, reliable pregnancy rates for transferred frozen-thawed embryos are available only for mice and cattle. In a study involving transfers of more than 500 mouse embryos that had been frozen for 6 to 65 months, about 50% developed into normal fetuses or live mice at birth [88]. In another study, 66% of almost 200 transferred mouse embryos developed into live fetuses [89].

Surgical transfers of 550 frozen-thawed cow embryos yielded a pregnancy rate of 54%; however, only the best quality embryos were frozen [90]. In another report, involving more than 800 cow embryos that varied in quality when frozen and which were transferred non-surgically, the pregnancy rate was 32% [91]. Leibo [82] concluded from extensive commercial experience that non-surgical transfers of frozen embryos can yield a pregnancy rate that is more than 80% of that obtained by the transfer of freshly recovered embryos.

A key factor affecting success of embryo transfer is the degree of estrous synchrony between the embryo donor and the recipient. In general, the survival of embryos is reduced if the estrous cycle of the recipient is more than one day out of phase with that of the donor at the time of embryo recovery. In the cow, pregnancy rates were equivalent when recipients were in estrus from 36 hours before to 12 hours after the donor [92]. In contrast, mice, rat, and pig embryos that are retarded in comparison to the estrous stage of the recipient are much less likely to survive than embryos that are synchronous or advanced by as much as 24 hours [93].

For non-domestic mammals in captivity, the problems of anestrus and detection of estrus greatly increases the difficulty in obtaining synchronous transfer of embryos. Development of methods for synchronizing estrus in non-domestic species with progestogens or prostaglandins could have considerable value for timing the recovery of embryos for maximum survival during freezing as well as the subsequent transfer into the host animal.

Applications in germplasm preservation

General considerations

Preservation of germplasm in the form of embryos, which are diploid, is considerably more efficient than is storage of germplasm in the haploid state as spermatozoa. The embryo provides the full genetic complement of the breed or species, whereas if females of the threatened breed or species are unavailable for insemination, the spermatozoa must be used for several generations of backcrossing to reconstitute a representative of the breed or species. In addition, unless semen from unrelated males is used in back-crossing, considerable inbreeding will result.

In the case of domestic mammals, one can safely assume that the rare and endangered breed will be sufficiently similar to the plentiful breeds that spermatozoa or embryos will be compatible for fertilization or development even if none of the rare breed is available. However, in the case of many endangered species, such females may not be available, and less desirable alternatives must be explored. One of these possibilities is to use related species. A number of interspecific pregnancies have been established among equine species, but most of these hybrids are sterile, the exception being Przewalski's horse (*Equus przewalskii*) crossed with domestic horse [94]. In Bovidae, fertile hybrids have been reported between *Bos taurus* and *B. indicus* and between American bison (*Bison bison*) with *Bison bonasus*, *Bos indicus*, *Bos taurus*, *Bos grunniens*, and other species [95]. Other examples of interspecies and intergenus hybridization among non-domestic mammals have been compiled [95]. Although interspecies fertilization may occur under natural mating or artificial insemination with fresh semen, one might expect reduced success with frozen-thawed semen.

In certain instances, an endangered species might be rescued by transferring its embryos into the uterus of a more numerous closely related genus or species. Notable examples are successful interspecies transfers between *B. taurus* and *B. indicus* [96], transfers of Gaur (*B. gaurus*) embryos into domestic cow [97], Przewalski's horse (*E. przewalskii*) embryos and Grant's Zebra (*Equus burchelli*) embryos into domestic horse [71], mouflon (*Ovis musimon*) embryos into domestic sheep [98], and Bongo antelope (*Taurotragus euryceros*) embryos into African eland (*Taurotragus oryx*) [99].

In the past few years, a number of embryo manipulations have resulted in establishment of true interspecific pregnancies between sheep and goats [100]. These manipulations result in the trophoblast being that of the host species with the embryo being that of the foreign species. Whether this strategy will be useful for other interspecies transfers remains to be determined.

References

1. Mantegazza, P. 1866. Sulla sperma umano. R.C. 1st Lombardo 3: 183–196.
2. Polge, C., Smith, A.U. and Parkes, A.S. 1949. Revival of spermatozoa after vitrification and dehydration at low temperatures. Nature (Lond.) 164: 666.
3. Whittingham, D.G., Leibo, S.P. and Mazur, P. 1972. Survival of mouse embryos frozen to $-196°C$ and $-269°C$. Science 178: 411–414.
4. Wilmut, I. 1972. The effect of cooling rate, warming rate, cryoprotective agent, and stage of development on survival of mouse embryos during freezing and thawing. Life Sci. 11: 1071–1079.
5. Mazur, P. 1963. Kinetics of water loss from cells at subzero temperatures and the likelihood of intracellular freezing. J. Gen. Physiol. 47: 347–369.
6. Mazur, P. 1976. Freezing and low temperature storage of living cells, pp. 1–12. In: O. Muhlbock (ed.), Basic Aspects of Freeze Preservation of Mouse Strains. Gustav Fischer Verlag, Stuttgart.
7. Ashwood-Smith, M.J. and Friedmann, G.B. 1979. Lethal and chromosomal effects of freezing, thawing, storage time, and X-irradiation of mammalian cells preserved at $-196°C$ in dimethyl sulfoxide. Cryobiology 16: 132–140.
8. Committee on Germplasm Resources. 1978. In: Conservation of germplasm resources: An imperative, pp. 79–91. National Academy of Sciences, Washington, DC.
9. Shikama, K. 1965. Effect of freezing and thawing on the stability of double helix of DNA. Nature 207: 529–530.
10. Ashwood-Smith, M.J., Trevino, J. and Warby. C. 1972. Effect of freezing on the molecular weight of bacterial DNA. Cryobiology 9: 141–143.
11. Russell, W.L. 1963. The effect of radiation dose rate and fractionation on mutation in mice, pp. 205–217. In: F.H. Sobels (ed.), Repair from Genetic Radiation-Damage and Differential Radiosensitivity in Germ Cells. Macmillan, NY, and Pergamon, London.
12. Watson, P.F. 1978. A review of techniques of semen collection in mammals. Symp. Zool. Soc. Lond. 43: 97–126.
13. Martin, I.C.A. 1978. The principles and practice of electroejaculation of mammals. Symp. Zool. Soc. Lond. 43: 127–152.
14. Seager, S.W.J., Wildt, D. and Platz, C. 1980. Semen collection by electroejaculation and

artificial vagina in over 100 species of zoo mammals. 9th Internat. Congr. Anim. Reprod. Artif. Insem. Madrid 2: 571–580.
15. Lake, P.E. 1978. The principles and practice of semen collection and preservation in birds. Symp. Zool. Soc. Lond. 43: 31–49.
16. Gee, G.F. and Temple, S.A. 1978. Artificial insemination for breeding non-domestic birds. Symp. Zool. Soc. Lond. 43: 51–72.
17. Serebrovski, A.S. and Sokolovskaja, I.I. 1934. Electroejakuljacia. Ptic. Prob. Zhivotu, Mosc. 5: 57 (Anim. Breed. Abstr. 3: 73–74).
18. Watanabe, M. 1957. An improved technique of the artificial insemination of ducks. J. Fac. Fish Anim. Husb., Hiroshima Univ. 1: 363–370.
19. Courtens, J.L., Nunes, J.F. and Corteel, J.M. 1984. Induction of the acrosome reaction in the spermatozoa of the goat by secretions of the male accessory glands and milk. Gamete Res. 9: 287–302.
20. Watson, P.F. 1981. The effects of cold shock on sperm cell membranes, pp. 189–218. In: G. J. Morris and A. Clarke (eds.), Effects of Low Temperature on Biological Membranes. Academic Press, London.
21. Pursel, V.G. 1979. Effect of cold shock on boar sperm treated with butylated hydroxytoluene. Biol. Reprod. 21: 319–324.
22. Fiser, P.S. and Fairfull, R.W. 1984. The effect of glycerol concentration and cooling velocity on cryosurvival of ram spermatozoa frozen in straws. Cryobiology 21: 542–551.
23. Colas, G. 1975. Effect of initial freezing temperature, addition of glycerol and dilution on the survival and fertilizing ability of deep frozen ram semen. J. Reprod. Fertil. 42: 277–285.
24. Cochran, J.D., Amann, R.P., Froman, D.P. and Pickett, B.W. 1984. Effects of centrifugation, glycerol level, cooling to 5 °C, freezing rate and thawing rate on the post-thaw motility of equine sperm. Theriogenology 22: 25–38.
25. Cristanelli, M.J., Amann, R.P., Squires, E.L. and Pickett, B.W. 1985. Effects of egg yolk and glycerol levels in lactose-EDTA-egg yolk extender and of freezing rate on the motility of frozen-thawed stallion spermatozoa. Theriogenology 24: 681–686.
26. Nishikawa, Y., Intani, A. and Shinomiya, S. 1972. Studies on the protective effects of egg yolk and glycerol on the freezability of horse sperm. 7th Internat. Congr. Anim. Reprod. Artif. Insem. 2: 1545–1549.
27. Corteel, J.M., Bril, G., Lebœuf, B. and Montigny, G. 1976. Fertilizing capacity of washed deep frozen goat sperm. 8th Internat. Congr. Anim. Reprod. Artif. Insem. (Krakow) 4: 981–985.
28. Waide, Y., Niwa, T. and Asanuma, R. 1977. Studies on preservation of liquid and frozen semen of domestic animals. 3. Viability and fertility of frozen goat spermatozoa. Jap. J. Anim. Reprod. 23: 129–137.
29. Clark, C.E. and Shaffner, C.S. 1960. The fertilizing capacity of frozen chicken sperm and the influence of related in vitro processes. Poultry Sci. 39: 1213–1220.
30. Sexton, T.J. 1980. Optimal rates for cooling chicken semen from 5 to minus 196 celsius. Poultry Sci. 59: 2765–2770.
31. Lake, P.E. and Ravie, O. 1984. An exploration of cryoprotective compounds for fowl spermatozoa. Brit. Poultry Sci. 25: 145–150.
32. Stoss, J. 1983. Fish gamete preservation and spermatozoan physiology, pp. 305–350. In: W. S. Hoar, D. J. Randall, and E. M. Donaldson (eds.), Fish Physiology, Vol. IX Reproduction Part B Behavior and Fertility Control. Academic Press, New York.
33. Ott, A.G. and Horton, H.F. 1971. Fertilization of chinook and coho salmon eggs with cryopreserved sperm. J. Fish. Res. Board Can. 28: 745–748.
34. Billard, R. 1978. Some data on gametes preservation and artificial insemination in teleost fish. Actes Colloq. Cent. Nat. Exploitation Oceans (CNEXO) 8: 59 – 73.

35. Jondet, R. 1964. Congelation rapide du sperme de taureau conditionne en paillettes. 5th Internat. Congr. Anim. Reprod. Artif. Insem. (Trento) 4: 463–467.
36. Nagase, H. and Niwa, T. 1963. Studies on the deep-freezing technique for bull semen. III. Deep freezing of bull semen in pellet form. Jap. J. Anim. Reprod. 9: 73–77.
37. Kozumplik, J. 1978. The size of pellets and the individual properties of boar sperm, as related to the number of motile spermatozoa after defreezing. Vet. Med., Praha 23: 457–463.
38. Robbins, R.K., Saacke, R.G. and Chandler, P.T. 1976. Influence of freezing rate, thaw rate and glycerol level on acrosomal retention and survival of bovine spermatozoa frozen in French straws. J. Anim. Sci. 42: 145–154.
39. Amann, R.P. and Pickett, B.W. 1987. Principles of cryopreservation and a review of cryopreservation of stallion spermatozoa. Equine Vet. Sci. 7: 145–173.
40. Pursel, V.G. and Park, C.S. 1985. Freezing and thawing procedures for boar spermatozoa, pp. 147–166. In: L. A. Johnson and K. Larsson (eds.), Deep Freezing of Boar Semen. Swedish University of Agricultural Sciences, Uppsala.
41. Lake, P.E. and Stewart, J.M. 1978. Artificial insemination in poultry. Bull. Min. Agric. Fisheries, Food: 213.
42. Blaxter, J.H.S. 1953. Sperm storage and cross-fertilization of spring and autumn spawning herring. Nature (London) 172: 1189–1190.
43. Mounib, M.S., Hwang, P.C. and Idler, D.R. 1968. Cryogenic preservation of Atlantic cod (*Gadus morhua*) sperm. J. Fish. Res. Board Can. 25: 2623–2632.
44. Graybill, J.R. and Horton, H.F. 1969. Limited fertilization of steelhead trout eggs with cryopreserved sperm. J. Fish. Res. Board Can. 26: 1400–1404.
45. Guest, W.C., Avault, J.W. and Roussel, J.D. 1976. Preservation of channel catfish sperm. Trans. Am. Fish. Soc. 105: 469–474.
46. Moczarski, M. 1977. Deep freezing of carp *Cyprinus carpio* L. sperm. Bull. Acad. Pol. Sci., Ser. Sci. Biol. 15: 187–190.
47. Hoyle, R.J. and Idler, D.R. 1968. Preliminary results in the fertilization of eggs with frozen sperm of Atlantic salmon (*Salmo salar*). J. Fish. Res. Board Can. 25: 1295–1297.
48. Mounib, M.S. 1978. Cryogenic preservation of fish and mammalian spermatozoa. J. Reprod. Fertil. 53: 13–18.
49. Hara, S., Canto, J.T. and Almendras, J.M.E. 1982. A comparative study of various extenders for milkfish, *Chanos chanos* (Forsskal), sperm preservation. Aquaculture 28: 339–346.
50. Larson, E.V. and Graham, E.F. 1973. Measurements of cooling rates and cell survival at different points within a sphere. Cryobiology 10: 515 (Abstr.).
51. Stein, H. and Bayrle, H. 1978. Cryopreservation of the sperm of some freshwater teleosts. Ann. Biol. Anim., Biochim., Biophys. 18: 1073–1076.
52. Piironen, J. and Hyvarinen, H. 1983. Cryopreservation of spermatozoa of the whitefish (*Coregonus muksun* Pallas). J. Fish. Biol. 22: 159–163.
53. De Montalembert, G., Bry, C. and Billard, R. 1978. Control of reproduction in northern pike. Am. Fish. Soc., Spec. Publ. 11: 217–225.
54. Bamba, K. and Cran, D.G. 1985. Effect of rapid warming of boar semen on sperm morphology and physiology. J. Reprod. Fertil. 75: 133–138.
55. Bamba, K. and Cran, D.G. 1988. Effect of rapid warming of bull and rabbit semen. J. Reprod. Fertil. 82: 501–507.
56. Bamba, K. and Cran, D.G. 1988. Further studies on rapid dilution and warming of boar semen. J. Reprod. Fertil. 82: 509–518.
57. Senger, P.L., Becker, W.C. and Hillers, J.K. 1976. Effect of thawing rate and post-thaw temperature on motility and acrosomal maintenance in bovine semen frozen in plastic straws. J. Anim. Sci. 42: 932–936.

58. De Abreu, R.M., Berndtson, W.E., Smith, R.L. and Pickett, B.W. 1979. Effect of post-thaw warming on viability of bovine spermatozoa thawed at different rates in French straws. J. Dairy Sci. 62: 1449–1454.
59. Crabo, B.G., Graham, E.F. and Larson, E.V. 1972. The influence of thawing methods on frozen boar spermatozoa. Cryobiology 9: 331 (Abstr.).
60. Wilmut, I. and Rowson, L.E.A. 1973. Experiments on the low-temperature preservation of cow embryos. Vet. Rec. 92: 686–690.
61. Bank, H. and Maurer, R.R. 1974. Survival of frozen rabbit embryos. Exp. Cell Res. 89: 188–196.
62. Willadsen, S.M., Polge, C., Rowson, L.E.A. and Moor, R.M. 1974. Preservation of sheep embryos in liquid nitrogen. Cryobiology 11: 560.
63. Whittingham, D.G. 1975. Survival of rat embryos after freezing and thawing. J. Reprod. Fertil. 43: 575–578.
64. Bilton, R.J. and Moore, N.W. 1976. *in vitro* culture, storage and transfer of goat embryos. Austr. J. Biol. Sci. 29: 125–129.
65. Yamamoto, Y., Oguri, N., Tsutsumi, Y. and Hachinohe, Y. 1982. Experiments in the freezing and storage of equine embryos. J. Reprod. Fertil. Suppl. 32: 399–403.
66. Kramer, L., Dresser, B.L., Pope, C.E. Dalhausen, R.D. and Baker, R.D. 1983. The nonsurgical transfer of frozen-thawed eland embryos. Proc. Amer. Assoc. Zoo Veterinarians: 104–105.
67. Trounson, A.O. and Mohr, L. 1983. Human pregnancy following cryopreservation thawing and transfer of an eight-cell embryo. Nature (London) 305: 707–709.
68. Pope, C.E., Pope, V.Z. and Beck, L.R. 1984. Live birth following cryopreservation and transfer of a baboon embryo. Fertil. Steril. 42: 143–145.
69. Hearn, J.P. and Summers, P.M. 1986. Experimental manipulation of embryo implantation in the marmoset monkey and exotic equids. Theriogenology 25: 3–11.
70. Balmaceda, J.P., Heitman, T.O., Garcia, M.R. Pauerstein, C.J. and Pool, T.B. 1986. Embryo cryopreservation in cynomolgus monkeys. Fertil. Steril. 45: 403–406.
71. Harvey, B. 1982. Cryobiology and the storage of teleost gametes. Proc. Internat. Symp. Reprod. Physiol. Fish, 1982: 123–127.
72. Harvey, B. and Ashwood-Smith, M.J. 1982. Cryoprotectant penetration and supercooling in the eggs of salmonid fishes. Cryobiology 19: 29–40.
73. Dukelow, W.R. 1980. Laparoscopy in small animals and ancillary techniques, pp. 95–105. In: R.M. Harrison and D.E. Wildt (eds.), Animal Laparoscopy. Williams and Wilkins, Baltimore.
74. Trounson, A.D., Willadsen, S.M. Rowson, L.E.A. and Newcomb, R. 1976. The storage of cow eggs at room temperature and at low temperatures. J. Reprod. Fertil. 46: 173–178.
75. Willadsen, S.M. 1977. Factors affecting the survival of sheep embryos during deep freezing and thawing, pp. 175–189. In: K. Elliott and J. Whelan (eds.), The Freezing of Mammalian Embryos. CIBA Symposium 52, Elsevier, Amsterdam.
76. Bouyssou, B. and Chupin, D. 1981. Two step freezing of cattle blastocysts with dimethysulphoxide (DMSO) of glycerol. Theriogenology 17: 159–167.
77. Massip, A., Van Der Zawlmen, P., Hanzen, C. and Ectors, F. 1982. Fast freezing of cow embryos in French strains with an automatic program. Theriogenology 18: 325–332.
78. Renard, J-P., Heyman, Y. and Ozil, J-P. 1982. Congélation de l'embryon bovin: Une nouvelle méthode de décongélation pour le transfert cervical des embryons conditionnés une seule fois en paillettes. Ann. Med. Vet. 126: 23–32.
79. Wilmut, I. 1986. Cryopreservation of mammalian eggs and embryos, pp. 217–247. In: R.B.L. Gwatkin (ed.), Developmental Biology: A Comprehensive Synthesis. Plenum, New York.

80. Rall, W.F. and Fahy, G.M. 1985. Ice-free cryopreservation of mouse embryos at − 196 °C by vitrification. Nature 313: 573–575.
81. Whittingham, D.G., Wood, M., Farrant, J., Lee, H. and Halsey, J.A. 1979. Survival of frozen mouse embryos after rapid thawing from − 196 °C. J. Reprod. Fertil. 56: 11–21.
82. Leibo, S.P. 1984. A one-step method for direct nonsurgical transfer of frozen-thawed bovine embryos. Theriogenology 21: 767–790.
83. Trounson, A. 1986. Preservation of human eggs and embryos. Fertil. Steril. 46: 1–12.
84. Schneider, U. and Maurer, R.R. 1983. Factors affecting survival of frozen-thawed mouse embryos. Biol. Reprod. 29: 121–128.
85. Mazur, P., Rigopoulos, N., Jackowski, S.C. and Leibo, S.P. 1976. Preliminary estimates of the permeability of mouse ova and early embryos to glycerol. Biophys. J. 16: 232a (Abstr.).
86. Wilmut, I., Polge, C. and Rowson, L.E.A. 1975. The effect on cow embryos of cooling to 20,0 and − 196 °C. J. Reprod. Fertil. 45: 409–411.
87. Willadsen, S.M., Polge, C., Rowson, L.E.A. and Moor, R.M. 1976. Deep freezing of sheep embryos. J. Reprod. Fertil. 46: 151–154.
88. Lyon, M.F., Glenister, P.H. and Whittingham, D.G. 1981. Long-term viability of embryos stored under irradiation, pp. 139–148. In: G.H. Zeilmaker (ed.), Frozen Storage of Laboratory Animals. Gustav Fischer, Stuttgart.
89. Renard, J-P. and Babinet, C. 1984. High survival of mouse embryos after rapid freezing and thawing inside plastic straws with 1–2 propanediol as cryoprotectant. J. Exp. Zool. 230: 443–48.
90. Pettit, W.H. 1985. Commercial freezing of bovine embryos in glass ampules. Theriogenology 23: 13–16.
91. Wright, J. 1985. Commercial freezing of bovine embryos in straws. Theriogenology 23: 17–29.
92. Hasler, J.F., McCauley, A.D., Lathrop, W.F. and Foote, R.H. 1987. Effect of donor-embryo-recipient interactions on pregnancy rate in a large-scale bovine embryo transfer program. Theriogenology 27: 139–168.
93. Wilmut, I., Sales, D.I. and Ashworth, C.J. 1985. The influence of variation in embryo stage and maternal hormone profiles on embryo survival in farm animals. Theriogenology 23: 107–119.
94. Short, R.V., Chandley, A.C., Jones, R.C. and Allen, W.R. 1974. Meiosis in interspecific equine hybrids. II. The Przewalski horse/domestic horse hybrid (*Equus przewalskii* x *E. caballus*). Cytogenet. Cell Genet. 13: 465–478.
95. Gray, A.P. 1972. Mammalian hybrids: Technical Communication No. 10 of Commonwealth Agricultural Bureau of Animal Breeding and Genetics. Edinburgh.
96. Summers, P.M., Shelton, J.N. and Edwards, J. 1983. The production of mixed-species *Bos taurus-Bos indicus* twin calves. Anim. Reprod. Sci. 6: 79–89.
97. Stover, J., Evans, J. and Dolensek, E.P. 1981. Inter Proc. Am. Assoc. Zoo Veterinarians: 122–124.
98. Bunch, T.D., Foote, W.C. and Whitaker, B. 1977. Interspecies ovum transfer to propagate wild sheep. J. Wildl. Manage. 41: 726–730.
99. Dresser, B.L., Pope, C.E., Kramer, L., Kuehn, G., Dahlhaussen, R.D., Maruska, E.J., Reece, B. and Thomas, W.D. 1985. Birth of bongo antelope (*Tragelaphus euryceros*) to eland antelope (*Tragelaphus oryx*) and cryopreservation of bongo embryos. Theriogenology 23: 190. (Abstr.).
100. Anderson, G.B. 1988. Interspecific pregnancy: Barriers and prospects. Biol. Reprod. 38: 1–15.

Invertebrate germplasm cryopreservation: potential, problems, and prospects

ROGER A. LEOPOLD and ROBERT R. ROJAS
Biosciences Research Laboratory, U.S. Dept. Agric., Agricultural Research Service, Fargo, North Dakota, 58105, U.S.A.

Abstract Although the demand for long-term storage of invertebrate germplasm has never been more evident than it is now, specific knowledge of such storage is sparse. To date, germplasm has been stored mainly by those concerned with plant and domestic animal breeding or with type culture collections for protozoans. In the research laboratory, the continuous rearing of numerous strains of insects and other invertebrates is costly and labor intensive. Catastrophic loss and genetic drift are also important considerations when an organism such as an insect is to be used as a biological control agent and is reared under factorylike conditions. Cryopreservation, or cold storage, provides a feasible alternative to continuous rearing while circumventing the aforementioned difficulties. However, the unique reproductive and environmental adaptations characteristic of invertebrates often preclude the use of much of the conventional cryobiological methodology. This report focuses on research needs for invertebrate cryopreservation and suggests some possible routes for gaining success in the long-term storage of invertebrate germplasm.

Key words: Metazoa, cold tolerance, cold storage, cold shock, cryoprotectants

Introduction

Invertebrates, and insects in particular, have a long history of being subjected to whole-body cold-tolerance experiments. The early work of Reaumur [1] in 1736 and the reviews of Urarov [2], Luyet and Gehenio [3], Ushatinskaya [4], Asahina [5], Baust and Rojas [6] are evidence of man's avid interest in the cryobiology of this group of animals. However, within this extensive record scant attention has been given to the development of methodology for the long-term cryopreservation of the germplasm of metazoan invertebrates. For some researchers, this may not be a startling revelation, since they cannot perceive the value or application of such an endeavor. For others, it may be mildly surprising, since they have a general awareness that many members of this group normally overwinter under subfreezing conditions and would seem to be readily adaptable to the techniques of cryopreservation. Both of these perceptions, although understandable, are at variance with the current needs of the research and agricultural com-

L. Knutson and A.K. Stoner (eds.). Biotic diversity and germplasm preservation, global imperatives.
 ISBN 0-7923-0178-1

munities and with the amount of existing cryobiological information that has been or perhaps can be used to develop germplasm storage technology.

Applications for germplasm cryopreservation

From an economic point of view, the need for increased research in germplasm preservation for metazoan invertebrates is best exemplified by the burden borne by U.S. research agencies involved in supporting the *Drosophila* stock centers at Bowling Green, Ohio and Pasadena, California. No less than 4000 fruitfly strains are being maintained by these two facilities, and this has been estimated to constitute less than half of the number of strains existing worldwide (M. Ashburner, pers. comm.). As with most laboratory animals that must be reared continuously to maintain their germlines, the need to start new generations of flies about every 30 days is costly and labor intensive. Because of the inherent costs associated with mass rearing numerous strains, economics rather than future research potential often dictates whether a particular strain will be retained. Within our laboratory alone, more than 175 genetic strains of house flies and over 50 geographical isolates of the cotton boll weevil have been discarded in the last 10 years with little or no chance of recovery.

In addition to the possibility of cross-contamination between strains and the threat of catastrophic loss caused by disease or accidents at large stock centers, colony domestication and genetic drift often result in undesired genetic and behavioral changes [7–10]. Genetic drift and laboratory adaptation also become important considerations when a mass-reared species is to be released in the field as a biological control agent (i.e. entomophagous, phytophagous, parasitic, and sexually sterile insects). The possible loss of field competitiveness of a species reared under large-scale factorylike conditions (Fig. 1) is a constant concern for those involved with biological control programs. It has been suggested that to maintain genetic diversity of the domesticated colony under such conditions, the founder strain should be saved and periodically backcrossed to it [11, 12]. Further, in situations where animals are to be reared and released in biocontrol programs, rearing needs often fluctuate with periodic outbreaks caused by migration, accidental introduction, or cyclical seasonal reproduction of the pest species. The availability of technology to stockpile cryopreserved biocontrol agents for later use in immediate colony build-up or direct release into the field would alleviate the need to construct large rearing facilities to combat periodic pest outbreaks.

In the realm of mariculture, the mass culture of marine invertebrates such as shrimp, oysters, and other shellfish currently requires a continous

Figures 1a, 1b. Factorylike conditions under which the screwworm fly, *Cochliomyia hominivorax* (Coquerel) is reared for use in the sterile insect release method. 1a. A bank of colony cages containing a portion of adult brood stock. 1b. A conveyor belt collector and larval/pupal separation apparatus.

maintenance of a brood stock or reliable access to a natural source of some stage of the species to be cultured. Ready access to a germplasm storage bank could alleviate problems such as disease and seasonal fluctuations in reproductive capacity that are associated with maintenance of a large brood stock and would allow for the possibility for improvement of strains through selective breeding of stored germlines.

Ultra-low temperature tolerance of invertebrates

With few exceptions, research on metazoan cryobiology has been conducted to elucidate how poikilothermic animals tolerate or avoid freezing in their native environment. An arbitrary gauge of an animal's potential (and perhaps the researcher's intent) for possible long-term cryopreservation is its survival at the ultra-low temperatures of liquid nitrogen, oxygen, or helium. Relatively few invertebrates, other than protozoans [13–17] have been examined for ultra-low temperature freezing tolerance. This is not meant to imply that storage of an animal's germplasm at temperatures above that of liquid nitrogen cannot be useful. Any low temperature storage regime that meets the goals of a particular situation and is economical should not be discounted (see below).

Nematodes were among the first multicellular animals to be shown to tolerate freezing to $-196°C$ [18, 19]. DeConinck [18] used a two-step freezing regime on the free-living nematode *Anguillula silusiae* (deMan) that entailed a cooling to $-30°C$ before plunging to $-196°C$. Parasitic forms of nematodes appear to be less tolerant to freezing, although Schiller *et al.* [20] were able to store immature stages of *Onchocerca volvulus* Leuckart within human cutaneous tissue and in a black fly vector for as long as 504 days after freezing *in situ* in a cryoprotectant at $-196°C$.

Animals such as rotifer and tardigrade cysts and chironomid fly larvae that tolerate extensive dehydration have been demonstrated to withstand freezing in liquid nitrogen or liquid helium and to continue to develop normally after thawing [21–23]. Hinton [23] reduced the water content of the larvae of the chironomid fly, *Polypedilum vanderplanki* Hinton, to about 8% before transferring to liquid helium. Dehydration tolerance is probably one of the attributes which also enables certain lepidopteran and hymenopteran insects to survive ultra-low temperature freezing in limited numbers. Tissue cells of insects typically can dehydrate concurrent with extracellular ice growth during a gradual fall in the freezing temperature [24]. Certain stepwise cooling procedures consisting of two to five steps can be used to mimic dehydration. These procedures typically incorporate a 'hold' of varying lengths of time at $-30°C$ or $-70°C$ before the material is transferred to a liquid gas. Use of such a procedure allowed post-thawing survival for pupae of the swallowtail butterfly, *Papilo machaon* L. [25]; prepupae of the slug caterpillar, *Monema flavescens* Walker [25] and the sawfly, *Trichiocampus populi* Okamoto [26]; and larvae of the European cornborer, *Ostrinia nubialis* (Hübner) [27–29]. However, even though significant numbers of these insects treated with multistep freezing methods survived for periods as long as 70 days [27], almost all of the insects experienced difficulty proceeding to the next respective stage, an indication of tissue damage.

Experimentation with exposing marine organisms to ultra-low temperatures has been limited, but the results so far show potential for long-term storage. Asahina (cited in [5]), using a two-step freezing procedure with the pluteus stage of the sea urchin, achieved low but significant survival after freezing. Baust and Lawrence [30] reported low survival of various penaied shrimp nauplii after exposure to $-196\,^{\circ}C$ but found that survival could be enhanced considerably by use of cryoprotective agents and choosing the most chill-tolerant stage [31, 32].

Of the ultra-low temperature tolerant organisms mentioned above, almost all were examined for tolerance during either an immature, drought-resistant, and/or quiescent 'hibernating' stage. This strongly suggests that survival under low temperatures, as displayed by invertebrates experiencing one or more of these stages, was enhanced by a natural ability to withstand some aspect of environmental stress that may not be directly related to freezing tolerance. Even rotifer species, which can normally survive ambient freezing during an active reproductive stage, experience better postfreezing survival when they have been desiccated and are in a drought-resistant state [21].

Inherent problems of invertebrate cryopreservation

Embryonic nutrition

Cryopreservation of invertebrate germplasm, and of arthropods in particular, poses some problems that are common to mammalian systems but that are unique in other respects. For example, the size and amount of yolk of comparable stages of a house fly and of a mammalian embryo differ (Fig. 2). In general, most invertebrate embryos develop within a closed system, as do many non-mammalian vertebrate embryos and all nutrients needed for embryonic development come from the egg yolk. For this reason, most invertebrate eggs are typically larger and richer in yolk than are mammalian eggs, which after fertilization derive nourishment for embryonic growth from the mother. Thus damage to the yolk would be highly detrimental to the invertebrate system because it is the sole source of nutrition. Yolk that is high in lipids may be especially vulnerable to disruption by cold and freezing [33]. In fact, lipovitellin, a common lipoprotein in the yolk of egg-laying animals, has been shown to be disassociated by freezing [34]. The membrane surrounding the yolk granule also may be damaged by chilling and/or freezing, which would lead to loss of the integrity of the essential com-

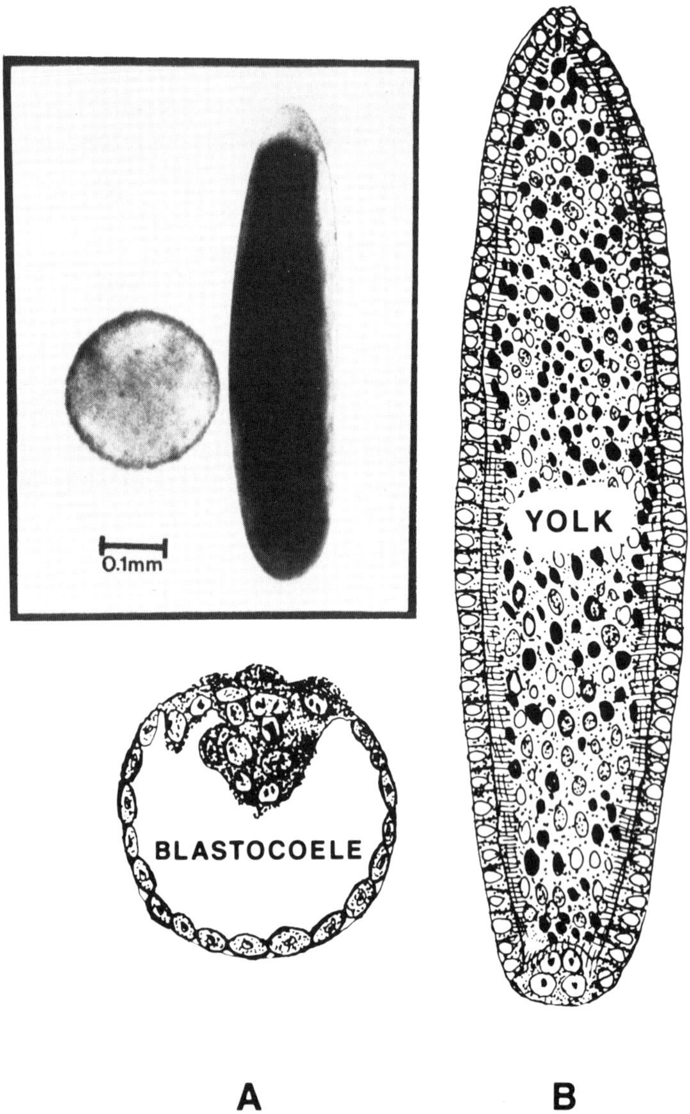

Figure 2. Comparison of relative size and amount of yolk between (A) an 8-day-old 'hatched' blastocyst of the pig and (B) a 1-hour-old embryo of the house fly. Inset shows size difference between the pig and house fly embryos.

ponents. This may be why the lipid-rich pig embryo cannot be cooled lower than a critical temperature between 10° and 15°C [35, 36]. Lovelock [37] has suggested that lipid-protein complexes may be the 'Achilles heel' of many cells subjected to chilling and freezing. These molecules are held together, lipid to protein, by weak associations and not by strong covalent bonds. Lovelock describes these complexes as inherently unstable and maintained in living cells by continous synthesis.

Heacox *et al.* [33] stated that the loss of viability of insect embryos as a result of freezing may occur because cell function is determined very early in development. The loss in ontogenetic plasticity in a system like this would mean that freeze damage to a few essential cells may result in loss or malformation of vital larval or adult tissues or organs.

Embryonic membrane systems as barriers to permeation by cryoprotectants

Since most invertebrate embryos develop outside of the mother, they are endowed with elaborate membrane systems to protect them from environmental hazards. These membranes present barriers to cryoprotectant-induced dehydration and permeation. The outer embryonic membranes of an insect and a mammalian embryo are contrasted in Fig. 3. The mammalian blastocyst at 7 to 8 days postovulation is relatively permeable, with only two outer membranes, the zona pellucida and the vitelline membrane. The typical insect embryo, in contrast, has a vitelline membrane, a hydrophobic waxy layer, and an elaborate chorion consisting of an innermost crystalline chorion, inner chorion, outer chorion, and a fibrous exochorion.

In addition to the problems inherent to invertebrates, there are also advantages to working with them. To this point, only egg and embryonic systems have been mentioned. For mammmalian systems, cryopreservation beyond the egg or embryonic stage of the whole organism would be untenable, but for the invertebrates this is not necessarily the case. In fact, post-embryonic stages of a variety of invertebrates have been frozen successfully or are naturally freezing tolerant (Table 1). Another advantage is that often a large number of individuals are available, since many invertebrates have a short generation time and are also often prolific producers of their progeny. Survival assessments following a chilling-warming regime or other treatment are facilitated by rapid development external to the mother. Furthermore, research with invertebrates is generally felt to be unencumbered by the ethical issues that may be involved with the cryopreservation of mammalian organisms.

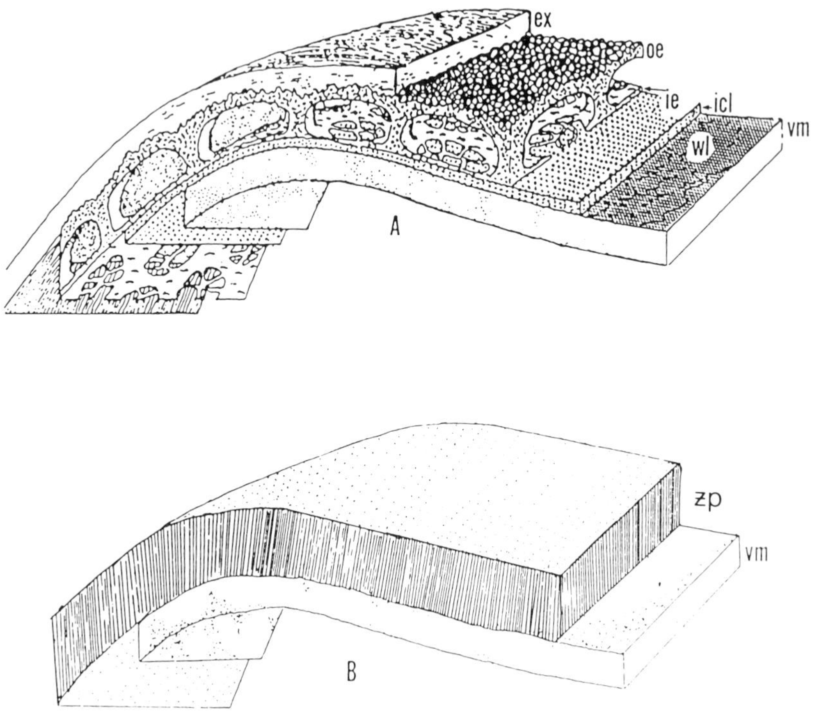

Figures 3a, 3b. Egg membrane systems of (a) an insect (modified from Margaritis [80]) and (b) a mammal. Abbreviations: vm-vitelline membrane, wl-waxy layer, icl-inner chorionic layer, ie-inner perforated endochorion, oe-outer chorion, ex-fibrous outer exochorion, zp-zona pellucida.

Approaches to cryopreservation of invertebrates

Determination of natural cold tolerance

In devising a protocol for cryopreservation of an invertebrate, the first determination to be made is to survey the natural cold tolerance of all developmental stages (i.e. eggs, immature stages, and adults) and at various ages within these stages. This will give the researcher a profile from which may be chosen the most cold-tolerant stage with the caveat that supercooling points do not necessarily reflect lower lethal limits. Also, the tolerance should be determined for short as well as long-term exposure to cold. This type of determination allows for the selection of the most cold tolerant stage and defines the limitations that are either to be overcome or worked within.

Strong and Leopold [38] assessed the cold tolerance of different developmental stages of the house fly (Fig. 4), a freeze-intolerant species. In this study the insect's capacity to survive supercooling was compared to its

Table 1. Examples of invertebrate species which tolerate freezing for a minimum of 24 hours

Species and Stage	Temperature tolerated for 24 hours	Reference
Insecta		
Lepidoptera		
Hyalophora cecropia (L.), pupae	Below −70 °C	Asahina & Tanno [71]
Coleoptera		
Dendroides canadensis Latreille, larvae	−28 °C	Duman [72]
Hymenoptera		
Bracon cephi (Gahan), larvae	Below −40 °C	Salt [73]
Diptera		
Eurosta solidaginis (Fitch), larvae	Below −40 °C	Salt [74]
Orthoptera		
Cryptocercus punctulatus Scudder, adults	−10 °C	Hamilton *et al.* [75]
Pelecypoda		
Filibranchia		
Mytilus edulis L., adults	−10 °C	Williams [76]
Gastropoda		
Pulmonata		
Melampus bidentatus Say, adults	−13 °C	Loomis [77]
Arachnida		
Scorpionida		
Centruroides vittatus Say, adults	−6 °C	Whitmore *et al.* [78]
Crustacea		
Anostraca		
Artemia salina Leach, cysts	−190 °C	Whitaker [79]
Nematoda		
Rhabditida		
Anguillula silusiae (deMan), adults	−196 °C	DeConinck [18]

ability to withstand continuous exposure to cold temperatures. As is the case with many organisms, innate supercooling capacity was shown not to be an accurate indicator of cold tolerance.

It is indeed fortunate if the organism to be preserved possesses a freezing-tolerant stage. Such organisms are able to withstand ice formation in extracellular spaces and perhaps intracellular as well. The ability to tolerate freezing often is associated with the accumulation of cryoprotectant(s) (either low molecular weight sugars and sugar alcohols [5, 39–41] and/or high molecular weight thermal hysteresis proteins [42]). These compounds afford protection to the organism in a variety of ways once freezing occurs [43, 44]. Potent extracellular, proteinaceous, ice-nucleating agents also are

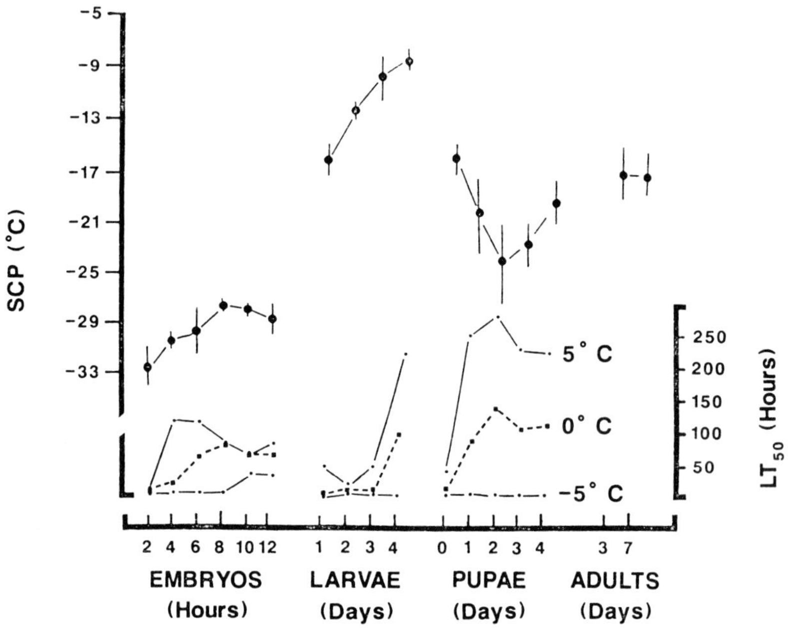

Figure 4. Ontogenetic variability of supercooling points (SCP) and cold tolerance (LT_{50}) for the house fly, *Musca domestica* (modified from Strong-Gunderson and Leopold [38]).

often present in the organism. They act to induce freezing at relatively high subzero temperatures, allowing sufficient time for water to efflux from the cells and thereby precluding the formation of potentially fatal intracellular ice [24, 45]. These compounds also act to keep the organism in the frozen state, where metabolic reactions are slowed and freeze-thaw cycles are avoided. Often these compounds are accumulated in response to exposure to low temperatures [46, 47]. If freezing tolerance is exhibited for extended periods of time, there is no need for the researcher to look further.

Cold shock: detection and avoidance

Cold tolerance is slight or nonexistent in many economically important species. These organisms may experience injury or may be killed by exposure to temperatures above the supercooling point. Cold shock occurs in a diverse range of organisms including bacteria, yeast, algae, fungi, protozoa, higher plants of tropical origin, fish eggs, mammalian spermatozoa, and mammalian somatic cells and embryos [48]. This phenomenon presents a formidable barrier to succesful cryopreservation. Avoidance of cold shock injury can be achieved only by recognizing its existence and its cause(s) in the organism. An example of cold shock injury occurring in an insect is illustrated in Table 2. In this case, significant numbers of diapausing sunflower moth larvae are able to withstand $-10\,^{\circ}C$ for seven days whereas their non-diaupausing counterparts are killed with less than a single day's exposure. Though the supercooling point (SCP) for the diapausing larvae were slightly lower than those for the non-diapausing larvae, the $30\times$ greater concentration of trehalose was suggested as having a beneficial effect on avoidance of cold shock [49]. In certain instances, cold shock may be recorded as 'prefreeze events' occurring prior to the SCP exotherm. Knight *et al.* [50] were able to detect prefreeze events that apparently coincided with the lethal temperature for an aphid (Fig. 5). The general consensus among researchers in this field is that cold shock injury results from the induction of phase transitions in membrane lipids and subsequent loss of membrane permeability [50, 51]. Possible ways to overcome cold shock injury are by addition of membrane stabilizers such as some of the polyols (glycerol, sorbitol, and trehalose [51, 52]) as well as inducing changes in membrane lipid saturation by prechilling [53].

Cryoprotectants: permeability, optimization, and selection

In all probability, cryoprotectants will be needed to confer to the organism some degree of freezing tolerance. Since one mode of action of non-penetrating cryoprotectants is dehydration one must determine if the organism is freely permeable to water as well as to 'penetrating' cryoprotectants whose site of

Table 2. Glycogen, trehalose, SCP, and survival of diapause and non-diapause larvae of *Homeosoma electellum* (Hulst)

	Diapause	Non-Diapause
Glycogen (μg/mg fresh wt.)	21.7 ± 1.5[a] 23.5 ± 1.7	11.4 ± 0.9
Trehalose (μg/mg fresh wt.)	9.0 ± 0.7 10.3 ± 0.8	0.3 ± 0.3
SCP (°C)	-21.9 ± 0.6 -23.8 ± 0.3	-15.3 ± 0.3
% survival to pupation	44% (3)[b] 33% (7)	0% (1)

[a] The values represent the means ± SEM.
[b] ( ) No. days exposure to $-10\,^{\circ}C$.

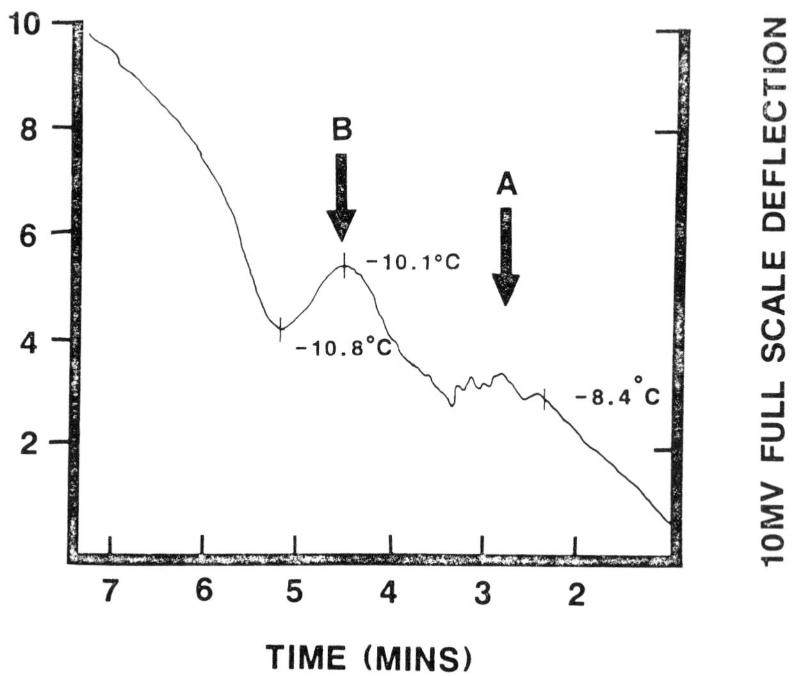

Figure 5. Pre-freeze exotherms (peaks A and B) produced by cooling an aphid at 1 deg min × $^{-1}$. Exotherm A (−8.4°C) is associated with the LT_{50} which occurs before the supercooling point exotherm for this aphid (ca. −20°C) (−8.4°C = midpeak of exotherm A, −10.1°C = midpeak of exotherm B, −10.8°C = endpoint of exotherm B) (Redrawn from Knight *et al.* [50]).

action may be both intra-and extracellular. Because desiccation is a major problem faced by terrestial invertebrates, they are often protected by external coats (i.e. chorion, waxy layers, and cuticle). These external coatings greatly decrease permeability to water and cryoprotectants, and it may be necessary to alter the organism's permeability to water and to the cryoprotectant of choice. An organism's permeability can be evaluated by microscopic measurement of size changes after exposure to hypertonic solutions and/or cryoprotectants. Shrinkage of the organism in hypertonic solutions would be indicative of water loss from the organism. Following shrinkage with a non-penetrating cryoprotectant, organisms reswell if an equal osmolal solution of a 'penetrating' cryoprotectant is substituted for the suspending solution and permeates the organism. Thus the organism can be considered permeable to the cryoprotectant.

If the stage selected is non-permeable to water and cryoprotectant, the researcher still has some options. One option is to render the organism permeable through the use of various organic solvents. Ta-Te Lin *et al.* [54]

reported successfully permeabilizing *Drosophila* embryos by removing the egg chorion in 50% liquid bleach (5.25% sodium hypochlorite by weight) and then treating the embryos with hexane followed by isopropanol. This treatment yielded eggs with 80–85% permeability (as determined by osmotic contraction in 1M sucrose) and 70–80% survival (hatching). Rojas and Leopold (unpublished), using the permeabilization technique of Heacox *et al.* [33], determined that that there is a volumetric swelling of 10 hr-old housefly embryos in hypo-osmotic solutions of sucrose and subsequent reswelling occurs after removal from a hyper-osmotic sucrose solution (2M) and placed in DMSO (2M) (Fig. 6). This does not occur when eggs are only dechorionated and the waxy layer still intact. Another option is to choose a stage of the organism that is more permeable, albeit perhaps less cold tolerant. Heacox *et al.* [33] reported that experiments using neutral red as an indicator of permeability showed that the vitelline envelopes of young (1–3 hrs) embryos are readily permeable to the dye, whereas older embryos become increasing less permeable. However, Strong and Leopold [38] reported that 10- and 12-hour-old embryos showed the greatest tolerance to the coldest temperature tested (−5°C).

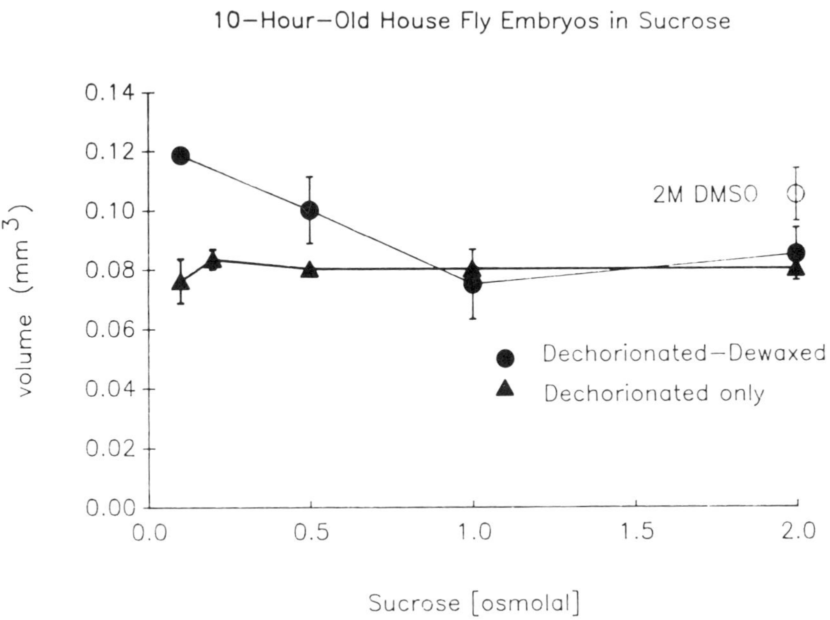

Figure 6. Volume change of dechorionated and dewaxed 10-hour-old house fly embryos. Swelling of dechorionated and dewaxed embryos occurs in hypo-osmotic sucrose solutions, reswelling following substitutions of 2 osmolal sucrose with 2 osmolal DMSO indicates replacement of water within the embryos. No volumetric changes occur in embryos that have been only dechorionated and not dewaxed.

The amount of dehydration that the organism can tolerate and the toxicity of the cryoprotectant will be determining factors as to the final concentration of cryoprotectant used. Stepwise addition and removal of cryoprotectants can be used to avoid undue osmotic stress to the organism. Exposing the organisms to the cryoprotectant just long enough for complete permeation will help to avoid cryoprotectant-induced toxicity [55]. Futhermore, Mazur [56] has shown for human red blood cells and eight-cell mouse embryos [57] that the unfrozen water fraction (U) of the suspending medium is crucial to survival following freezing. The eight-cell mouse embryos were injured when U dropped below 5–7.5%. By determining U, the researcher can choose a concentration of cryoprotectant-NaCl that will be above the critical U value at a given subfreezing temperature.

Multicomponent cryoprotectant systems are found in nature. Their use in the laboratory also should be considered as a way to reduce toxicity elicited by a single cryoprotectant. A multicomponent system takes advantage of synergism by combining the dehydrating effects of non-penetrating cryoprotectants with the 'water replacing' effects of penetrating cryoprotectants. A natural freezing tolerance is conferred to the third-instar larva of the gall fly *Eurosta solidaginis* (Fitch) through seasonal accumulation of a multicomponent cryoprotectant system (Fig. 7). High glycerol and trehalose concentrations are specific to this stage of development, whereas elevated sorbitol levels are induced by low-temperature acclimation [40, 58]. Thus acclimation is yet another option that might be used to increase the level of an innately produced cryoprotectant and its associated concomitant increase in cold-tolerance.

Organisms and cells often exhibit distinct optima in survival when presented with different cooling and warming rates [6, 56]. Miller [59] has shown that the cooling/freezing rate optima of the beetle *Upis cerambroides* L. was below 0.32°C/min; the percentage of beetles surviving undamaged

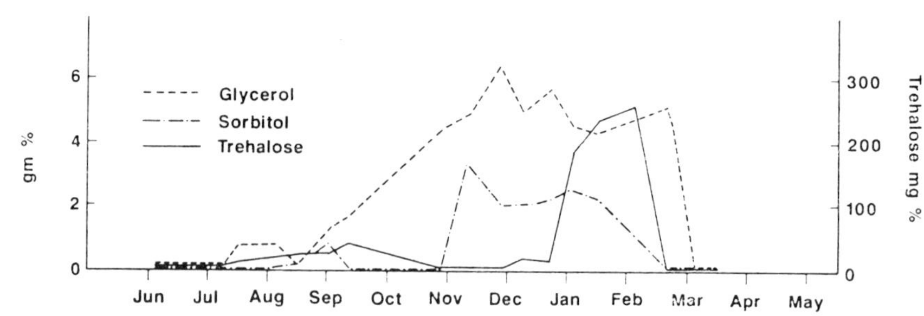

Figure 7. Seasonal accumulation of a multi-component cryoprotectant system in the freeze-tolerant third-instar larva of the gall fly, *Eurosta solidaginis* (Fitch) (modified from Morrissey and Baust [58]).

dropped from 100 to 4 with a change in rate from 0.29° to 0.35°C/min. Similarly, the small wingless chironomid midge *Belgica antarctica* (Jacobs) has a cooling/freezing rate optimum of 0.1°C/min, but it is not as sharp as that of *U. cerambroides* [60]. Heacox *et al.* [33] found that when house fly embryos were cooled to −20°C in the presence of DMSO, a cooling rate of 0.6°C/min afforded maximum survival and optimizing the warming rate with the freezing process can be of value.

Instrumentation

Several instruments with potential to elucidate some of the phenomena discussed earlier in this review (cold shock, yolk lipoprotein dissociation, intracellular vs. extracellular freezing, prefreeze events, and cryoprotectant permeation) are currently gaining increased use in cryobiological research. For example, differential scanning calorimetry (DSC) is becoming more widely used in the field of cryobiology. Although relatively expensive, this instrumentation can provide valuable insight into freezing damage and cold shock injury. DSC also can be used to detect such events as membrane-phase transitions and protein denaturation associated with cold shock. Other exothermic and endothermic events such as freezing, melting, glass transition points, recrystallization, and supercooling points all can be determined routinely by the use of DSC.

Cryomicroscopes (optical microscopes fitted with freezing stages having precisely controlled cooling rates) also can be used to determine such events as freezing, melting, and supercooling as well as extracellular vs. intracellular ice formation. For example, Leibo *et al.* [61] used a cryomicroscope to determine cooling rates that allowed formation of intracellular ice in mouse ova.

A more sophisticated and quantitative means of determining solute penetration is through the use of proton imaging with simultaneous, spatially resolved, high resolution nuclear magnetic resonance (NMR). This kind of instrumentation would allow for both the determination of solute penetration and its differential distribution within the organism. Gassner and Lohman [62] used an NMR imager and its submicroliter cube resolution to access the distributional changes in water and lipid content during embryonic development of the desert locust, *Schistocerca gregaria* (Forskål) (Fig. 8).

Finally, high performance liquid chromatography can expedite the screening of naturally occurring cryoprotectants in invertebrates, and the accumulation and turnover can be followed with temperature acclimation. The method of Hendrix *et al.* [63] allows for rapid simultaneous screening of low molecular weight carbohydrates and polyols. Figure 9 shows a typical chromatograph of a standard of sugars and sugar alcohols and a sample prepared from the freezing-tolerant third-instar larva of the gall fly *E. solidaginis*. The accumulations of glycerol, sorbitol, and trehalose in this organism are associated with freezing tolerance and are influenced by temperature acclimation.

Summary: research protocol and prospects

Like most aspects of scientific endeavor, there are logical, sequential processes through which a researcher can proceed to evaluate cryopreservation for the long-term storage of invertebrate germplasm. However, before embarking upon such research involving development of germplasm-storage capabilities, some thought should be given as to the best avenue and to what is adequate for the particular organism. For most invertebrates, the essential value to mankind is that of being part of our biosphere. Their germplasm is best preserved as a resource for future generations by protection and maintenance within their natural habitat. For those species of more immediate usefulness, germplasm preservation under a liquid nitrogen system may not be possible or desireable. Depending upon the organism's biology, physiology, and adaptability to cryopreservation methods, the

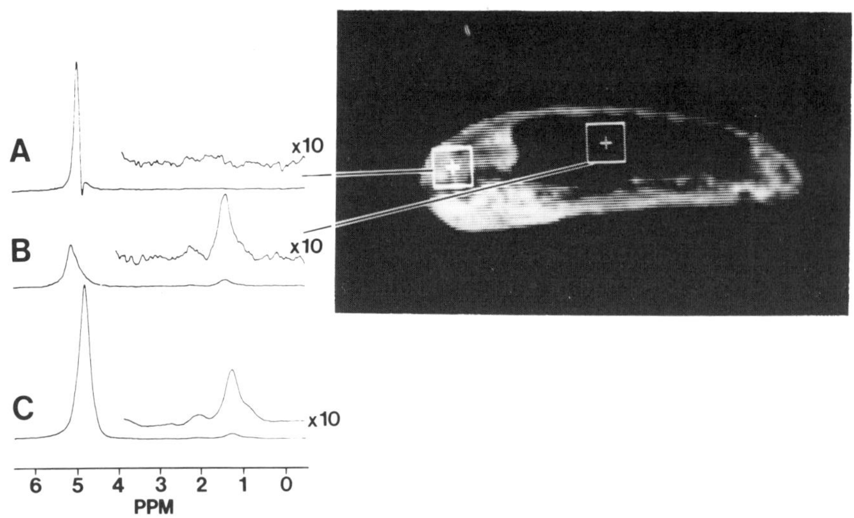

Figure 8. NMR spectroscopy of a locust embryo showing lipid and water signals. (A) region devoid of lipid (B) homecoel showing lipid presence (C) overall spectrum showing water and lipid resonances (Gassner and Lohmann [62]; reprinted by permission of Proc. Natl. Acad. Sci. U.S.A.).

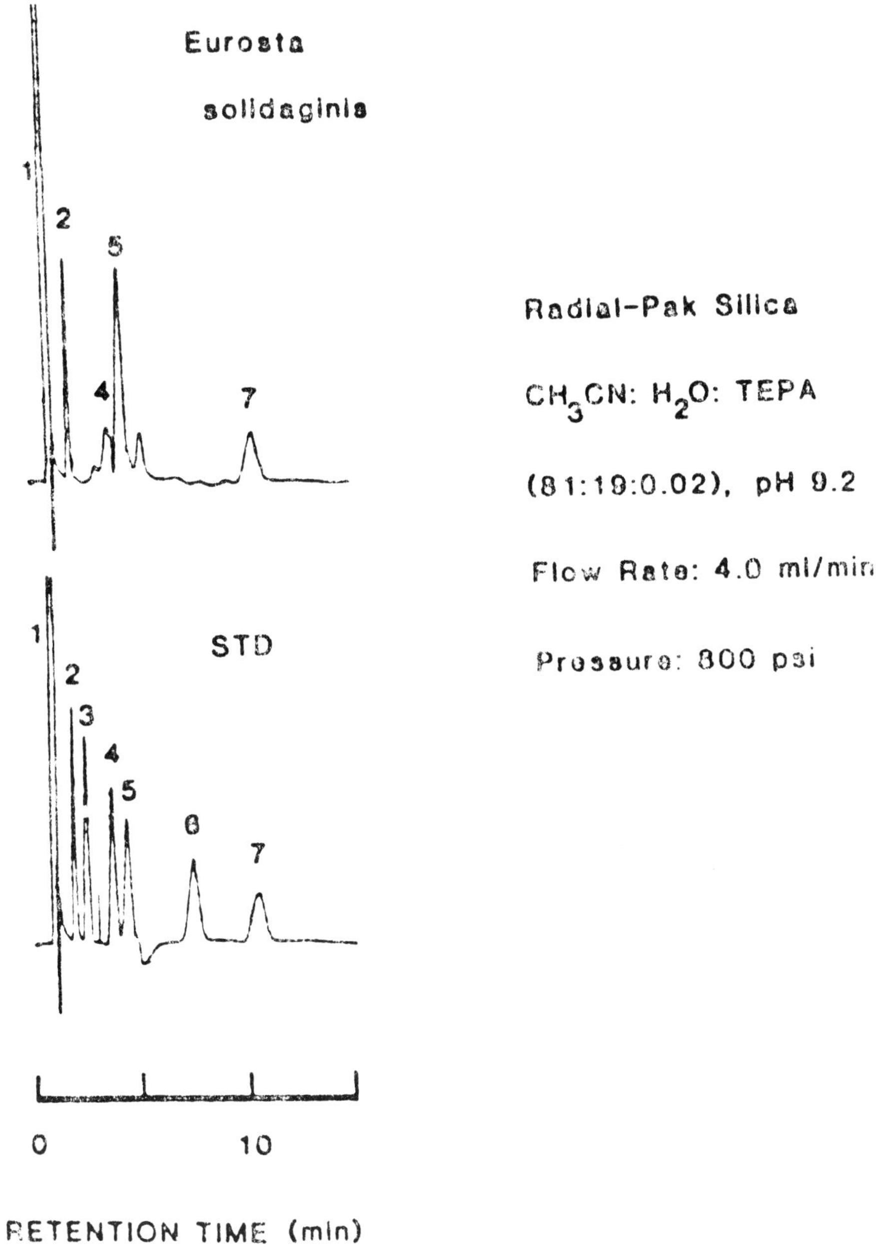

Figure 9. Sample high performance liquid chromatographs of polyols and sugars separated by use of a 101 silica cartridge (10 cm x 100 mm I.D.). Upper chromatograph represents a typical separation of an extract of the freeze-tolerant third instar larvae of the gall fly *Eurosta solidaginis* (Fitch). Lower chromatograph is of a standard (STD). (1) water front, (2) glycerol, (3) erythritol, (4) fructose, (5) sorbitol, (6) sucrose, (7) trehalose.

storage methodology must be constructed on a framework of practicality, economy, and perhaps on insights to exploit the animal's own adaptive responses to environmental stress or stresses. For example, Asahina [64, 65] found that most overwintering prepupae of *Monema flavescens* Walker survived a 435-day storage period at 0 °C, whereas supercooling to − 10 °C or freezing at − 20 °C lowered survival considerably and that storage for 70 days at − 190 °C interferred with normal emergence. Thus the researcher must carefully weigh the practicality and economy of trying to extend the potential storage time to a level of geological time as opposed to accepting what the organism is already capable of doing.

Having made a tentative decision that some type of germplasm cold storage is desirable for a particular organism, the decisions of how to implement and maintain such a program then must be made. Fig. 10 represents our concept of a protocol that could be used to evaluate an organism's suitability for cryopreservation. It is, of course, obvious that consideration must be given to the availability of resources needed to mount this type of research effort, such as adequate cryogenic equipment and trained personnel. Initial start-up costs may diminish the long-term benefits that could be obtained from cryopreservation of an organism deemed to have low economic, research, or aesthetic value. If the invertebrate is a terrestial inhabitant of the temperate zones, existing knowledge about its natural cold tolerance and overwintering strategy can be an obvious benefit to those interested in its cryopreservation. Information on which life stage overwinters and whether it is freezing tolerant or intolerant can reduce or eliminate some of the preliminary testing.

Determining how to prevent cold shock is one of the key issues in this process. The existence of cold shock requires that application of some type of chemical cryoprotection will be a necessary adjunct to implementing a cold storage technique. However, in those organisms that innately produce cryoprotectants as part of a cold tolerance strategy, a period of cold acclimation is often required to induce resistance to cold shock [66]. Thus the possibility of using an acclimation regime should be considered as an alternative or as an aid to the application of exogenous cryoprotectants.

Some invertebrates undoubtedly will be found to be intractable in that all of their developmental stages will resist whole-body cryopreservation methods. These sensitive stages cannot be permeabilized without damage or are unable to withstand the toxic effects of the cryoprotectants. In these situations, other techniques of germplasm storage involving the preservation of reproductive tissues and cells may be applicable. If the organism employs internal fertilization in its reproduction, artificial insemination using frozen-stored sperm (as with honey bees [67]), may aid the researcher in a selective breeding program. Gonadal transplants, either of the larval organ [68] or

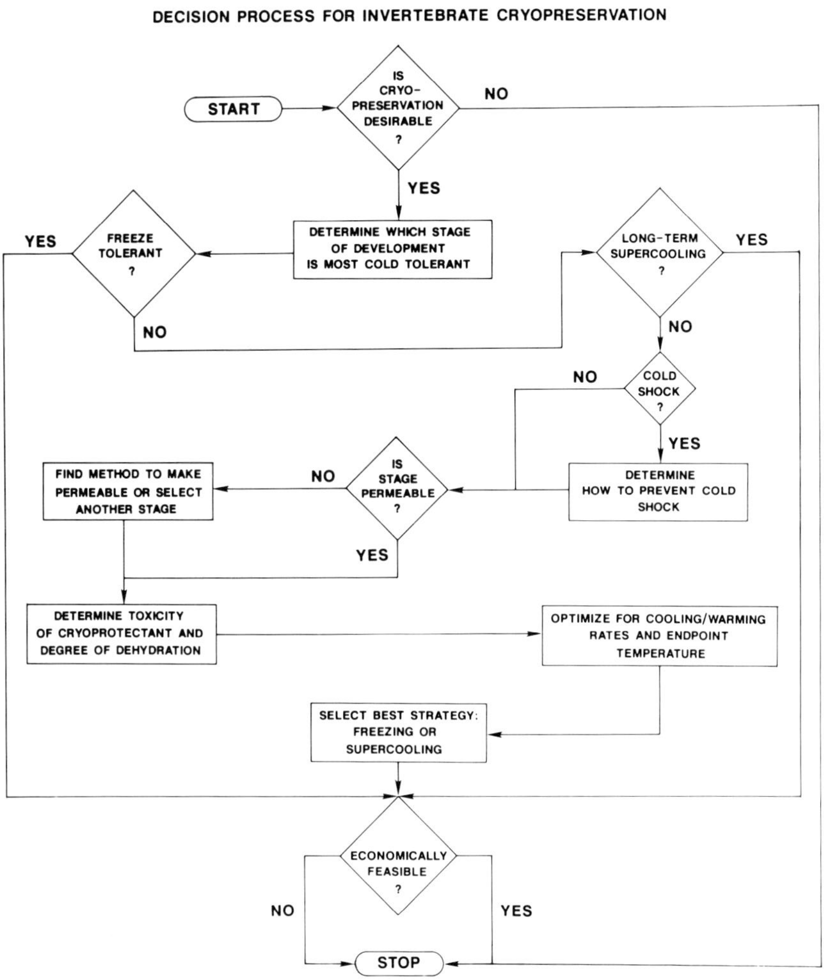

Figure 10. Decision process for invertebrate cryopreservation.

of primordial germ cells [69, 70] also may have applications in the long-term storage of germplasm of laboratory research animals. The obvious drawbacks of all three methods include:

1. A requirement for microsurgical skills,
2. Preservation of only one-half of the genome,
3. The limited numbers of species amenable to this type of micromanipulation, and
4. The labor intensiveness of all three methods.

As with the methods involving cryopreservation of the whole organism, practicality, efficacy, and economy of the technique has to be weighed against continuous laboratory rearing or promoting *in situ* preservation in the natural habitat. It is expected that when the *in situ* method of preservation is no longer adequate, whole-organism cryopreservation will become the method of common useage while the germ cell/reproductive tissue storage techniques will be reserved for high impact situations where risks of loss of invertebrate germplasm are unacceptable at any cost.

References

1. Réaumur, R. A. F. 1736. Mémoires pour Servir a l'Histoire des Insectes 2: 141–147. L'Imprimerie Royale, Paris, 170 pp.
2. Uvarov, B. P. 1931. Insects and climate. Trans. Roy. Entomol. Soc. Lond. 79: 1–247.
3. Luyet, B. J. and P. M. Gehenio. 1940. Life and death at low temperatures. Monograph No. 1. Biodynamica, Normandy, Missouri. 341 pp.
4. Ushatinskaya, R. S. 1957. Principles of Cold Resistance in Insects. Acad. Sci. USSR Press, Moscow. 314 pp.
5. Asahina, E. 1966. Freezing and frost resistance in insects. Pp. 451–485 in H.T. Meryman (ed.), Cryobiology. Academic Press, London. 775 pp.
6. Baust, J. G. and R. R. Rojas. 1985. Review Insect cold hardiness: Facts and fancy. J. Insect Physiol. 31: 755–759.
7. Lopez-Fanjul, C. and W. G. Hill. 1973. Genetic differences between populations of *Drosophila melanogaster* for a quantitative trait. I. Laboratory populations. Genet. Res. 22: 69–78.
8. Bush, G.L., R. W. Neck, and G. B. Kitto. 1976. Screwworm irradiation: Inadvertant selection for noncompetitive ecotypes during mass rearing. Science 193: 419–423.
9. Richerson, V. and E. A. Cameron, 1974. Differences in pheromone release and sexual behavior between laboratory-reared and wild gypsy moth adults. Environ. Entomol. 3: 475–481.
10. Boller, E. F. 1979. Behavioral aspects of quality in insectary production. Pp 153–160 in M. A. Hoy and J. J. McKelvey (eds.), Genetics in Relation to Insect Management-Working Papers. Rockefeller Foundation, N.Y. 125 pp.
11. Craig, G. B., Jr. 1964. Applications of genetic technology to mosquito rearing. Bull. W.H.O. 31: 469–473.
12. Crystal, M. M. Reproductive behavior of laboratory-reared screwworm flies. J. Med. Entomol. 4: 443–450.
13. Levine, N. D. and W. C. Marquardt. 1955. The effect of glycerol and related compounds on survival of *Tritichomonas foetus* at freezing temperatures. J. Protozool. 2: 100–107.
14. Smith, A. U. 1961. Biological Effects of Freezing and Supercooling. Williams and Williams, Baltimore. 462 pp.
15. Hwang, S. W. 1964. Freezing and viability of *Tetrahymena pyriformis* in dimethylsulfoxide. Science 144: 64–65.
16. Diamond, L.S. 1964. Freezing preservation of protozoa. Cryobiol. 1: 95–102.
17. Simon, E. M. 1971. *Paramecium aurelia*: Recovery from − 196°C Cryobiol. 8: 361–365.
18. DeConinck, L. 1951. On the resistance of the free-living nematode *Anguillula silusiae* to low temperatures. Biodynamica 7: 77–84.
19. Asahina, E. 1959. Frost resistance in a nematode, *Aphelenchoides ritzemabosi*. Low Temp. Sci. Ser. B. 17: 51–62.

20. Schiller, E. L., V. M. Turner,, H. F. Marroquin, and R. D'Antonio. 1979. The cryopreservation and in vitro cultivation of larval *Onchocera volvulus*. Am. J. Trop. Med. 28: 997–1009.
21. Koehler, J. K. and L. K. Johnson. 1969. Food supply as a factor in the survival of frozen and thawed rotifers. Cryobiol. 5: 375–378.
22. Pennak, R. W. 1953. Fresh-water invertebrates of the United States. Ronald Press, New York. 769 pp.
23. Hinton, H. E. 1960. Cryptobiosis in the larva of *Polypedilium vanderplanki* Hinton (Chironomidae) J. Insect Physiol. 8: 155–163.
24. Zachariassen, K.E. and H.T. Hammel, 1976. Nucleating agents in the haemolymph of insects tolerant to freezing. Nature, Lond. 262: 285–287.
25. Asahina, E. and K. Aoki. 1958. A method by which frost-hardy caterpillars survive freezing at a super-low temperature. Low Temp. Sci. Ser. B. 16: 55–63.
26. Asahina, E. and K. Tanno. 1964. A large amount of trehalose in a frost resistant insect. Nature, Lond. 204: 1222.
27. Takehara, I. and E. Asahina. 1960. Frost resistance and glycerol content in overwintering insects. Low Temp. Sci. Ser. B. 18: 57–65.
28. Losina-Losinsky, L. K. 1962. Survival of insects at super low temperatures. Dokl. Akad. Nauk. SSSR 147: 1247–1249.
29. Losina-Losinsky, L. K. 1963. Resistance of some insects to the temperature of liquid helium (−269°C) under conditions of intracellular freezing in absence of antifreezes. Cytologya, Akad. Nauk. 5: 220–221.
30. Baust, J. G. and A. L. Lawrence. 1977. Rapid freezing: an essential criterion for the successful cryopreservation of immature larval shrimp. Cryobiol. 14: 705.
31. Baust, J. G. and A. L. Lawrence. 1980. Ontogenetic variability of chill tolerance in larval *Artemia salina* I. Single type cryoprotectants. Aquaculture 20: 305–11.
32. Baust, J. G. and A. L. Lawrence. 1980. Ontogenetic variability of chill tolerance in larval *Artemia salina* II. Dual cryoprotectants. Aquaculture 20: 313–21.
33. Heacox, A. E., R. A. Leopold, and J. D. Bramer. 1985. Survival of housefly embryos cooled in the presence of dimethylsulfoxide. Cryo-Letters 6: 305–312.
34. Lea, C. H. and J. C. Hawke. 1952. 2. The influence of water on the stability of lipovitellin and effects of freezing and drying. J. Biochem. 52: 105–114.
35. Wilmut, I. 1972. The low temperature preservation of mammalian embryos. J. Reprod. Fertil. 31: 513–514.
36. Edidin, M. and V. A. Petit, 1979. The effect of temperature on the lateral diffusion of plasma proteins. Pp. 155–174 in K. Elliot and J. Whelan (eds.), The Freezing of Mammalian Embryos. Excerpta Medica, Amsterdam. 330 pp.
37. Lovelock, J. E. 1957. The denaturation of lipid protein complexes as a cause of damage by freezing. Proc. Roy. Soc. B. 147: 427–433.
38. Strong-Gunderson, J. M. and R. A. Leopold. 1989. House fly cryobiology: Determination of supercooling capacity and low-temperature tolerance of *Musca domestica* L. Environ. Entomol. (in press).
39. Salt, R. W. 1961. Principles of insect cold-hardiness. Annu. Rev. Entomol. 6: 55–74.
40. Rojas, R. R., R. E. Lee, Jr., T. A. Luu, and J. G. Baust. 1983. Temperature dependence-independence of antifreeze turnover in *Eurosta solidaginis* (Fitch). J. Insect Physiol. 29: 865–869.
41. Baust, J. G. 1973. Mechanisms of cryoprotection in freezing tolerant animal systems. Cryobiol. 10: 197–205.
42. Duman, J. G. 1979. Thermal-hysteresis-factors in overwintering insects. J. Insect Physiol. 25: 805–810.

43. Baust, J. G. 1981. Biochemical correlates to cold hardening in insects. Cryobiol. 18: 81–83.

44. Knight, C. A. and J. G. Duman. 1986. Inhibition of recrystallization of ice by insect thermal hysteresis proteins: a possible cryoprotective role. Cryobiol. 23: 256–262.

45. Duman, J. G. and J. L. Patterson. 1978. The role of ice nucleators in the frost tolerance of overwintering queens of the bald faced hornet. Comp. Biochem. Physiol. 59A: 69–72.

46. Baust, J. G. and L. K. Miller. 1972. Influence of low temperature acclimation on cold hardiness in the beetle *Pterostichus brevicornis*. J. Insect Physiol. 18: 1935–1949.

47. Baust, J. G. and R. E. Lee. 1982. Environmental triggers to cryoprotectant modulation in separate populations of the gall fly, *Eurosta solidaginis* (Fitch). J. Insect Physiol. 28: 431–436.

48. Morris, G. J., G. Coulson, M. A. Meyer, and M. R. McLellan. 1983. Cold shock a widespread cellular reaction. Cryo-Letters 4: 179–192.

49. Rojas, R. R., J. G. Riemann, and R. A. Leopold. 1989. Diapause and overwintering capabilities in the larva of the sunflower moth (*Homoeosona electellum*) (Hulst). Environ. Entomol. (in press).

50. Knight, J. D., J. S. Bale, H. Gleave, and R. A. Stevenson. 1986. A multi-channel micro-bead thermocouple thermometer for the detection of low thermal energy changes in small invertebrates. Cryo-Letters 7: 299–304.

51. Quinn, P. J. 1985. A lipid-phase separation model of low-temperature damage to biological membranes. Cryobiol. 22: 128–146.

52. Crowe, J. H. and L. H. Crowe. 1984. Preservation of membranes in anhydrobiotic organisms: The role of trehalose. Science 223: 701–702.

53. Hazel, J. R. and C. C. Prosser. 1974. Molecular mechanisms of temperature compensation in poikilotherms. Physiol. Rev. 54: 620–677.

54. Lin, T. T., D. V. Lynch, S. P. Myers, R. E. Pitt and P. L. Steponkus. 1987. Volumetric behavior and hydraulic conductivity of *Drosophila* embryos. Cryobiol. 24: 542.

55. Fahy, G. M. 1984. Cryoprotectant toxicity: Biochemical or osmotic? Cryo-Letters 5: 79–90.

56. Mazur, P. 1984. Freezing of living cells: Mechanisms and implications. Am. J. Physiol. 247: C125–C142.

57. Schneider, U. and P. Mazur. 1987. Relative influence of unfrozen fraction and salt concentration on the survival of slowlyfrozen eight-cell mouse embryos. Cryobiol. 24: 17–41.

58. Morrissey, R. E. and J. G. Baust. 1976. The ontogeny of cold tolerance in the gall fly, *Eurosta solidagenis* (sic) J. Insect Physiol. 22: 431–437.

59. Miller, L. K. 1978. Freezing tolerance in relation to cooling rate in an adult insect. Cryobiol. 15: 345–349.

60. Miller, L. K. 1969. Freezing tolerance in an adult insect. Science 166: 105–106.

61. Leibo, S. P., J. J. McGrath, and E. G. Cravalho. 1978. Microscopic observation of intracellular ice formation in mouse ova as a function of cooling rate. Cryobiol. 15: 257–271.

62. Gassner, G. and J. A. B. Lohman, 1987. Combined proton NMR imaging and spectral analysis of locust embryonic development. Proc. Natl. Acad. Sci. USA. 84: 5297–5300.

63. Hendrix, D. L., R. E. Lee, J. G. Baust, and H. James. 1981. Separation of carbohydrates and polyols by a radially compressed high performance liquid chromatographic silica column modified with tetraethylenepentamine. J. Chromatography 210: 45–53.

64. Asahina, E. 1955. Freezing and supercooling as a method of storage of a mobile animal, a preliminary experiment. Zool. Mag. (Tokyo) 64: 280–285.

65. Asahina, E. 1969. Frost resistance in insects. Pp. 1–49 in J. W. L. Beament, J. E. Treherne

and V. B. Wigglesworth (eds.), Advances in Insect Physiology. Academic Press, New York. 308 pp.

66. Lee, R. E., Jr., C.-P. Chen, and D. L. Denlinger. 1987. A rapid cold-hardening process in insects. Science 238: 1415–1417.
67. Harbo, J. 1983. Survival of honey bee (Hymenoptera: Apidae) spermatozoa after two years in liquid nitrogen (– 196 °C). Ann. Entomol. Soc. Amer. 76: 890–891.
68. Bruschweiler, W. and W. Gehring. 1973. A method for freezing living ovaries of *Drosophila melanogaster* larvae and its application to the storage of mutant stocks. Experientia 29: 134–135.
69. Allis, C. D., E. M. Underwood, J. H. Caulton, and A. P. Mahowald. 1979. Pole cells of *Drosophila melanogaster* in culture. Dev. Biol. 69: 451–465.
70. Strong, J. M., A. E. Heacox, and R. A. Leopold. 1985. Successful transplantation of pole cells between the wild type (FW_2) and a mutant strain of the house fly. Proc. N. Dakota Acad. Sci. 39: 62.
71. Asahina, E. and K. Tanno. 1966. Freezing resistance in the diapausing pupa of the *Cecropia* silkworm at liquid nitrogen temperature. Low Temp. Sci. Ser. B. 24: 25–34.
72. Duman, J. G. 1980. Factors involved in overwintering survival of the freeze tolerant beetle, *Dendroides canadensis*. J. Comp. Physiol. 136: 53–59.
73. Salt, R. W. 1959. Role of glycerol in the cold-hardening of *Bracon cephi* (Gahan). Can. J. Zool. 37: 59–69.
74. Salt, R. W. 1957. Natural occurrence of glycerol in insects and its relation to their ability to survive freezing. Can. Entomol. 89: 491–494.
75. Hamilton, R. L., D. E. Mullins and D. M. Orcutt. 1985. Freezing-tolerance in the wood roach *Cryptocercus punctulatus*. Experimentia 41: 1535–1537.
76. Williams, R. J. 1970. Freezing tolerance in *Mytilus edulis*. Comp. Biochem. Physiol. 35: 145–161.
77. Loomis, S. H. 1985. Seasonal changes in the freezing tolerance of the intertidal pulmonate gastropod *Melampus bidentatus* Say. Can. J. Zool. 63: 2021–2025.
78. Whitmore, D. H., R. Gonzalez, and J. G. Baust 1985. Scorpion cold-hardiness. Physiol. Zool. 58: 526–537.
79. Whitaker, D. M. 1940. The tolerance of *Artemia* for cold and high vacuum. Jour. Exp. Zool. 83: 391–399.
80. Margaritis, L. H. 1985. Structure and physiology of the egg shell. Pp. 153–230 in G. A. Kerkut, and L. I. Gilbert (eds.), Comprehensive Physiology, Biochemistry, and Pharmacology. Pergamon Press, Oxford, England. 487 pp.

Techniques for preserving of plant germplasm

LEIGH E. TOWILL and ERIC E. ROOS
Plant Germplasm Preservation Research Unit, USDA, ARS National Seed Storage Laboratory, Fort Collins, Calorado 80523, U.S.A.

Abstract Preservation of genetic diversity in plants requires effective application of storage technologies. Active and base collections may require different storage conditions. Plant germplasm is stored as seeds and vegetative plants. Preservation strategies are categorized as *in vivo* (greenhouses, plantations, orchards, etc.), *in vitro* (plants in vessels under normal or restrictive growth conditions), cold (seeds and *in vitro* plants; some vegetative, *in vivo* propagules), or cryogenic (seed, excised embryos, pollen and buds/shoot tips from *in vivo*- or *in vitro*-grown materials). Desiccation tolerant seeds are stored at about 5 °C (longevities of 5 to 20 + years) or about − 10 °C to − 20 °C (20 to 50 + years). Long-term storage data for most species are sparse and longevities vary considerably for seed lots within a species. Cryogenic storage (in liquid nitrogen) should give extremely long periods of viability (100 + years). Species with desiccation sensitive seed must be maintained as vegetative plants because of short seed longevity.

In vitro maintenance of some species is feasible, but not extensively practiced. Clones of some species are maintained if there is some desirable attribute of the line or if considerable time is required for the plant to reach maturity. The germplasm of clonal species is preserved as vegetative plants, mainly *in vivo* (greenhouses, orchards, botanical gardens and arboreta). *In vitro* storage may supplement other techniques, but it requires research and development for application to diverse genotypes. Implementation of cryogenic storage of shoot-tips/buds and of desiccation sensitive seed requires further study.

Key words: seeds, clones, cryopreservation, desiccation tolerance, desiccation sensitive, liquid nitrogen, seed storage

Introduction

Plant germplasm comprises the total array of living plant species, subspecies, genetically defined stocks, genetic variants, and mutants whose continuing availability is important for the present and future health and economic welfare of mankind [1]. The need to preserve germplasm of diverse species has been amply stressed [2, 3, 4, 5, 6]. The extent of materials considered to be germplasm depends upon the crop. Thus, potato germplasm consists of some 160 or more species because many of the tuber-bearing *Solanum* species intercross, whereas maize germplasm consists of only a few species but comprises thousands of adapted land races.

L. Knutson and A.K. Stoner (eds.). Biotic diversity and germplasm preservation, global imperatives.
 ISBN 0-7923-0178-1

Advances in the knowledge of gene structure, function, and manipulation, as well as an increased understanding of incompatibility barriers and ways to overcome them, herald a new era for crop improvement by use of genes from taxonomically distant species. We need to adopt the broadest possible definition of germplasm for collection and preservation purposes in order to ensure the availability of these species.

The object of this review is to address the following questions: What storage technologies are presently used, and what alternative technologies exist that may allow for more efficient preservation of germplasm maintained as seeds and as clones?

Active and base collections

Active collections need to be contrasted to base collections because different storage technologies may be applied to each. An active collection functions to obtain, maintain, evaluate, and distribute diversity [7]. Stocks are depleted by distribution, by testing for viability estimates, and by decreases in viability during storage. When stocks are low, replenishment is by prescribed regeneration procedures that differ among species.

A base collection is essentially a duplicate of an active collection, but it is held in long-term storage as a back-up should there be a loss in the active collection. Base collection stocks generally are not distributed but do suffer viability decline and losses as a result of testing. Thus, for preservation of diversity and for minimizing costs, storage strategies that give the greatest longevity are desired for base collections.

Accordingly, different storage technologies are highly desirable for active and base collections. Two examples are illustrative. Cryogenic storage of seed may give extremely long lifetimes useful for a base collection. However, active collections hold large numbers of seed. Space considerations and the extent of distribution may dictate that storage in vaults at 5 °C is sufficient. For woody clonally propagated materials, growth in greenhouses, orchards, or *in vitro* plant cultures are preferred for the active collection because budwood, cuttings, or cultures can be supplied promptly. However, a field collection is not a satisfactory base collection because of its susceptibility to loss from pathogens, adverse weather and other factors. Cryogenically stored shoot-tips or buds would be useful for a base collection because they preserve the genetic integrity of the clone; however, they are not readily used forms and must be grafted or cultured *in vitro*.

Since research and development goals differ for seed and vegetatively maintained germplasm, each will be discussed separately.

Seed-maintained germplasm

Seeds are the most common storage propagule for germplasm preservation and are classified as being orthodox (capable of retaining viability after being dried to less than about 5% moisture content – also termed 'desiccation tolerant' seed) and recalcitrant (those losing viability after being dried below a critical limit, usually about 12–30% moisture – also termed 'desiccation intolerant' seed) [8]. Desiccation-intolerant seeds are also termed 'short-lived' seeds. Most temperate crop species have orthodox seeds.

Orthodox seed storage

Bewley and Black [9], Priestley [10], Roos [11] and Toole [12] have recently reviewed the topic of seed longevity. Earlier reviews by Barton [13], Biasutti Owen [14], Crocker [15], Harrington [16], Justice and Bass [17], and Roberts [18] all contain excellent information and special insight on this topic. The classic compilation of data on seed longevity by Ewart [19] is still a primary source of information for many species. Bass [20] compiled tables of seed longevities for many agricultural and horticultural species stored under various conditions.

Many studies have shown the importance of seed moisture and storage temperature for retaining viability in orthodox seeds [21, 22]. Longevity also can be affected by genotype and by environmental factors during ripening, harvesting, and drying of the seed before storage [21, 23, 24], but these factors will not be discussed here. Lowering the seed moisture content to 5–7% and lowering the storage temperature increases longevity for many species, but data on seed stored under these conditions are often limited, especially with storage at subfreezing temperatures. Freeze drying has also been used to reduce seed moisture content and improve seed storability [25, 26, 27].

Short-term seed storage usually is conducted under ambient conditions, while medium-term storage is conducted from +1 °C to +4 °C, and long-term storage is conducted from −10 °C to −20 °C. For example, longevities for rice seed under short-, medium-, and long-term storage are about 5, 30, and estimated to be greater than 70 years, respectively [28].

Lowering of the storage temperature to approximately −20 °C should greatly increase longevity, thus reducing the need for costly germination tests and regeneration cycles. Approximations based on a number of assumptions suggest that seed from many species may have longevities of about 100 years when stored at −20 °C [22, 23].

Hundreds of papers have been published on seed storage [10, 29], and a large number of them have included information on storage at cool (5 °C) or cold ($<0\,^{\circ}$C) temperatures. The storage times reported usually have been 5 to 10 years. Roos [11] listed only two papers that contained data on subfreezing storage of seeds for 25 years. Seeds of *Lobelia cardinalis* L. stored at $-5\,^{\circ}$C and 4.7% seed moisture content had a germination of 64% after 25 years versus an original germination of 73% [30]. *Hibiscus cannabinus* L. seeds retained their full germination (88%) after 25 years at $-12\,^{\circ}$C and 10.4% seed moisture content [20].

Rincker [31, 32, 33, 34] examined the effects of subfreezing storage on germination and forage production of forage grass and legume seed. In one study, 291 seed lots representing seven species were stored for as long as 20 years at $-15\,^{\circ}$C and 60% RH [32]. In general, seed lots of original high quality retained high germination potential. For example, 14 lots of birdsfoot trefoil (*Lotus corniculatus* L.) stored for 18 years had 95% germination both before and after storage. Germination declined more quickly for seed lots with lower initial seed quality. For example, 20 lots of bromegrass (*Bromus inermis* Leyss.) with an initial germination of 87%, fell to 55% after 18 years. In a second study, 260 lots of nine forage species were placed in subfreezing storage directly after harvesting [33]. After 20 years only three seed lots had lost 20% or more of their viability. Upon the retirement of Clarence Rincker from USDA, ARS, over 3000 seed lots were transferred from Prosser, Washington, to the National Seed Storage Laboratory (NSSL) in Fort Collins, Colorado, in order to continue this valuable, long-term seed storage study [35]. All seed lots were tested for viability and seed moisture content. Analyses were done on the effects of year and location of production on germination [36]. While location of production appeared to have minimal affect on germination after storage, year of production was found to have significantly affected storability. Cryogenic storage refers to storage of propagules at very low temperatures. Ideally, temperatures less than about $-130\,^{\circ}$C are desired because of the absence of liquid water, very low molecular kinetic energies, and extremely slow diffusion [37]. Chemical reactions (e.g., metabolism) thus occur very slowly, and storage longevities are postulated to be extremely long and limited only by the buildup of genetic lesions resulting from background irradiation [38]. Storage in liquid nitrogen (LN, $-196\,^{\circ}$C) or in the vapor above LN (ca. $-150\,^{\circ}$C to $-180\,^{\circ}$C) is usually envisioned because it is a relatively cheap cryogen. Long periods of storage are desired to reduce problems associated with seed regeneration schemes, including cost of regeneration and risks of loss caused by disease and environmental problems [39]. Where an adequate cryogen source is available, cryogenic seed storage is cost-competitive with low temperature ($-20\,^{\circ}$C) storage [40]. A number of assumptions are used in these calculations, but when costs associated with regeneration cycles and germination testing are included, cryogenic storage is economical.

Orthodox seeds at low moisture levels usually retain the same percentage germination after LN treatment as before; hence, extended storage may be possible [41]. Some limitations on LN use for orthodox seed may exist and may relate to the seed coat, seed oil content, and seed moisture content. Some orthodox seeds are desiccation tolerant but are sensitive to low temperatures. Some temperate fruit and nut seeds (*Corylus* spp., *Juglans* spp., and some *Prunus* spp.) exhibit these characteristics, although the extent of sensitivity and the mechanisms of injury are not known.

Cooling of orthodox seed requires no special equipment. Seed is placed into the vapor phase within a LN storage refrigerator. Warming of seed from LN storage is not controlled. Samples usually are placed directly at room temperatures after removal from the LN refrigerator.

Recalcitrant seed storage

The previous discussion concentrated on orthodox seed longevity. Recalcitrant seeds have short lifespans, ranging from a few weeks to several months [8, 16, 42]. Harrington [16] compiled a list of plant species with short-lived seeds, many of which we would now call recalcitrant (the term recalcitrant had not come into general usage at that time). A list of species suspected of having recalcitrant seeds was compiled by King and Roberts [43]. Three factors may contribute to the short longevity of stored recalcitrant seeds; sensitivity to desiccation, chilling injury, and problems such as microbial contamination and germination during storage that are associated with seeds of high-moisture content.

Recalcitrant seeds are produced by aquatic species, large-seeded species, species native to tropical areas, and some temperate zone species of trees (e.g., oaks) [8, 44]. Coconut, cocoa, mango, nutmeg, and rubber plants are examples of economically important species with recalcitrant seed. Adequate tests should be made to determine if seed of a species displays true recalcitrant behavior [45]. For example, seeds from citrus, formerly thought to be recalcitrant, were shown to be dormant, requiring a longer period of germination than undried seeds [22].

To retain viability, recalcitrant seeds are stored at as low a temperature as possible under conditions that retain relatively high seed moisture levels and assure a supply of oxygen for respiration. Seed longevities range from weeks to months under these conditions and are too short for adequate germplasm preservation [46]. Therefore, these species usually are stored as vegetative plants. If diversity is considerable, many individuals must be conserved, resulting in high space and maintenance costs.

More research is necessary to define the mechanisms of desiccation and chilling injury and to investigate methods of alleviating it. Cryogenic treatments involving cryoprotectants [44] or the use of excised embryos [47] may prove to be useful, but no procedure has been developed so far for the practical, reproducible cryogenic preservation of recalcitrant seeds. Axillary shoot tip or bud cryopreservation from *in vivo* or *in vitro* plants might be useful for base-collection storage of these species.

Seed deterioration

It is becoming increasingly apparent that the life span of a seed or seed lot can be extended greatly by low temperature storage. However, seeds eventually lose the capability to germinate and die. Even under relatively good storage conditions, as maintained at the USDA National Seed Storage Laboratory in Fort Collins, Colorado, deterioration still occurs.

Attempts to quantify seed deterioration generally have used the seed germination percentage and, more specifically, the standard germination test as prescribed in the official rules for testing seeds [48, 49]. These rules prescribe that only 'normal' seedlings are included in the percentage of germinated seeds. This practice may introduce errors in the germination value as a result of the subjective classification of abnormals.

Predicting storability. Roos [11] has reviewed some of the early attempts to predict seed longevity. Harrington's 'rules of thumb' related the effects of seed moisture content and storage temperature to the rate of seed aging [50]. The rules state that for each 1% increase in seed moisture (between 5 and 14%), the lifespan of the seed is halved; and for each 5°C increase in temperature (between 0 and 50°C), the lifespan is halved independently of the seed moisture content.

The accelerated aging (AA) test [51, 52] was originally developed to predict the relative storability of seed lots. In this test, seeds are placed under high temperatures (40–50°C) and high relative humidities (>90%) for short periods of time (2-5 days) and are then subjected to the standard germination test. Results are compared with those of unaged seeds tested at the same time. Although this test now has become an accepted vigor test [53], its original purpose was to predict which seed lots should be carried over to the following year by seed companies. Since this test was developed in Mississippi, it further was assumed that carry-over seed would be subjected to the hot, humid conditions prevailing in that State.

While the basic principle of the AA test – namely that seeds having low vigor (see below) will not store as well as those having high vigor – is probably valid, it is not necessarily true that seeds stored under gene bank conditions will behave in the same manner. The mechanism for loss in germinability under the two sets of conditions probably is not the same. Evidence for this comes from three observations:

1. Under AA, seed moisture is elevated to a point at which microorganisms can grow readily,
2. Under normal (slow) aging, the percentage of abnormal seedlings increases as germination (normal seedlings) drops, whereas under AA this large increase in abnormal seedlings is seldom seen, possibly because the seeds pass quickly through this stage, and
3. The frequency of chromosomal aberrations in root tips of germinating seeds, which is normally correlated with germination percentage [54] is lower in seeds subjected to AA.

An alternate approach, similar to accelerated aging, to predict which seed lots might deteriorate first in gene banks involves the use of artificial aging techniques [55, 56]. The significant differences between these two approaches lie in the use of lower seed moisture contents and storage temperatures and longer storage periods. Seed moisture contents typically should be below levels where active fungal growth can occur (<15%), storage temperatures are less than 35 °C, and storage periods should be of the order of weeks or months. The objective is to achieve slow deterioration which more closely approximates that of seeds in long-term cold storage. Moore and Roos [57] compared different artificial aging tests to evaluate rates of deterioration and statistically analyze the quantitative responses. Artificial aging tests which employed high humidities and high storage temperatures (90% RH, 32 °C) resulted in high data heterogeneity and concomitant lack of reliability for predicting seed survival. More moderate aging conditions (70% RH, 21 °C) resulted in low data heterogeneity. A computer program was developed [58] to calculate the deterioration rates and apply the statistical analysis. This program was used by Priestley *et al.* [59] to determine differences in seed longevity at the species level.

Mathematical models of deterioration. A mathematical model of seed deterioration has been proposed [18, 60, 61]. The 'basic viability equation' can be used to predict germination after any storage period at any temperature and moisture content (within reasonable limits). The equation depends upon the determination of several constants, including the initial seed quality (specific for each seed lot) and four other constants for the effects of temperature, seed moisture content (determined for each species), and species. Once the constants have been determined for a given species and seed lot, predictions can be made as to longevity under any number of different storage conditions. An inherent problem with this and with all predictive models is that the equation may be invalid if it is extrapolated beyond the limits of the data (temperature and seed moisture conditions) used in developing the various constants.

Loss of vigor. Detecting loss of seed vigor before germination actually declines presents a much more difficult problem. Early detection of deterioration (vigor loss) is of crucial importance to germplasm curators as they must monitor seeds in storage and regenerate them before a significant loss in viability has transpired in order to prevent genetic changes from occurring [11]. Numerous reviews of vigor testing have been published [53, 62, 63, 64, 65, 66, 67]. Some of the more common vigor tests include accelerated aging, controlled deterioration, cold test, conductivity test, seedling growth rate, and the tetrazolium test. The AA test, which already has been discussed, is widely used for predicting storability and field performance as it can be applied to a wide range of species. The chief difficulty in getting repeatable results appears to depend upon problems associated with non-uniform water uptake during the exposure to high humidity [68, 69].

The controlled deterioration test [70] is similar to the AA test except that seeds are partially imbibed on moistened filter paper to pre-determined moisture levels. Moisture content is checked by frequent weighings during imbibition. The seeds are then sealed in containers and held for 24 hr at 10 °C to equilibrate before being incubated in a 45 °C water bath for 1 day. Seeds are then germinated (as detected by radicle emergence) at 20 °C.

The cold test, which is primarily used for predicting field emergence of corn seed, is routinely applied by major seed companies for evaluating seed lots prior to planting. This test was developed in the 1940s [71] and utilizes soil which contains various microorganisms from the field. Thus, the interaction of low temperature, microorganisms, and seed quality is measured. The applicability of the cold test for predicting the success of seed storage is minimal.

The conductivity test measures electrolytes leached from the seed upon soaking. It was proposed by Matthews and Bradnock [72, 73] as a seed vigor test for predicting field emergence of peas. This test assumes that seed deterioration is manifested as a loss of membrane integrity. If this is, in fact, the mechanism of deterioration, this test should have wide applicability for predicting seed storability.

Seedling growth rate is a comparison of root and/or shoot length or seedling weight after germination. Thus, it measures the metabolic efficiency of mobilization of storage reserves by the seed. It is applicable to essentially all seed species. However, precise control of environmental conditions is needed in order to achieve reliable results [53]. Once again, this test measures the overall capacity of the seed to germinate and grow and thus should have wide application to seed storage.

Tetrazolium testing (TZ) was originally developed as a quick alternative to the standard germination test, particularly when seed dormancy was encountered [13]. The principle of the test is the measurement of dehydrogenase activity, which results in a reduction of the soluble, colorless tetrazolium salt to an insoluble formazan, which stains the seed tissues red. A critical evaluation of the staining pattern is necessary for reliable test results. Although the TZ test has wide application, its subjectiveness makes it of questionable value for detecting small differences in vigor in stored seeds.

A new approach is needed to evaluate seed deterioration. Germination testing will continue to be the standard method for the near future; however, this is a time consuming and labor intensive endeavor. Most vigor tests are also time consuming and, if anything, are even more expensive than the germination test. Ideally, a non-destructive method of scanning a seed sample could be developed. Such a method could be automated, and a seed sample could be subjected to repeated analyses over time. Such a technique could involve infra-red analyses, or perhaps some form of nuclear magnetic resonance. Delfini *et al.* [74] used ^{31}P NMR spectroscopy to follow phosphate metabolites directly in intact, germinating lettuce seeds. Alternatively, the head gas above a sealed storage sample could be analyzed for volatiles which might be given off during extended storage [75]. For these types of approaches to be developed, more information is needed on the basic mechanisms of seed deterioration.

Seed storage guidelines

Methods for seed handling, packaging, germination testing, and shipping have been evaluated and somewhat standardized. Orthodox seeds are usually dried to moisture contents of 5–7% and then are sealed in moisture-proof foil-laminated bags for storage at temperatures between 5°C and −20°C [7, 17, 76]. Care must be taken to ensure seals are complete and the bags are free of any punctures. Other storage containers may suffice [7, 77]. Accurate seed viability tests and methods of breaking dormancy are necessary for evaluating longevity [17, 78].

Seed germination testing is costly in terms of manpower and in sacrificing valuable seeds for use in viability estimates. Viability testing should be less frequent if storage is at lower temperatures, but data are needed on exactly what viability testing schedule is desired. Viability tests using fewer seeds have been proposed but are not widely used [78, 79].

Statistically based improvements in this area are known and should be developed and incorporated into the testing programs.

The recommendation from the International Board for Plant Genetic Resources (IBPGR) for retaining maximum diversity within an accession is to regenerate seed whenever germination drops about 15% below what was initially present before storage [7]. Practical considerations of available labor, space, and time, however, often delay regeneration until viability has dropped 20–50% or even much more from the original value. This delay in regeneration may lead to loss of diversity in heterozygous and heterogeneous accessions.

Loss of viability during storage is associated with mutation, notably accumulations of chromosome damage and point mutations. Cells with chromosome change are usually selected against during growth of the seedling (diplontic selection), but point mutations may be passed to succeeding generations [22, 23, 80]. Thus, genetic changes do occur during storage.

Intertwined with the question of tolerable percentage drop in viability before regeneration is the question of how many individuals are used to produce the next generation of seed [81]. The proposal to regenerate an accession whenever germination drops about 15% (as opposed to the more practical 20–50% decrease), tends to increase the frequency of regeneration and associated costs [79]. The balance between loss in diversity from delaying regeneration cycles and hence using seed samples with lower viability [82], and the loss due to genetic drift from more frequent regeneration cycles [83] if the 15% viability decrease is adopted is not easy to evaluate unless adequate characterization of the diversity is available. Genetic drift could be reduced if larger numbers were grown out for regeneration, but practical considerations (e.g. space and manpower) usually limit the number of individuals used. Population studies along with knowledge of rare allele frequencies are necessary in order to estimate numbers of parents needed to retain diversity.

Clonal preservation

Active collections

Germplasm is also maintained vegetatively as clones. The reasons for retaining clonal lines relate to maintenance of heterozygosity and adapted complexes, inability to set seed easily, long juvenile stages (especially in woody lines), and production of recalcitrant and/or large seed [84, 85].

In vivo. Clones usually are maintained *in vivo* in greenhouses or field plantings, such as orchards or plantations. Plants parts such as cuttings, budwood, root divisions, and tubers from some species can be stored for about one year under refrigeration, and are used for propagation and distribution. The useful storage time varies considerably depending upon species. Propagules from chilling sensitive species require storage at warmer temperatures. Field and greenhouse maintenance requires considerable space, and high costs are associated with pruning, disease and weed control, and propagation. Field materials are susceptible to loss from disease or disaster. Phytosanitary practices may be more stringent in greenhouses and screenhouses, and disease occurrence and spread are thus reduced. Tremendous adaptive characteristics associated with survival under different climatic conditions often exist within a species. Growth of all lines at a single location may place some lines under environmental stress and decrease fecundity. Germplasm could be lost.

Perennial ornamental species often are preserved as clones in botanical gardens and arboreta, but the number of lines stored are few relative to those required to maintain diversity. Gardens and arboreta have high maintenance costs, problems with disease, pests, environmental stress, and have priorities as to types of materials displayed to their constituents [1]. Such institutions usually do not distribute large numbers of materials. Safe preservation is best accomplished if multiple sites are used. Cold or cryogenic storage of seed from perennial ornamentals might be a feasible route for long-term preservation.

In vitro. In vitro plants are useful for preservation of active collections for a wide array of species [84, 85, 86, 87]. The advantages include maintenance of disease-free state, more rapid multiplication rates, ease of rooting (especially for woody materials), reduced maintenance requirements, possibly reduced labor input, and ease of germplasm exchange and shipment both within and between countries.

The term *in vitro* may also refer to cells, tissues, or organs in culture, but their use for germplasm storage is not feasible at the present time due to cytological instabilities in culture systems and somaclonal variation in regenerated plants [88, 89]. If reproducible plant regeneration were possible from a range of genotypes, and if somaclonal variation were restricted (or controlled), use of cells or protoplasts for storage of some species may be feasible.

For germplasm storage, *in vitro* plants are initiated from meristem tips, buds, or stem tips and are propagated through divisions; thus, lines are maintained as clones [90, 91]. Shoot multiplication *in vitro* must occur by axillary bud proliferations to avoid selection of somaclonal variants that might arise from adventitious buds produced within the tissue mass. Virtually any species can be introduced into culture and maintained as an *in vitro* plant, but indiscriminate application is neither practical nor desirable for germplasm preservation. Care must be taken to apply this technology to species that absolutely require clonal maintenance.

In vitro storage of plants is conducted either under normal growth conditions (such as 20–25 °C under light), reduced temperatures (– 3 to 12 °C, with or without light), or on a growth-restrictive nutrient medium at ambient or reduced temperatures [92]. Storage in this manner gives longevities of weeks, months and, in some cases, years before subculture is needed. Examples of storage periods include 16 months for garlic [93], 4 years for apple [94], one year for kiwifruit [95], and 18 months for pear [96]. Information is scant describing whether a single protocol can be used for diverse genotypes within a species or whether regenerants after storage are normal and easily acclimatized. A single procedure worked well for potato cultivars [97], but was not useful for diverse *Chrysanthemum* lines [98].

Potential problems with the application of *in vitro* plant preservation include difficulties in culture establishment, micropropagation, rooting, and acclimatization [85, 99, 100, 101, 102]. Many of the details for these stages are cultivar specific. Explants from mature woody plants often are more difficult to establish in culture than are those from juvenile plants [101, 102]. Diverse genotypes may not survive in the same physical and chemical environment. Growth abnormalities may occur [103, 104]. There is an uncertainty about the occurrence of somaclonal variants. Adequate characterization of each *in vitro* line (e.g., electrophoretic profiles: isozymes; restriction fragment length polymorphisms) must be established and lines periodically examined to avoid identification errors.

Experience with *in vitro* maintenance for germplasm preservation is limited; thus, time and cost analyses are lacking. An international database on *in vitro* conservation has been established to improve information flow among workers with common interests [105]. Combinations of *in vivo* and *in vitro* maintenance are being used in some U.S. Fruit and Nut Clonal Repositories and other collections and should produce information on effectiveness, acceptability, and cost. Development and implementation are impeded by inadequate information on optimal storage conditions, on application to diverse genotypes and species, and on genotypic stability. At present there are inadequate data to provide answers to questions on the extent of duplicate samples, duplicate locations, and handling needs. Until acceptability is gained though experience, *in vitro* storage is now best utilized as an adjunct to *in vivo* storage [106].

Many developed and developing countries are exploring *in vitro* storage technology for clonal lines (e.g., cassava, potatoes, sweet potatoes, strawberries) and have relied upon the international centers and the IBPGR for guidance and for information exchange to apply *in vitro* culture to genetic conservation more directly. The IBPGR advocates the use of *in vitro* plants for active collections and has published recommendations for *in vitro* genebank operations [87, 107, 108]. A IBPGR-CIAT collaborative project on cassava should provide practical knowledge [109].

Base collections
The need for base collections is well established for seed germplasm but is not well defined for clonal germplasm. Species diversity could be preserved through storage of seed populations for some crops. Long-term preservation of the clone could be accomplished cryogenically with isolated shoot-tips or buds. Pollen preservation could supplement seed storage or clonal preservation.

1. *Seed.* Seeds could be used to preserve diversity for clonal crops if they can be easily produced, and if they are orthodox [110]. The main advantage in preserving seed is the simplicity of long term storage under $-20°C$ or cryogenic conditions. Considerations for storage are discussed above.

Pollen. Pollen is not a conventional vehicle for germplasm preservation. At present pollen is stored mainly to facilitate in hybridizing materials that flower at different times. However, pollen storage may also be useful for germplasm preservation [39, 111]. A population of pollen grains collected from genetically different individuals would contain the nuclear genes within that population. Some cytoplasmic genes, however, would be lost by exclusion during subsequent use for fertilization.

Pollen, as with seed, may be grouped into types that are desiccation-tolerant and desiccation-sensitive [112]. Desiccation tolerant types store best when dried to a low moisture content and when held at low temperatures, analogous to orthodox seed. Dried pollen survives cryogenic exposure. Desiccation-sensitive pollen (for example, pollen of many members of the Poaceae) loses viability during drying and, therefore, cannot tolerate temperatures where its cell water freezes.

Extensive or systematic studies defining pollen desiccation resistance and sensitivity are few. Empirical information derived from exposure to drying and low temperatures has shown that pollen from some fruit crops of temperate and subtropical climates is desiccation-resistant [113]. Examples include grape [114], olive [115], some *Prunus* spp. [116], avocado [117], pistachio [118], and walnut [119].

Some success has been reported for cooling desiccation sensitive pollen to low temperatures, notably maize pollen [120], but data are sparse. Others have not been able to confirm the information, but it appears that careful adjustment of the moisture content to a specific range may be crucial [121, 122]. Other difficult-to-store pollens, such as from pearl millet [123] and sugarbeet [124], have survived LN exposure.

Storage of pollen requires little space and could be useful as a supplement for a base collection of clonal lines, for example, from species with recalcitrant seed. Species with recalcitrant seed do not necessarily have desiccation-sensitive pollen. Pollen alone, however, is not a satisfactory base collection because some cytoplasmic genes may be lost during transmission. Information is also needed on stability and longevity at subzero temperatures and on the development of handling systems before practical storage is achieved.

Shoot tips and buds. Cryopreservation of meristem tips, shoot tips, or buds is potentially useful for clonal base collections since it provides a cheap, spacing-saving method for long-term storage [125, 126, 127, 128, 129]. Since many species are held clonally, it is desirable to develop unified procedures that are applicable to groups of these species. Two general strategies are apparent. The first utilizes the ability of some species to cold-acclimate to very low temperatures, while the second utilizes addition of cryoprotectants to enhance survival of species that are less, or are not cold hardy.

The extreme cold hardiness that develops in some species is well documented. If twigs are collected in the correct physiological state, subjected to controlled cooling to LN and thawed, the excised buds survive and can be grafted. Several investigations have shown that hardy twigs, when cooled under controlled conditions, survive LN exposure [127, 128, 130, 131, 132, 133, 134, 135]. This strategy should be useful for preserving germplasm of apple, pear, blueberry, *Prunus* spp., gooseberries, and currants, as well as others. Treated buds are grafted to suitable rootstocks to generate a shoots.

Alternatively, meristem tips can be dissected from LN-treated buds and cultured to produce a shoots [136]. Any method must take into account that genotypes within a species differ in hardiness. Reducing the moisture content in twigs prior to cooling may enhance survival in tender lines.

Cryopreservation of species that cold acclimate to a lesser extent and of species that do not acclimate at all will require the application of cryoprotectants, such as dimethylsulfoxide (DMSO) or combinations of DMSO with sucrose, glucose, polyethylene glycol, or proline [129, 137]. Buds or shoot tips from field, greenhouse, growth chambers, or *in vitro* plants are excised for treatment. In some cases cold treatment of the stock plant is beneficial [138, 139]. After excision, the sequence of steps in a cryogenic protocol involve preculture (a few hours to 2 days) in growth medium or with low levels of cryoprotectants; controlled exposure to a defined concentration of cryoprotectant; ice nucleation of the sample just below the freezing point of the solution; controlled cooling (about 0.25 to 1 °C/min) to about −35 °C to −40 °C followed by transfer directly into LN; rapid warming; and post thaw treatments prior to culturing to determine viability. The variations on the basic protocol are too numerous to discuss here. Details for several of these steps have been reviewed [125, 128, 129, 140]. Shoot-tips from herbaceous species (examples: carnation, cassava, chickpea, *Digitalis*, pea, peanut, potato, strawberry) and from woody species (apple) can survive LN treatment. Many questions remain, for example, whether a single procedure works well for a diverse array of genotypes, and if plant regenerants are completely normal after treatment.

A persistent problem in the cryogenic storage of *in vitro* cultures is the maintenance of organization within the shoot tip and regeneration of the shoot without (or with minimal) callus formation or adventitious bud development. With current procedures, shoot tips from herbaceous lines exposed to LN often form either a small amount of callus before shoots are regenerated [126, 141] or form adventitious shoots at different locations within the tips [142]. Both probably increase the chance for somaclonal variation. LN-treated buds from hardy or acclimated woody lines may develop directly into a plant, but because of limited studies it cannot be stated that all species develop similarly.

An alternative to ice formation during low temperature exposure is the process of vitrification which refers to solidification of the system (glass formation) during exposure to low temperatures without actual ice crystallization. Use of cryoprotectant solutions and of cooling and warming rates that promote and stabilize glasses within cells are necessary and may prove to be a useful strategy to enhance cryopreservation, particularly in multicellular systems [132, 143, 144, 145, 146]. Vitrification could eliminate certain injury causing events during cooling/thawing and increase the allowable size of the propagule being treated. High concentrations of cryoprotectants are necessary and toxicity may be a problem. Vitrification is being studied actively in many animal systems, and is an intriguing and as yet unexplored method for cryogenic storage of plants.

The genetic stability of materials exposed to cryogenic protocols has been questioned. One issue is whether extremely long-term storage exposes samples to background levels of irradiation that could produce genetic change or lethality since DNA repair processes are not functioning at low temperatures [147, 148, 149]. There are few data in higher plant systems that provide answers. Projections from studies with animal cells exposed to high levels of radiation while cryogenically stored suggest that lethality would occur only after several thousand years with background radiation levels [38, 147, 148, 149]. Storage clearly is not indefinite at low temperatures, but the results suggest a few centuries of storage is realistic.

A second issue for hydrated materials which are processed with cryoprotectants is whether treatments occurring during the cryogenic protocol are mutagenic [150, 151]. Microbes are the experimental system of choice for examining mutagenesis, but no clear answer has emerged. The few studies performed suggest that mutation rates are not increased. A related process, freeze-drying, is mutagenic, but the mutagenicity is usually attributed to events during the drying phase and not to cooling and thawing *per se* [152, 153]. Other studies suggest that DNA is damaged during cryogenic exposure [150, 154]. The use of some cryoprotectants increases survival, and in these cases less DNA damage is observed. There are no comparable studies with cells or tissues from higher organisms; such information is needed.

The opinion, therefore, of most cryobiologists is that cryogenic storage is a safe method for retaining viability for many years. Cryogenic storage has been pragmatically used for microbes and for animal cell cultures, semen, and embryos. Storage times reported range from only a few months to a few years. The American Type Culture Collection is now storing plant callus from some species in LN [155]. Cryogenic storage of buds and shoot-tips from clonal lines requires additional research before practical storage can be initiated. Few laboratories, however, are engaged in such work [156]. As far as we are aware, no institute or program has routinely stored clonal germplasm under cryogenic conditions. Thus, no consensus or information is available on storage systems, retrieval, and testing procedures, and distribution, nor on the number of replicates required per accession and extent of collection duplication.

Future

Two other technologies may be useful in the future for supplementing the preservation of materials which are difficult to preserve with conventional techniques. These are the production and preservation of synthetic seeds and the preservation of nucleic acids.

The occurrence of somatic embryogenesis has not been reported as frequently as organogenesis in plant tissue cultures, but it has been observed in an increasing number of crop species [157, 158]. With improved techniques for induction and the subsequent encapsulation of somatic embryos with polymers, the generation of synthetic seeds has been possible [159]. Thus, clones could be preserved through the generation of synthetic seeds and their subsequent conventional low temperature storage. Several problems are apparent, among which are whether somatic embryogenesis could eventually be easily induced in a range of genoypes, whether somatic embryos are desiccation resistant, and whether the extent of somaclonal variation in synthetic seeds is significant. Since these are areas of active research, an increased understanding of the induction and control of somatic embryogenesis should be forthcoming. At that time the potential for use in germplasm preservation may be more easily assessed.

Even more speculative is the possibility of using nucleic acids, particularly DNA, for preservation. Preservation could be in the form of extracted DNA or in bacteria as genomic or cDNA libraries. The rapid increase in understanding of gene manipulation and function suggests that molecular approaches for improving crops for many different traits is forthcoming. Storage of nucleic acids will not substitute for preservation of the organism for several reasons, including inability to perform crosses in a conventional manner and inability to characterize or evaluate the donor for traits. It may be of benefit under certain situations. For example, DNA may be extracted from nonviable seeds and thus preserve at least some of the genes within the otherwise useless materials. Such recovery has been reported from older materials, but the DNA was extensively degraded [163, 164, 165]. Whether larger DNA fragments can be extracted from younger, but nonviable, tissues is unknown.

Another potential use for nucleic acid storage may be for conservation of diversity of little known wild species, many of which occur in the tropics. Since germplasm collections are not generally equipped to deal with the maintenance of large numbers of these species, the genetic material could at least be stored for later use with minimal maintenance. These samples could later be probed for specific, probably single gene, traits and thus some of the genetic information could be utilized. Of course, complete evaluation of the species would be impossible.

Nucleic acid storage, therefore, has the potential to supplement conventional techniques for the preservation of biological diversity. Such application needs careful thought and is dependent upon increased understanding of gene manipulation and the control of gene expression.

References

1. National Academy of Sciences. 1978. Conservation of Germplasm Resources: an Imperative. Committee on Germplasm Resources, Division of Biological Sciences, Washington, DC.
2. Frankel, O.H. and M. Soulé. 1981. Conservation and evolution. Cambridge University Press, Cambridge, England.
3. Wilkes, G. 1983. Current status of crop plant germplasm. CRC Critical Reviews of Plant Science 1: 131–181.
4. Frankel, O.H. and A.H.D. Brown. 1984. Plant genetic resources today: a critical appraisal, pp. 249–257. In: J.H.W. Holden and J.T. Williams (eds.). Crop Genetic Resources: Conservation and Evaluation. George Allen and Unwin, London.
5. Holden, J.H.W. and J.T. Williams (eds.). 1984. Crop Genetic Resources. Conservation and Evaluation. George Allen and Unwin, London.
6. Oldfield, M.L. 1984. The Value of Conserving Genetic Resources. U.S. Dept. Interior, Washington, DC.

7. Cromarty, A.S., R.H. Ellis and E.H. Roberts. 1982. The Design of Seed Storage Facilities for genetic Conservation. Intern. Board for Plant Genetic Resources. IBPGR Secretariat, Rome.
8. Chin, H.F. and E.H. Roberts. 1980. Recalcitrant crop seeds. Tropical Press SDN. BHD. Kuala Lumpur, Malaysia.
9. Bewley, J.D. and M. Black 1982. Physiology and Biochemistry of Seeds in Relation to Germination. Springer-Verlag, New York.
10. Priestley, D.A. 1986. Seed Aging: Implications for Seed Storage and Persistence in the Soil. Cornell University Press, Ithaca, New York.
11. Roos, E.E. 1986. Precepts of successful seed storage, pp. 1–25. In: M.B. McDonald, Jr. and C.J. Nelson (eds.). Physiology of Seed Deterioration. Spec. Pub. 11, Crop Science Society of America, Madison, Wisconsin.
12. Toole, V.K. 1986. Ancient seeds: Seed longevity. J. Seed Technol. 10: 1–10.
13. Barton, L.V. 1961. Seed Preservation and Longevity. Interscience, New York.
14. Biasutti Owen, E. 1956. The Storage of Seeds for Maintenance of Viability. Bull.43, Commonwealth Bureau of Pastures and Field Crops. Commonwealth Agricultural Bureaux, Farnham Royal, Bucks, England.
15. Crocker, W. 1938. Life-span of seeds. Bot. Rev. 4: 235–274.
16. Harrington, J.F. 1972. Seed storage and longevity, pp. 145–245. In: T.T. Kozlowski (ed.). Seed Biology, Vol. 3. Academic Press, New York.
17. Justice, O.L. and L.N. Bass. 1978. Principles and Practices of Seed Storage. USDA Handb. 506. U.S. Government Printing Office, Washington, DC.
18. Roberts, E.H. 1972. Viability of Seeds. Chapman and Hall, London.
19. Ewart, A.J. 1908. On the longevity of seeds. Proc. Roy. Soc. Victoria 21: 1–210.
20. Bass, L.N. 1981. Storage conditions for maintaining seed quality, pp. 239–321. In: E.E. Finney, Jr. (ed.). Handbook of Transportation and Marketing in Agriculture, Vol. 2. Field Crops. CRC Press, Boca Raton, Florida.
21. Bass, L.N. 1980. Seed viability during long-term storage, pp. 117–141. In: J. Janick (ed.). Horticultural Reviews. Avi Publi., Westport. Connecticut.
22. Roberts, E.H. 1983. Loss of seed viability during storage, pp. 9–34. In: J.R. Thompson (ed.). Advances in Research and Technology of Seeds, Part 8. Pudoc, Wageningen.
23. Roberts, E.H. and R.H. Ellis. 1984. The implications of the deterioration of orthodox seeds during storage for genetic resources conservation, pp. 18–37. In: J.H.W. Holden and J.T. Williams (eds.). Crop Genetic Resources: Conservation and Evaluation. George Allen and Unwin, London.
24. Smith, R.D. 1985. Maintaining viability during collecting expeditions. Appendix III, pp. 13–22. In: IBPGR Advisory Committee on Seed Storage: Report of the Third Meeting. IBPGR Secretariat, Rome.
25. Woodstock, L.W. 1975. Freeze-drying as an alternative method for lowering seed moisture. Proc. Assoc. Offic. Seed Anal. 65: 159–163.
26. Woodstock, L.W., J. Simkin and E. Schroeder. 1976. Freeze-drying to improve seed storability. Seed Sci. Technol. 4: 301–311.
27. Woodstock, L.W., S. Maxon, K. Paul and L. Bass. 1983. Use of freeze-drying and acetone impregnation with natural and synthetic anti-oxidants to improve storability of onion, pepper, and parsley seeds. J. Amer. Soc. Hort. Sci. 108: 692–696.
28. Chang, T.T. 1984. The role and experience of an international crop-specific genetic resources center, pp. 35–45. In: Conservation of Crop Germplasm: an International Perspective. Spec. Publ. 8, Crop Science Society of America, Madison, Wisconsin.
29. Barton, L.V. 1967. Bibliography of Seeds. Columbia University Press, New York.
30. Barton, L.V. 1960. Storage of seeds of *Lobelia cardinalis* L. Contrib. Boyce Thompson Inst. 23: 267–268.

31. Rincker, C.M. 1980. Effect of long-term subfreezing storage of seed on legume forage production. Crop Sci. 20: 574–577.
32. Rincker, C.M. 1981. Long-term subfreezing storage of forage crop seeds. Crop Sci. 21: 424–427.
33. Rincker, C.M. 1983. Germination of forage crop seeds after 20 years of subfreezing storage. Crop. Sci. 23: 229–231.
34. Rincker, C.M. and J.D. Maguire. 1979. Effect of seed storage on germination and forage production of seven grass cultivars. Crop Sci. 19: 857–860.
35. Roos, E.E., P.C. Stanwood and C.M. Rincker. 1986. Long-term sub-freezing storage of forage seeds – a second look. Agron. Abstr. pp. 128–129.
36. Roos, E.E., P.C. Stanwood and C.M. Rincker. 1987. Effect of year and location of production on long-term storage of forage grass and legume seeds. Agron. Abstr. pp. 130–131.
37. Mazur, P. 1966. Physical and chemical basis of drying in single celled micro-organisms subjected to freezing and thawing, pp. 214–315. In: H.T. Meryman (ed.). Cryobiology. Academic Press, New York.
38. Ashwood-Smith, M. and E. Grant. 1977. Genetic stability in cellular systems stored in the frozen state, pp. 251–272. In: K. Elliot and J. Whelan (eds.). The Freezing of Mammalian Embryos. Elsevier, Amsterdam.
39. Roberts, E.H. 1975. Problems of long-term storage of seed and pollen for genetic resources conservation, pp. 269–296. In: O.H. Frankel and J.G. Hawkes (eds.). Crop Genetic Resources for Today and Tomorrow. Cambridge University Press, Cambridge.
40. Stanwood, P.C. and L.N. Bass. 1981. Seed germplasm preservation using liquid nitrogen. Seed Sci. Technol. 9: 423–437.
41. Stanwood, P.C. 1985. Cryopreservation of seed germplasm for genetic conservation, pp. 199–226. In: K.K. Kartha (ed.). Cryopreservation of Plant Cells and Organs. CRC Press, Boca Raton, FL.
42. Roberts, E.H. 1973. Predicting the storage life of seeds. Seed Sci. Technol. 1: 499–514.
43. King, M.W. and E.H. Roberts. 1980. Maintenance of recalcitrant seeds in storage, pp. 53–89. In: H.F. Chin and E.H. Roberts (eds.). Recalcitrant Crop Seeds. Tropical Press SDN. BHD. Kuala Lumpur, Malaysia.
44. King, M.W. and E.H. Roberts. 1979. The Storage of Recalcitrant Seeds – Achievements and Possible Approaches. IBPGR Report Secretariat, Rome.
45. Roberts, E.H., M.W. King and R.H. Ellis. 1984. Recalcitrant seeds: Their recognition and storage, pp. 38–52. In: J.H.W. Holden and J.T. Williams (eds.). Crop Genetic resources: Conservation and evaluation. George Allen and Unwin, London.
46. Roberts, E.H. and M.W. King. 1981. Storage of recalcitrant seeds, pp. 39–48. In: L.A. Withers and J.T. Williams (eds.). Crop Genetic Resources, the Conservation of Difficult Material. IUBS Series B 42.
47. Grout, B.W.W., K. Shelton and H.W. Pritchard. 1984. Orthodox behavior of oil palm seed and cryopreservation of the excised embryo for genetic conservation. Ann. Bot. 52: 381–384.
48. Association of Official Seed Analysts. 1984. Rules for testing seeds. J. Seed Technol. 6: 1–126.
49. International Seed Testing Association. 1985. International rules for seed testing. Seed Sci. Technol. 13: 300–513.
50. Harrington, J.F. 1963. Practical advice and instructions on seed storage. Proc. Intl. Seed Test. Assoc. 28: 989–994.
51. Delouche, J.C. 1965. An accelerated aging technique for predicting relative storability of crimson clover and tall fescue seed lots. Agron. Abstr. p. 40.

52. Delouche, J.C. and C.C. Baskin. 1973. Accelerated aging techniques for predicting the relative storability of seed lots. Seed Sci. Technol. 1: 427–452.
53. Association of Official Seed Analysts. 1983. Seed Vigor testing handbook. Contribution no. 32. Association of Official Seed Analysts, Boise, Idaho.
54. Murata, M., E.E. Roos and T. Tsuchiya. 1981. Chromosome damage induced by artificial seed aging in barley. I. Germinability and frequency of aberrant anaphases at first mitosis. Can. J. Genet. Cytol. 23: 267–280.
55. Murata, M., E.E. Roos and T. Tsuchiya. 1980. Mitotic delay in root tips of peas induced by artificial seed aging. Bot. Gaz. 141: 19–23.
56. Roos, E.E. and C.M. Rincker. 1982. Genetic stability in 'Pennlate' orchardgrass seed following artificial aging. Crop Sci. 22: 611–613.
57. Moore, F.D., III and E.E. Roos. 1982. Determining differences in viability loss rates during seed storage. Seed Sci. Technol. 10: 283–300.
58. Moore, F.D., III, A.E. McSay and E.E. Roos. 1983. Probit analysis: A Computer Program for Evaluation of Seed Germination and Viability Loss Rate. Colorado State Univ. Exp. Stn. Tech. Bull. 147, 7 p.
59. Priestley, D.A., V.I. Cullinan and J. Wolfe. 1985. Differences in seed longevity at the species level. Plant Cell Environ. 8: 577–562.
60. Ellis, R.H. and E.H.Roberts. 1981. The quantification of ageing and survival in orthodox seeds. Seed Sci. Technol. 9: 373–409.
61. Roberts, E.H. 1986. Quantifying seed deterioration. pp. 101–123. In: M.B. McDonald, JR. and C.J. Nelson (eds.). Physiology of Seed Deterioration. Spec. Pub. 11, Crop Science Society of America, Madison, Wisconsin.
62. Heydecker, W. 1972. Vigour. pp. 209–252. In: E.H. Roberts (ed.) Viability of Seeds. Chapman and Hall, London.
63. Pollock, B.M. and E.E. Roos. 1972. Seed and seedling vigor, pp. 314–388. In: T.T. Kozlowski (Ed.). Seed Biology, Vol 1. Academic Press, New York.
64. Woodstock, L.W. 1973. Physiological and biochemical tests for seed vigor. Seed Sci. Technol. 1: 127–157.
65. Ching, T.M. 1973. Biochemical aspects of seed vigor. Seed Sci. Technol. 1: 73–88.
66. McDonald, M.B., Jr. 1975. A Review and evaluation of seed vigor tests. Proc. Assoc. Offic. Seed Analysts 65: 109–139.
67. Perry, D.A. (Ed.). 1981. Handbook of Vigor Test Methods. Intern. Seed Test. Assoc., Zurich.
68. McDonald, M.B., Jr. 1977. The influence of seed moisure on the accelerated aging vigor test. J. Seed Technol. 2(1): 18–28.
69. Tao, K-L.J. 1979. An evaluation of alternative methods of accelerated aging seed vigor test for soybeans. J. Seed Technol. 3(2): 30–40.
70. Matthews, S. 1980. Controlled deterioration: A new vigour test for seeds, pp. 647–660. In: P.D. Hebblethwaite (ed.). Seed Production. Butterworths, Boston.
71. Rice, W.N. 1960. Development of the cold test for seed evaluation. Proc. Assoc. Offic. Seed Analysts 50: 118–123.
72. Matthews, S. and W.T. Bradnock. 1967. The detection of seed samples of wrinkled-seeded peas (*Pisum sativum* L.) of potentially low planting value. Proc. Intl. Seed Test. Assoc. 32: 553–563.
73. Matthews, S. and W.T. Bradnock. 1968. Relationship between seed exudation and field emergence in peas and French beans. Hortic. Res. 8: 89–93.
74. Delfini, M., R. Angelini, F. Bruno, F. Conti, M.E. Di Cocco, A.M. Giuliani and F. Manes. 1985. Phosphate metabolites in germinating lettuce seeds observed by ^{31}P NMR spectroscopy. Cellular Molecular Biol. 31: 385–389.

75. Benson, E.E. and L.A. Withers. 1987. Gas chromatographic analysis of volatile hydrocarbon production by cryopreserved plant tissue cultures: A non-destructive method for assessing stability. Cryo-Letters 8: 35–46.
76. Bass, L.N. 1984. Germplasm preservation, pp. 55–67. In: Conservation of Crop Germplasm – an International Perspective. Spec. Pub. 8, Crop Science Society of America, Madison, Wisconsin.
77. Ellis, R.H., T.D. Hong and E.H. Roberts. 1985a. Handbook of Seed Technology for Genebanks. Vol. I. Principles and Methodology. Handbooks for Genebanks: No. 2. International Board for Plant Genetic Resources, IBPGR Secretariat, Rome.
78. Ellis, R.H., Hong, T.D. and Roberts, E.H. 1985b. Handbook of Seed Technology for Genebanks. Vol.II. Compendium of Specific Germination Information and Test Recommendations. Handbooks for genebanks 3. International Board for Plant Genetic Resources, IBPGR Secretariat, Rome.
79. Ellis, R.H. and Whitehead, J. 1987. Open, truncated and triangular sequential seed testing procedures. Seed Sci. Technol. 15: 1–17.
80. Roos, E.E. 1982. Induced genetic changes in seed germplasm during storage, pp. 409–434. In: A.A. Khan (ed.). The Physiology and Biochemistry of Seed Development, Dormancy and Germination. Elsevier Biomedical Press, New York.
81. Oka, H.I. 1975. Maintenance and storage 9: Considerations on the population size necessary for conservation of crop germplasm, pp. 57–63. In T. Matsuo (ed.). JIBP Synthesis 5, Gene Conservation. University of Tokyo Press.
82. Roos, E.E. 1984. Genetic shifts in mixed bean populations. I. Storage effects. Crop Sci. 24: 240–244.
83. Roos, E.E. 1984. Genetic shifts in mixed bean populations. II. Effects of regeneration. Crop Sci. 24: 711–715.
84. de Langhe, E.A.L. 1984. The role of *in vitro* techniques in germplasm conservation, pp. 131–137. In: J.H.W. Holden and J.T. Williams (eds.). Crop Genetic Resources: Conservation and Evaluation. George Allen and Unwin, London.
85. Towill, L.E. 1988. Genetic considerations for germplasm preservation of clonal materials. HortSci. 23: 91–97.
86. Withers, L.A. 1984. Germplasm conservation in vitro: present state of research and its application, pp. 138–157. In: J.H.W. Holden and J.T. Williams (eds.). Crop Genetic Resources: Conservation and Evaluation. George Allen and Unwin, London.
87. Withers, L.A. and J.T. Williams. 1986. *In vitro* conservation. IBPGR, Rome.
88. Bayliss, M.W. 1980. Chromosomal variation in plant tissues in culture. Intl. Rev. Cytol Suppl. 11A: 113–144.
89. D'Amato, F. 1985. Cytogenetics of plant cell and tissue cultures and their regenerates. CRC Crit. Rev. Plant Sci. 3: 73–112.
90. Styer, D.J. and C.K. Chin. 1983. Meristem and shoot-tip culture for propagation, pathogen elimination and germplasm preservation, Hort. Rev. 5: 221–227.
91. Hu, C.Y. and P.J. Wang. 1983. Meristem, shoot tip and bud culture, pp. 177–227. In: D.A. Evans, W.R. Sharp, P.V. Ammirato, and Y. Yamada (eds.). Handbook of Plant Cell Culture. Vol. 1. Techniques for Propagation and Breeding. Macmillan, New York.
92. Aitken-Christie, J. and A.P. Singh. 1987. Cold storage of tissue cultures, pp. 285–304. In: J.M. Bonga and D.J. Durzan (eds.). Cell and Tissue Culture in Forestry, vol. 2. Martinus Nijhoff, Dordrecht.
93. El-Gizawy, A.M. and Ford-Lloyd, B.V. 1987. An *in vitro* method for the conservation and storage of garlic (*Allium sativum*) germplasm. Plant Cell, Tiss. and Org. Cult. 9: 147–150.
94. Druarte, Ph. 1985. *In vitro* germplasm preservation technique for fruit trees,

pp. 167–171. In: A. Schafer-Menuhr (ed.). *In vitro* Techniques: Propagation and Long Term Storage. Martinus Nijhoff/Dr. W. Junk Publ. Boston.

95. Monette, P.L. 1987. Organogenesis and plantlet regeneration following *in vitro* cold storage of kiwifruit shoot tip cultures. Sci. Hort. 31: 101–106.
96. Wanas, W.H., J.A. Callow and L.A. Withers. 1986. Growth limitations for the conservation of pear genotypes, pp. 285–290. In: L.A. Withers and P.G. Alderson (eds.). Plant tissue culture and its Agricultural Applications. Butterworths, London.
97. Mix, G. 1985. Preservation of old potato varieties, pp. 149–153. In: A. Schafer-Menuhr (ed.). *In vitro* techniques. Propagation and Long Term Storage. Martinus Nijhoff, Dordrecht.
98. Preil, W. and M. Hoffmann. 1985. *In vitro* storage in Chrysanthemum breeding and propagation, pp. 161–165. In: A. Schafer-Menuhr (ed.). *In vitro* Techniques: Propagation and Long Term Storage. Martinus Nijhoff/Dr W. Junk, Dordrecht.
99. Stushnoff, C. and C. Fear. 1985. The potential use of *in vitro* storage for temperate fruit germplasm: A status report. IBPGR. 21 p. Rome.
100. Ziv, M. 1986. *In vitro* hardening and acclimatization of tissue culture plants, pp. 187–196. In: L.A. Withers and P.G. Alderson (eds.). Plant Tissue Culture and its Agricultural Applications. Butterworths, London.
101. Bonga, J.M. 1987. Clonal propagation of mature trees: problems and possible solutions, pp. 249–271. In: J.M. Bonga and D.J. Durzan (eds.). Cell and Tissue Culture in Forestry, Vol. 1, General Principles and Biotechnology. Martinus Nijhoff, Dordrecht.
102. Hutchinson, J.F. and R.H. Zimmerman. 1987. Tissue culture of temperate fruit and nut trees. Hort. Rev. 9: 273–349.
103. Kevers, C.M., M. Coumans, M.-F. Coumans-Gilles and Th. Gaspar. 1984. Physiological and biochemical events leading to vitrification of plants cultured *in vitro*. Physiol. Plantarum 61: 69–74.
104. Paques, M. and Boxus, Ph. 1987. 'Vitrification': review of the literature. Acta Hort. 212: 155–166.
105. Wheelans, K. and Withers, L.A. 1984. The IBPGR international database on *in vitro* conservation. IBPGR Plant Genetic Resources – Newsletter 60: 33–38.
106. van Sloten, D.H. and Holle, M. 1988. Temperate fruit genetic resources and the IBPGR. HortScience 23: 73–74.
107. International Board for Plant Genetic Resources. 1986. IBPGR Advisory Committee on *in vitro* Storage. Design, Planning and Operation of in vitro Genebanks. IBPGR, Rome.
108. International Board for Plant Genetic Resources. 1988. IBPGR Advisory Committee on *in vitro* Storage. Conservation and Movement of Vegetatively Propagated Germplasm: *in vitro* Culture and Disease Aspects. IBPGR, Rome.
109. Chavez, R., W.M. Roca and J.T. Williams. 1987. IBPGR-CIAT collaborative project on a pilot *in vitro* genebank. FAO/IBPGR Plant Genetic Resources Newsletter 71: 11–13.
110. International Board for Plant Genetic Resources. 1985. Long-term Seed Storage of Major Temperate Fruits. IBPGR, Rome.
111. Schoenike, R.E. and C.F. Bey. 1981. Conserving genes through pollen storage, pp. 72–73. In: E.D. Franklin (ed.). Pollen Management Handbook, USDA Handbook 587, Washington, DC.
112. Towill, L.E. 1985. Low temperature and freeze-/vacuum drying preservation of pollen, pp. 171–198. In: K.K. Kartha (ed.). Cryopreservation of Plant Cells and Organs. CRC Press, Boca Raton, Florida.
113. Akihama, T. and M. Omura. 1986. Preservation of fruit tree pollen, pp. 101–112. In: Y.P.S. Bajaj (ed.). Biotechnology in Agriculture and Forestry, 1. Trees I. Springler-Verlag, Berlin.

114. Parfitt, D.E. and A.A. Almehdi. 1983. Cryogenic storage of grape pollen. Am. J. Enol. Vitic. 34: 227–228.
115. Parfitt, D.E. and A.A. Almehdi. 1984a. Cryogenic storage of olive pollen. Fruits Var. J. 38: 14–16.
116. Parfitt, D.E. and A.A. Almehdi. 1984b. Liquid nitrogen storage of pollen from five cultivated *Prunus* species. HortSci. 19: 69–70.
117. Sedgley, M. 1981. Storage of avocado pollen. Euphytica 30: 595–599.
118. Vithanange, H.I.M.V. and D. Alexander. 1985. Synchronous flowering and pollen storage techniques as aids to artificial hydridization in pistachio (*Pistacia* spp.). J. Hort. Sci. 60: 107–113.
119. Juza, J.G. and V.S. Polito. 1988. Cryopreservation of English walnut (*Juglans regia* L.) pollen. Euphytica 37: 141–148.
120. Barnabas, B. and E. Rajki. 1981. Fertility of deep-frozen maize (*Zea mays* L.) pollen. Ann Bot. 48: 861–864.
121. Hoekstra, F.A. 1986. Water content in relation to stress in pollen, pp. 102–122. In: A.C. Leopold (ed.). Membranes, Metabolism and Dry Organisms. Cornell University Press, Ithaca, New York.
122. Kerhoas, C., G. Gay and C. Dumas. 1987. A multidisciplinary approach to the study of the plasma membrane of *Zea mays* pollen during controlled dehydration. Planta 171: 1–10.
123. Hanna, W.W., G.W. Burton and W.G. Monson. 1986. Longterm storage of pearl millet pollen. J. Hered. 77: 361–362.
124. Hecker, R.J. P.C. Stanwood and C.A. Soulis. 1986. Storage of sugarbeet pollen. Euphytica 35: 777–783.
125. Sakai, A. 1984. Cryopreservation of apical meristems. Hort. Rev. 6: 357–372.
126. Kartha, K.K. 1985a. Meristem culture and germplasm preservation, pp. 115–134. In: K.K. Kartha (ed.). Cryopreservation of Plant Cells and Organs. CRC Press, Boca Raton, Florida.
127. Sakai, A. 1986. Cryopreservation of germplasm of woody plants, pp. 113–129. In: Y.P.S. Bajaj (ed.). Biotechnology in Agriculture and Forestry, vol. 1: Trees I. Springer-Verlag, Berlin.
128. Chen, T. and K.K. Kartha. 1987. Cryopreservation of woody species, pp. 305–319. In: J.M. Bonga and D.J. Durzan (eds.). Cell and Tissue Culture in Forestry, vol. 2. Martinus Nijhoff, Dordrecht.
129. Withers, L.A. 1987. The low temperature preservation of plant cell, tissue and organ cultures and seed for genetic conservation and improved agricultural practices, pp. 389–409. In: B.W.W. Grout and G.J. Morris (eds). The Effects of Low Temperatures on Biological Systems. Arnold Press. London.
130. Sakai, A. and Y. Nishiyama. 1978. Cryopreservation of winter vegetative buds of hardy fruit trees in liquid nitrogen. HortSci. 13: 225–227.
131. Katano, M., A. Ishihara and A. Sakai. 1983. Survival of dormant apple shoot tips after immersion in liquid nitrogen. HortSci. 18: 707–708.
132. Hirsh, A.G., R.J. Williams and H.T. Meryman. 1985. A novel method of natural cryoprotection. Plant Physiol. 79: 41–56.
133. Hirsh, A.G. 1987. Vitrification in plants as a natural form of cryoprotection. Cryobiology 24: 214–228.
134. Stushnoff, C. 1987. Cryopreservation of apple genetic resources. Can J. Plant Sci. 67: 1151–1154.
135. Tyler, N., C. Stushnoff and L.V. Gusta. 1988. Freezing of water in dormant vegetative apple buds in relation to cryopreservation. Plant Physiol. 87:201–205.

136. Moriguchi, T., T. Akihama and I. Kozaki. 1985. Freezepreservation of dormant pear shoot apices. Jpn. J. Breed. 35: 196–199.
137. Finkle, B.J., M.E. Zavala and J.M. Ulrich. 1985. Cryoprotective compounds in the viable freezing of plant tissues, pp. 75–113. In: K.K. Kartha (ed.). Cryopreservation of Plant Cells and Organs. CRC Press, Boca Raton, Florida.
138. Caswell, K.L., N.J. Tyler and C. Stushnoff. 1986. Cold hardening of *in vitro* apple and Saskatoon shoot cultures. HortSci. 21: 1207–1209.
139. Reed, B.M. 1988. Cold acclimation as a method to improve survival of cryopreserved *Rubus* meristems. Cryoletters 9: 166–171.
140. Kartha, K.K. (ed.) 1985. Cryopreservation of Plant Cells and Organs. CRC Press, Boca Raton, Florida.
141. Towill, L.E. 1983. Improved survival after cryogenic exposure of shoot-tips derived from *in vitro* plant cultures of potato. Cryobiology 20: 567–573.
142. Haskins, R.H. and K.K. Kartha. 1980. Freeze preservation of pea meristems: Cell survival. Can J. Bot. 58: 833–840.
143. Fahy, G.M., D.R. MacFarlane, C.A. Angell and H.T. Meryman. 1984. Vitrification as an approach to cryopreservation. Cryobiology 21: 407–426.
144. Boutron, P. and P. Mehl. 1986. Nouveaux cryoprotecteurs pour la cryopreservation des cellules par vitrification totale. Bull. Soc. Bot. Fr. Actual Bot. 133: 27–39.
145. Fahy, G.M., D.I. Levy and S.E. Ali. 1987. Some emerging principles underlying the physical properties, biological actions and utility of vitrification solutions. Cryobiology 24: 196–213.
146. Rall, W.F. 1987. Factors affecting the survival of mouse embryos cryopreserved by vitrification. Cryobiology 24: 387–402.
147. Lyon, M.F., D.G. Whittingham and P. Glenister. 1977. Long-term storage of frozen mouse embryos under increased background irradiation, pp. 273–281. In: K. Elliot and J. Whelan (eds.). The Freezing of Mammalian Embryos. Elsevier, Amsterdam.
148. Ashwood-Smith, M.J. and G.B. Friedmann. 1979. Lethal and chromosomal effects of freezing, thawing, storage time and x-irradiation on mammalian cells preserved at $-196\,^{\circ}$C in dimethylsulfoxide. Cryobiology 16: 132–140.
149. Glenister, P.H., D.G. Whittingham and M.F. Lyon. 1984. Further studies on the effect of radiation during the storage of frozen 8-cell mouse embryos at $-196\,^{\circ}$C. J. Reprod. Fert. 70: 229–234.
150. Calcott, P.H. and A.M. Gargett. 1981. Mutagenicity of freezing and thawing. FEMS Microbiol. Lett. 10: 151–155.
151. Ashwood-Smith, M.J. 1985. Genetic damage is not produced by normal cryopreservation procedures involving either glycerol or DMSO: a cautionary note however on the possible effects of DMSO. Cryobiology 22: 427–433.
152. Ashwood-Smith, M.J. and E. Grant. 1976. Mutation induction in bacteria by freeze-drying. Cryobiology 13: 206–213.
153. Tanaka, Y., M. Yoh, Y. Takeda and T. Miwatane. 1979. Induction of mutation in *Escherichia coli* by freeze-drying. Appl. Envir. Microbiol. 37: 369–372.
154. Calcott, P.H. 1986. Cryopreservation of microorganisms. CRC Crit. Rev. in Biotech. 4: 279–297.
155. American Type Culture Collection. 1985. New plant tissue culture collection. American Type Culture Collection Quarterly Newsletter 5: 1.
156. International Board for Plant Genetic Resources. 1986. IBPGR Advisory Committee on *in vitro* Storage. Report of the third meeting. IBPGR, Rome.
157. Ammirato, P.V. 1983. Embryogenesis, pp. 82–123. In: D.A. Evans, W.R. Sharp, P.V. Ammirato, and Y. Yamada (eds.). Handbook of Plant Cell Culture. Vol. 1. Techniques for Propagation and Breeding. Macmillan, New York.

158. Tulecke, W. 1987. Somatic embryogenesis in woody perennials, pp. 61–91. In: J.M. Bonga and D.J. Durzan (eds.). Cell and Tissue Culture in Forestry. Vol. 2 Specific Principles and Methods: Growth and Development. Martinus Nijhoff, Dortdrecht.
159. Redenbaugh, K., D. Slade, P. Viss and J. Fujii. 1987. Encapsulation of somatic embryos in synthetic seed coats. Hort. Sci. 22: 803–809.
160. Gray, D.J. 1987. Quiescence in monocotyledonous and dicotyledonous somatic embryos induced by dehydration. HortSci. 22: 810–814.
161. Vasil, I.K. 1986. Relative genetic stability of embryogenic cultures of the Gramineae and uniformity of regenerated plants, pp. 108–116. In: J. Semal (ed.). Somaclonal Variations and Crop Improvement. Martinus Nijhoff, Dordrecht.
162. Maheswaran, G. and E.G. Williams. 1987. Uniformity of plants regenerated by direct somatic embryogenesis from zygotic embryos of *Trifolium repens*. Ann. Bot. 59: 93–98.
163. Rollo, F. 1987. Nucleic acids in mummified plant seeds: screening of twelve specimens by gel-electrophoresis, molecular hybridization and DNA cloning. Theor. Appl. Genet. 73: 501–506.
164. Rogers, S.O. and Bendich, A.J. 1985. Extraction of DNA from milligram amounts of fresh, herbarium and mummified tissues. Plant Mole. Biol. 5: 69–76.
165. Paabo, S. 1985. Molecular cloning of ancient mummy DNA. Nature 314: 644–645.

Plant germplasm acquisition and exchange

GEORGE A. WHITE[1] and JAMES A. BRIGGS[2]
[1] *Germplasm Services Laboratory, Plant Sciences Institute, Agricultural Research Service, U.S. Department of Agriculture, Beltsville, Maryland 20705, U.S.A.*
[2] *National Plant Materials Center, Soil Conservation Service, U.S. Department of Agriculture, Beltsville, Maryland 20705, U.S.A.*

Introduction

Plant germplasm is an essential element in crop production and improvement programs worldwide. In the U.S., this fact quickly became a stark reality for the first settlers to the east coast from Europe. Subsequently, government officials, recognizing a dearth of native crops, initiated modest efforts to introduce and test seeds and plants from abroad. In 1898, the U.S. Department of Agriculture established the Seed and Plant Introduction Section to provide for the worldwide acquisition of plant materials [1]. In the same year, new introductions were assigned Plant Introduction (PI) numbers and distributed to research stations and universities for testing. Data compilation for USDA Plant Inventories also started in 1898 and continues to the present. During the past 90 years, more than 518,000 PI numbers have been assigned and data have been compiled for 196 published inventories.

For more than four decades after the establishment of the Seed and Plant Introduction Section, limited facilities existed for maintenance of introduced materials. As a result, a high percentage of the early introductions were lost. This major problem persisted with few exceptions until the passage of legislation in the 1940s and 1950s that established the four Regional Plant Introduction Stations, the National Seed Storage Laboratory (NSSL), and the Interregional Potato Station.

In 1972, a reorganization within the Agricultural Research Service of the U.S. Department of Agriculture (ARS-USDA) gave momentum to the gradual consolidation of a national program for plant germplasm. The current program with its many activities, locations, and complexities is called the National Plant Germplasm System [2, 3].

L. Knutson and A.K. Stoner (eds.). Biotic diversity and germplasm preservation, global imperatives.
 ISBN 0-7923-0178-1

Flow channel for exchange of plant germplasm

The National Plant Germplasm System (NPGS) policy is free exchange with all countries of small experimental quantities for research purposes. This policy is enhanced through a plant germplasm flow channel for plant germplasm (Fig. 1). The Plant Introduction Office (PIO) in the Germplasm Services Laboratory, Beltsville Agricultural Research Center, is responsible for coordinating plant germplasm exchanges between the U.S. and other countries. Close liaison with officials of the Animal Plant Health Inspection Service (APHIS) of USDA is maintained so that the quarantine requirements of the importing country can be met. The flow channel basically consists of sending shipments through the USDA Plant Germplasm Quarantine Center, Beltsville, Maryland. Each shipment is examined for possible pests. Imported materials are distributed to germplasm curators and/or researchers, usually by the PIO, if no restrictions exist or if required permits are in order. If the importing countries' requirements can be met, a phytosanitary certificate is issued. If not met, the PIO requests import permits, waivers to additional declarations (ADs), or special permission to allow shipment to the intended recipient.

Plant materials for exchange with foreign countries are provided by working collections in the NPGS, individual scientists, cooperating agencies such as the Soil Conservation Service (SCS) and the Forest Service, or commercial companies. Because of the magnitude of U.S. agriculture and

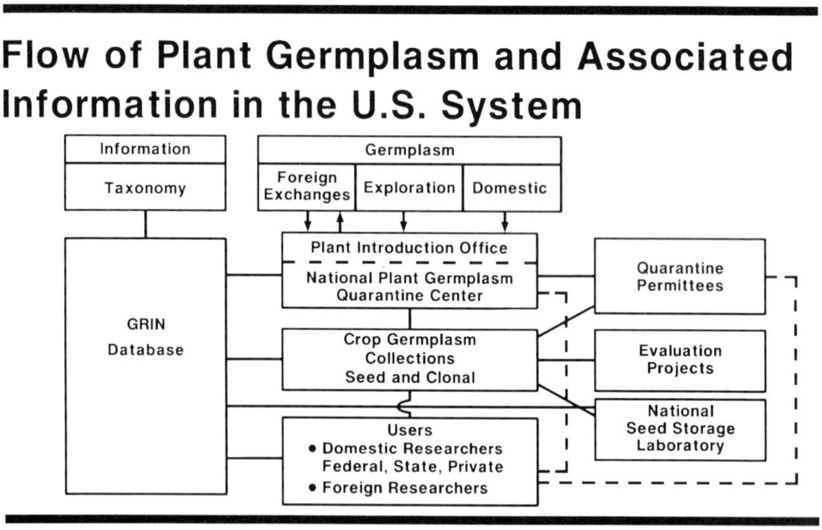

Figure 1. The primary flow channel in the U.S. for imports and exports of plant germplasm and associated information.

its research programs, there are many plant germplasm sources for a wide diversity of crops. Curators of germplasm collections are responsible for the proper maintenance, evaluation, increase, distribution, deposition of seed samples into the NSSL, and for documentation of evaluation and inventory data. All data are maintained in the Germplasm Resources Information Network (GRIN) database, located at the Agricultural Research Center, Beltsville, Maryland.

Magnitude of germplasm exchanges

The U.S. has a long history of generosity with its holdings of plant germplasm. This history may in part reflect our dependence on other countries for germplasm from centers of genetic origin and diversity. Collections made years ago in other countries and maintained in U.S. collections are available to the original source country as well as other countries. U.S. cultivars are also available. Annual average distributions for 1983-1987 are summarized in Table 1.

Plant materials introduced from other countries may or may not be incorporated into the NPGS. Generally, more items are 'passed through' (mainly to individual scientists) than are assigned PI numbers and sent to germplasm collections (Table 2). The 10-year (1978-1987) average number of assigned PI numbers is 9,347 annually.

The SCS, in close cooperation with the NPGS, provides germplasm, especially accessions having forage and conservation traits, to other countries. During 1985-1987, seeds of approximately 800 accessions, including the major SCS cultivars, were sent through the NPGS to 23 countries.

Acquisition of plant germplasm

The various sources of plant germplasm are summarized in Table 3. Accurate and complete passport information, such as scientific, cultivar, and local names, locality and date of collection, collector names and numbers, and plant characteristics about each item should accompany the material. Foreign introductions should be identified by scientific names so that their quarantine status can be determined readily. Unidentifiable materials are considered to be in the 'prohibited category' (requires an official permit from the Animal Plant Health Inspection Service) of quarantine. Likewise, materials for overseas shipment are identified to determine the requirement of the importing country. A brief discussion of the primary sources of plant germplasm follows.

Table 2. Annual acquisition of plant germplasm through exchanges and other sources by the U.S., including passport data-documented and undocumented accessions.

Year	Items without PI numbers[a]	No. of PIs	Origin of PIs in %		
			Foreign	U.S.	Unknown
1983	14,082	7,263	96.3	3.6	–
1984	16,597	11,030	97.0	2.7	0.3
1985	7,878	6,020	88.6	9.1	2.4
1986	12,954	6,069	83.3	15.0	1.7
1987	4,421	8,056	82.5	5.2	12.3
Ave.	11,006	7,688	90.4	6.3	3.3

[a] Foreign introductions. Many of these accessions will be assigned PI numbers later and incorporated into NPGS.

and local names, locality and date of collection, collector names and numbers, and plant characteristics about each item should accompany the material. Foreign introductions should be identified by scientific names so that their quarantine status can be determined readily. Unidentifiable materials are considered to be in the 'prohibited category' (requires an official permit from the Animal Plant Health Indspection Service) of quarantine. Likewise, materials for overseas shipment are identified to determine the

Table 3. Primary sources of plant germplasm for the National Plant Germplasm System.

Source	Comments
Exchange	Most important source over an extended period of time.
Exploration	Foreign and domestic; requires extensive preparation and cooperation.
Cooperation with U.S. agencies	Soil Conservation Service assembles collections and exchanges materials with other countries.
Agreements	Bilateral agreements between U.S. and other countries. IBPGR base collection allocations. Crop Science Registration program.
Special projects	Latin American Maize Project. Plant and Seed Materials Project. Other AID projects.
Breeding programs	Breeding & genetic materials with known traits.
Other significant sources	Travelers, Crop hobbyists.

Evaluation and inventory data are added to the GRIN database by the specialists at the working collections. Taxonomic and geographic data are also maintained on the database to provide current and accurate information.

Exchanges. The two-way movement of plant germplasm provides a permanent access channel between countries. The PIO provides national coordination for the NPGS and U.S. scientific exchanges. In the case of highly perishable materials such as plants, cuttings, tubers, etc., direct shipment from the U.S. source to the intended recipient is recommended when the import requirements can be met. The U.S. annually exchanges plant germplasm with 120 or more countries (Table 1).

Explorations. Because of the limited number of native crop species, the U.S. initiated plant germplasm explorations. Most early collections, however, have been lost because proper maintenance facilities did not exist. The first official exploration was conducted by D.J. Brown who collected seeds in Europe during 1854-55 [4]. Efforts up to 1898 resulted in the introduction of the navel orange, flax, olive, persimmon, sorghum, wheats and other cereals, and especially fruit varieties from Europe. Since 1898, plant-collecting expeditions supported by the USDA-ARS, state universities, botanic gardens, special grants, and international groups, especially the International Board for Plant Genetic Resources (IBPGR), have covered much of the world.

A systematic protocol exists within the NPGS for handling foreign and domestic exploration proposals. Proposals must be submitted by the lead scientist to the Coordinators of the Regional Plant Introduction Stations and must have Crop Advisory Committee (CAC) approval. The Coordinator arranges for reviews by Regional Technical Committees. Approved proposals by the Regional Committees are reviewed by the Plant Germplasm Operations Committee (PGOC) and ranked based on their national urgency and available funds. Finally, National Research Program Leaders review and recommend approval by the ARS administrator.

The Germplasm Services Laboratory, Plant Sciences Institute, USDA-ARS, Beltsville, assists collectors in obtaining permits, meeting quarantine requirements, identifying contacts, documenting data, and distributing the collected materials.

The CACs study germplasm gaps, review sources of germplasm, establish acquisition and exploration priorities, and determine crop genetic vulnerability. Such activities are directed toward the development of national priorities for plant exploration and germplasm acquisition.

Plant explorations require extensive planning and must address specific needs. Questions that should be answered before preparation of the proposal

Table 4. Representative collections of plant germplasm obtained through field exploration for incorporation into the National Plant Germplasm System.

Collector(s)	Year	Country	Crop	PI numbers
Doney, Whitney	1987	England, Scotland, Ireland	*Beta*	518298–518440
Jefferson	1986	Taiwan, Korea, Japan	*Prunus* (ornamental)	510764–510892
Seiler, Roath, Skoric	1985	NE United States[a]	*Helianthus*	503202–503291
Kaiser, Muehlbauer	1985	Turkey	*Cicer*, *Lens*	509117–509432
Percival, Schwendiman, Belot	1985	Caribbean Islands	*Gossypium*	501301–501480
Nuez	1984	Spain[b]	Vegetables	512339–513136
Rumbaugh, Taylor	1984	Romania	*Astragalus*, *Trifolium*	494615–494750
Rumbaugh, Graves	1983	Morocco	*Medicago*, *Trifolium*, other forages	516479–517193
Tarn, Gomez, Hjerting, Ross	1983	Mexico[c]	Tuber-bearing *Solanum* spp.	498011–498062
Whitaker, Provvidenti	1982	Turkey, Greece	*Lactuca*	490992–491263
Stewart	1981	Australia	*Gossypium*	464838–464863
Simpson, Pietrerelli, Zurita	1981	Peru[d]	Rust resistant *Arachis* (cultivated)	476143–476210
Dewey, Sunderman, Thompson, Orton	1980	People's Rep. of China	Cereals, vegetables, forages	447272–447422 478310–478397 499380–499564
Craddock, Barclay	1979	Chile	Cereals	436007–436526
Oakes, Brewbaker	1978	Mexico	*Leucaena*	443555–443721
Winters, Knight	1978	Brazil, Argentina, Paraguay	Oil palms, fruits, ornamentals	424618–424925
Fogg	1977	Taiwan	*Setaria italica*	433382–433481

Table 4. (Continued).

Collector(s)	Year	Country	Crop	PI numbers
Van der Zwet	1976	Czech., Romania, Yugoslavia	*Pyrus*	483379–483404
Oakes, Bashaw	1976	South Africa	Warm-season grasses	409104–410406
Voigt, El Lakany	1975 1977	Yemen Arab Republic	*Sorghum bicolor*	464942–465964 474582–475743 485721–486126 495917–496250 508105–508248
Remaily	1975	NE United States	*Vitis* spp.	483128–483190
Dewey, Plummer	1975	USSR	Forage grasses	439881–440784
White, Lessman, Meyer	1974	Mediterranean	*Crambe* spp. (new oilseed crop)	388700–388743 388753–388877

[a] Jointly sponsored by United States Department of Agriculture-Agricultural Research Service and International Board for Plant Genetic Resources.
[b] Sponsored by IBPGR.
[c] Cooperative efforts by Mexico, U.S., Canada, and Denmark.
[d] Financially supported by the U.S. Agency for International Development.

include: What germplasm is already available in U.S. programs? Can the needs be met by introducing germplasm from other countries through exchanges? Does the exploration fit national priorities? Can the objectives be met successfully?

Materials are processed in various ways depending on the nature of the materials, quarantine status, need for increase before documentation, etc. The crop curators are responsible for increases, evaluation, maintenance, deposition of seed into NSSL for long-term storage, distribution, and computerization of evaluation and inventory data. The PIO documents the passport data, assigns PI numbers, and distributes the materials to the proper curator.

Table 4 lists representative collections from different parts of the world.

Cooperation with U.S. agencies. SCS cooperates with the NPGS through its ambitious assemblage of domestic and foreign collections of plant materials. These genetically rich collections are sources of superior genotypes that are selected for use in soil and water conservation programs. While minimal efforts are made to preserve non-selected materials for further distribution, substantial numbers of accessions are normally available for some time after the initial screening and selection process.

Domestic field collections are important because SCS screens and identifies superior genotypes. Collection and evaluation information on approximately 25,000 domestic items is documented in the SCS national database. This number represents about 50% of all domestic collections since 1935. Accession documentation consists of plant morphological and environmental parameters and evaluation data. Field collection emphasis is placed on sampling the potential area of intended use as thoroughly as possible and then on rigidly evaluating materials for superior genotypes.

Foreign germplasm is important in soil conservation programs. Currently, 18,900 SCS accessions have NPGS PI numbers. An additional 14,900 accessions have been obtained directly by SCS through NPGS flow channels. Foreign accessioning and seed increases are coordinated by the National Plant Materials Center, Beltsville. The highly diverse collection consists of 2,572 species within 559 genera, and obtained from 137 countries. The SCS contributes plant germplasm to NPGS primarily in two ways:

1. All materials selected for advanced evaluation are assigned PI numbers. Seeds are deposited in appropriate working collections and in the NSSL for long-term storage. The NPGS assumes the maintenance responsibilities for these materials.
2. NPGS curators and other scientists can obtain accessions not included

above from the SCS for inclusion in their programs. As appropriate, these items can be added to the NPGS.

The results of acquisition and evaluation efforts by SCS programs have been spectacular. Since 1941, more than 275 cultivars have been developed and released for commercial use. Of these, 36% were derived from foreign introductions. Releases include grasses, legumes, forbs, and woody species. Intended uses of these and future releases are shown in Fig. 2. Cultivars developed through the intensive acquisition and testing process have had a major impact on programs on worldwide conservation and seed and plant trade programs.

Seeds are sold through both domestic and international markets. The total estimated retail value of seeds of SCS-developed cultivars in 1986 was $35 million. Acreage, production, and value of selected cultivars are listed in Table 5. Many of the SCS-developed cultivars are now major components of our environment. For example, the first seedings of the cultivar 'A-68' of Lehmann lovegrasses (*Eragrostis lehmanniana* Nees) were in the southwestern United States following introduction from South Africa in the period 1937-1940. Today 'A-68' is found on about 355,000 acres and pro-

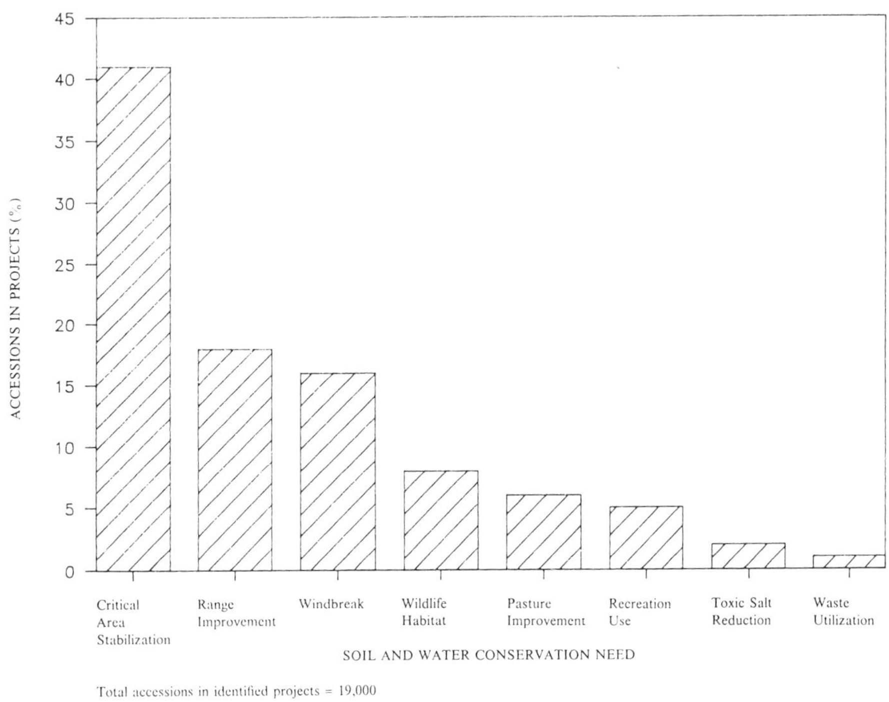

Figure 2. Identified soil and water conservation needs and current Soil Conservation Service plant development projects devoted to each area.

Table 5. Acreage and commercial value of selected cultivars developed by the Soil Conservation Service–USDA.

Cultivar	Scientific name	Source	Seed production (000s of pounds)	Retail sales value (000 dollars)	Land area protected (acres)
Cape	*Ammophila breviligulate* Fern.	Domestic	6,625[a]	918	56[b]
Cardinal	*Elaeagnus umbellata* Thunb.	Introduced	1,640[a]	317	1,355
Caucasian	*Bothriochloa bladhii* S. T. Blake	Introduced	4	84	2,000
Cochise	*Eragrostis lehmanniana* Nees × *E. trichophora* Cosson & Dur.	Introduced	30	97	30,000
Corto	*Atriplex semibaccata* R. Br.	Introduced	50	825	6,200
Critana	*Elytrigia dasystachya* (Hook.) A. Love & D. Love	Domestic	240	1,194	59,000
Lutana	*Astragalus cicer* L.	Introduced	289	1,402	29,000
Pennsacola	*Paspalum notatum* Fluegge	Domestic	1,160	1,061	290,000
Rosana	*Elymus smithii* (Rydb.) Gould	Domestic	115	548	16,400
Secar	*Elytrigia spicata* (Pursh) D. R. Dewey	Domestic	80	515	13,300
Selection 75	*Panicum coloratum* L.	Introduced	60	390	30,000
Streamco	*Salix purpurea* L.	Domestic	258[a]	19	78[b]
Zorro	*Vulpia myuros* (L.) C. Gmelin	Domestic	60	630	1,500

[a] Number of plants in thousands.
[b] Linear distance in miles.

vides 0.6–1.0 tons of forage per acre. In addition, commercial producers have made substantial plantings of 'A-68' in northern Mexico [5, 6]. The many cultivars of wheatgrasses (*Agropyron* spp.) and other genera provide a similar story within the high desert areas of the U.S. Currently, 107 grass accessions are in the advanced stages of cultivar development by SCS. Who knows but that their impact may be equal or greater than already made by *Eragrostis* and *Agropyron*.

Other government agencies such as the Office for International Cooperation and Development (OICD-USDA), Foreign Agriculture Service (FAS-USDA), and The Agency for International Development (AID) provide support to international projects that lead to exchanges, in some cases to field collections, and to contacts that contribute to exchanges.

Agreements. This source of plant germplasm has become increasingly important during the last 20 years. Some examples of agreements are:

1. Bilateral Agreements Between the U.S. and Other Countries – Such agreements include plant germplasm explorations, exchange of scientific teams, and exchange of specific germplasm. The U.S.-U.S.S.R. agreement has resulted in U.S. field explorations in the U.S.S.R. for forage species and soybean accession exchanges. One U.S.S.R. team in cooperation with U.S. scientists made a large collection of *Helianthus* throughout extensive areas of the U.S.
2. IBPGR Base Collection Allocations – The U.S. has a well-established facility, the National Seed Storage Laboratory (NSSL), for long-term storage of plant germplasm. IBPGR staff designate sites around the world for base storage and backup base storage. Upon request from the IBPGR, USDA-ARS agreed to provide base storage for the following crops: *Allium, Amaranthus, Citrus, Cucurbita, Cynodon, Solanum* (eggplant), *Leucaena, Abelmoschus* (okra), *Paspalum, Pennisetum, Phaseolus, Oryza sativa* L., *Sorghum, Glycine, Ipomoea* (sweet potato-seed), *Saccharum, Lycopersicon, Triticum, Vigna* (cowpea), and *Zornia*. The NPGS has proceeded to incorporate these collections into the working collections as well as depositing seed in the NSSL for base storage.
3. Crop Science Registration Program-Officials of the NPGS and the Crop Science Society have agreed to incorporate all cultivars, germplasm, parental lines, and genetic stocks registered by Crop Science into the NPGS [7]. During 1987, 323 items, consisting of 29 crop groups, were assigned PI numbers and distributed to NSSL and to the appropriate working collections. A special Crop Science Registration (CSR) designation is entered into the database, which allows easy retrieval of information about registered materials. PI numbers are assigned at the time manuscripts describing the materials to be registered are approved for

publication in Crop Science (Table 6). This agreement between the Crop Science Society and the NPGS allows the systematic inclusion of germplasm into the national program.

4. Other Sources of Plant Germplasm – A number of other important sources also contribute plant germplasm to the NPGS and exchange germplasm with other countries. Contributions by breeders and other researchers (Federal, State, private, and non-profit) are increasingly being formalized through releases and registrations. However, breeding materials from scientific programs that are threatened because of retirement, program changes, and death, should be carefully reviewed for possible inclusion in the NPGS.

The Latin American Maize Project (LAMP) was established to increase, evaluate, and provide long-term storage for maize germplasm collections. Funded by a private U.S. company, this program involves the active participation of Bolivia, Brazil, Chile, Colombia, Guatemala, Mexico, Paraguay, Peru, Uruguay, U.S.A., and Venezuela.

The Plant and Seed Materials project of the PIO that is funded by AID has resulted in the inclusion of sesame from Thailand and forage grasses from Morocco into the NPGS. AID funds supported extensive sorghum collecting in the Yemen Arab Republic (see Table 4). Travelers with broad

Table 6. Representation Plant Introduction Number data record for a *Crop Science*-registered germplasm of alfalfa (*Medicago*).

PI 506239. *Medicago sativa* L. FABACEAE Alfalfa.

Donated by Erwin, D. C., Dept. of Plant Pathology, Univ. of California, Riverside, California. United States. Received January 15, 1987.

PI 506239	Donor id.: A76–18. Origin: United States. Pedigree: Derived from 34 resistant plants. Other id.: GP-195. Group: CSR-Alfalfa. Remarks: Plant nondormant type. Special Attributes: Resistant to *Phytophthora* root rot. Breeding material. Seed. Lit. ref.: Crop Sci. 27:820. 1987.
Destination:	Western Regional PI Station, 59 Johnson Hall, Washington State Univ., Pullman, Washington 99164, United States. National Seed Storage Laboratory, Colorado State Univ., Ft. Collins, Colorado 80523, United States.
Records:	Erwin, D. C., Dept. of Plant Pathology, Univ. of California, Riverside, California 92521, United States.

interests in various plants contribute germplasm. Last, but by no means least, are the hobbyists. This enthusiastic and diverse group of individuals assembles (and tests) collections of fruit, heirloom beans, vegetables, etc. Frequently, they are important sources of low value crops or obsolete cultivars for which there is little or no research effort.

While the acquisition and exchange of plant germplasm are essential, without preservation they are incomplete. The following comments sum up the NPGS philosophy concerning plant germplasm preservation:

> 'We are fully committed to the conservation of biological diversity for the long term' says Dr. Nyle Brady of USAID.
>
> 'The earth's diversity of plants and animals and the genetic heritage they represent are humanity's endowment for future survival.'
>
> The Environment, 1987.

References

1. Hyland, H.L. 1977. History of U.S. Plant Introduction. Environ. Rev. J. 4: 26–33.
2. Parliman, B.J. and White, G.A. 1985. The plant introduction and quarantine system of the United States. Pp. 361–434 *in* Plant Breeding Reviews 3, AVI Publishing Co. Westport, Connecticut.
3. Wilkes, G. 1985. Current status of crop plant germplasm. Pp. 133-181 *in* CRC Critical Reviews in Plant Science, CRC Press Inc. Boca Raton, Florida.
4. Anonymous. 1971. The national program for conservation of crop germplasm. Univ. Printing Dept., Univ. Georgia, Athens. 73 pp.
5. Cox, J.R., Martin-R, M.H., Ibarra-F, F.A., Fourie, J.H. Rethman, N.F.G. and Wilcox, D.G. 1988. The influence of climate and soil on the distribution of four African grasses. J. Range Mgt. 41(2): 127–139.
6. Cox, J.R., Ruyle, G.B., Fourie, J.H. and Donaldson C. 1988. Lehmann lovegrass – Central South Africa and Arizona, USA. Rangeland 10(2): 53–55.
7. White, G.A., Eberhart, S.A., Miller, P.A. and Mowder, J.D. 1988. Plant materials registered by Crop Science incorporated into NPGS. Crop Sci. 28: (716–717).

Plant germplasm information systems

JIMMIE D. MOWDER[1] and ALLAN K. STONER[2]

[1]*Database Manager, Germplasm Resources Information Network*

[2]*Research Leader, Germplasm Services Laboratory, Agricultural Research Service-USDA, Beltsville, Maryland 20705, U.S.A.*

Abstract The ability of modern information systems to simplify the keeping of extensive, detailed records makes them invaluable for the effective management and use of plant germplasm collections. A feasability study conducted by the U.S. Department of Agriculture, Agricultural Research Service, determined that a centralized computer database should be developed to provide and store detailed information on collection, maintenance, distribution, and use of plant germplasm held by the U.S. National Plant Germplasm System (NPGS). As a result, the Germplasm Resources Information Network (GRIN) was created. GRIN, which is accessible to the entire scientific community, serves as a central repository for genetic resources information, provides an incentive for the standardization of crop descriptors and evaluation information, and provides a mechanism by which germplasm curators can manage inventories and distribution records. Taxonomic, passport, characteristic, and evaluation data on available germplasm can be reviewed and retrieved through the use of a menu-driven system. Plant scientists can use these data to determine the existence of germplasm that could be useful in their research programs. Requests for germplasm then can be created and forwarded to the appropriate germplasm maintenance site via GRIN. Although GRIN was designed to meet the specific requirements of the NPGS, portions of it could serve as a prototype for other, similar systems.

Key words: computer, crop descriptors, database

For many decades, considerable attention has been devoted by the scientific community to the collection and preservation of plant germplasm. Much less attention, however, has been given to the concurrent collection and preservation of information pertinent to the stored seeds or plant material. Today the realization is greater than ever before of the importance of accumulating and recording as much information as possible about the germplasm maintained.

Availability of data describing preserved germplasm aids administrators and collection curators in monitoring inventories and seed germination records, tracking requests for and shipping germplasm, assessing the genetic diversity represented in collections, and coordinating activities between germplasm collections. Scientists who use germplasm in plant-improvement programs or other types of research benefit by knowing where specific germplasm can be obtained and by having rapid access to detailed informa-

L. Knutson and A.K. Stoner (eds.). Biotic diversity and germplasm preservation, global imperatives.
 ISBN 0-7923-0178-1

tion about the origin of specific genotypes and of what traits they possess.

The need for an efficient information system increases as the size of germplasm collections increase and as more data are accumulated on accessions in the collections. To be maximally effective, an information system should:

1. Assist the curator in managing the collection,
2. Facilitate the use of diverse germplasm by providing users with rapid access to botanical, horticultural, and agronomic information about specific accessions [1],
3 Facilitate the acquisition of germplasm by collection or exchange, and
4. Enhance coordination of plant germplasm maintenance activities within and between countries.

Information on plant germplasm for single germplasm collections or maintenance sites can be managed by simple manual or computerized systems. At the other extreme, large numbers of collections or collection sites can be linked by a complex computerized network. Several information systems for plant germplasm collections are in use or in various stages of development around the world [2]. In most cases these systems serve small geographic areas or specific gene banks. Recent communication by one of us (JDM) with scientists from India, Canada, Mexico, and Australia indicate that these countries have developed or are in various stages of developing computerized information systems to maintain data on their national plant germplasm collections.

The International Board for Plant Genetic Resources (IBPGR) gathers data on existing germplasm collections around the world and publishes general information on the contents of these collections in its *Directories of Collections*. More detailed information on specific crops is contained in central databases that IBPGR supports through contracts with individual institutions. For example, the European Association for Research on Plant Breeding has received support for a European apple database, and the Office of Wine and Vine has been funded to initiate development of a global database on grape collections [3]. The IBPGR also has issued descriptor lists to aid in characterization and evaluation of most major crops. These lists help to ensure that attributes of germplasm stocks are described in internationally standardized ways.

In 1964, the International Council of Scientific Unions established the International Biological Programme to gain a better understanding of the environment and to facilitate improved management of the world's natural resources. At the conclusion of the Programme in 1974, a report identified the TAXIR computerized information system as a possible base for development of an information system for managing plant germplasm data [4].

Japan's Genetic Resource Information Management System for Center of Germplasm Seed Storage is a computerized plant germplasm information system containing some of these concepts and data structures.

In the U.S., germplasm of agriculturally important food and fiber crops and their relatives is acquired, maintained, evaluated, and distributed by the National Plant Germplasm System (NPGS) (See Shands, Fitzgerald, and Eberhart, this volume). The NPGS is a coordinated network of federal, state, and private institutions, agencies, and research units. Because of the national scope of the NPGS, the U.S. Department of Agriculture (USDA), Agricultural Research Service (ARS) has been designated to serve in the leadership role. The NPGS maintains approximately 350,000 accessions of germplasm as seed or vegetatively propagated stocks in working and base collections. These accessions represent nearly 170 plant families, 1,000 genera, and 6,000 species. New accessions are added to the system at a rate of 7,000–15,000 per year. The germplasm collection sites include four Regional Plant Introduction Stations (RPIS) at Griffin, Georgia; Ames, Iowa; Geneva, New York; and Pullman, Washington. Important commodity and special collections are located at Aberdeen, Idaho (small grains); Urbana, Illinois and Stoneville, Mississippi (soybeans); College Station, Texas (cotton); Fargo, North Dakota (flax); Oxford, North Carolina (tobacco); Logan, Utah (forage and range grasses); Tifton, Georgia (millet); Byron, Georgia and Mayaguez, Puerto Rico (bamboo); Sturgeon Bay, Wisconsin (potatoes); Griffin, Georgia; Stillwater, Oklahoma; and Stephensville, Texas (peanuts); Madison, Wisconsin (*Brassica* species); and Raleigh, North Carolina and Homestead, Florida (*Tripsacum*). Special genetic stock collections are located at Fort Collins, Colorado (barley); Urbana, Illinois (maize); Geneva, New York (pea); Davis, California (tomato); and Columbia, Missouri (wheat). Since 1980, clonal repositories have been established at Corvallis, Oregon; Davis, California; Geneva, New York; Miami, Florida/Mayaguez, Puerto Rico; Brownwood, Texas; Hilo, Hawaii; Orlando/Leesburg, Florida; and Riverside/Brawley, California. Finally, duplicate samples of material in working and commodity collections are held in a base collection at the National Seed Storage Laboratory (NSSL), Fort Collins, Colorado.

In the U.S., a list of introduced germplasm and associated descriptive information has been continuously published since 1896 as USDA plant inventories. The most recent inventory covers the period July-December 1986 [5]. Although the inventories contain much valuable information, they often are not readily available to scientists, they are not very timely, and they cannot be updated. During 1976 and 1977, the ARS conducted a study to examine the adequacy of the NPGS information management system. The study concluded that 'An information system exists within the plant genetics resources community of the U.S., but this system lacks the organization, communication techniques, trained personnel, and funding to satisfy the requirements of the NPGS community.' As a result of this finding, the ARS initiated the development of a comprehensive system that would:

1. Serve as a central repository for genetic resources information accessible by the entire scientific community,
2. Begin standardization of crop descriptors and evaluation information, and
3. Provide a mechanism by which collection sites can better manage their collections.

A cooperative agreement was established with the Laboratory for Information Science in Agriculture (LISA) at Boulder and later at Ft. Collins, Colorado, to develop a computer-based information system for the NPGS. An initial analysis of the needs of the NPGS community led to identification of two basic groups of information users – those who supply information and those who demand it – and to identification of the specific needs of the two groups. The supply group includes those who collect, maintain, evaluate, and distribute germplasm to researchers, whereas the demand group includes plant breeders and other scientists who use plant germplasm.

A detailed survey also was conducted of the overall structure and operating procedures of the NPGS. This survey included an analysis of the specific methods to be used system-wide and by individual operational components (i.e. NSSL, RPIS) to define, store, and retrieve data, as well as a determination of the type and quality of data to be stored. The results of the analysis made it clear that the computer software for the system would have to be infinitely more complex than was any commercially available product. Operational prototypes of a system were subsequently developed and tested at several NPGS collection sites. Evaluation of the prototypes led to development of a user-oriented design that enables persons unfamiliar with computer terminology and hardware to access easily all available data on plant germplasm.

The 5-year ARS/LISA Germplasm Resources Information Project (GRIP) resulted in the production of a functional information system for which the ARS assumed managerial responsibility in June 1983. The new system was named the Germplasm Resources Information Network and soon became known by the acronym 'GRIN'.

GRIN was designed around a central computer to optimize accessibility and operational speed for all users. A database 'network' concept was adopted to enhance information by reducing redundancy and by 'linking' together most of the pertinent information about a particular germplasm accession (i.e. from its native habitat to the most recently acquired characteristic and evaluation data). This centralization also allows scientists access to a more extensive collection of samples from which to choose, thereby reducing the possibility of their overlooking a potentially valuable germplasm accession.

The GRIN database contains three major types of data – passport/accession, evaluation/observation, and inventory. Passport data (sometimes referred to as accession data) include the basic information that accompanies the plant material at the time it first 'passes' into the National Plant Germplasm System. Passport data provide an historical record of the sample and rarely are changed once they have been entered into the system. They also provide general information that may be useful to a researcher in identifying and locating specific genetic material appropriate for a project. Passport data include the following:

- Appropriate plant identification numbers or identifiers and designations of the sites where a supply of the germplasm is available.
- Life form (annual, spreading shrub, vine, etc.) and the form in which the germplasm was received or exists (seed, tuber, plant, root, cell suspension).
- Common name and taxonomic classification (family, genus, species).
- Name of the original collector, when and where it was collected (country, town, latitude, longitude) and specific conditions of the collection site (hot climate, arid region, alkaline soil, etc.).
- Crop category (cereal, forage, fiber, vegetable, ornamental, etc.) and pedigree or information about its parentage.
- General references to any reported attributes (susceptibility or resistance to environmental stresses or conditions, etc.).

Evaluation or observational data categorize information about the growth characteristics of a specific accession, disease and insect resistance or susceptibility, quality of the fruits or grain, time required for a crop to mature, and other traits of the germplasm. Several observation records may be associated with a single accession of germplasm, each of which may describe growth characteristics under a variety of conditions.

Evaluation or observation records are available as individual items or as units in selective searches. Usually at least one observation record is provided when the plant material first enters the GRIN databank; however, some accessions lack an initial observation record or it is incomplete.

Observation data are recorded in a series of categories called Descriptors, which include such information as plant height, leaf size, petiole length, days to maturity, flower color, fruit quality, reactions to pathogens, etc. Information about the number and types of evaluation categories and the manner in which they are evaluated is available in the Catalog of the GRIN System [6].

Inventory data are created, maintained, and frequently updated by the curators of the germplasm collection sites. Although these data are part of the GRIN system, they are not available to public users except by contacting the curator of the site at which the plant material is stored.

Inventory data include the following: locally assigned identifiers; availability for distribution; quantity of propagules on hand; hundred-seed weight; replenishment and germination levels; germination history; and maintenance techniques.

The data in the GRIN are interconnected either by the taxonomic name or by a unique primary identifier that may be a Plant Introduction (PI) number or a maintenance-site assigned number. The germplasm collection sites are considered the 'owners' of the germplasm and the accompanying information about it. Thus personnel at the maintenance sites are responsible for ensuring that all data are entered in the GRIN in a timely and accurate manner. Maintenance of GRIN data is supported through automatic updates that are easily and quickly available to everyone. Much of the passport data associated with an accession typically is entered in the GRIN or supplied to the collection site by the Plant Introduction Office, Beltsville, Maryland. Correct scientific nomenclature is monitored and maintained by plant taxonomists. All evaluation data gathered at the collection sites or from ARS-supported projects must be entered into the GRIN database by collection-site personnel.

The entire GRIN database is backed up on magnetic tape three times a week and has a safeguard feature that records all transactions so that data are protected up to the moment of a system problem. This assures that data are protected even if a catastrophic event destroys the equipment or the data. System software (computer programs) protects a site's data-area in the database from accidental or malicious contamination.

Access to the GRIN and retrieval of information from the database can be achieved through telecommunications by publicly and privately employed scientists in the U.S., Mexico, and Canada. In addition, retrieval-only access to the GRIN is available to scientists affiliated with the 13 international agricultural centers under the aegis of the Consultative Group on International Agricultural Research (CGIAR). Scientists from any country may also contact the Database Managment Unit (DBMU), Germplasm Services Laboratory (GSL), Plant Sciences Institute, ARS/USDA, Beltsville, Maryland, to have the database queried and the results mailed to them. The DBMU has responsibility for the continued development and operation of the GRIN.

GRIN has been designed to meet the specific needs of the NPGS and is intended to store data only on germplasm maintained in that system; thus it is not intended to be an international system. However, portions of the GRIN can serve as a prototype or model for other national systems or for a worldwide plant germplasm information storage and retrieval system.

The input of Crop Advisory Committees (CAC's) was important to the implementation of the GRIN. The Committees are national working groups of specialists who provide analysis, data, and advice about germplasm within a specific crop or group of related crops of present or future economic importance. Input by CAC's to the GRIN includes the development of descriptors for specific crops and standards for evaluation of germplasm. The CAC's also provide a link to the 'user community' and in many instances to international organizations (i.e. IBPGR) concerned with the collection and preservation of plant germplasm.

The NPGS is embarking on the inclusion of several new technologies to assist plant scientists in recording and disseminating their research data. For example, the DBMU is experimenting with a video image digitization system. In cooperation with the NSSL at Ft. Collins, Colorado, and the North Central RPIS at Ames, Iowa, the DBMU is investigating the feasibility of exchanging digitized images of seeds. If successful, this procedure could help determine whether similarly named accessions at Ft. Collins and Ames are, in fact, identical. Video imaging also can be useful in defining descriptors and in describing evaluation results.

Bar coding is being used as a rapid and accurate mechanism for identifying seed packets, rows, plots, and individual accessions growing in greenhouses, screenhouses, orchards, and growth chambers. Researchers at the National Plant Germplasm Quarantine Center, Glenn Dale, Maryland, use bar coding to label plants and to track their location and progress through a lengthy disease-indexing procedure. Researchers at the National Small Grains Collection, Aberdeen, Idaho, use bar coding and electronic balances connected to a computer to identify and record the amount of seed available for approximately 110,000 accessions.

Digitized geographic and political boundary information now enables the production of maps of previous plant collection sites. Accurate outlines of a country or continent can be drawn, and the collection point of each accession for which the location is stored as latitude and longitude coordinates can be marked. The degrees and minutes of coordinates can be stored, allowing sites to be designated with an accuracy of within one mile. Maps such as these permit the user to view an entire region to check for areas where collections for a specific species have not been attempted.

Hand-held data entry devices are becoming popular for recording observational data. These devices can be carried easily in the field or greenhouse for data entry. The data then can be transferred easily (up-loaded) to a microcomputer or to the GRIN Prime[R] computer for permanent storage. The use of hand-held devices can reduce the time between data capture and its availability to the public and reduce potential errors by eliminating the key-entry step required if data are recorded in longhand on data sheets.

Summary

A critically important aspect of the management of genetic resources is concurrent management of information pertinent to preserved seeds or plant material. An effective information system should:

1. Aid managers and curators of collections by providing accurate information on inventories, seed requests, seed germination results, etc., and
2. Provide germplasm users with easy and rapid access to descriptive information about the preserved germplasm, thereby encouraging more effective use of plant germplasm.

Modern computer technology makes it possible to design an information management system to meet the specific needs of a country or organization. The Germplasm Resources Information Network has been designed specifically for the National Plant Germplasm System, but it can serve as a model for other plant germplasm information storage and retrieval systems.

References

1. Perry, M., A. K. Stoner, and J. D. Mowder. 1988. Plant Germplasm Information Management System: Germplasm Resources Information Network. HortScience 23(1): 57–60.
2. Konopka, J. and J. Hanson (eds.). 1985. Information handling systems for genebank management. Proc. of Workshop at Nordic Gene Bank, Alnarp, Sweden, 21–23 Nov. 1984. Internat. Board for Plant Genetic Resources, Rome.
3. Van Sloten, D. H. and M. Holle. 1988. Temperature fruit genetic resources and the IBPGR. HortScience 23(1): 73–74.
4. Frankel, O. H. and J. G. Hawkes (eds.). 1975. Crop Genetic Resources for Today and Tomorrow. Cambridge University Press, Cambridge, Massachusetts.
5. Sterns, S. and G. A. White. 1986. Plant Inventory No. 195. Plant materials introduced from July 1 to Dec 31, 1986. Agric. Res. Serv., U.S. Dept. Agric., Beltsville, Maryland.
6. Anonymous. 1986. A manual for public users of the germplasm resources information network. The GRIN databank. Agric. Res. Serv., U.S. Dept. Agric., Pl. Genet. Germplasm Inst., GRIN Database Manage. Unit, Beltsville, Maryland.

[R] Mention of a trade name does not constitute a guarantee of the product or an endorsement by the U.S. Department of Agriculture over other products not mentioned.

Animal germplasm information systems

REX. L. POWELL and H. DUANE NORMAN
Animal Improvement Programs Laboratory, Livestock and Poultry Sciences Institute, Agricultural Research Service, USDA, Beltsville, Maryland 20705, U.S.A.

Abstract Information has been published by breed, research, and conservation organizations on numbers and locations of animals and, in some cases, measures of phenotypic performance. Literature searches can identify reports on characteristics of specific breeds, but published information is scant about many endangered breeds. Comprehensive information would be more readily available from data banks designed to document animal germplasm resources. Recent efforts to establish data banks have been made by European and African groups and by the United Nations. Information on the genetic merit of endangered livestock breeds generally is lacking because small numbers of animals preclude accurate genetic evaluations and because refined analytic procedures may be unavailable in the country where the animals are located. Information on previous research needs to be consolidated to avoid duplication of effort and promote further research. Care needs to be taken so that breeds are preserved with their true genetic uniqueness, not in name only. Increasing availability and application of computers should improve recording, preserving, and sharing of information. Ironically, accurate and objective information on genetic merit may hasten the demise of breeds or strains apparently less competitive under current conditions unless conservation measures are taken.

Key words: data bank, gene bank, gene preservation, genetic evaluation, registry society

Introduction

An animal germplasm information system can be defined as all collections of data and descriptions that provide guidance on animal numbers, locations, and characteristics of animals. This includes national and international inventories, data and gene banks, and other summarizations. The goal of an information system is to provide a single source for all data related to a specific subject. One problem with such systems is that not all information is current; information from one source may have been superceded by a later report of data or interpretations published in another medium. Although all information systems could be developed for all breeds of all species worldwide, this discussion will be limited to information systems developed for domestic animals or livestock.

The interest of the United Nations' Food and Agriculture Organization (FAO) in gene preservation dates back to the 1950's [1]. Conferences in

L. Knutson and A.K. Stoner (eds.). Biotic diversity and germplasm preservation, global imperatives.
 ISBN 0-7923-0178-1

succeeding decades led to joint efforts by FAO and the United Nations' Environmental Program (UNEP), which were authorized in 1972.

The First World Congress on Genetics Applied to Livestock Production in 1974 included the roundtable session 'The Conservation of Animal Genetic Resources.' The introductory paper by Mason [2] set the stage for a powerful session. Bowman [3] described the development of the Rare Breeds Survival Trust in the United Kingdom. Maijala [4] presented a perspective on gene conservation that still is timely. Turton [5] addressed collection and dissemination of information on livestock breeds.

Effective use of the 'best' animals for optimum food production under existing circumstances necessitates setting aside less productive animals, strains, breeds, and even species. For most species, some, if not most, producers place a high value on purity, which may result in discarding of a breed or strain that is only slightly less worthy than another [6]. At least half of the British cattle breeds listed by Housman [7] are approaching extinction[6], and in Norway, the 30 cattle breeds present in the 1930's had been reduced to only 3 less than 40 years later. Liberal use of crossbred bulls in artificial insemination (AI) has permitted at least some of the genes of the lost breeds to be retained in the general population. Numbers of Finnish landrace sheep (Finnsheep) decreased from over 1 million in 1950 to 150,000 in 1967 [6] because other breeds were favored for growth rate and muscling. Only after numbers of the breed had dropped markedly were its high fertility traits recognized as valuable. Visscher [8] reported on a project that involved incorporating Finnsheep into a synthetic line.

Decrease in numbers of breeds is directly related to decrease in numbers of breeders. For example, intensive commercial breeding programs in poultry have resulted in a few companies (and, therefore, a few breeds/strains) that dominate the enterprise of a country or a continent. However, many types of poultry have been preserved by hobby breeders of fancy breeds. Shows and exhibitions promote and recognize this effort in this and many other species.

Loss of specific breeds or strains is only of academic interest if they possess no genes that would be useful in the future. Unfortunately, the needs of the future are unknown. A generation ago, the preferred meat animals in the U.S. were blocky, short, and squat by today's standards. Belated recognition of efficiency of meat animals with larger, leaner frames has resulted in new selection goals and breed changes. The beef cattle industry is permeated with foreign beef and dual-purpose breeds, with even some infusion from dairy breeds. To the extent that traditional British breeds exist (although in a different form), valuable genes probably have been retained.

Maintenance of genetic diversity is supported by the likelihood of changes in needs or demands. Market changes in terms of real or perceived health concerns can alter selection among and within breeds. New or increased frequency of certain diseases may favor breeds or strains that were previously overlooked, as illustrated by the leukosis-resistant Fayoumi-hen from Egypt [6]. Regional or national increases in affluence may generate a market for a product that was not competitive simply as food. Changing environments may require selection for different genotypes. Much of the dairy production in developed countries is characterized by a high level of grain (concentrate) feeding and requires little foraging on the part of the animal. An increase in grain cost relative to pasture and roughage could result in a much different preferred genotype.

Genetic diversity may be lost because of selection or by random drift. Decay of genetic variation by drift is generally slower than by directional selection [6]. Genetic diversity can be conserved by:

1. Maintaining several pure breeds or strains,
2. Establishing a common gene pool, or
3. Establishing banks for semen and eggs.

Techniques of AI, embryo transfer, and cloning amplify the possibility for loss of genetic diversity. With more intense selection or smaller effective population size, the expected loss in variation is greater.

Cloning, in particular, leads farmers to dream of a whole herd or flock that is genetically identical to their 'top' animal. However, determining what constitutes a truly outstanding animal is an imperfect process, especially for females. Genetic progress has resulted largely from use of available tools and knowledge that sound decisions are made overall even if some animals are overevaluated. 'Putting all the eggs in one basket' seldom has been successful and is not recommended. In addition and more importantly, genetic progress depends on variation. Within a genetically uniform herd, progress ceases, and improvement must come from superior genes outside the herd (or other pool from which the cloning selection was made). However, if outside genes are superior, the advisability of the initial cloning is questionable. Cloning is not a proper tool for gene preservation because it further narrows an already small population.

Is there a need?

Maijala [9] enumerated the reasons for conservation and grouped these reasons into three main categories:

1. Economic-biological,
2. Scientific, and
3. Cultural-historic.

Although an economic-biological consideration may be stated as the primary reason for conservation, a cultural-historic interest often is the stronger underlying concern. Most people feel a responsibility to maintain existing nature for its own worth or for its link to their heritage.

Selection pressure in many breeds and species rapidly is changing breed frequency and the genetic nature of breeds. Infusion of North American Holstein genes into practically every corner of the world is one example. While this has resulted in higher milk yields, increased size, and improved udder conformation, traits such as concentration of milk constituents and carcass quality have suffered [10]. Another dramatic example is the change in body composition of pigs. The change from fatness to leanness has been accompanied by problems of meat quality and susceptibility to stress. The latter has been especially troublesome for landrace populations in which the 'Halothane gene' has reached a frequency of near 0.9 [11].

Societies established to promote a particular breed (breed associations) provide a mechanism for exchange of information among members, promotional effort, and a system for recording ancestry. Major goals usually do not include maintaining genetic diversity but rather, to improve stock, which entails removing animals (and, therefore, genes) that are deemed less desirable. More than 92% of the dairy cows on production testing programs in the U.S. are Holsteins [12]. The other breeds offer diversity but are being selected in essentially the same direction as are Holsteins. Lerner and Donald [13] suggest that Friesians and Jerseys may contain all the needed genes for dairy cattle. Among all breeds, they are near the largest and smallest in frame and near the extremes in both volume and composition of milk.

On a short-term basis, the most efficient food production clearly conflicts with maintenance of diversity in the form of breeds that seemingly are inferior under current conditions for the primary trait. If superior stock is available, it is unrealistic to expect a country in which a substantial portion of the population is undernourished to devote additional resources to maintaining marginal native livestock. Improvement of inferior breeds would be slow and without guarantee that the resulting stock would be competitive. After consideration of the propriety of preservation, there is the difficult question of who will pay the cost. Society as a whole will be the long-term beneficiary of any good that accrues. Momm [14] concluded that the government is responsible for genetic resources and offered four principles on how the government can discharge its responsibilities.

Conservation of genetic resources has been of popular interest in the last decade and a half. Each of the three world congresses on genetics applied to livestock production have included a session on the topic. The annual meetings of the European Association for Animal Production (EAAP) generally have such a session, too. The presentations usually presume the cause is noble and restate the needs and methods, but true progress is slow. The topic might be termed a 'motherhood' issue; i.e., one assumed correct from the outset with little questioning expected or tolerated.

A shift in the distribution of a quantitative trait should not be confused with loss of genetic variation. Although a reduction in genetic variation is expected for a selected trait, the trend in dairy data from 1965 through 1982 was for a slightly increasing heritability for milk yield [15]. Land [16] questioned whether genetic variation actually is being lost or whether the change can be described more accurately as a shift in average performance with no significant loss of variation. He viewed as remote the likelihood that uniform selection practices would produce a narrow genetic base and that conditions (environmentally or marketwise) suddenly would change such that conserved breeds would be required. However, that is not to say that specific genes (alleles) could not be lost. Land [17] also questioned whether conserved populations could be used effectively and suggested selection of different strains toward divergent biological goals as a means for having the desired genetic variation available when needed.

Smith [18] suggested that conservation efforts may be adequate in developed countries and that the real need is to conserve germplasm in developing countries. He estimated costs of conservation by maintaining live breeding stock or freezing semen or embryos, and concluded that the possible benefit was far greater than the cost. Because the consumer is the ultimate recipient of benefits, the government should provide the financial support for conservation practices.

Individual animal and breed information

Information on the genetic merit of individual animals may not exist or may be sophisticated, depending on species and country. Genetic evaluations are calculated by various methods and may differ across breeds within a country and species. Furthermore, definitions of traits may differ within and across countries. Information on the genetic merit of livestock breeds in danger of extinction is often lacking because small numbers of animals preclude accurate genetic evaluations. Refined analytic procedures frequently are unavailable in the country where the animals are located. In many cases, no attempt has been made to partition phenotypes into genetic and environmental effects. Rare breeds may be exposed to extremely nonrandom environments, both positive and negative such that raw measurements may be misleading. Objective data are needed for comparison of breeds and strains.

Information on individual animals may come from test stations or on-farm measurements. In either case, supervision usually is by some objective person. An example of such a program is the on-farm production testing of dairy animals in the U.S. initiated in 1906. This program has grown to be available in all states and encompass over 4.5 million cows in nearly 58,000 herds [12]. Sophisticated procedures are available to estimate genetic merit for animals by use of their own performance (if applicable) and information from a number of relatives. Such programs include data from small populations and could include rare breeds.

Information on dairy populations from many countries and methods used for genetic evaluation have been collected by the International Bull Evaluation Service [19]. Evaluations resulting from programs in various countries provide estimates of genetic merit for individual dairy bulls. These estimates are not comparable across countries, however. Scales of measurement and definitions of traits vary, but the critical unknown is the genetic base. This is not the genetic base connected with maintenance of genetic diversity, but a definition of the zero point for a genetic evaluation system. This genetic base is important because evaluations are on a deviation basis, and differences between bases is an essential factor in comparisons of genetic evaluations across countries. The International Bull Evaluation Service represents an international approach to determining the most accurate way to convert dairy sire evaluations from one country to another.

While direct comparison of meat animal breeds has been common, choice of replacement dairy breeds is subjective even now. Objective knowledge of the merits of different breeds is scarce. For example, in the U.S., relative milk production per cow is known for the various breeds. However, that knowledge provides only general guidance as to the true ranking for merit of the seven dairy breeds because cows of different breeds are subjected to different environmental qualities. A cow that produces 6,000 kg of milk in a lactation in one herd might produce 9,000 kg in another herd. Scientific projects to compare breeds likely would be unpopular, at least among advocates for most breeds. Even comparing production per cow in a common environment would not measure total efficiency, although it would provide more objective information than is currently available. Thus, determining an objective ranking of breeds is not a trivial task and may not even be advisable. Uncertainty may passively promote continuance of a number of breeds, strains, or lines that preserve genetic diversity.

Data banks

Availability of information on the performance of breeds in vastly different environments is important. Interactions between genotype and environment often have been found to be of little or no importance across developed countries. However, the degree of interaction may depend on the degree of difference in environment. Maijala [9] stated that although significant interactions seldom had been found for cattle, environmental differences may have been small for many reports. Manipulation of conditions to provide an environment compatible with the country of origin may be possible in some situations, but in others may not be technologically, economically, or culturally feasible. Collaborative work involving the Meat Animal Research Center, Clay Center, Nebraska suggests genotype-environment interactions are important (Gregory, personal communication). Resistance to disease (including insect vectors) and tolerance for temperature in the host country are important and cause significant interactions [20]. Crosses may permit increased performance while retaining sufficient environmental compatibility from native stock. Once the optimal percentage of foreign genes is determined, recordkeeping is necessary to manage breeding to produce the desired product. The increased performance may require improved (or at least different) management.

Characterization of breeds is best accomplished if breed performance is measured under a common specified environment. Comparison of national strains has been limited. A few comparisons of limited scale have been conducted with strains of Holstein-Friesians. The most extensive study is the trial supervised by FAO in Poland [21]. Such trials provide much more definitive comparisons than are possible from data collected in varying environments and which may be reported subjectively. Trials of indigenous breeds or crosses of indigenous and exotic tropical and temperate breeds against reference breeds could identify useful breeds or strains from among those with high potential, especially if the experience of the FAO International Black and White Strain Comparison in Poland were used [22]. However, such trials are expensive and cannot be short-term if they address longevity or lifetime efficiency.

Data-collection systems may not exist in areas with breeds or strains in danger of loss. Without objective data (such as milk records or growth rate), characterization of breeds is hampered, and genetic improvement is unlikely. The International Livestock Center for Africa has developed a computer data bank for recording performance of individual animals [20]. This data bank will allow summarization in several ways that include breed comparisons. The recording and analysis system is offered to national research groups in Africa at no charge. Individual animal databases are available concerning reproduction, weight, dairy yield, wool, draft ability, and health.

Potential genetic improvement is a function of population size. Genetically similar groups of animals may exist in different countries and under different names, and could be linked together to promote progress not possible with each indigenous population. Maijala *et al.* [23] reported on answers from 22 countries to a survey of an EAAP working party on breeds, endangered breeds, and conservation activities. Cattle, horses, pigs, sheep, and goats were included. Estimates were made for eight other countries from the literature. Mason's [24] dictionary was cited as being valuable in determining which breeds were essentially the same but referred to in different languages. Of the 1,263 country populations, 130 existed in at least two countries. Criteria for endangered status varied by species. Of 241 endangered populations, 26 had nonendangered counterparts in other countries, a demonstration of the need for an international role in determining which breeds can and should be preserved. Available sources likely will not permit all breeds or strains to be maintained, and there is little room for waste or duplication of effort.

Maijala and Simon [25] described recent European efforts toward establishing two data banks. One is in progress for cattle, goats, horses, pigs, and sheep. Data for a breed include general information, development, appearance, main use, and performance relative to standard breeds. The other data bank will be for reindeer, farmed fur-bearing species, poultry, rabbits, and bees. This second data bank will include detailed characteristics for breeds and for individuals conserved alive or frozen.

The Commonwealth Bureau of Animal Breeding and Genetics has published *Animal Breeding Abstracts* since the early 1930's [26]. Since 1973, data have been computerized such that online searches may be conducted by remote telecommunication with the main site in Edinburgh or with locations in the U.S., Italy, or West Germany. The Bureau will conduct searches for those not wishing to conduct their own. Additionally, journal tapes may be leased. Searches of this database may be conducted in many ways to assemble the particular set of information desired.

In the 1960's, FAO began a series of Expert Consultations on the evaluation, conservation, and use of animal genetic resources. In 1973, FAO and UNEP joined in a pilot project on the conservation of animal genetic resources. One of the joint efforts was in the area of data banks, with a goal of providing a description of breeds and strains in a common format for various purposes, including identification of endangered breeds. Early findings for Asia, Africa, and Latin America have been published by FAO/UNEP in the Animal Production and Health series. Further work still is awaiting funding. One important finding was that only 25 to 30% of all research reports published in local languages had been included in *Animal Breeding Abstracts* by the Commonwealth Bureau of Animal Breeding and Genetics [22]. Some of these reports have limited distribution, and translation from other than major languages is a problem. Regardless of the reasons for exclusion from *Animal Breeding Abstracts*, the Commonwealth

Bureau largely may be serving only the needs of the developed world [22]. Lack of access to research reports was found to result in duplication of research; thus a more comprehensive system is needed.

Trials by FAO/UNEP in Africa, Asia, and Latin America from 1983 through 1985 provided experience upon which recommendations were made relative to data banks [27]. Two resulting approaches were referred to as the Asian approach and the Latin American approach. The Asian approach used fixed-length records in which trait information was (or could be) identified by position within the fixed record. All traits were defined initially and, thus, were addressed by each record, if only by omission. By contrast, the Latin American approach associated a code with each trait such that new traits could be added. Record length was variable, and positions were not tied to any trait. Each approach had advantages and disadvantages, and record size was a consideration. With the Asian approach, unused fields wasted considerable space, whereas with the Latin American approach, trait codes were significant consumers of space. The recommendation was for application of a commercial software package that followed the Asian approach but which recognized null fields such that they required little or no disk space. This may be accomplished by use of codes, which may be invisible to the user, or through a very flexible storage algorithm.

Comparisons also were made of the use of a central (global) center versus regional sites. In spite of the efficiency of a global center, the recommendation was for regional data banks because of the greater likelihood of success from a greater feeling of active involvement.

The FAO/UNEP descriptor lists provide a format for coding data gleaned from scientific papers, annual reports, theses, and any other source, whether published or not [28]. Descriptor lists are provided for cattle, buffalo, pigs, sheep, and goats [28] and for poultry [29]. The formatting system is based on 'master' and 'slave' records. Master records, one for each breed or separate strain, contain information on physical characteristics that may come from a number of sources. Master records are intended to be descriptive and qualitative. Slave records are from a single source for each breed and contain objective data on performance and on the environment under which the performance was measured. Thus a particular source that addresses more than one breed will have a slave record for each breed. Slave records should indicate whether information is a collation of existing data or the result of a comparison of breeds under a common environment. The descriptor list for slave records includes a code for reliability of the source.

Extraction and summarization of data require more than clerical expertise. The team leader for the building of a data bank must be an animal geneticist with experience with the particular species, have the ability to judge the authenticity and reliability of sources, and have an appreciation for computerization and statistical summarization.

Although computers tend to get deserved attention for enabling establishment of data banks, infrequent flow of new information and the appeal of color pictures to aid in description of breeds make traditional manual, visual methods worthy of consideration [11].

Gene banks

Locations and contents of gene banks may be part of the information within data banks, but their special nature warrants separate discussion.

Preservation through cryogenic storage has been addressed by Pursel and Johnson (this volume). Such storage of germplasm may be more feasible in the long-term than maintenance of breeding herds of live animals, but it is not without serious difficulties. Hodges [22] lists five considerations:

1. Location of breeds in an isolated area with poor facilities for collecting semen from males not accustomed to the procedure,
2. Difficulty in testing for disease as well as in crossing international borders to the storage site,
3. Maintenance of a constant supply of liquid nitrogen,
4. Location of the bank(s) and ownership of germplasm, and
5. Financial responsibility.

The question of who will be responsible financially for conservation of germplasm is of general importance because those areas with the greatest opportunities for conservation usually are the areas with the most limited abilities to afford conservation activities.

The FAO defines a gene bank as a repository where samples of animal genetic resource populations that are being preserved are kept [30]. Parez [31] suggested that gene banks already are frequent in developed countries through stores of frozen semen. However, the spirit of the definition does not seem to be met unless an effort is made to maintain semen well beyond the normal marketing needs. Existence of these banks does testify to the capability to collect, process, and store frozen genetic material. A number of banks are operating in developing countries in Africa, Latin America, and Asia [31]. In many reports, the recommendation is that at least 25 individuals be represented in frozen semen or embryos, a figure that has been attributed to Gowe [32]. The Milk Marketing Board has established a semen bank for bulls of rare breeds in the United Kingdom [33]. The American Minor Breeds Conservancy (AMBC) is establishing semen banks for Milking Devon and Florida Native/Pineywoods cattle and plans to include other breeds.

Maijala *et al.* [34] reported on Scandinavian efforts toward conservation of animal genetic resources. The Nordic Council of Ministers supported an expert group to collect information on threatened breeds and the situation of semen banks. Although a Nordic Gene Bank for Farm Animals was to be created [23], the function was to collect and disperse information; actual gene banks would have been within each country.

Conservation projects

The FAO [30] defines preservation as the 'aspect of conservation by which a sample of an animal genetic resource is designated to an isolated process of maintenance, by providing an environment free of the human forces which might bring about genetic change.' Thus, mere continuation of a breed, selected in any way, would not fit the definition of preservation of the breed. However, it would fit the FAO [30] definition of conservation by management, which is 'planned genetic change with the aim of sustaining, utilizing, restoring, or enhancing the animal genetic resource.'

France, Spain, Hungary, Poland, West Germany, Italy, Britain, and The Netherlands have active conservation programs [11], most with government support. Attitudes and actions by country were summarized by Maijala *et al.* [11], and applicability of conservation methods were scored for various aspects in regard to five species.

A report on conservation activities in Hungary was combined with descriptions of the breeds involved in the program [35]. Conservation began in the late 1950's, when the last state herds of some breeds were retained. In 1973, the government program was initiated. Conservation activities in the U.S.S.R. were described by Dmitriev [36].

Various organizations are involved in conservation programs in Great Britain [33, 37]. The best known activity is that of the Rare Breeds Survival Trust, which promotes conservation in a number of ways, including financial support for breeders willing to maintain qualifying breeds.

Various historical working farms and similar operations maintain a few animals of breeds representative of earlier times. However, the small numbers of animals per location make inbreeding likely, with the resulting loss of heterozygosity. Some committed individuals are working diligently to maintain larger breeding units of rare breeds.

At the 1976 BARC Workshop on Germplasm Preservation, Miller [38] suggested that conservation efforts for cattle through farm parks and living museums are on too small a scale to succeed. The small herd sizes lead to a loss of 2–7% of heterozygosity per generation. Thus a program intended for preserving genes may retain the breed but fail to provide potential genetic diversity. Miller likened efforts for gene preservation to purchasing an insurance policy. The event insured against may be highly unlikely but still should be considered and insured against.

The Board on Agriculture of the U.S. National Research Council initiated a 3-year study, Management of Global Genetic Resources: Agricultural Imperatives, in 1986. This project has involved approximately 50 scientists in an examination of the complex technical and socioeconomic issues affecting germplasm conservation and use. The anticipated reports produced by this study will include strategies for management of animal germplasm (both globally and in the U.S.) and will briefly discuss animal germplasm information systems.

Presently, the major organized conservation efforts in livestock in the U.S. are through the AMBC and the National Council on Gene Resources [39]. Both are recent organizations: AMBC was formed in 1977 and the National Council in 1980.

Bodo [40] stresses that conservation interest must be in preserving genetic uniqueness, not just the breed itself. He illustrates this point with the case of the Hannoveraner horses from the Holstein region of Germany. These horse were light draft horses just after World War II but now are famous sport horses with good jumping ability. The breed still exists but primarily in name only. Even the most diverse breeds will tend toward a common animal if subjected to the same selection pressures as more common breeds. Maintaining only the breed characteristics of color, length of hair, uniqueness of horns, etc., is not really maintaining meaningful genetic diversity.

Nearly extinct breeds may appear to be rejuvenated through popular effort or the incentive of funds to support conservation, but the real genetic diversity has been lost. Glodek [41] cited the Angeln Saddleback pig as an example of a breed that had extremely restricted numbers (69 sows in six herds in 1985). Although the current breed has the external appearance of the original breed, the diversity in both performance and marker gene frequencies indicates that the current breed is markedly different from animals in the 1960's. Genetic drift or unrecorded upgrading with other breeds may have resulted in loss of unique genes.

Inventories and information

A number of articles in *Animal Genetic Resources Information*, a semi-annual newsletter published by FAO and UNEP beginning in 1983 and available from FAO, Rome, describe endangered breeds and conservation efforts. The series *FAO Animal Production and Health Papers* provides reports presented at expert panel meetings and recommendations. Considerable information may be gleaned from the 34 articles in 'Strategies for Improved Use and Conservation' [30].

Information on livestock in many parts of the world is not generally available in journals in the West. Some information from China was compiled by Hoefer and Tsuchitani [42] in *Animal Agriculture in China*. A major FAO publication described livestock breeds of China in considerable detail [43]. A survey has been made in the U.S.S.R. in cooperation with the Center for International Projects in Moscow and the All Union Academy of Agricultural Sciences [44]. The National Research Council prepared the book *Little-Known Asian Animals With a Promising Economic Future* [45], in which for each of 15 species or subspecies of bovine and swine, characteristics and potential advantages were discussed.

The Beltsville Agricultural Research Center Workshop on Germplasm Preservation was held in 1976. At that workshop, Bereskin [1] cited the work of the Jackson Laboratory at Bar Harbor, Maine, for maintenance of many inbred lines of mice. Bereskin's report also referred to FAO interest in the topic of gene preservation in the 1950's. The University of Nebraska approach to preserving swine germplasm was to develop a gene pool from 14 different breeds. Crawford and Somes were credited with compiling registers of poultry stocks.

A primary source of information on numbers and locations of livestock in developed countries are breed-registry organizations. For operational efficiency, some breed organizations have combined operations while maintaining separate organizational entities. Costs are lowered by sharing or contracting office, computer, field staff, and publication resources with other breed organizations. In the U.S., the American Guernsey Association has moved its headquarters to coincide with the American Jersey Cattle Club in Columbus, Ohio. The American Milking Shorthorn Society has moved to Beloit, Wisconsin, to make use of resources of the Brown Swiss Cattle Breeders' Association located there. The Ayrshire Breeders Association has continued at its headquarters in Brandon, Vermont, but has arranged for computer processing by the Holstein Association.

Supporters of some breeds are too few to make a breed organization feasible. In the United Kingdom, Countrywide Livestock Limited provides a registry program for breeds that lack their own registry organization. In the U.S., animals of these breeds can be registered with AMBC. For example, AMBC provides registry programs for Jacob, Navajo-Churro, Wiltshire Horn, and Gulf Coast Native sheep and the Exmoor pony, and likely will develop a program for Florida Native/Pineywoods cattle. In addition, AMBC surveys rare livestock populations, organizes rare breed shows and sales, publishes a newsletter, and is setting up a semen bank for rare breeds of cattle. In 1985, AMBC published its *North American Livestock Census* [46], which identified endangered breeds of cattle, sheep, goats, pigs, horses, donkeys, mules, poultry, dogs, and other animals. The *1987 AMBC*

Poultry Census and Sourcebook provided descriptions for 15 endangered breeds [47]. A separate and more extensive list is in *International Registry of Poultry Genetic Stocks*, which describes 630 breeds and varieties of chicken, quail, and turkey as well as 310 specialized lines and strains [48]. Sources for these stocks are provided as are linkage maps and locations of mutant stocks.

Irony of information

Momm [14] began the summary of his paper by stating that intensive animal breeding not only produces genetic gains, but also results in losses of genetic resources. Thus, progress has a price. Not only does selection, the keystone of progress, have the side effect of removing some genes from existence, but information in general is likely to work against conservation. Increase in accurate and accepted evaluations of breeds and strains appears to be correlated positively with demise of those ranking lower for the trait or index employed.

Philipsson [10] provided estimates of the degree to which North American Holstein genes were present in other populations of black and white cattle. For bulls being sampled in 1986, Czechoslovakia, Denmark, France, Israel, Italy, Japan, Netherlands, South Africa, Switzerland, West Germany, and Yugoslavia all reported proportions of North American genes of 80% or more.

Before the advent of genetic evaluations in a country, each farm or region could feel that they had the best livestock. If statistical analyses demonstrate otherwise, most breeders want to incorporate the best genetics and to discard what they had been content with (if not proud of) previously. Thus information is a two-edged sword. Information is important to preservation efforts by identifying endangered breeds and their characteristics and locations. However, with information that truly characterizes a breed relative to another, the profit-oriented (survival-oriented) producer will tend to want the superior breed. Such is the irony of information.

Conclusions

Although sometimes unpopular, periodic examination of the justification for conservation or at least the direction of activities is appropriate. Also important is determining to what degree a breed might be preserved in name only rather than with its presumed genetic uniqueness. How successful have been conservation efforts and specifically information systems is a matter of perspective. Is the container half full or half empty? Over the past few years, progress has been made toward the establishment of useful data banks and gene banks. Other efforts also are ongoing to document data that can be entered into data banks. However, most efforts, especially the major ones, are not complete. Their success in documenting and retaining genetic material is still a matter of speculation. Because the goal is to be prepared for an unknown future that is ever moving away, perhaps we can never measure success in this area.

References

1. Bereskin, B. 1976. Preservation of animal germ plasm–an overview. Mimeo, U.S. Dept. of Agric., Beltsville, Maryland. 13 pp.
2. Mason, I.L. 1974. Introduction presentation. Proc. 1st World Congr. Genet. Appl. Livest. Prod. II: 13–21.
3. Bowman, J.C. 1974. Conservation of rare livestock breeds in the United Kingdom. Proc. 1st World Congr. Genet. Appl. Livest. Prod. II: 23–29.
4. Maijala, K. 1974. Conservation of animal breeds in general. Proc. 1st World Congr. Genet. Appl. Livest. Prod. II: 37–46.
5. Turton, J.D. 1974. The collection, storage and dissemination of information on breeds of livestock. Proc. 1st World Congr. Genet. Appl. Livest. Prod. II: 61–74.
6. Maijala, K. 1970. Need and methods of gene conservation in animal breeding. Ann. Genet. Sel. Anim. 2: 403–415.
7. Housman, W. 1905. Cattle: breeds and managements. Live Stock Handbooks, 5th ed. Winston, London. 382 pp.
8. Visscher, A.H. 1987. Development of a synthetic dam line in a reciprocal cross between Finnish Landrace and Ile de France in the Netherlands. Livest. Prod. Sci. 17: 77–87.
9. Maijala, K. 1986. Motives, possibilities and methods of maintaining numerically small cattle breeds. World Rev. Anim. Prod. XXII(1): 43–50.
10. Philipsson, J. 1988. Prospects and problems in the international use of Friesian cattle strains. Proc. 7th World Holstein Conf., Nairobi. 15 pp.
11. Maijala, K., A.V. Cherekaev, J.-M. Devillard, Z. Reklewski, G. Rognoni, D.L. Simon and D.E. Steane. 1984. Conservation of animal genetic resources in Europe, final report of an E.A.A.P. working party. Livest. Prod. Sci. 11: 3–22.
12. Wiggans, G.R. and C.A. Ernst. 1988. NCDHIP participation as of January 1, 1988. Natl. Coop. Dairy Herd Impr. Prog. Handb., Fact Sheet K-1. 4 pp.
13. Lerner, I.M. and H.P. Donald, 1966. Modern Developments in Animal Breeding. Academic Press, New York, 294 pp.
14. Momm, H. 1987. Governmental responsibility for genetic resources in animal breeding. 38th Annu. Mtg., Euro. Assoc. Anim. Prod. 1:12 (Abstr.)
15. DeVeer, J.C. and L.D. Van Vleck. 1986. Comparison of heritability estimates from daughter on dam regression with three models to account for production level of dam. J. Dairy Sci. 69:2890.
16. Land, R.B. 1986. Genetic resource requirements under favourable production marketing systems: priorities and organization. Proc. 3rd World Congr. Genet. Appl. Livest. Prod. XI: 486–491.
17. Land, R.B. 1981. An alternative philosophy for livestock breeding. Livest. Prod. Sci. 8: 95–99.

18. Smith, C. 1984. Estimated costs of genetic conservation in farm animals. FAO Anim. Prod. Hlth. Pap. 44/1: 21–30.
19. International Bull Evaluation Service. 1986. Sire evaluation procedures for dairy production traits practised in various countries, 1985. INTERBULL Bull. 2. 76 pp.
20. Brumby, P.J. and J.C.M. Trail. 1986. Animal breeding and productivity studies in Africa. Bull. Internat. Livest. Ctr. Africa 23: 23–27.
21. Jasiorowski, H.A., M. Stolzman and Z. Reklewski. 1987. International FAO black and white cattle strain comparison (1974–1984). World Anim. Rev. 62: 2–15.
22. Hodges, J. 1986. Animal genetic resources in the developing world: goals, strategies, management and current status. Proc. 3rd World Congr. Genet. Appl. Livest. Prod. XI: 474–485.
23. Maijala, K. 1984. Scandinavian activities on the conservation of animal resources. Anim. Genet. Resources Info. 1/84: 20–26.
24. Mason, I.L. 1969. A world dictionary of breeds of livestock. 2nd ed. Commonw. Agric. Bur. Tech. Commun. 8. 272 pp.
25. Maijala, K. and D. Simon. 1987. Plans for European data banks on animal genetic resources. 38th Ann. Mtg., Euro. Assoc. Anim. Prod. 1:8 (Abstr.)
26. Turton, J.D. 1984. The Commonwealth Bureau of Animal Breeding and Genetics and the provision of information for data banks on animal genetic resources. FAO Anim. Prod. Hlth. Pap. 44/1: 142–147.
27. Food and Agriculture Organization, United Nations. 1986a. Animal genetic resources data banks: 1. Computer systems study for regional data banks. FAO Anim. Prod. Hlth. Pap. 59/1. 18 pp.
28. Food and Agriculture Organization, United Nations. 1986b. Animal genetic resources data banks: 2. Descriptor lists for cattle, buffalo, pigs, sheep and goats. FAO Anim. Prod. Hlth. Pap. 59/2. 150 pp.
29. Food and Agriculture Organization, United Nations. 1986c. Animal genetic resources data banks: 3. Descriptor lists for poultry. FAO Anim. Prod. Hlth. Pap. 59/3. 111 pp.
30. Hodges, J. (ed.). 1987. Animal genetic resources: strategies for improved use and conservation. FAO Anim. Prod. Hlth. Pap. 66. 316 pp.
31. Parez, M. 1984. Harvesting, processing, storage and subsequent use of animal cells in developing countries. FAO Anim. Prod. Hlth. Pap. 44/2: 67–87.
32. Springmann, K., W. Schutz and H. Krausslich. 1987. Factors affecting the utilisation of genetic resources for future animal production. 38th Annu. Mtg., Euro. Assoc. Anim. Prod. 1:24 (Abstr.)
33. Alderson, L. 1986. Mobilization of the forces of society for the conservation of animal genetic resources. Anim. Genet. Resources Info. 5/86: 1–5.
34. Maijala, K., C. Ehrenberg and N. Kolstad. 1982. Scandinavian report on animal genetic resources and their conservation. FAO Anim. Prod. Hlth. Pap. 24: 35–136.
35. Bodo, I. 1985. Hungarian activities on the conservation of domestic animal genetic resources. Anim. Genet. Resources Info. 4/85: 16–22.
36. Dmitriev, N.G. 1987. Conservation of animal resources in the USSR. Anim. Genet. Resources Info. 6/87: 13–16.
37. Alderson, L. 1985. The conservation of animal genetic resource in Great Britain. Anim. Genet. Resources Info. 4/85: 26–31.
38. Miller, R.H. 1977. The need for and potential application of germ plasm preservation in cattle. J. Hered. 68: 365–374.
39. Brooke, C. 1984. Organizational activities for the conservation of farm animals in the United States. Anim. Genet. Resources Info. 3/84: 31–32.
40. Bodo, I. 1987. Importance of type and variation within the breed in preservation of domestic animals. 38th Annu. Mtg., Euro. Assoc. Anim. Prod. 1:30 (Abstr.).

41. Glodek, P.H. 1987. Is the Angeln Saddleback pig a genetic resource for future animal production? 38th Annu. Mtg., Euro. Assoc. Anim. Prod. 1:34 (Abstr.).
42. Hoefer, J.A. and P.T. Tsuchitani, eds. 1980. Animal Agriculture in China. National Academy Press, Washington, DC. 197 pp.
43. Peilieu, C. 1984. Livestock breeds of China. FAO Anim. Prod. Hlth. Pap. 46. 217 pp.
44. Food and Agriculture Organization. (In press). Animal genetic resources of the USSR. FAO Anim. Prod. Hlth. Pap. 65.
45. Ruskin, F.R. (ed.) 1983. Little-Known Asian Animals With a Promising Economic Future. National Academy Press, Washington, DC. 131 pp.
46. Henson, E. 1985. 1985 North American Livestock Census. The American Minor Breeds Conservancy, Pittsboro, NC. 34 pp.
47. Christman, C. and L. Heise. 1987. 1987 AMBC Poultry Census and Sourcebook. American Minor Breeds Conservancy, Pittsboro, North Carolina, 34 pp.
48. Somes, R.G., Jr. 1984. International registry of poultry genetic stocks. Storrs Agric. Exp. Sta. Bull. 469. 96 pp.

Accessibility of world germplasm resources: animal health constraints

JOHN A. ACREE
Veterinary Services, Animal and Plant Health Inspection Service, U.S. Department of Agriculture, Hyatlsville, Maryland, 20782, U.S.A.

Abstract Germplasm is not distributed evenly about the globe for a variety of natural and artificial reasons. This uneven distribution creates a demand for germplasm movement into new areas. Animal diseases are not distributed evenly about the globe either, and for a variety of natural and artificial reasons also. This uneven distribution creates a demand for animal health barriers to germplasm movement. In response to demands from livestock breeders we have developed methods to overcome these barriers with a minimal risk of spreading disease.

Concern for the preservation of genetic diversity elevates the previous demands for access to isolated germplasm into imperatives. An equal imperative, of course, is a regard for world animal health. Fortunately, research has provided technologies such as embryo transfer, which, when combined with other safeguards, may enable us to move germplasm from area to area with minimal risk of disease transmission.

Key words: germplasm exchange, animal health, embryo transfer, semen, animal disease distribution

Germplasm is not evenly distributed about the globe

Germplasm (and here I am speaking of 'valued' or 'desirable' germplasm from the livestock breeders' point of view) is not evenly distributed throughout the world. Some of the reasons for this uneven distribution are natural. The drive to disperse may be opposed by the drive to specialize, and so a subspecies evolves to fill a limited niche. Migration is sometimes blocked by deserts, seas, or mountains. A variety of complex interactions with other species and/or the environment can combine to define where we will find certain biotypes.

Other reasons for uneven distribution are entirely artificial. Man's preferences for certain species, or for breeds within a species, may account entirely for their existence in certain areas. Our impact on, and alterations to, the environment can favor the local survival of one species over another.

For whatever reason, uneven global distribution of germplasm represents a powerful potential for genetic migration. There is natural dispersal pressure upon a species to colonize distant territory. This may serve to insure the survival of its kind against local calamity. There is the artificial pressure that we supply through our fascination with, or appreciation of, exotic breeds and species.

L. Knutson and A.K. Stoner (eds.). Biotic diversity and germplasm preservation, global imperatives.
 ISBN 0-7923-0178-1

Disease is not evenly distributed about the globe

Certain animal diseases are caused by microbiologic agents. Each of these agents has its own germplasm. Therefore, it is no surprise that 'undesirable' germplasm, from animal health's point of view, also is unevenly distributed about the world. Again, some reasons for this distribution are natural. A pathogen's range may be limited because nature confines its host, reservoir, or vector to certain areas. The disease agent may have its own environmental demands, or it may be dependent upon a certain host density that occurs in only a few areas.

As with the 'desirable' germplasm, some reasons for this uneven distribution are artificial, the most significant of which are man's campaigns to eradicate animal diseases. These efforts have eliminated certain disease agents from large geopolitical areas of the world. Other artificial reasons, almost equally important, are the restrictions man has imposed on the international movement of animals. These restrictions have excluded the more serious animal disease agents from all the advanced countries. We should note that most of these disease agents are those that infect domestic ungulates and poultry. We should also note that their uneven distribution represents a powerful potential for spread.

The world distribution of foot-and-mouth disease, for example, testifies to the enormous effort expended by certain nations in eliminating and excluding the virus. Almost 40 countries are completely free of this disease and, of course, take pains to remain so. Many countries are free of African swine fever, rinderpest, and contagious bovine pleuropneumonia. However, as yet only a handful of countries enjoy freedom from bovine brucellosis, tuberculosis, and leukosis. It is understandable, therefore, that once a nation has expended time, effort, and other resources in freeing its livestock of an animal disease, it must impose severe restrictions upon the introduction of certain species from areas not free of that disease. This is because there has always been a close association between the movement of animals and the movement of animal pathogens.

Concomitant migration of genes and germs

A dramatic example of the intercontinental movement of an animal pathogen in association with animal movement was the rinderpest pandemic of Africa that began in 1889. The virus was introduced into Ethiopia by cattle that the invading Italians brought from Aden and India. By late 1896, the disease had devastated all susceptible species in the sub-Saharan portion of Africa. Even today, it remains a significant animal health problem for some African countries. As far as we know, rinderpest has never spread into new areas except by the movement of live animals.

Other famous examples of costly animal importations include the single cow from a British ship that introduced contagious bovine pleuropneumonia into New York in 1843 and the four cows imported from Holland that introduced the same disease into Boston in 1859. From these two beach-heads, pleuropneumonia spread south and west and crossed the Appalachian mountains in the early 1880's. We did not eradicate it from the U.S. until 1892. The first three introductions of foot-and-mouth disease into the U.S. were traced to imported breeding cattle. The appearance of foot-and-mouth disease in Mexico in 1946 coincided with the importation of bulls from Brazil. In more recent times, African horse sickness was diagnosed in Spain, perhaps introduced by zebra from Namibia.

In view of this history, it is not surprising that most livestock owners in countries in which good animal health is enjoyed demand that their regulatory veterinarians take steps to prevent the introduction of devastating livestock diseases via animal importation.

Unfortunately, some of the steps demanded in the past completely prohibited the movement of certain species between many countries and geopolitical areas. This was because disease reporting of that time was such that we could be certain only that a disease agent had, or had not, expressed itself clinically in animals of a given country. If we had reason to believe that all frontiers of a country were well policed as far as animal movement, we considered the entire livestock population of the country as a unit of uniform health status. Therefore, this unit either had evidence of the disease in question or it did not. If we had reason to believe that the frontiers of a country, or a number of its neighboring countries, were not well controlled, the dimension of this unit was expanded until a geographical barrier, such as a mountain, an ocean, or a desert provided a reliable border. Therefore, the known existence of a disease anywhere within certain large areas of the world could deny access to all the livestock germplasm of those areas. An example of this is the U.S. Tariff Act of 1930 which prohibited the importation of ruminants and swine from any country in which outbreaks of foot-and-mouth disease or rinderpest were known to occur.

These absolute restrictions invariably resulted in an increased demand for the prohibited germplasm. Experts in animal breeding began to assure other livestock owners, as well as animal health authorities, that the infusion of this exotic germplasm into their national herds or flocks would increase productivity substantially above what they could expect from exploiting the domestic gene pool.

In response to this demand, animal health authorities of the world developed ways to import desirable livestock and poultry from many prohibited areas with a minimal risk of importing pathogens. For the sake of simplicity, I will continue to use the generic terms 'animal health authorities,' while admitting that I am most familiar with the actions and attitudes of the authorities of North America, particularly those of the U.S. When I say 'serious animal disease' I usually am referring to the model of foot-and-mouth disease in cattle.

Response to the demand for breeding livestock

The terminology that developed in association with this geographical prohibition concept was somewhat misleading. Entire countries or continents were referred to as 'infected' with foot-and-mouth disease or 'infected' with rinderpest. This implied, to some people that all animals in those countries or on those continents were capable of transmitting the disease agent. This, of course, was not true. It is rare for an infectious agent to convert every host in a population into a carrier. In fact, there is seldom a necessity, from the disease agent's point of view, to do so. Therefore, in any 'infected' country, there are subpopulations of livestock that are not infected and have never been exposed. These nonexposed subpopulations are the safest sources of export germplasm.

This approach, however, is based upon the critical assumption that the members of the nonexposed subpopulations are the genetic equivalents of those from the exposed. In other words, 'low-risk' germplasm can be found in an 'infected' country which will suit breed improvement purposes just as well as any of the 'high-risk' germplasm.

Having made this assumption, the animal health authorities attempt to locate and verify the 'low-risk' germplasm sources by using a series of proven procedures. From these sources, germplasm units, in the form of live breeding animals, are obtained. These individual animals are then subjected to a process designed to further isolate them from exposure and to verify their freedom from previous exposure. Some of the steps commonly used in this process are described below.

Step 1 – Certification of herds and areas

The animal health authorities of the importing country usually require affirmation from the animal health authorities of the exporting country that specified herds are located within areas where the disease in question has not been observed for a certain length of time. The reliability of this procedure depends upon the nature of the disease, the accuracy of the diagnosis, and the completeness of the reporting. However, for those diseases that usually express themselves in a dramatic clinical fashion, and in countries with good disease surveillance, this step identifies herds with a low probability of exposure.

Step 2 – Initial serological sampling of the herd of origin
The species in which we are interested usually will develop circulating antibodies when exposed to disease agents. We have developed sensitive serologic tests to detect these antibodies. Invariably, there will be some exposed individuals that yield false negative results, but if we test a representative sample of the entire herd, we will have a high probability of detecting any exposure that it might have received.

Step 3 – Initial serological testing of the export candidate
This test usually is the first screening of the individual animals considered for international movement. If the herd has been properly certified and tested, we expect very few of these individuals to have detectable levels of antibody. Therefore, to reduce the possibility of false negative results we use the most sensitive serological test available. Because a test's specificity is lessened when used in a population having a low incidence of the disease, the number of false positives obtained will be high. This is unimportant however, because of our assumption that we do not eliminate the only source of a desired germplasm when we reject any one animal or any one herd.

Step 4 – Initial agent isolation tests of the export candidate
We may attempt to isolate the disease agent of concern from the candidate animals. Although our ability to detect a disease agent is seldom as good as our ability to detect antibodies to the agent, the use of this step in combination with step 3 provides a high probability of revealing exposure.

Step 5 – Pre-embarkation quarantine of an export cohort
The animals that qualify on certification and initial testing are usually placed in cohorts and isolated from all other animals in the country or region of origin. The length of quarantine is at least twice the median incubation period of the disease in question. During this time, the export animals are retested and observed for signs of disease.

This isolation period serves several purposes. First, it permits animals in the incubatory stage of the disease to develop clinical signs or detectable levels of antibody. Because of the prior steps few, if any, inapparent carriers will remain in such a cohort. However, if one of the cohort is a carrier this procedure ensures its prolonged, close, contact with animals that should be fully susceptible. Therefore, the cohort functions as a sentinel system in that the appearance of disease signs or antibodies in any member indicates the presence of a carrier animal. If nothing else, the time in quarantine may permit some of the carrier animals to overcome their infections or reduce them to a level below that needed for disease agent transmission.

Step 6 – Destination quarantine of the import cohort

The imported animals often are isolated after arrival in the country of destination but before their release into the national herd. The cohort is tested and observed and allowed to mingle with fully-susceptible domestic sentinels of several species. At the end of this quarantine, all imports and sentinels are subjected to a final diagnostic testing.

Step 7 – Post release observation

After the animals are released, they usually become part of a known herd until their germplasm can be amplified and marketed. In some cases, as a final precaution, these herds are observed for a number of weeks or months after the addition of the exotics. Animal health authorities of the U. S. have limited authority under which they can employ this option.

Importation schemes seldom use all the above steps. The safety of the system comes from the *combination of the probabilities*, provided by each step, of detecting and eliminating a carrier animal. The probability is small that an exposed herd will be certified as nonexposed. The probability is smaller that a representative sample of the animals in an exposed herd will have no detectable antibody to the disease. The probability is still smaller that an animal selected as an export from this herd will be a carrier of the disease agent. The probability is extremely remote that a carrier will have no detectable antibodies to the agent, not be able to infect sentinel animals and yet transmit the disease agent after release. Therefore, this system can only fail if there is an unbroken chain of rare events, each dependent upon the one before.

By using schemes such as this, it was possible to obtain germplasm with minimal risk from many of the 'infected' countries and regions of the world. These schemes were productive in that large numbers of breeding animals and thousands of doses of bovine semen have been imported into North America from countries in which foot-and-mouth disease exists. These schemes were successful in that the importations produced no outbreaks of the disease. These schemes were instructive in that we acquired considerable experience in obtaining 'low-risk' germplasm. But the cost and logistical complexity of these methods still imposed considerable constraint on the worldwide accessibility of breeding livestock. Other methods were sought in the hope that their incorporation into the import schemes would simplify and economize the process.

Incorporation of semen into import schemes

The techniques developed for preserving the semen of several species have provided another tool for obtaining 'low risk' male genes from certain countries. The introduction of germplasm in the form of semen inherently is safer than the movement of live animals for the reasons listed below.

1. The larger internal and external metazoan parasites will not be transmitted by this route,
2. The only animal that the semen contacts in the country of origin is its donor,
3. The only disease agents that need concern us are those that can come in contact with the donors testes, other primary reproductive organs, accessory organs, urethra, penis, and prepuce,
4. The semen can be diluted and treated with antibiotics,
5. A representative sample can be obtained from an ejaculate and tested, in a fairly reliable fashion, for the presence of the disease agent, and
6. The preserved semen can be quarantined while its donor is isolated for a certain period of time and retested.

Because the health of the semen *donor* is of utmost importance, the animal may be subjected to Steps 1 through 5, listed in the Previous Section.

The above features, however, have allowed the importation of bovine germplasm, even from countries in which foot-and-mouth disease is endemic, without the application of step 6, which is the very expensive destination quarantine.

Incorporation of embryo transfer into import schemes

It is fortunate that the techniques of embryo transfer have now been developed for many species. We can preserve the embryos of some of these species in the frozen state for considerable periods of time. Although an embryo is just as well equipped genetically as a mature animal, the risk of disease transmission associated with international movement of frozen embryos is less than that with live animals or even semen. This is because the embryo offers the following unique dimensions to host-agent interactions.

1. As with semen, an embryo cannot be involved in the transmission of the larger internal and external metazoan parasites,
2. An embryo, at the age it is usually collected, has had a very limited exposure to disease agents. It contacts only two other individuals of its species before transfer; its sire and its dam. Its contact with its sire is

unimportant, with the rare exception of a gamete-associated disease agent. For all practical purposes, an embryo can acquire an infectious agent only from its dam. We can assume, therefore, that it will have been exposed only to the agents to which its dam has been exposed. Therefore, we can identify embryos that have a low probability of being exposed by applying, to the embryo's dam, those proven techniques we have used to identify individuals that have a low probability of being exposed,

3. Furthermore, because of an embryo's location within the dam, it can be exposed only to the agents we might find in the reproductive and circulatory systems. This aspect greatly limits the number of agents we must consider,
4. Even if an embryo is exposed to a disease agent, its unique anatomy protects it against infection. The noncellular zona pellucida that surrounds the embryo at the time of collection seems to exclude most pathogens. Therefore, if the intact embryo is passed through a series of sterile rinses, agents acquired in the uterus of the dam may be removed from its surface,
5. Within the zona pellucida, the embryo's cellular make-up at the time of collection is not likely to successfully host any disease agent. Most research suggests that agents penetrating the zona will kill the embryo. Dead embryos will usually be discarded at the evaluation stage and not be a concern in international movement,
6. Additionally, in those species in which we are able to obtain acceptable recipient pregnancy rates from the transfer of frozen embryos, we have another valuable tool. As in the case of frozen semen, we can keep the preserved embryos in quarantine while their isolated dam is observed and retested.

Most research on disease transmission by embryos has been conducted in cattle, swine, and sheep. These studies have involved small numbers of replications but, for the most part, they support our intuitive assessment that disease agents are unlikely to be transmitted by the embryo. However, these studies have demonstrated the necessity to study disease transmission by embryos on an agent-by-agent and species-by-species basis.

We know that we can never expect to prove that a disease agent *cannot* be transmitted by the embryo. With many replications of attempts to transmit agents by this route, however, we can estimate a *probability* of such transmission occurring. Unfortunately, we have sufficient replications to estimate this probability with very few agents and in very few species. Therefore, the studies we need, for many agents and many species, are field trials in which large numbers of embryos, sired by naturally exposed fathers and collected from naturally exposed dams, are transferred into susceptible recipients. When the recipients and the resultant offspring are tested, we will have an estimate of the probability of the agent's being transmitted by an *exposed* donor. This probability will provide an idea of how many additional steps we will have to take to increase the likelihood that the donor of embryos we intend to import has *not* been exposed.

Research now is being conducted with bovine embryos using the agents of foot-and-mouth disease and rinderpest. Foot-and-mouth disease also is being investigated in swine. When this work is complete, we hope to develop a safe but economical method of importing previously inaccessible germplasm of these species.

The demand for breeding livestock is not the only demand that animal health authorities have met. Animal exhibitors early argued the benefits of access to wild species in prohibited areas. They cited not only the entertainment value of having these animals for display, but also the education, conservation, and research benefits.

Response to the demand for wildlife species

The wild species that were inaccessible because of animal health constraints were, by and large, ungulates and certain avian biotypes. The equivalent of a good herd history seldom can be provided for wildlife and there is often a question as to the interspecific accuracy of diagnostic tests. Consequently, these animals were a serious health concern. However, animal health officials had more latitude in developing methods to accommodate wild species because no U. S. law specifically prohibited their importation. Therefore, these officials took advantage of the fact that zoological specimens would seldom be involved in domestic commerce and that their use required almost daily observation to develop an eighth step in the import process termed the 'Permanent Post-Entry Quarantine.' This step required the imported specimens to spend their lifetimes in one of several approved zoological parks. Authorities had to be notified when the imports were moved from one approved zoo to another. Any unusual illness had to be reported and the imported wildlife were not permitted to contact domestic livestock.

The additional imperative

It is essential that we take steps to preserve animal genetic diversity. This idea is not contrary to animal health purposes if the preservation is accomplished *in situ*. However, it is implied that certain animal germplasm can be preserved only if it is moved from its country of origin to another, safer, place. Germplasm movement is an animal health concern, especially if it is of a species that can transmit serious disease agents of livestock. This is because there can be no such movement without the risk of spreading disease. However, animal health authorities justified the taking of this risk when the aim of the importers was to exploit the exotic germplasm. The argument for taking risks when the aim is to save a breed, species, or type from extinction should be even more compelling.

Possible response to the additional imperative

The first and most obvious way to respond to this imperative is to use some of the tools we have employed in responding to the previous demands. With the familiar species, especially livestock, we shall use frozen embryos more and breeding animals less.

Other, unfamiliar species, however, will need to be imported. The best response to this imperative usually will involve frozen embryos or frozen semen. We hope that someone is working to develop the techniques that are needed for this approach.

One of the great advantages of being able to freeze the gametes of a species is that it provides us an option in the face of urgency. If a breed or subspecies is in immediate threat of extinction but satisfactory methods have been developed to preserve its embryos or semen, we can hold its germplasm in any reasonably secure facility until we evaluate the health hazard it presents.

It would seem wise to place priority on studies to develop techniques of embryo transfer in species needing preservation. Once this is done, we can turn to developing embryo freezing techniques. Then, perhaps while the critical germplasm is in storage, we can test the species for its ability to transmit disease agents via the embryo.

We also have reason to hope that research in other areas, such as the newer biotechnology, will provide more sensitive diagnostic methods. The focus here should be on tests to determine more accurately the health status of the *donors* of embryos and semen. Epidemiology and disease ecology have enabled us to assess the risk of importing livestock. We expect the assistance of these disciplines in the importation of critical species.

Conclusion

1. There is good reason for the existence of the animal health constraints that have limited our access to world germplasm.
2. Animal health authorities, in response to client demands, have worked within these constraints to develop low-risk methods by which we can access this germplasm.
3. Animal health authorities recognize the imperative of preserving biotic diversity and will, within the limits of our laws and with due regard for all our client interests, work closely with those who must achieve this preservation through the international movement of animal germplasm.

SECTION FIVE

International issues and linkages

Plant genetic resource initiatives in international development*

JOEL I. COHEN[1] and ROBERT BERTRAM[2]

[1] *Biotechnology Specialist, Office of Agriculture, Bureau for Science and Technology, Agency for International Development, Washington, DC, U.S.A.*

[2] *Agricultural Research Specialist, Consultative Group on International Agricultural Research Staff, Directorate for Food and Agriculture, Agency for International Development, Washington, DC, U.S.A.*

Abstract Governments both in developed and developing countries are exploring means to support comprehensive plant genetic resource programs interactive with the international network coordinating global distribution and conservation of germplasm. National plant genetic resource programs can serve as an effective interface between the conservation of a country's biological resources and their availability for use in crop improvement. Recent developments in the global exchange of germplasm, molecular characterization of genetic diversity, and conservation biology, and the increasing size of many collections have focused renewed attention on national germplasm capabilities.

Within the context of international development, genetic resource initiatives have been supported through projects designed to improve crop efficiency and productivity in developing countries. These research programs, many of which are based at international agricultural research centers, have also provided a means to fund crop-specific plant genetic resource activities. This funding includes support for germplasm conservation and evaluation, conventional breeding, and, more recently, biotechnology. Support for plant genetic resource initiatives at the national level is becoming more appropriate for development assistance. However, for this support to become more institutionalized, a number of steps must be taken and suitable policies determined.

Introduction

International agricultural assistance traditionally has supported research to improve crop efficiency and productivity in developing countries. Support of these research programs, many of which are based at international and national agricultural research centers, also has provided a means to fund crop-specific plant genetic resource activities. This funding has included support for germplasm conservation and enhancement, conventional breeding, and, more recently, biotechnology.

---

* The opinions expressed are those of the authors and not necessarily those of the Agency for International Development.

L. Knutson and A.K. Stoner (eds.). Biotic diversity and germplasm preservation, global imperatives.
 ISBN 0-7923-0178-1

A new project proposed jointly between the U.S. Agency for International Development's mission in New Delhi and the Indian Council of Agricultural Research links development objectives with a comprehensive plant genetic resource system. This proposed project builds upon the Government of India's seventh 'Five Year Plan' which identified the conservation of plant, animal, and fish germplasm as a priority area in the field of agricultural research. In accordance with this statement, the Fourth and Fifth Meeting of the Indo-US Subcommission on Agriculture identified genetic resource conservation as an area of priority for joint collaboration [1].

Establishing national priorities for the conservation of genetic resources is but one example of recent developments which help create a framework supportive of collaborative germplasm initiatives. Other examples include: increased attention given to global exchange of germplasm, increased emphasis on creating a sustainable agricultural resource base, increasing progress in the molecular characterization of genetic diversity, new emphasis being placed on conservation biology, and a growing need for a national role in genetic resource programs. These developments have lent importance and urgency to national genetic resource initiatives, allowing them to achieve independent project status as opposed to remaining minor activities subsumed within broad development objectives.

Recent proposals considering genetic resource initiatives in a system and development project context have helped explore national program opportunities. For these activities to continue and become more institutionalized, a number of steps must be taken and suitable policies be determined. To understand these steps and policies, we first describe the context that relates genetic resources and international development activities.

International context for genetic resources and development projects

Genetic resource issues, especially in recent years, often have been characterized in terms of germplasm-rich countries providing and germplasm-poor countries using. In many cases, these terms have been used interchangeably with developing and developed countries, or South and North, respectively. However, these terms are not entirely reflective of the actual occurrence and provenance of genetic resources.

From the aspect of international development activities, especially those that relate to agricultural research and development, a very different situation would appear to be operative. From this perspective, a large, decentralized network of genetic resource activities is evident in both developed and developing countries. This global network, as described by the International Board for Plant Genetic Resources (IBPGR), is composed of a consortium of national, regional, and international institutes cooperating in diverse activities and various segments of germplasm systems [2].

The coordination required for germplasm exchange within the global network illustrates the interdependence that already exists between participating genebanks [3]. This interdependence is helping to erase the artificial distinction between 'germplasm users' and 'germplasm providers.' Whether germplasm represents materials native to a given area has far less consequence than what it is needed for, how much is available, its viability, and national abilities for storage, exchange, and use. The emerging global network uniting users and providers is so interdependent that scientists often assume both roles.

Contrasted with the importance of this global network, genetic resource-specific activities seem to have a relatively low status among development priorities. Germplasm conservation is a long-term activity, the initial investment being subject to considerable recurrent costs. Argument for this type of investment is hampered because few demonstrated economic linkages, such as jobs, production, and income – all important criteria for both donors and developing country ministries – have been ascribed to national germplasm systems. Funds for development must compete with a host of other potential uses; within agriculture, emphasis has been placed on cultivar development and release, extension, and other high-payoff activities.

The sense that germplasm activities are being attended to elsewhere – as in the case of the Agency for International Development (AID) through central funding of the IBPGR, International Agricultural Research Centers (IARCs), or through activities supported by the U.S. Department of Agriculture – must certainly play a role in the lower priority accorded to activities specifically designed to enhance germplasm conservation or use.

The low priority ascribed to germplasm development often is supported by plant breeders burdened with the incessant demand for release of new cultivars. A similar negative attitude can be encountered from development administrators and ministry officials faced with a wide range of pressing opportunities. Many breeders do not utilize exotic material because of the extensive back-crossing required, or they rely on material provided by the IARCs or from pre-breeding programs, if it is used at all. Many breeders do not familiarize themselves with material available from genebanks because of its non-adapted nature, poor combining ability, and their belief that variability in elite germplasm is far from exhausted [4].

Fortunately, the IARCs have the resources and means to assess and use exotic materials and can make the resulting lines available through international nurseries and yield trials distributed to national programs. They cannot, however, carry this responsibility alone, particularly for crops outside their research mandates.

Linking national genetic resource programs to international development

Users and providers of germplasm, as well as conservationists in general, have anticipated the need to further develop capacities in field collection, conservation, and evaluation [5]. However, these needs have not been given adequate attention by either development practitioners or by developing country officials. Mutual collaboration between industrialized and developing countries, derived from improvement of national germplasm systems, has yet to be completely realized.

As suggested by T.T. Chang, cooperation between countries could begin around the following generalities: developed countries could serve as sources of funding, preservation and distribution sites, and training centers, while developing countries could serve as centers of diversity, rejuvenation of seed in home environment, and *in situ* conservation [6]. We recognize that these distinctions currently exist; however, we expect them to become less dramatic as national programs of developing countries evolve and their capacity for preservation and distribution increases.

There are several concerns that development agencies may wish to have addressed before supporting national plant genetic resource programs. First, they will need to be convinced that the proposed activity does not duplicate germplasm research being conducted elsewhere. This issue of duplication will have to be balanced by:

1. The concept of functional redundancy within the larger international system, and
2. Relationships established between Genetic Resource Units (GRUs) located at particular international centers and the proposed national germplasm system.

The relationships between national germplasm systems and germplasm activities at international agricultural centers will become even more important as national programs evolve. A working balance between the two systems will depend on the specific role each IARC determines for its germplasm responsibilities and national priority established for a related genetic resource system (Fig. 1). Future support for national genetic resource activities will require better coordination so that clear responsibilities are established between the two systems.

While most of the IARCs have sizeable GRUs, each center determines the priority and role that such units have within their mandates. Certain points of commonality between the GRUs have been established, including their overall goal to conserve and safeguard samples of extant diversity from their mandate crops and to make these freely available to all users [7]. However, the resources available at each center for its GRU vary greatly.

Figure 1. Balance of responsibilities for genetic resource activities divided between the international agricultural research centers and enhanced national programs.

The International Rice Research Institute (IRRI) has accepted the mandate for conserving the full spectrum of rice germplasm and has placed this responsibility within its International Rice Germplasm Center [8]. Other centers, such as the International Center of Improvement of Maize and Wheat (CIMMYT), are seeking to facilitate greater national abilities in germplasm systems and have informal agreements with IBPGR regarding germplasm storage responsibilities for their mandated crops [9].

These understandings between national and international genetic resource programs should evolve in a complimentary manner, similar to those established by the IARCs and national programs regarding responsibilities for germplasm development and subsequent cultivar release [10] (Fig. 2).

A second area of concern to potential donors is the degree to which a proposed activity complements existing agricultural research and development activities in the host country. Development programs only recently have considered projects for the management and construction of national germplasm programs. However, indefinite support of national systems by donors is not feasible, so each country must be prepared to assume recurrent costs and plan for future expansions eventually. Projects to develop national germplasm systems must complement existing development priorities and include host country commitment to assume recurrent costs on an agreed upon timetable.

Development projects traditionally provide incremental, focused assistance through host country governments and donor missions for specific priorities of national economic importance [11]. Development agencies have focused on enhancing national capabilities determined by host country

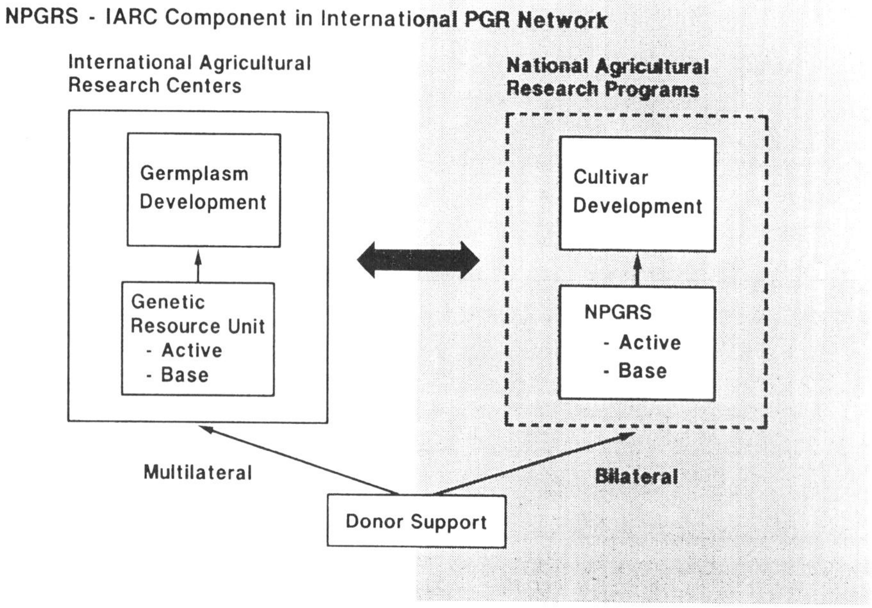

Figure 2. Division of responsibilities for germplasm and cultivar development integrated with division between responsibilities for genetic resource activities.

projections and donor strategy statements. Projects developing national capabilities in germplasm will have to be included in future documents of this type while gaining acceptability at the mission, country, central, and regional bureaus and at Congressional levels. Efforts of this type will link genetic resource activities with vital economic and production strategies and policies. This linkage, focusing on utility, is a vital step for all nations establishing a national germplasm system.

Another consideration involves the relation of proposed genetic resource systems to conventional plant breeding programs. Support provided to enhance national capabilities in germplasm management has a different goal and purpose than does support provided for cultivar development. However, for genetic resource systems to be effective, they must be an independent yet integrated part of a country's overall agricultural research program. The linkage between genetic resource activities and plant breeding can be fostered through active collections and pre-breeding, or enhancement; however, such activities alone will not ensure collaboration and flow of genetic materials. This will depend upon initiatives of both breeders and germplasm managers.

It is through such linkages that obstacles to using new and unusual germplasm in breeding programs can be partially overcome. Breeders can make themselves familiar with available variability by visiting genetic resource nurseries. Germplasm curators can produce and distribute catalogs of available accessions, noting traits of special interest to breeders.

Linkage between cultivar development and germplasm systems are best developed through interdisciplinary crop advisory committees able to advise on development priorities simultaneously (Fig. 3). This will provide justification of support at a more basic level by determining to what degree breeders are able to use unadapted or unimproved germplasm. No simple answer exists for this question. In countries where resources are severely limited, any diversion of effort from more urgent problems may meet with opposition.

It is fair to say, though, that plant genetic resource activities have been neglected in many countries, with traditional agricultural research programs receiving most of the support given to plant sciences. We think that some shift in this balance, or lack of it, is warranted at this time. First, from the point of view of genetic resource programs, a significant international system now exists to back up and benefit from the operation of national programs. National programs add depth to the international network and greatly increase accessibility of germplasm to potential users. On the plant breeding side, genetic resource programs remain a longterm activity, but their importance and potential contribution cannot be ignored. Interest in genes from germplasm collections increases as crops require additional stress tolerance or disease resistance characters.

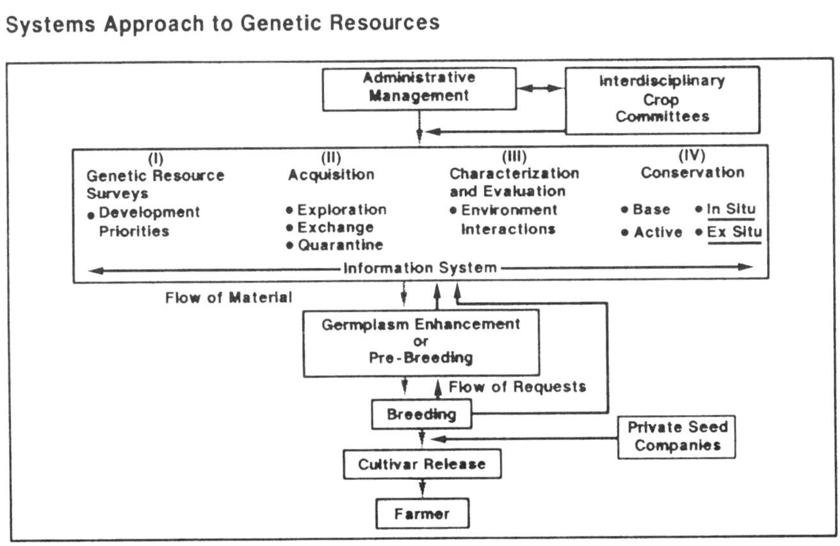

Figure 3. National genetic resource system components and flow chart with linkages to interdisciplinary crop advisory committee, breeding, and seed companies.

System and project context for national germplasm programs

Development support for a systems approach to plant genetic resources (Fig. 3) represents modifications based upon previous examples of genetic resource impact chains [12]. Donors will decide on a case-by-case basis which components of this comprehensive system will be supported. Administrative management of a complex germplasm system is a job not to be underestimated. It is in this area of human resource development that donor agencies have extensive experience. However, decisions as to where and how to develop such management will require coordination with other national systems.

Within the main body of the system itself are four germplasm-oriented disciplines: genetic resource surveys, germplasm acquisition, characterization and evaluation, and conservation. A fifth area, computer-based information systems, connects the other four. Each component has its own requirement for training, research, and management, in addition to its operational responsibility.

Major components within each of these areas are listed in Fig. 3. It is recognized that these may vary, depending upon the system required. The ability to plan for genetic resource surveys and conservation sites with development perspectives in mind is an important feature. Included within acquisition are exploration, exchange, and a fully operative plant germplasm quarantine system. Characterization and evaluation must include plans for localized trials to evaluate genotype by environment interactions and should correspond with needs of breeders. Conservation includes not only base and active collections but also reflects relative national priorities extended to *ex situ* and *in situ* efforts.

Once material has been processed through this internal system, it is available for use by plant breeders, pathologists, entomologists, and other researchers. It either can be used directly or enter pre-breeding programs for germplasm enhancement, with agronomically fit material advanced to breeding programs for further introgression. Enhancement can include combining or pyramiding of desired genes for disease resistance or other special characters. The exact path taken will vary from crop to crop.

The connections established here between users, both public and private, are key linkages between national germplasm priorities and conventional development priorities of cultivar improvement. Linkages with private seed companies, both national and international, can be particularly valuable as a means to increase support for germplasm activities while increasing the role of genetic resources in commercial product development.

Potential goals for projects supporting a national germplasm system could include the following: conservation of biological diversity, preservation of a country's sustainable agricultural resource base, and modernization of scientific institutes. The first purpose, at the project purpose level, would be to provide the means to complete a comprehensive, national plant genetic resource system which will strengthen, enhance and coordinate the facilities and research activities required for conserving, collecting, evaluating, and exchanging plant genetic resources. The second purpose would be to enhance a nation's role in the global exchange network, including the operation of a safe and efficient germplasm quarantine system.

Once goal and purpose have been established, actual objectives or accomplishments must be determined. This entails deciding on the level of support for each component of a national system targeted for development. Determining appropriate support for components necessitates decisions based in part upon items presented in Table 1. Based upon these considerations, specific accomplishments for each individual national project will be determined. Potential accomplishments include examples listed in Table 2.

Not all countries or national germplasm systems will warrant such complete investment. However, donors may find it useful to develop priorities for potential funding requests from countries enlisting their support. Criteria for screening such requests should take into account items enumerated in Table 1.

Advantages provided by efficient national germplasm programs

Modern and effective national germplasm programs can provide for species-specific research and evaluation that are often beyond the scope of the IARCs. This is particularly important to non-mandate crops, which are

Table 1. Supporting components of national germplasm systems.

Decisions should be based upon the following criteria:
1. Current operational status of national system
2. Number of accessions held at present and number envisioned at completion of project
3. Number and type of species (crops, primitive cultivars, wild and weedy relatives) to be conserved
4. Potential number of users
5. Present status of base and active collections
6. Willingness and ability to participate in the reasonable, free, and open exchange of germplasm
7. Potential designation by International Board for Plant Genetic Resources as a base collection site
8. Regional importance placed upon development of particular national system and its potential role in the international system for germplasm exchange

Table 2. Accomplishments attributable to projects supporting national germplasm system projects.

Potential project accomplishments would include:
1. A comprehensive inventory of national base and active collections
2. Upgrading the quality of localized active collections
3. Strategic design of future collaborative plant exploration and collection trips
4. Priorities for new *ex situ* collections and for *in situ* reserves
5. Standardized germplasm data and information base
6. Enhanced base collection facilities
7. Establishing interdisciplinary crop advisory committees
8. Management support and human resource development
9. Enhancing germplasm quarantine facilities and procedures

often of considerable importance to local economies and which form an integral part of indigenous farming systems. National programs concerned with these indigenous crops provide valuable support for conservation and evaluation activities that might otherwise be neglected.

An important component of this research includes evaluation in crop-specific environments to give better estimates of genotype by environment interactions. Local evaluation solidifies the tie between national programs and regional, active collections. Support to active or working collections makes them an integral part of development efforts by linking genetic resource systems with conventional breeding programs and assisting in the characterization and evaluation of germplasm (Fig. 4). These and other advantages derived from enhanced national programs are listed in Table 3.

Nationally supported active collections generally handle materials for

Table 3. Advances derived from enhanced national germplasm systems.

Potential advances derived from development assistance to national systems:
1. Ability to participate in international germplasm exchange
2. Provide for species-specific evaluation often beyond the scope of the international agricultural research centers
3. Evaluation in agreement with plant breeding objectives, facilitated through support to active collections
4. Better estimates of genotype by environment interactions
5. Coordinated user information network
6. Centralized control over germplasm quarantine procedures
7. Catalyst for collaborative collection and conservation
8. National systems will develop a broader base of support in host-country scientific community

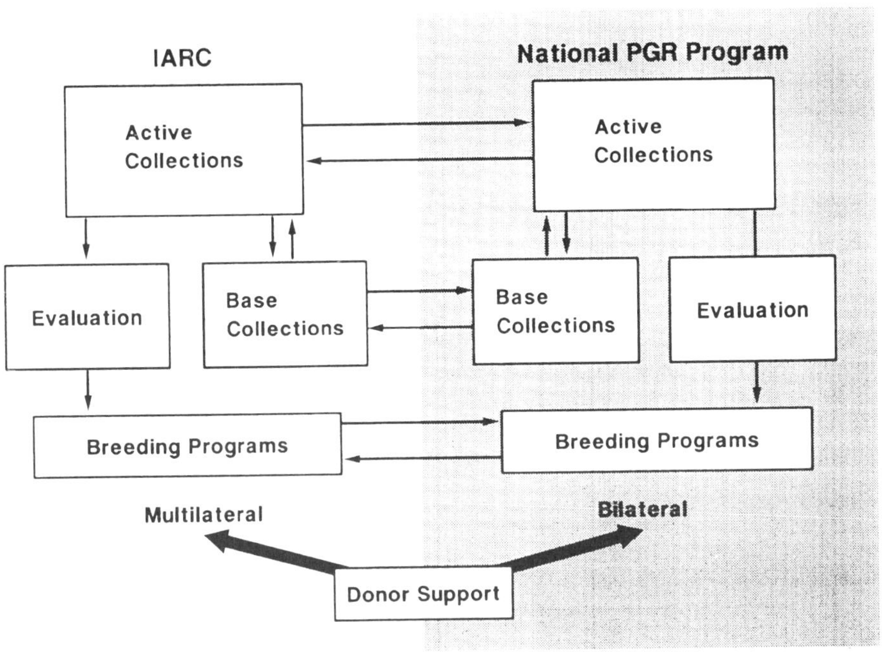

Figure 4. Linkage between collection (base and active), evaluation responsibilities, and breeding programs.

short to intermediate periods of storage and have storage and handling requirements that are less exacting and comprehensive than those of base collections. They generally turn materials over faster and are better situated in decentralized locations than long-term base collections. These regional locations provide breeders with easier access to materials and provide for observations in environments similar to those in which they would be used. Field space will be needed for such grow-outs, and this may require linkage to university or agricultural research centers. Recurrent costs associated with maintaining and evaluating germplasm will need to be calculated, but this is an area in which many developing countries have a distinct advantage, since labor costs are relatively low.

Not surprisingly, the IBPGR has made the development of active genebanks an institutional priority, in part because they help bridge conservation and use [13]. Active genebanks substantially contribute to the documentation and information available to base collections through their regeneration, characterization, multiplication, and distribution of genetic materials. Active linkage favors the development of national systems driven by demand and interest of the local research community, increasing the probability that resources will be provided to conduct the activity beyond the period of donor support.

Just as is the case with active genebanks, developing countries have considerable comparative advantage in relation to field genebanks. Field genebanks maintain vegetatively held material and are especially useful in handling perennial and woody species. Many tropical species cannot be handled as dried, cold-stored seeds and instead must be held for short periods or maintained in the field. Spices, herbs, and medicinal and plantation crops are among the plants that may best be handled in this way. In addition, literally hundreds of species of fruits and nuts may be best suited to management near their native habitats. A growing interest in perennial crops can be envisioned as farmers and researchers in humid and sub-humid tropical regions search for environmentally and economically sustainable multi-canopy production systems.

Recent developments affecting plant genetic resources

A number of developments – scientific advances, policy and priority shifts, and accomplishments – have brought about changes that prompt us to reconsider the role of plant genetic resource activities within the international development area. First, it must be noted that for some of the major food crops, a significant portion of the genetic diversity in primitive cultivars and land races has been collected. As a result, increased emphasis now is being placed on efforts toward their characterization and use which are often time consuming, labor intensive, and costly. The ability is desired to divide these responsibilities among national and international genetic resource systems efficiently and effectively.

A second factor increasing the use and exchange of plant genetic resources is biotechnology. These technologies presently consist primarily of *in vitro* tissue culture procedures designed for germplasm activities. Only now is the potential for transferring genes and chromosomes through recombinant DNA technologies being realized. These technologies will make more possible the selective use of wild relatives to crop species or even completely alien genes without the rigorous backcrossing required by conventional selection techniques. Other potential applications of biotechnology to germplasm research are presented in Table 4. The need to integrate these and other applications of biotechnology with conventional plant breeding has been established [14].

In some developing countries, especially those in Asia, remarkable gains have been made in the production of rice and other staple foods – a major accomplishment in a region only 20 years ago thought by many to have insurmountable food shortage problems. This progress has engendered interest in and demand for diversification in agricultural production. High

Table 4. Potential applications of biotechnology for integration with genetic resource activities.

Components of conventional germplasm-based technologies	Conventional Timespan	Potential biotechnology contributions
Acquisition/Exchange	1 year	*In vitro* culture, disease indexing, and eradication
Conservation	Ongoing	*In vitro* conservation
Evaluation	2 seasons	Molecular diagnostics
Germplasm Enhancement	3–5 seasons	Embryo rescue, molecular diagnostics, selection in tissue culture, somaclonal variation
Wide Hybridization	2 years	Embryo rescue, somaclonal variation, anther culture, protoplast fusion

value horticultural crops are taking on new importance, including some that may have had little improvement, especially at the local level. Scientists in developing countries increasingly may become interested in considering a broader variety of crops, some of which may have needs that require support from or interaction with an accessible germplasm collection.

In 1986, Congress enacted legislation that annually earmarks a certain portion of foreign assistance dollars for conservation of biological diversity. While specifically targeting tropical nations as priority areas for conservation activities, this legislation moreover has enhanced the connection between conservation, biological diversity, and plant genetic resources. The AID has responded to this congressional request by soliciting funding proposals from appropriate missions, bureaus, and offices.

This legislation has proved to be a valuable impetus in looking beyond genebanks for effective conservation of plant genetic resources, possibly through *in situ* reserves, although few working examples of such systems remain. New efforts are needed linking the need for conservation and quantification of genetic resources with conservation biology [15]. For example, documentation is urgently needed of areas in greatest need of conservation and which contain primitive, wild, or weedy species suited for *in situ* preserves.

Genetic resource programs and international crop improvement

Despite a lack of support for the reasons described above, activity in this area has been considerable. In the case of the AID, such activity has been undertaken mainly through central funding channels with support provided to institutions directly associated with plant genetic resources. Ongoing support has been provided to the IBPGR since its founding. In fact, for most of its history, the AID has been the largest contributor. The IBPGR not only has helped to ensure that genetic diversity is collected, catalogued, and conserved, but also actively promotes unrestricted international exchange of plant germplasm in these collections. In many cases the IBPGR provided direct assistance to national programs in genetic resources, through training, provision of equipment, or critically needed funding associated with collecting or conserving important germplasm.

In recent years, the IBPGR has sought more actively to encourage the use of scientific techniques to minimize the loss or destabilization of genetic diversity represented in collections. To this end, IBPGR is sponsoring research programs in collaboration with a variety of institutions regarding strategic areas relevant to plant genetic resource programs. Current research thrusts include seed physiology, *in vitro* conservation, pathology and germplasm exchange, analysis of genetic variation patterns on an ecogeographical basis, and preservation of genetic integrity during regeneration of accessions [16]. The overall contribution of IBPGR in all of these areas has been far out of proportion to its size [17].

Through its contribution to the IARCs, the AID has supported varying degrees of germplasm activities on an array of important food crops. At some centers, the GRU primarily constitutes an active collection directly supporting breeding programs. Other centers have developed a full world collection for their mandated crops, including wild relatives. In most cases, because of proximity, available resources, and the nature of the breeding objectives, breeding programs at the centers have been able to tap into a wide variety of materials contained in their collection.

To encourage the exchange of germplasm, the Plant and Seed Materials project was initiated by the AID in 1973. This project operates under the leadership of Dr. George White at the Plant Introduction Office at Beltsville, Maryland. The project provides limited quantities of plant and seed materials for evaluation in developing countries. Since its initiation, more than 100,000 varieties or selections of virtually all economically important crops have been sent to 100 + countries.

Through its Office of the Science Advisor, the AID supports small research grants, both directly and through the National Academy of Science, on a wide array of topics related to development. The AID program seeks to build research linkages and training opportunities with scientists from developing countries and is complementary to the larger program thrusts of the Agency. Biological diversity is one of the program's disciplines, and a number of these grants support plant genetic resource-related research. The program includes characterization and systematic evaluation of wild relatives of crop species. Examples of supported projects include assessment of potential sources of disease resistance, techniques to sample and characterize germplasm of important palm species, and techniques to broaden collections by use of remote sensing and ecological data.

The above activities have, in our view, been very effective. Each has been evaluated for relevance and contribution to overall objectives of agricultural development. We anticipate that these or similar activities will long be needed. However, it is neither our contention nor the main objective of this paper to argue that such activities continue to suffice if plant genetic resource opportunities and needs are to be fully addressed. This is especially true as the scope of plant genetic resource work expands from collection and conservation to include characterization and use more prominently. Just as in the developed countries, national systems responding to national priorities, capabilities, and comparative advantages will need to be developed in developing countries in the vital work of germplasm conservation and use.

Policy implications

We have presented herein the rationale for how a national plant genetic resource system could be appropriate for certain developing countries. There are clearly policy-level impacts from this (Table 5). Support for plant genetic resource activities will hasten the development of national strategies for germplasm conservation and use. The activity could garner new interest, particularly from development practitioners. Such attention reflects the fact that genetic resource initiatives can stand alone among activities in portfolios of developing country and donor activities and will help to ensure that a higher priority be accorded such activities in the future.

Regarding the scientific community of a single country, we have noted that purposeful efforts to integrate genetic resource activities into national agricultural research will help ensure broader support and interest (Fig. 5).

Table 5. Policy implications of support provided to national germplasm systems

Policy considerations for donors to consider:
1. Provide opportunity to develop and pursue national germplasm strategy
2. Interdependence of collections provides rationale for free exchange
3. Shared responsibility of collections minimizes ownership concerns
4. Ability to link agricultural and natural resource researchers through common conservation goals

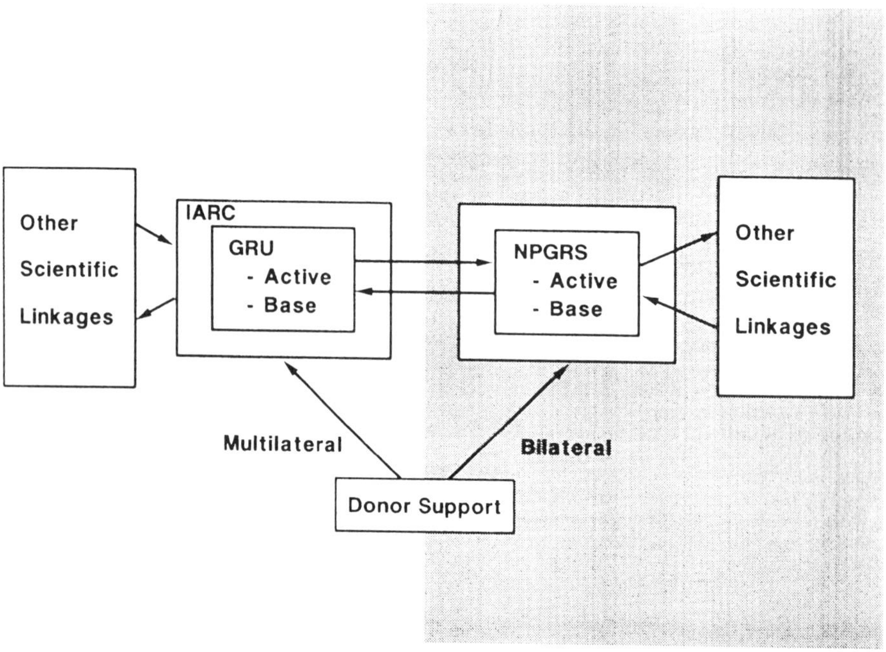

Figure 5. Linkages provided to other scientific communities involving either national or international genetic resource units.

Plant genetic resource programs are unusual in that they link diverse disciplines – botany and ecology – with natural resource interests, and link plant breeders with pathologists, entomologists, and other agricultural scientists. Such linkages need to be fostered so that agricultural priorities are no longer viewed as being at odds with conservation priorities. Genetic resource work is, after all, applied biodiversity conservation and can help underscore areas where conservation and agricultural priorities intersect.

Returning to the operations level of genetic resource initiatives, a decentralized approach such as advocated in this paper has important implications for how germplasm activities should be designed. A decentralized system emphasizes the interdependence of all germplasm holders, users, evaluators, and collectors. An extension of this concept is that of free exchange – a system in which all users are mutually dependent – hardly is workable without a global commitment to the unrestricted distribution of materials to all bona fide workers, regardless of institutional or political location or affiliation. Departure from the concept of interdependence also means retreat from the doctrine of free exchange and advocating a doctrine of ownership. This is most frequently encountered in the case of tropical cash crops, but even in those cases it represents a limited and unduly protective viewpoint.

Increased involvement of developing countries in the international germplasm system can only strengthen the mutual goodwill of this network [17]. The shared responsibility required to manage each national system will build a constituency to ensure that the benefits and advantages of the global network are broadly understood, whereas the shared responsibility required to manage the international system will help alleviate concerns over questions of ownership, accessibility, and benefit from germplasm activities. Developing a clear role for national systems within the context of the global network can help ensure that scientific cooperation and progress, not politics, determine policy.

References

1. Indo-U.S. Subcommission On Agriculture. 1988. Report of the Fifth Meeting, December 7–11, 1987. Government of India, New Delhi.
2. IBPGR. 1986. Annual Report 1985. International Board for Plant Genetic Resources, Rome. 92 pp.
3. Plucknett, D.L., N.J.H. Smith, J.T. Williams and N.M. Anishetty. 1987. Gene Banks and the World's Food. Princeton University Press, Princeton, New Jersey.
4. Cohen, J.I. and W.C. Galinat. 1984. Potential use of alien germplasm for maize improvement. Crop Science 24: 1011–1015.
5. Esquinas-Alcazar, J.T. 1987. Plant genetic resources: a base for food security. Ceres 118: 39–45.
6. Chang, T.T. 1987. The availability of crop germplasm. Pp. 225–231. In Crop Exploration and Utilization of Genetic Resources. Proceedings of International Symposium, Changhua, Taiwan. Taichung District Agricultural Improvement Station, Taiwan, R.O.C.: 6–12.
7. CGIAR Center Director's Statement. 1987. Statement of Common Policies on Plant Genetic Resource Activities in the Consutative Group on International Agricultural Research. International Food Policy Research Institute, Washington, DC.
8. Chang, T.T. 1984. The role and experience of an international crop-specific genetic resources center. Pp. 35–45. In Conservation of Crop Germplasm-An International Perspective. Crop Science Society of America, Madison, Wisconsin.
9. Hendry, P. 1988. An interview with Donald L. Winkelmann. Ceres 116: 40–44.
10. CGIAR. 1985. 1984 Annual Report. Consultative Group on International Agricultural Research Secretariat, Washington, DC.
11. International Development Cooperation Agency. 1987. Development Issues: U.S. Actions Affecting Developing Countries. Washington, DC.
12. Hawkes, J.G. 1985. Plant Genetic Resources The Impact of the International Agricultural Research Centers. Consultative Group on International Agricultural Research Study Paper 3. World Bank, Washington DC.
13. IBPGR. 1986. IBPGR Programme Structure. International Board for Plant Genetic Resources, Rome.
14. Cohen, J.I., D.L. Plucknett, N.J.H. Smith, and K.A. Jones. 1988. Models for integrating biotech into crop improvement programs. Bio/Technology 6: 387–392.

15. Strauss, M.J., J.I. Cohen, and J. Pino. 1988. Quantification of diversity in *ex situ* plant collections. Abstracts of Papers, 1988 AAAS Annual Meeting, p. 52. Am. Assoc. Adv. Sci. Washington, DC.
16. IBPGR. 1987. Annual Report 1986. International Board for Plant Genetics. Rome, Italy.
17. Wilkes, G. 1983. Current status of crop plant germplasm. CRC Critical Rev. Pl. Sci. 1(2): 133–181.

Threatened habitats and germplasm preservation: a Central American perspective

RODRIGO GÁMEZ
CIBCM, Universidad de Costa Rica, Ciudad Universitaria 2060, Costa Rica

Abstract Central America and Mexico, historically a focal point in the development of agriculture, have long been recognized as one of the most important centers of biotic diversity in the world. Yet deforestation, soil erosion, over-exploitation of natural resources by traditional and modern methods, and related environmental destruction are severely threatening the rich biodiversity of the region. Disruption of the wide variety of ecosystems and habitats results mainly from a rapidly growing population, complex political and socio-economic pressures, and lack of knowledge of ways to carry out sustainable agriculture and forestry. Present concerns about the disappearance of irreplaceable germplasm are based on the interest of preserving the several hundred different species of plants and animals of economic importance, organisms that have been domesticated or harvested by indigenous cultures from wild or semi-wild populations. These populations still harbor equally valuable genetic resources of potential use in modern agriculture and industry, as do the tens of thousands of other species that occur with them. The preservation of Central American germplasm will depend largely on the conservation of the wide range of habitats and cultural traditions typical of the area. This task will be attainable only if it is perceived as a shared responsibility of the Central American and developed nations.

Key words: Central America, habitat destruction, tropical biodiversity, tropical conservation, cultural conservation, socio-economic development

Introduction

The destruction of natural resources in Central America (Belize, Guatemala, Honduras, El Salvador, Nicaragua, Costa Rica and Panamá) is receiving increasing public attention as the effects of the rapid deterioration of the environment anticipated by biologists and conservationists become evident to politicians, economists, planners, and other sectors of the population of this region [1, 2]. One of the many reasons for concern about the conservation of germplasm of direct agricultural interest, and the protection of wildlands in Central America and Mexico, is that this area has been long recognized as a historical focal point in the development of agriculture. It is one of the most important centers of biodiversity and genetic resources of economic interest in the world [3–11]. Several crops of worldwide importance, such as maize, beans, squash, peppers and cacao, are only a

L. Knutson and A.K. Stoner (eds.). Biotic diversity and germplasm preservation, global imperatives.
 Printed in the Netherlands. ISBN 0-7923-0178-1

few of the several hundred different species of plants domesticated or harvested from wild populations in the region by pre-Columbian indigenous cultures. Deforestation, soil erosion, overexploitation of natural resources by traditional or modern technologies, and related environmental destruction are severely disrupting the wide variety of Central American ecosystems and habitats typical of the region. An additional concern is that the economies of the Central American countries, as well as the vast majority of the population, depend heavily upon the natural resources of the region. Yet, these resources are being rapidly degraded [1].

Here I discuss how habitat destruction affects the preservation of genetic resources in Central America, and what I perceive as some of the main socio-economic forces driving this process. These ideas are derived from a variety of published and unpublished observations by a number of people, including myself, with different backgrounds, experiences, and viewpoints. The fundamental concern is the conservation of the tropical world and its indigenous cultures, with a parallel feeling of urgency for action because... 'we can see clearly what is happening, what will be the irreversible consequences for biology and humanity, and how the solutions must be constructed' [7]. If our generation cannot save the tropical world, it will not be there for the next.

Natural characteristics of Central America

A detailed description of the geographic characteristics and natural resources of Central America was recently published [1], and only the main features of this account are outlined here. The Central American isthmus is a relatively narrow land bridge, situated in the northern Neotropics, extending from approximately 7° N latitude to about 18° N latitude. The vegetation that originally covered most of this land mass was forest. It is a land of extremes. Complex plate tectonics and active volcanism have shaped the distinctive geographical features of the region. Most of the landscape is extremely rugged terrain, with nearly three-fourths of the area being hilly or mountainous, with highland valleys lying above 500 m in altitude. Aside from the narrow Pacific and Caribbean coastal plains, there is little flatland in Central America. As a result of its altitudinal diversity, its location between two large oceanic weather regimes, and its being perpendicular to the trade winds, the isthmus has a highly variable climate within and between years. While hot and humid climates prevail in the lowlands, the highlands are cooler. The range of rainfall is extremely wide: Costa Rica is one of the rainiest countries on earth (1,000–7,500 mm per year) but the Motagua valley of Guatemala (500 mm per year) is as arid as the deserts of the southwestern United States. Total average rainfall tends to increase from north to south along the isthmus. The length of the dry season is more important in distinguishing the climatic variations of Central America than is the total amount of precipitation. In some areas such as the Atlantic watersheds, rain falls almost year-round. However, in the Pacific coastal plains, the dry season is more severe and may include from five to eight rain-free months. That is to say, the eastern side of Central America was primarily rain forest, while the western side was dry forest. Local climates vary widely between these extremes.

Because of its volcanic origins, Central America is known to have some of the most fertile soils on earth. Yet for every hectare of these rich soils, there are probably about two hectares of very low quality soils, soils that are shallow, saline, arid, lateritic, waterlogged, leached, extremely steep, or very rocky.

Due to its geographical position as a bridge between two continental land masses and its geological, topographic, and climatic characteristics, Central America is a mosaic of ecosystems. Its forests provide a huge assortment of habitats containing millions of species of organisms. Viewed as repositories of biological diversity, these forests are among the world's most species-rich ecosystems [8, 9]. About 1300 ha. of forest at La Selva Biological Station in the Atlantic lowlands of Costa Rica are thought to support nearly 1.5 times the number of plant and animal species known for the entire state of California [10]. Santa Rosa National Park, 108 km^2 in area in the dry forests of northwestern Costa Rica, contains as many as 13,000 species of insects, 700 species of plants, and 400 species of vertebrates [11].

Ethnic and cultural diversity in Central America is almost as great as that of its flora and fauna [1, 12]. The majority of its present population is a mixture of European, Mesoamerican Indian, African, and Caribbean stocks. Although Spanish is the predominant language, numerous local Indian languages as well as English and Creole are used, at times interchangeably. The bulk of the population has traditionally inhabited the mountainous highlands and the Pacific slopes and lowlands.

Genetic resources and agriculture

The diversity of genetic resources of direct interest to humans has remained high only in countries and regions where traditional agricultural methods are still practiced; modern agriculture, based largely on uniform crop varieties, ignores the heterogeneity typical of the indigenous tropical agroecosystems and leads to their extinction [3, 13]. This is generally the case throughout Central America. Traditional farming systems in the isthmus certainly have been characterized by diverse use of plant species and genetic diversity in space and time, both in the form of polycultures or agroforestry systems [14]. For example, autochthonous Indian populations in southern Costa Rica still use over 90 different species of cultivated, semi-cultivated, or wild plants, including trees for food, medicine, construction materials and other basic needs [15]. Similar practices are still followed in Mexico and Guatemala [16]. These diverse systems are designed to fulfill complex nutritional, agronomic, and socio-economic needs [13] rather than to fuel an international market community.

Useful wild species such as *Lycopersicum esculentum* var. *cerasiform* (Dunal) Hort may be allowed to grow as weeds in cultivated fields or backyards; fruit from wild trees (*Pouteria* spp., *Licania platypus* (Hemsl.) Fritsch, *Inga* spp., *Hymenaea courbaril* L., *Cassia grandis* L., *Manilkara zapota* (L.) Royen) are harvested from secondary forests. Planting semi-wild as well as domesticated fruit trees around houses to provide food and shade is a common practice in the Pacific lowlands of Central America. Probably for agronomic, ecological, or cultural reasons, many species of economic interest are highly restricted in distribution. 'Tacaco' (*Frantzia tacaco* (Pittier) Wunderlin), a vegetable fruit, is only known as a cultivated species in the highlands of Costa Rica; 'loroco' (*Fernaldia pandurata* (A.DC.) Woodson) and 'pito' (*Erythrina* spp.) flowers are only consumed in El Salvador and Guatemala; young stems of 'chipilin' (*Crotalaria* spp.) and 'ixtlán' (*Solanum* spp.), as well as the inflorescence of 'pacayas' (*Chamaedorea* spp.), are used as occasional food. Some commercial products are still harvested from wildlands; these include 'chicle' (*Manilkara zapota* (L.) Royen) and 'ame' (*Dioscorea* spp.) for cortisone, collected from Guatemalan and Mexican forests; 'ipecacuana' (*Cephaelis ipecacuanha* (Brot) A. Rich.) is directly harvested from Costa Rican and Nicaraguan rain forests. Central American forests still remain a major source of wild species of ornamental orchids, bromeliads, ferns, and foliage plants [6].

As in other areas where agriculture was developing, the hunting-gathering activities of the early populations were followed by a gradual process of domestication and selection of primitive plant cultivars. After several millennia, Central American cultures were using combinations of several hundred plant species. Recent inventories [6] have identified over a hundred different species of these plants, many of which are still commonly used in the area as cereals, dyes, stimulants, fibers, animal feed, fruits, grains, vegetables, medicine, oil sources, and ornamentals. Although many of these plants are not used outside Mesoamerica, some constitute staple foods of local or regional interest. They include several species of worldwide importance such as maize, beans, cucurbits, cacao, tomatoes, peppers, cotton, and orchids. In addition, genetic variability is wide within each cultivated species. Much of this variability is still found in the wild, both in the same or in closely related species. In the last few years, botanical explorations in Costa Rica and Guatemala have even revealed the existence of several previously unknown species of *Phaseolus* [17].

While not stressed here, the same people who harvested wild plants also harvested a wide variety of vertebrates as meat animals, and aquatic systems were for them a fountain of species-rich invertebrates. Such traditions still exist throughout the area [1, 12].

Germplasm in wildlands

Above I touched on species, mainly plants, that came into use by pre-Columbian cultures. They were chosen by indigenous people several millennia ago, obviously according to their nutritional, agronomic, and socio-economic needs and the ability of the cultures to process technically and propagate food. The species that they chose represent only a small fraction of the millions of species still surviving in tropical wildlands and represent perhaps the least exploited of the biological resources of the area [1, 7]. Although seemingly of limited economic value today, many of those organisms are undoubtedly capable of providing useful products to fulfill man's modern needs. For example, how could one predict the potential value of a foul-tasting seed from a rainforest understory shrub, otherwise known as coffee? Wild organisms offer potential sources of pharmaceuticals, natural chemical products, and germplasm for the improvement of existing crops and the development of new domesticated species for food and biochemical processing. Realization of this potential requires, among other things, saving the wildlands, imagination, gathering basic information about the traits of the organisms, data storage and retrieval systems for the accumulated information, financial resources, selective breeding, biotechnology and genetic engineering, gene and species banks, and basic understanding of tropical biology [7].

While there is a rising lament over the loss of wild germplasm, the fact is that the agricultural industry is extremely conservative about exploring its potential. Examples of new uses or crops appear infrequently, examples that would stimulate an avalanche of awareness of the human loss as wild tropical biodiversity is destroyed. For example, Central American wildlands are also known to be rich depositories of wild genotypes of presently cultivated species and/or their close relatives of *Phaseolus* [17] and *Zea* [4]. Nevertheless, with a few notable exceptions this genetic information remains unused.

In spite of the current accelerated destruction of natural habitats in

Central America, fragments or remnants of the original wildlands still remain, either within reserves or interspersed among crops and ranch lands. The reason for this is that until recent years (and from a European perspective), agricultural exploitation of the tropics has been 'inefficient'. Highly 'efficient' agriculture destroyed much of the biological diversity of North America. Little remains to allow restoration of extra-tropical habitats on good soils, as for example in the eastern United States. On the contrary, the Guanacaste province of Costa Rica, after 400 years of cultivation since the Spanish colonization, still retains enough remnants of the tropical dry forest that once covered the area to allow a process of restoration of the original species-rich ecosystems and habitats [18].

Threats to the natural habitats

The deterioration of the natural environment of Central America has recently accelerated rapidly as a result of large-scale political, military, social, and economic problems. Although it is not within the scope of this presentation to treat in detail these complex issues, they cannot be ignored in any analysis of the future of the rich ecosystems of this area in particular or of the world in general [19, 20].

Political and socio-economic context

The development of the seven Central American countries faces critical political, economic, and social problems. A detailed analysis of the prevailing political and socio-economic trends and their impact on the natural resources of the region was recently published by Leonard [1]. Here I summarize some relevant aspects of this account.

Nicaragua, Honduras, El Salvador, Guatemala and Panamá have suffered decades of political turmoil and uncertainty associated with military actions, social unrest and unstable democratic processes. On the contrary, Costa Rica, a country that constitutionally eliminated its army several decades ago, has remained as an island of stability, peace, democracy, and social justice.

Population growth is one of the major problems facing Central America. There were 26 million Central Americans in 1986; this is a doubling of the population since 1960 and a fivefold increase from 1920, when 5 million people inhabited the region. The population is growing at an annual rate of 2.8%, only slightly less than that of Africa. The growth rates of nearly 3.5% per year for Nicaragua, Honduras, and Guatemala are extraordinarily high, and these countries represent 60% of the region's population. El Salvador, Belize, Costa Rica, and Panamá have more moderate growth rates

(2.0–2.7%), but in the past only Costa Rica and Panamá have been able to reduce population growth rates. The geographic distribution of the population is also very uneven. El Salvador is one of the most densely populated countries in the world (245 persons per km^2), while Belize is only sparsely populated (6 persons per km^2). But except for El Salvador, population distribution is also highly uneven within each country. Nearly 80% of the total population of Central America lives in a small number of densely populated areas located in the lowlands, slopes, or highlands of the Pacific watersheds. These areas of dense populations comprise much of the land with the most fertile volcanic soils and favorable climates, as exemplified in Guatemala, El Salvador, Nicaragua, and the Central Valley of Costa Rica. The competition for available land for agricultural production and urban development is becoming a major issue. But the real tragedy is that if we expect a moderate to high standard of living for each Central American, the area is already overpopulated, given the productive potential of these countries's natural resources.

Among the numerous demographic issues that directly affect the preservation of natural resources in Central America, two aspects related to current migratory movements must be emphasized. One is the accelerating movement of people into urban areas. The other is the movement, either directed by governmental institutions or by private initiative, into the less populated and still forested areas of the Caribbean watersheds. The impact of these movements on the natural resources of the region will be treated in the following sections.

Most of the population of Central America is a mixture of the Spanish settlers and the native Indian population. Except for Guatemala, where nearly half of the population are Indians who maintain to widely different degrees their autochthonous languages and culture, only small endogenous pockets of the original ethnic groups remain in the rest of the countries. Both Indian and mixed rural populations still practice traditional farming systems as part of the remnants of their cultural-ethnic traditions. This sector of the Central American population, mostly living under marginal socio-economic conditions, is the primary refuge of much of the crop germplasm derived through centuries of cultural development.

Striking differences in general health conditions are also found in the isthmus. While Costa Rica, and to a lesser degree Panamá and Belize, have health standards that compare favorably to those of the United States and Canada, the rest of the region lags well behind. Serious health and nutritional problems affect many rural and urban sectors of the population, to a degree approaching that of the poorest countries in Asia or Africa.

Global economic problems, the burdens of external debt services, and the weakness of their internal economies have led to marked declines in per-capita income and serious fiscal crises in the seven countries of the region. On a per-capita basis, gross national product (GNP) also shows wide disparities among countries; Panamá and Costa Rica have almost three times the GNP per person of El Salvador and Honduras. Except for Nicaragua, income distribution also shows striking disparities within countries throughout the region, where the richest 20% of the population has traditionally controlled between 49 and 66% of the national income.

Central American countries have traditionally depended upon exportation of a few and similar agricultural products and raw materials to generate their foreign currency. The great dependence on a few agricultural commodities renders Central American economies extremely vulnerable to fluctuations in climate, international market prices, and many socio-economic factors that are largely beyond their control. The possibilities of economic diversification in the near future are very bleak, partly because of substantial dependence on imports of petroleum and many other basic materials. The pressures to increase exports of natural commodities are being maintained and are leading to even greater exploitation of natural resources. This export increase implies increased deforestation to raise timber exports (while it remains) and intensification of livestock ranching, and development of coffee, banana, cotton, rice, sorghum, oil palm, and other types of highly commercial plantations. These trends continue in spite of the fact that.... 'the large jumps in international prices during the 1970s seriously exacerbated chronic balance-of-trade problems that had resulted from the general pattern of exporting basic, low value-added, unrefined agricultural commodities and importing high value-added, expensive technological and capital goods' [1]. Strong policies to stimulate export-oriented agriculture have discouraged commercial agricultural efforts in the production of basic foods for domestic consumption, which has hardly improved over subsistence agriculture. Subsistence farming has low production largely because it is practiced on marginal farmland and because it is practiced by the poorest and least politically- and economically-influential sectors of the population.

Another element of singular importance relates to energy supply and use. All motorized transportation energy used in the isthmus is based on oil (60% of the total energy used is supplied by renewable natural resources, including biomass, hydropower or geothermal power). Because all seven countries must import their petroleum, the great increase in oil prices in the late 1970s has meant that the region's export earnings are primarily consumed in purchasing oil.

Two-thirds of the electric energy currently produced in Central America comes from hydroelectric power plants. Costa Rica generates nearly 99% of its electricity from this source, but Belize generates all of its electricity from diesel power plants. Wood and charcoal remain traditional sources of energy, particularly in rural areas. Households and many small rural industries use wood as their main source of energy. Honduras, Guatemala, El Salvador, and Nicaragua are the countries most dependent on firewood. Fulfillment of requirements for firewood represents an important pressure on the remaining forest, particularly on a local scale in the drier and heavily populated Pacific watersheds of the isthmus. Uncontrolled firewood harvest is a major impediment to forest recovery once the forest been partly cleared (e.g., by a lumbering operation).

Throughout Central America, agricultural activities in general are thus divided between commercial and subsistence sectors. In addition to these commercial and subsistence sectors, in many parts of Central America exist large farms or ranches that are operated by owners who limit investments and production costs and hold the land in part for other purposes beside basic production. They are often absentee land owners who maintain other sources of income in urban areas. This group constitutes the vestiges of a colonial aristocracy, whose primary interest to control land is based on cultural, social, or political reasons and not on the long-term maximization of agricultural production for themselves or the country. The underutilization of some of the best agricultural lands held in this manner contrasts dramatically with the overexploitation of the much more common subsistence-farming sector of the population.

As in the rest of the world, access to land and its resources is thus quite unequally distributed in Central America. In Costa Rica, a country considered to have the most equitable socioeconomic structure, 1% of the total number of landholdings account for 36% of the land in the form of farms of 500 hectares or more. In Guatemala, this same percentage of land is accounted for by 0.2% of the landholdings, while in El Salvador and before the present agrarian reform, 1.5% of the landholdings controlled 50% of the agricultural land. Mainly for the purpose of reducing population and land pressures in the Pacific slopes, most Central American countries have devised land-development schemes for the less populated and exploited Atlantic lowlands. Soils in these very humid areas are generally inadequate for sustained agricultural purposes utilizing contemporary agrobusiness technologies and administrative structures.

Environmental implications of socio-economic trends

The Central American landscape is changing at a very fast rate as a result of complex interrelated forces that include a rapidly growing population that demands more land on which to grow its food and cash crops, and from which to harvest timber for start-up capital [1].

Forest destruction

The rapid conversion of forest to pastures and agricultural production constitutes the most important environmental change presently occurring in Central America [1], with the accompanying and direct loss of huge numbers of species representing invaluable genetic resources.

According to FAO data [1], in 1980 only 40% of the isthmus was in forest, 13% in annual or permanent crops, and 22% in pastures, with the remaining 25% devoted to other uses. While the total area devoted to cultivated crops has not changed much, the major trend is reduction in forest cover and increase in pasture lands. In Costa Rica, Honduras, and Nicaragua not less than 30% of the land is now devoted to cattle ranching.

Less than 40% of the forest in Central America remains today. What is more striking is the rate at which the forest has been disappearing. While approximately one-third of the forests has been cleared in the 450 years since the Spanish settlement of the area, about one-third was destroyed in less than three decades, between 1950 and 1985. The rate of deforestation has been steadily increasing in recent decades. In Costa Rica nearly 20% of the land was cleared between 1970 and 1980, at a rate of nearly 50,000 hectares per year. Excluding the protected areas, only 250,000 hectares remain today of wild forest containing trees big enough to harvest [1, 2, 21, 22].

The destruction of the rainforests located mostly in the Atlantic watersheds of Central America has attracted considerable attention. This type of habitat represents most of the total forest cover remaining in the isthmus. A much less publicized fact is that the dry forest of the Pacific side is by far the most threatened of the tropical habitats of the region. In pre-Columbian times, dry forests extended from north-central Mexico to Panamá, covering nearly 550,000 km^2, on the Pacific side alone, a similar or larger area than that occupied by rainforest in the Mesoamerican lowlands. It is estimated that less than 2% of relatively undisturbed dry forest exists today, and only 0.09% of it is under some type of protection [18, 23].

The causes of deforestation are numerous, complex, and interrelated, but direct evidence indicates that the present major force is the quest for unowned forested land that can be cleared and put into production. For many years, demand for timber was not the main threat. In Costa Rica alone, nearly 50% of the timber (approximately 2.2 million m^3 per year) was burned or allowed to rot *in situ* [21, 22]. Colonization of remote areas is motivated by the shortage of available arable land in the densely populated areas of the Central American Pacific lowlands. Cut trees provide a source of badly needed capital, and land owners convert wild vegetation to crops, including pasture grasses, to obtain a cash return in a relatively short time (D. Janzen, personal communication).

Land degradation

A direct consequence of the extensive removal of the forest cover for cattle ranching and agricultural colonization has been soil erosion. In the absence of adequate soil conservation and management technologies, this problem has attained critical dimensions, particularly in areas where topographic, edaphic, and climatic conditions are inadequate for such types of land utilization. With the exception of Belize, erosion and soil degradation in the seven countries range from approximately 10% to nearly 50%, as it was in the case of El Salvador in 1980. But the degree of land degradation in the Pacific watersheds of the other countries is not much different from that of El Salvador and may be as high as 40% for all the area. This erosion is a result of the intense exploitation, without adequate soil conservation, that the area has experienced for several centuries, a process that has been accelerated in the last few decades for the reasons discussed above [1].

Rapid, disorganized urban development in the cooler highlands of Central America is continuing on the most fertile and best agricultural soils. This process is creating not only a serious shortage of agricultural land but is also causing the loss of the best soils in these countries, as no other areas exist with soils of similar quality [1, 10, 24]. The conflict between urban expansion and maintenance of agricultural land is best exemplified in the Central Valley of Costa Rica, where two-thirds of the population occupy 6% of the national territory. Soil, topographic, and climatic conditions have determined that this valley is the agricultural heartland of the country, occupied mostly by coffee, vegetable, and dairy farms. Of the approximately 680 km^2 of this valley, nearly 240 km^2 are situated on some of the best agricultural volcanic soils in the world. Recent estimates indicate that over 150 km^2 of this prime land already has been urbanized, and the current trend indicates that the metropolitan area will occupy practically all of the Central Valley by the year 2000 [1, 10, 24].

Agricultural and urban development not only destroyed the original habitats that existed in these highland areas, where population is densely concentrated, but are also rapidly destroying a substantial amount of indigenous germplasm of economic interest that remained under domestic cultivation. Numerous species of vegetables, fruits, or medicinal plants were cultivated around houses in gardens or backyards, but sprawling urbanization and modern patterns of consumption are rapidly eliminating these areas, as well as the cultural traditions that motivated their use (J. León, personal communication).

Agricultural practices

Certain modern agricultural practices also are responsible for the loss of germplasm of economic interest. Large farming operations that concentrate in commodity crops such as bananas, rice, sorghum, coffee, sugar cane, cotton, or beef production rely heavily on technology-intensive practices and external inputs such as pesticides, fuel, and fertilizers. Formerly these plantation-type operations, mostly based on the use of exotic tropical species, used relatively 'inefficient' technologies and allowed many edible or medicinal species of plants that were harvested by surrounding communities to grow as weeds in the fields. For example, a few years ago coffee was cultivated in most of Central America under the shade of numerous species of *Inga* spp. or *Erythrina* spp. These and other species of trees used as live fences include some important sources of fruits. In addition to the nitrogen-fixing capacity of the shade trees, these species of legumes provided firewood for domestic consumption as well as a favorable environment for the growth of edible and medicinal weeds (J. León, personal communication). The recent introduction of high-yielding coffee varieties demanding intensive use of fertilizer and herbicides brought along the removal of the shade and fence trees and the disappearance of numerous weedy species of economic interest to the rural communities.

The Central American Pacific lowlands have been ranched and farmed for hundreds of years in a manner that, as pointed out before, has allowed the survival of patches of the original species and wildlands that once covered the area. Dramatic impacts on these fragments of the indigenous vegetation are visible now when highly efficient corporate agriculture changes the landscape, eliminating topographic, soil, and habitat diversity, introducing a highly manipulated crop species, such as African oil palm, pineapple, rice, sorghum, or melons. Extensive areas of these pesticide-rich agroecosystems now occupy much of the Guanacaste and Puntarenas provinces of Costa Rica as well as other similar areas of Central America. Cattle ranching continues to be a threat to the remaining wildlands on the dry-forested Pacific watersheds, both in protected areas and in the patches of original vegetation. The introduction of cattle to the region also brought along the introduction of exotic species of grasses, mostly of African origin. But pasture lands are gradually invaded by indigenous species of weeds or trees that are of little use to cattle. Particularly in the Pacific lowlands of the isthmus, burning of pasture land during the dry season to eliminate these invaders is a common practice. Such fires frequently get out of control, burning not only the grasses but also the patches of indigenous vegetation as well as the neighboring remnants of forest and protected wildlands. They represent a permanent major threat to the remaining endangered tropical dry forest [18].

The value of conventional uses of Central American germplasm from domesticated, semi-domesticated, or wild species of organisms is incontestable. The highly-threatened dry forest in particular contains numerous species that represent genetic resources for the development of sturdy, drought resistant varieties capable of tolerating environmental fluctuations that frequently devastate cultivated fields. These wild lands similarly could provide the genes for the development of species able to grow in low-fertility soils or to support the numerous pests and diseases that prevail in tropical lowlands. Rainforests also can provide the genetic resources for the development of crop species that could grow successfully in humid places where present conventional agricultural technologies have failed. Equally unappreciated is that if the right measures to protect tropical wildlands are not taken now, this same potential for new crops could spell the end of the remaining wildlands. With new, 'engineered' crops, small farmers and corporations could then use areas presently considered as marginal.

Cultural preservation and restoration

When the problems of *in situ* preservation of germplasm of human interest, both in wildlands or agroecosystems are examined, it is not possible to separate the biological problems from the cultural or socio-economic context of the region where such biological materials exist. Conservation of plant genetic resources in germplasm banks represents an important and valuable strategy, but its limitations are numerous [25]. Among other considerations, gene banks freeze evolutionary processes: plants are no longer allowed to respond to environmental selective pressures whereas *in situ* preservation guarantees the continual and dynamic interaction of plants with the environment [25, 26, 27].

Consequently, and in particular under Central American circumstances, *in situ* conservation of plant genetic resources must preserve the agroecosystems of which these resources are an integral part. On the other hand, traditional agroecosystems cannot be preserved independently of the cultural and socio-economic organization of the local people [27]. These farming systems represent centuries of practical knowledge and experiences with the local environment, transmitted through generations by the indigenous farmers. Without access to external information and agronomic or economic inputs and markets, such systems have evolved as sustained-yield agroecosystems [27–29]. Peasant farmers thus have a key active role to play in the *in situ* preservation of germplasm [16], a role that has not been acknowledged in most current efforts of preservation of natural resources. It has been very properly stated that within this social and cultural context, natural resources cannot be perceived simply as genes to be conserved in a germplasm bank, usually managed by people with a very limited knowledge or understanding of the ecological conditions under which the desirable traits of those materials are expressed. These organisms lose a substantial part of their value when isolated from the indigenous cultures that selected them and which have the practical knowledge of the conditions under which they are to be used, and for what purposes. Much of their cultural and historical meaning thus is lost [27, 30–31].

The importance of the relationship between cultural diversity and conservation is clear, although it is not very frequently taken into consideration.

'The most frequently ignored or overlooked aspect of conservation of natural resources and economic development is human cultural diversity. Or to put it another way, people are generally overlooked even though it is obvious that without human beings there could not be such phenomena as resource conservation and resource development. But when the plans are made to conserve or develop natural resources the starting point of such deliberations is seldom if ever the diverse needs and aspirations of all the cultural elements that will be affected [12].'

The information on the socio-economic conditions of Central America clearly points out that current development trends are rapidly leading to the destruction of the rich environmental resources of the region. The disruption, since colonial times, of the indigenous cultural and socio-economic structure and the subsequent development of the prevalent pattern of economic development and political organization has not favored the rational and sustained use of natural resources. Central American people, particularly those in rural areas, have lost, for the most part their cultural richness and the modest but sustained agricultural production of their predecessors [7]. In contrast with the traditional farming systems for which they have substituted, modern agroecosystems are dysfunctional [5] and serve the interests of only a small sector of the local population and the extra-tropical markets they subsidize. Such trends need to be changed if the extraordinary biodiversity of the region is to be saved. Functional agroecosystems must be preserved or reconstructed, a task that has to depart from traditional unidisciplinary-oriented perspectives and which must integrate such specialists as ecologists, ethnobotanists, agriculturists, sociologists, economists, planners, and others with the peasant farmer. The preservation of biodiversity in wildlands also faces biological and social challenges, but what is evident is that the protected wildlands that do not become adequately integrated into society will not survive, as they should be instruments for material gains and the intellectual development of the communities in which they are embedded [7].

Few efforts are being made by international agencies operating in Central America to help improve the region's natural resource system [1]. It is striking to compare the minute amount of external aid spent in the protection or adequate management of natural resources with the huge amount that foreign agencies have invested in the exploitation of natural resources. But even more striking is the fact that in the last few years foreign powers have spent billions of dollars in military aid to Central American countries, except Costa Rica, a trend that has only exacerbated the internal problems of the region.

The future of the extraordinary biological and cultural diversity of Central America lies in the hands of these nations, and the foreign industrialized countries that have traditionally harvested the region's natural resources.

Acknowledgments

I am grateful to María E. Bozzoli, G. Budowski, D. Janzen, and J. León for sharing with me many of the ideas and concerns presented in this paper. Special thanks are due to D. Janzen for his stimulating discussions on the subject and critical review of the manuscript. Part of this work was supported by a grant to Fundacion Neotrópica of Costa Rica from the The John D. and Catherine T. MacArthur Foundation.

References

1. Leonard, J. H. 1987. Natural Resources and Economic Development in Central America. Transaction Books; New Brunswick & Oxford. 279 pp.
2. Parsons, J. J. 1976. Forest to pasture: Development or destruction?. Rev. Biol. Trop. 34 (Suppl. 1): 121–138.
3. Harris, D. R. 1972. The origins of agriculture in the tropics. Am. Scientist. 60: 180–193.
4. Wilkes, H. G. 1979. México and Central America as a centre for the origin of agriculture and the evolution of maize. Crop. Improv. 6: 1–18.
5. Janzen, D. 1973. Tropical agroecosystems. Science 182: 1212–1219.
6. León, J., H. Goldbach, and J. Engels. 1979. Los Recursos Genéticos de las Plantas cultivadas en América Central. Centro Agronómico Tropical de Investigación y Enseñanza: Turrialba, Costa Rica 32 pp.
7. Janzen, D. 1986. The future of tropical ecology. Annu. Rev. Ecol. Sys. 17: 305–324.
8. Hartshorn, G.S. 1983. Wildlands conservation in Central America. Pp. 423–444 in S. L. Sutton, T. C. Whitmore and A. C. Chadwick (eds.), Tropical Rain Forest Ecology and Management. Blackwell Sci. Publ.: Oxford 448 pp.
9. Janzen D. (ed.). 1982. Costa Rican Natural History. Univ. Chicago Press: Chicago 816 pp.
10. Hartshorn, G. 1982. Costa Rica. Perfil Ambiental: Estudio de Campo. Tropical Science Center: San José, Costa Rica 149 pp.
11. Janzen, D. 1987. Insect diversity of a Costa Rican dry forest: Why keep it, and how?. Biol. J. Linn. Soc. 30: 343–356.
12. Bennett, C. F. 1976. Cultural diversity in Central America and Panamá: Its relationship to conservation and planning. Rev. Biol. Trop. 34 (Suppl.1): 5–12.

13. Altieri, M. A. and M. K. Anderson. 1986. An ecological basis for the development of alternative agricultural systems for small farmers in the Third World. Am. J. Altern. Agric.1: 30–38.
14. Soria. J. 1976. Los sistemas de agricultura en el istmo Centroamericano. Rev. Biol. Trop. 34 (Suppl.1): 57–68.
15. Hazlett, D. L. 1986. Ethnobotanical observations from Cabécar and Guaymí settlements in Central America. Econ. Bot. 40: 339–352.
16. Alcorn, J. B. 1984. Development policy, forests, and peasant farms: Reflections of Huastec-managed forests' contributions to commercial production and resource conservation. Econ. Bot. 38: 389–406.
17. Debouk, D. G., R. Araya Villalobos, R. A. Ocampo Sánchez, and W. G. González Ugalde. (In press). *Phaseolus* germplasm collections in Costa Rica. Plant Gen. Res. Newsl.
18. Janzen, D. 1986. Guanacaste National Park: Tropical ecological and cultural restoration. Editorial Universidad Estatal a Distancia: San José, Costa Rica 103 pp.
19. Erlich, A. H. and P. R. Erlich. 1987. Earth. Franklin Watts, Inc.: New York 258 pp.
20. Raven, P. 1987. We Are Killing Our World: The Global Ecosystem in Crisis. The John D. and Catherine T. MacArthur Foundation: Chicago 24 pp.
21. Dirección General Forestal. 1985. Situación Actual Forestal de Costa Rica. Ministerio de Agricultura y Ganadería: San José, Costa Rica 40 pp.
22. Gámez, R. and Ugalde, A. 1988. Costa Rica's national park system and the preservation of biological diversity: Linking ecology with socio-economic development. pp. 131–142. In F. Almeda and K. Pringle (eds.), The Diversity and Conservation of Tropical Rainforests. California Academy of Sciences: San Francisco, 306 pp.
23. Janzen D. 1988. Tropical dry forests: The most endangered major tropical ecosystem. pp. 130–137 in E. O. Wilson (ed.), Biodiversity. National Academy Press Washington 521 pp.
24. Fournier, L. A. 1976. Efecto del urbanismo sobre el futuro desarrollo agrícola de Costa Rica. Rev. Biol. Trop. 34 (Suppl. 1): 49–55.
25. Prescott-Allen, R. and C. Prescott-Allen. 1982. The case for *in situ* conservation of crop genetic resources. Nature and Resources 23: 15–20.
26. Simmonds, N. W. 1962. Variability in crop plants, its use and conservation. Biol. Rev. 37: 422–465.
27. Altieri, M. A. and L. C. Merrick. 1987. *In situ* conservation of crop genetic resources through maintenance of traditional farming systems. Econ. Bot. 41: 86–96.
28. Altieri, M. A. and L. C. Merrick. 1988. Agroecology and *in situ* conservation of native crop diversity in the Third World. pp. 361–369 in E. O. Wilson (ed.), Biodiversity. National Academy Press: Washington 521 pp.
29. Wilken, G. C. 1977. Integrating forest and small-scale farm systems in middle America. Agro-ecosystems 3: 291–302.
30. Nabhan, G. P. 1985. Native crop diversity in Aridoamerica: conservation of regional gene pools. Econ. Bot. 30: 387–399.
31. Sarukhan, J. 1985. Ecological and social overviews of ethnobotanical research. Econ. Bot. 39: 431–435.

Ownership of plant genetic material

HOWARD J. BROOKS and CHARLES F. MURPHY
Agricultural Research Service, U.S. Department of Agriculture, Beltsville, Maryland, 20705, U.S.A.

Abstract Free international exchange of germplasm has been essential to the remarkable record of success achieved by U.S. plant breeders. The application of intellectual property rights to plant materials has evolved through four identifiable time periods. Prior to 1930, there were no intellectual property rights for plant materials. In 1930, plant patents became available for most vegetatively propagated material and in 1970, the Plant Variety Protection Act offered final product protection for most seed crops. In neither of these cases, however, was germplasm exchange seriously inhibited. Since 1985, utility patents have been applied to plants – although this interpretation of use of the utility patent has yet to be challenged in court. The current manner in which utility patents are being applied to plant material is estimated to have a significant impact on the free exchange, use, and control of plant germplasm worldwide.

Key words: plant patents, genetic material, plant germplasm, intellectual property, germplasm exchange, germplasm ownership.

Since the dawn of man, plants have been collected, transported, manipulated, shared, exchanged, and used to the benefit of all mankind. Prior to 1930, when the U.S. Plant Patent Act became law, plants and plant genetic material were considered common property of the world. In this paper we will discuss existing U.S. laws or interpretation of the laws which now allow ownership of plant materials and some possible implications of these laws to U.S. agriculture.

For about 100 years prior to the Plant Patent Act of 1930, U.S. explorers and scientists collected a wide array of useful plant material from around the world. It is well that they did, as the continental U.S. was largely void of native germplasm of commercial crops that we now grow for domestic use and export. Other developed countries did the same because no single country had, or has, all desired plant germplasm within its borders. Fortunately for us all, plant germplasm was freely exchanged and shared. There were no legal constraints to free access to plant genetic material, no apparent ethical constraints, and world agriculture advanced very rapidly.

Under the Plant Patent Act of 1930, those who invented or discovered asexually propagated plants that were novel and distinct could prevent others from selling them for a period of 17 years. As with the General Patent

L. Knutson and A.K. Stoner (eds.). Biotic diversity and germplasm preservation, global imperatives.
 ISBN 0-7923-0178-1

Act of 1836, which authorized patents on reproducible inventions, the intent of Congress was to help get new and novel plants into the marketplace, presumably for the benefit of U.S. agriculture and the general public.

About 6000 patents have been issued under the Plant Patent Act, with most being on roses, fruits, and other horticultural crops. While there was no research exclusion in the law, patent holders did not deny others from using patented plants as parents in breeding programs, and there was no thought of ownership of plant genetic material. Individual asexually propagated plants were protected but not individual genes, gene combinations, breeding systems, or plant products. It is our view that the Plant Patent Act promoted commercialization of improved plants developed or discovered by nurseries or private individuals and that the law proved to be beneficial to U.S. agriculture and to the general public. The Agricultural Research Service (ARS) of the U.S. Department of Agriculture (USDA) has never used this law to apply for plant patents. Under the Plant Patent Act, the USDA has the responsibility to review plant patent applications for the Patent Office. Thus, there would be a conflict of interest if ARS were to apply for plant patents. The USDA had made a policy decision prior to the passage of the Plant Patent Act not to apply for plant patents.

The 40-year period between 1930 and 1970 proved to be a good period for the development of improved varieties and for the advancement of U.S. and world agriculture. To our knowledge, no attempt was made by anyone to claim ownership of plant genetic material. Germplasm continued to be freely exchanged and used worldwide for development of improved varieties. The U.S. National Plant Germplasm System (NPGS) was established during this period to help facilitate worldwide collection and exchange of germplasm and to help preserve the world's diminishing resources.

In 1970, a second U.S. law was established to provide plant protection. This law, the Plant Variety Protection Act (PVPA), allowed patentlike protection on seed crops similar to the protection allowed on vegetatively propagated crops under the Plant Patent Act of 1930. Wisely, there was a research exemption in the law that allowed a protected variety to be used by anyone as a parent in breeding programs. Thus, it was U.S. law that helped to assure that plant genetic material remained common property and was freely available to plant scientists worldwide.

It is our view that the Plant Variety Protection Act (PVPA) was and is still a good law. It helped stimulate the expansion of the U.S. seed industry, it helped promote commercialization of improved varieties developed by commercial seed companies, and it helped the U.S. and other countries to continue to have free and unlimited access to improved plant genetic material from which additional improved varieties could be developed.

ARS has never individually applied for protection under PVPA, but we have joined with State cooperators to get Plant Variety Protection certificates on seed varieties jointly developed and released.

The 15-year period between 1970 and 1985 also proved to be reasonably good for world agriculture. More countries became self sufficient in the production of major food crops. We like to think that the policy of worldwide free exchange and use of plant germplasm helped contribute to this increase in world food production. But also during this period, international concerns began to surface about the ownership of plant genetic material and world germplasm. While world plant genetic resources have continued to be lost, some less developed and emerging countries have suggested that native plant germplasm be considered a national resource of individual countries and that it should not be freely shared with the rest of the world.

During this 15-year period, the U.S. and other developed countries made additional commitments to the collection, evaluation, and preservation of plant germplasm worldwide. The NPGS continued to be strengthened, as was the U.S. commitment to the principle of worldwide free exchange and use of plant genetic material. And perhaps as a consequence of potential advances in genetic engineering, international organizations purchased many seed companies in the U.S. and elsewhere. The ownership of the world seed industry now appears to be concentrated in fewer hands. This has been perceived by some persons as having the potential to monopolize world seed crops and plant genetic materials.

So who owns plant genetic material? Up to 1985, it was U.S. policy to consider plant genetic material as common property of the world. A similar policy and practice was accepted worldwide. There was no law, to our knowledge, that prevented the use of any genetic material or variety as a parent in a breeding program. Certainly, there was no U.S. law.

Now we come to September 18, 1985. On that date, the U.S. Patent and Trademark Office Board of Patent Appeals and Interferences ruled that plants and plant genetic material can be patented under the General or Utility Patent Act of 1836. This ruling is known as the Hibberd decision. The Hibberd decision reverses a position held in the Patent and Trademark Office for nearly 150 years. While under the Plant Patent Act or PVPA one could make only a single claim, following the Hibberd decision one can use a utility patent to make many claims of ownership. One can make claims on individual genes, gene combinations, characteristics, and products. It is also possible to claim rights to a process or breeding procedure.

After the 1980 Supreme Court ruling in the Chakrabarty case that microorganisms could be patented, most people were not surprised by the 1985 Hibberd decision. It seemed only a matter of time before the utility patent would be used to patent plants. More recently, the Patent Office has ruled that animal life created through genetic engineering techniques can also be patented.

We are of the opinion that there will be many applications to the U.S. Patent office for utility patents on plants. Since this appears to be the first law anywhere in the world to allow ownership of plant genetic material, we anticipate that many foreign scientists and seed companies also will apply for U.S. patents.

Unless the rulings of the Patent Office are challenged in the courts and reversed, the utility patent can be used in the U.S. to patent genes, gene combinations, and plant products. With this new interpretation of U.S. law, the future ownership of plant genetic material becomes less clear. We can anticipate some benefits to U.S. agriculture. We can also anticipate some possible serious disadvantages.

We are of the opinion that the Plant Patent and Plant Variety Protection Acts benefited U.S. agriculture by stimulating increased plant breeding by the private sector. We believe that there could be a similar benefit from the use of the utility patent on plants. As genetic engineering techniques are improved and used to introduce specific DNA segments into plants where they never existed before, we can visualize the expression of many new plant characteristics that were never possible before. Even if an individual or a company in the private sector were to receive patent rights and royalties on some of these manmade inventions, there could be long-range benefits to everyone. Once these genetic breakthroughs are made, there could be cross licensing by patent holders of the genetic material involved and a net gain to U.S. agriculture and consumers even with the cost of royalty payments. There could be a wealth of privately owned new and improved genetic material that could be exchanged or sold in the world community. The possibility of ownership of genetic material could continue to stimulate interest in plant breeding and variety development.

But there are some potential serious disadvantages as well, specifically where patents are granted for plant varieties and germplasm lines derived largely through coventional means. Considerable legal opinion suggests that a patent holder can control commercialization of plant material derived from a patented parent. The likely result could be a reduction in the free exchange of data and unpatented genetic material between scientists, organizations, and countries. Our concern is that the continued development of improved varieties might be made more difficult with less accessibility to improved germplasm. If this happens, there could be added costs to farmers for seed of patented varieties, which would be passed on to consumers. With a reduction in exchange of genetic material, there could be a corresponding reduction in genetic diversity. The expansion of the NPGS could be disrupted. It is unlikely that commercial seed companies or State universities desiring to patent genetic material would put their best germplasm in a national germplasm repository where plant material is freely shared with others.

What is the position of ARS on proprietary rights on plant genetic material? ARS maintains that the private sector should have proprietary rights on plants with unique characteristics, but free exchange of germplasm is encouraged. ARS has a policy to patent improved plant material developed by our scientists, following a review on a case-by-case basis, but only when it would benefit U.S. agriculture. In the event that ARS patents plant germplasm, we anticipate that we generally will make the patented material freely available to the world community. This would be a defensive position, but it would be consistent with the USDA position that genetic material should be freely available worldwide.

In conclusion, we recognize that U.S. policy on ownership of plant genetic material under U.S. law now appears to have been reversed as a result of the 1985 Hibberd decision. To the extent possible, and in spite of this recent new interpretation of a 150-year-old U.S. law, we ask that plant genetic resources continue to be considered as common property of the world.

We are excited by the potential benefits to plant improvement from modern technologies. Likewise, we recognize the financial and intellectual inputs necessary to achieve them. These inputs deserve reward, and utility patents offer an effective mechanism to protect inventions and reward the developers.

We also recognize that the vast store of plant germplasm was created not by the hand of man but by the hand of nature. We are committed to the protection and preservation of this natural resource and its free access and utilization by scientists of the world.

Plant germplasm and the economics of agriculture

DONALD N. DUVICK[1] and WILLIAM L. BROWN
[1]*Pioneer Hi-Bred International, Inc., 700 Capital Square, 400 Locust, Des Moines, Iowa 50309, U.S.A.*

Abstract Modern agriculture, particularly as practiced in the developed world, is largely dependent upon highly select sources of germplasm. The present germplasm base of some important U.S. crops is briefly described herein. Sources of new germplasm and methods of most appropriate germplasm use are discussed. Modern methods for using genetic diversity to increase yields and production stability are described and compared to earlier methods and strategies. New concepts are briefly explored, of protection of intellectual property as applied to plant germplasm, and subsequent implications for genetic diversity. Also examined are the influence of farmer and grower preferences on genetic diversity, and consequences of those preferences for developed and developing countries. A serious need exists today for research on germplasm management, use, and introgression into existing cultivars.

Key words: germplasm preservation, germplasm sources, genetic diversity, variability, selection, plant breeding, production stability, patents, intellectual property protection

Introduction

The economic success of crop agriculture depends on two fundamentals: the availability of appropriate germplasm and its appropriate deployment. Appropriate germplasm is that which reliably provides the greatest amount of product with the smallest expenditure of energy and land. Appropriate deployment of germplasm entails use of ecologically sound crop-growing practices to ensure maximum profitable production, within the limits of agricultural sustainability and with proper regard for human health. Available sources of useful germplasm and techniques for its manipulation have determined the nature of agriculture and its varying degrees of success through the millennia; they do so today and will continue to do so in the future.

Sources of plant germplasm and ways of manipulating it have changed markedly during the past century. Prior to the practical application of Mendelian genetics and expanded use of synthetic fertilizers, increases in production usually were dependent upon expansion of land areas under cultivation. Needed genetic variability was provided by the nature of the heterogeneous cultivars and numerous species cultivated by individual farmers. In modern agriculture, increases in crop production have come

L. Knutson and A.K. Stoner (eds.). Biotic diversity and germplasm preservation, global imperatives.
 ISBN 0-7923-0178-1

about largely through manipulation of germplasm by use of appropriate breeding techniques and through use of fertilizer, agricultural chemicals, and intensive monocultural practices [1]. Needed germplasm diversity is provided by extensive global interchange of basic and advanced breeding stocks and by a rapid turnover, on the farm, of leading cultivars as provided by plant breeders [2].

To be successful, modern plant breeding thus must have access to genetic variability in the form of highly select germplasm. In the absence of genetic variability, selection cannot progress, whereas in the absence of elite sources of variability, selection cannot move populations in the direction of higher productivity. Thus breeders in their efforts to improve plants usually rely on select cultivars as sources of new traits. Primitive cultivars and wild relatives are chosen as donors of new genes only when they are thought to be unique sources of such genes. The use of elite germplasm in preference to unselected sources results from the close linkages frequently observed between desirable and undesirable traits in unimproved germplasm. The use of unimproved (i.e. unadapted) germplasm requires additional generations of segregating progeny to achieve the desired results and thus decreases breeding efficiency.

Despite the advantages of the continued use of elite rather than exotic germplasm sources in plant improvement, the use of such material tends to narrow more rapidly the genetic base on the farm, thereby increasing potential genetic vulnerability. To counter this risk, it is essential to maintain and use global germplasm resources and to deploy rapidly and extensively the new products of breeding to the farm. The principle of 'genetic diversity in time' thus is practiced [2].

Today's practice of utilizing genetic diversity in time (genotype replacement) is more or less equivalent to the older agricultural practice of utilizing genetic diversity in place (use of heterogeneous cultivars and numerous species). But on a slower time scale, genetic diversity in time also occurred in primitive agriculture, and in nature as well. Charles Darwin noted this in his remarks on variation under domestication (remarks made as part of the foundation for his theory of evolution, 'Descent with Modification'). Darwin stated, 'Our oldest cultivated plants, such as wheat, still yield new varieties...'[3]. He also said, 'With our domestic productions, when an improved variety has been raised, it first supplants the less improved varieties in the same neighborhood; when much improved it is taken far and near, like our short-horn cattle, and takes the place of other breeds in other countries. Thus, the appearance of new forms and the disappearance of old are bound together.' The chief task of plant breeder, therefore, is to find and use efficiently desired genetic variation as they strive to improve humankind's economic well-being. Various strategies have been employed toward this goal.

Sources of useful genetic variation, and ways of using it

Over the years, breeders have sought and sometimes identified heterotic patterns in some crops that have been useful in augmenting performance. Corn belt maize arose from crosses of southern dents and northern flints [4]. Increased yield and other advantages associated with combining these two diverse sources of germplasm have been recognized since colonial times. Even today, maize breeders continue to exploit, via first-generation hybrids, the heterotic responses derived from combining segments of these two germplasms, which still exist in corn belt maize.

In breeding maize for the tropics and subtropics, a similar heterotic combination, of Tuxpeno Dent and Coastal Tropical Flint, has been used successfully, worldwide. Maize breeders now are finding ways to combine elite U.S. corn belt dents with elite tropical and subtropical strains.

In grain sorghum, 'Milo' and 'Kafir' are two distinct races that have been converted to dwarf stature and early maturity, thereby permitting their use in modern mechanized agriculture in temperate zones, such as the U.S. High Plains. Considerable heterosis results from the hybridization of these two races. Thus, most modern grain sorghums in use today are first-generation hybrids of lines derived from Milo and Kafir. Fortunately, cytoplasmic male sterility and male fertility restoration – conditions necessary for the exploitation of commercial hybrids in sorghum – are found in these two races. Milo provides the cytoplasmic factors necessary for male sterility, as well as nuclear fertility restorer genes; nuclear genes needed for expression of cytoplasmic male sterility are present in Kafir [5].

As is usually the case, this breeding success has resulted in concentration by farmers on a few highly successful genotypes and a consequent reduction in on-farm genetic diversity. All sorghum hybrids have Milo cytoplasm and thus its mitochondrial and chloroplastic genotype. U.S. sorghum hybrids also have a narrow base of nuclear genotypes; as just noted, inbred parents typically have had either Milo or Kafir nuclear background, with only rare additions of germplasm from other races.

The challenge to sorghum breeders is to use other sources of cytoplasmic sterility and other sources of nuclear variability. Exotic strains often can contribute useful disease-, insect- and stress-tolerance traits, but they usually present them in tight linkage with undesirable traits such as those that reduce yield under modern agronomic practices. Other sources of cytoplasmic sterility now are known, but they are less convenient to use than the Milo type.

Prior to the emergence of interest in growing hybrid wheat, which arose in the U.S. some twenty years ago but now appears to be waning, wheat breeders seemed to be of the opinion that the heterotic response in crosses of wheat cultivars was minimal. Field experience with hybrid breeding has demonstrated conclusively that significant amounts of heterosis are exhibited by some wheat hybrids [6].

Unfortunately, the most significant heterotic responses were obtained in crosses between different classes of wheat rather than within classes. Such differentiation, of course, would be expected on the basis of empirical knowledge that stringent selection for grain quality traits has greatly narrowed the genetic base of each class of wheat, and that the greatest amounts of heterosis typically are exhibited by crosses of genotypes derived from distinctly different germplasm pools.

Soybean breeding in the U.S. has been based upon a few original accessions from the Orient. Many accessions were examined during the initial introduction period, but eventual commercial success derived from only about a dozen cultivars of similar phenotype, obtained largely from a small area of northeastern China. These original selections have been sequentially hybridized and reselected to give rise to successive generations of improved new pureline cultivars [7].

In the U.S., two more or less separate soybean germplasm pools have arisen. Cultivars in one pool (northern) have indeterminate plant habit with flowering adapted to the relatively long day lengths of the U.S. corn belt. Cultivars in the other pool (southern) have determinate plant habit with flowering adapted to the shorter days of the U.S. midsouthern, southeastern, and Mississippi delta regions.

Although basic sources of germplasm for all U.S. soybean breeding are narrowly derived, breeders from time to time have introduced germplasm from exotic cultivars, using the exotics as essential sources of new genes for disease and nematode resistance. Backcross techniques have been used, however, to eliminate, as much as possible, all exotic genes except those with desired resistance or tolerance traits.

On the other hand, during the past 10 years a cooperative university/industry program systematically has explored previously untested exotic soybean cultivars for the presence of desirable agronomic traits, especially yield. Selected exotic cultivars then have been introgressed into elite U.S. stocks [C. Jennings, pers. comm.]. The aim is to introduce maximum amounts of exotic germplasm while maintaining sufficient proportions of elite germplasm to provide agronomic suitability of the new breeding stocks.

Additionally, soybean breeders now are hybridizing determinate with indeterminate U.S. soybean lines, and are selecting new cultivars adapted either to the north or the south, with useful traits from the opposing gene pool.

Sunflower now is grown commercially as first-generation hybrids in most parts of the world. This crop in the U.S., although derived from a species native to North America and with a wealth of available genetic variability in its wild relatives, nevertheless has been dependent upon a relatively narrow germplasm base. The crop was greatly improved agronomically in the area that is now the Soviet Union. A few cultivars from that country were the basis for development of a small number of inbred lines that were used in the first successful U.S. hybrids, in the 1970s and early 1980s. A very few hybrid combinations, often involving closely related sister lines, were used at first to plant most U.S. hybrid acreage. (Also, as with sorghum, all hybrids have been developed with a single source of male sterility-inducing cytoplasm.) Now, however, breeders are introgressing selections from related species into elite germplasm stocks. Such introgression will bring in needed new sources of disease and insect tolerance, useful additions to hybrid vigor, and potentially useful new forms of cytoplasmic male sterility. It is likely that results of this introgressive breeding will expand the useful genetic base of sunflower hybrids grown in the U.S. and in the rest of the world as well [T. Heaton, pers. comm.; 8].

Hybrid grain sorghum has gone through a similar narrowing and broadening cycle. The first hybrid, of a commercial Milo cultivar and a commercial Kafir cultivar, dominated the U.S. sorghum belt in the early years (the early 1960s). For many years thereafter, the sorghum germplasm base remained remarkably narrow; hybrid parents were hardly more than variations on the original two parents. But in recent years, purposeful introgression with tropical-adapted germplasm of several races of sorghum has given rise to productive, adapted, but genetically different hybrids with new levels of yield stability as well as greatly broadened tolerance to diseases and insects. Successful development of the tropical-introgressed lines owes much to pioneering genetic studies of the 1950s that clarified the genetics of flowering date and plant height in sorghum [9]. Such knowledge has made it relatively easy to transform (via backcrossing techniques) tall, late-flowering tropical cultivars into short, early-flowering types without much change in their other traits. The converted cultivars then can be evaluated in temperate regions of the U.S. without distortion caused by excessively late flowering and very tall stature; the best of the conversions are used as parents for introgression into adapted elite germplasm.

Wheats provide another example of the enhancing powers of exotic germplasm and the inter-related nature of breeding programs worldwide. The semi-dwarf wheats, famous as one of the two crops involved in the Green Revolution, attracted worldwide notice when CIMMYT semi-dwarf wheats, developed in Mexico with funding from the Rockefeller Foundation, transformed the food production capacities of India and Pakistan. The CIMMYT wheats were bred from crosses involving U.S. selections out of hybrids between U.S. standard-height cultivars and Japanese semi-dwarfs.

The Japanese wheats, in turn, traced back to Korean and 19th-century U.S. cultivars of quite different lineage and adaptation [10].

The importance, therefore, of sources of genetic diversity and wide ranging germplasm exchange in the development of new cultivars is evident. Genetic diversity not only promotes stability of production by helping growers to avoid epidemic catastrophe, but it also is essential for genetic improvements in cultivar performance, including yield. The two topics, 1. genetic diversity for promotion of stable production, and 2. genetic diversity for promotion of increased yields, merit further discussion.

Genetic diversity, production stability, and yield

Stability of production is desired by all producers of crop plants; even those with maximum yield as their primary goal must achieve such yields on a regular basis, for their high costs of production will force them out of business if crops fail or yields are abnormally low too frequently.

As noted previously, the early agriculturalists and the peasant farmers of today have depended on in-field or between-field genetic diversity to provide a significant portion of needed production stability. The underlying principle is that at least some genotypes will withstand or escape each year's disease, insect, and weather-stress problems.

But modern breeding, intended to provide cultivars for intensive farming in industrial countries, tends increasingly to work on the principle that all possible genetic sources of tolerance and resistance are built into one or a small number of high performance genotypes. Pyramiding of genes for rust resistance in individual wheat cultivars is one example of this principle in action. Individual maize hybrids with bred-in ability to produce the best possible yields with either insufficient or maximal amounts of soil nitrogen are another example. (Modern maize hybrids commonly are of this broadly-adapted dual purpose type, although knowledge is not widespread that they have this capability.) Sorghum hybrids likewise now combine resistance to heat and drought with the ability to make high grain yields when water is plentiful.

Individual modern cultivars thus are more stable in performance than were their landrace forebearers – if one defines stability as ability to avoid disastrous yield failures combined with ability to make very high yields when environments are maximally favorable. One also can define stability as ability to avoid disastrous yield failures combined with *lack* of ability to make very high yields in favorable environments. In some societies, this second kind of stability has been sufficient and even desirable, but such a goal now is unsatisfactory to agriculturalists in most parts of the world because of an increase in incidence of crop production for cash (or for subsistence plus cash) instead of for subsistence alone.

The key to bred-in stability (of the first type) is incorporation of large amounts of ancestral genetic diversity, carefully chosen and sequentially bred into increasingly stable high-yield cultivars that are tested repeatedly – proven to be reliable – across a wide range of environments.

An erroneous but nevertheless now widespread belief is that modern cultivars are less stable than their ancestors. This belief seems to be based on the logic that because high value inputs such as commercial fertilizers and pesticides are used on modern cultivars, the old cultivars, obviously sufficiently productive and successful before such inputs were used, must have been tougher than today's 'pampered' cultivars. Additionally, recent disease or insect epidemics or broad-scale adverse reactions to weather stresses are sometimes cited as proof of the increased delicacy of modern cultivars.

The facts do not bear out either conclusion, however. Epidemic- or weather-based famines were common in ancient agriculture. The early Romans sacrificed a red dog to the grain god, Robigus, hoping that he then would hold back the red rust, an ever-present danger to their life-supporting wheat crop.

Usually lacking are side-by-side comparisons of new cultivars with old ones, under identical growing conditions. When such side-by-side comparisons are made, it is seen that new cultivars are much more stable and much more stress tolerant than their fondly remembered predecessors [11]. This is because modern high input agriculture, monocultural in nature, introduces greater danger of certain kinds of biological stress than did the older, low input types of agronomic practice; modern agriculture requires use of cultivars with high levels of stress resistance. The new cultivars, therefore, are bred to stand up to the rigors of modern high input/high output agriculture. Old cultivars, when subjected to modern high input stresses, actually are much more tender than the new ones, showing stress symptoms such as excessive lodging, heat-induced grain abortion, and susceptibility to disease and/or insect damage. And even under the old style of farming, with its less stressful growing conditions (thinner sowing rates, less excessive use of nitrogen, use of polycultures), the old cultivars still are less productive than the new ones in most comparisons. Under the old style, however, the disadvantages of old cultivars typically are considerably less.

It should be clear from this discussion that breeding for stable production and breeding for improvement in yielding ability are in many respects the same process, and that steady progress in such breeding depends on a continuing supply of genetically diverse parents, adequately characterized for desirable traits of agronomic importance. Biotic diversity and its preservation and characterization are vital to the economic health of plant-based agriculture.

Needed sources of new germplasm traditionally have included primitive landraces (or wild relatives) and also elite cultivars, as noted earlier. New techniques involving *in vitro* DNA manipulation now allow potentially useful sources of germplasm to include not only plants of all degrees of relationship but also bacteria and animals. A *Bacillus thuringiensis* toxin, virulent to lepidopterous insects, is now produced by transgenic tobacco and tomato plants; an anti- microbial polypeptide produced by frog skin might, in theory, be produced in plants containing the frog genetic system responsible for production of the anti-microbial product. Biotechnology – the all-inclusive term for the new technologies centered around *in vitro* DNA manipulation – is giving new perspective to plant breeding and genetic diversity. It gives new perspective to the ownership of plant germplasm and its relationship to genetic diversity as well.

Genetic diversity and ownership of plant germplasm

The concept of ownership of plant products is as old as trade in seeds or in clonally reproduced plants such as tulips, fruit trees, or, roses. Certain native American cultures developed a concept of seed stock ownership. Increasingly precise definitions of ownership of proprietary plant materials have been devised in the industrial countries, especially during approximately the past 50 years.

In the U.S., starting with plant patents on clonally reproduced materials in 1930, then with the specially crafted Plant Variety Protection Act of 1970, and most recently with the 1980 Chakrabarty decision, proprietary rights to whole organisms increasingly and undirectionally have been enlarged and strengthened. Even more recently, the 1985 *ex parte Hibberd et al.* ruling (by the U.S. Patent and Trademark, Office Board of Patent Appeals and Interferences) sanctioned use of utility patents to protect ownership of appropriately derived plant parts and processes, self-reproducing cultivars, and hybrids. Products of biotechnology were the basis of the application on which this ruling was made.

Although court cases may bring about refinement or even reversal of some of the most recent decisions or rulings, it seems probable that in the U.S. and other industrial nations, the scope and effectiveness of intellectual property protection applied to plant germplasm will increase and will bring about changes in the ways in which germplasm is exchanged for breeding purposes. Since the subject of intellectual property protection applied to plant breeding is covered in another paper in this symposium, we do not need to treat it in depth, but we do need to comment on one aspect: the prospect that utility patents can prevent unauthorized individuals or organizations from freely using patented, released commercial cultivars as germplasm resources for breeding new lines during the life of the patent. This protection, afforded in the U.S. by utility patents as affirmed by the *ex parte Hibberd* decision, differs from that given by the U.S. Plant Variety Protection Act, which stipulates that protected items may be used for breeding once they are released.

We have noted several times herein that elite cultivars are a prime source of useful genetic variation. If breeders can no longer use such cultivars at will, must genetic progress in plant breeding come to a standstill? We think not, for the following reasons. The example of maize breeding – specifically, development of inbred lines in the U.S. – will illustrate what probably can happen if the utility patent is widely used.

Private seed companies, each with their own proprietary maize inbred lines (lines maintained as trade secrets and therefore not legally available to others), have made considerable progress in maize breeding over the past 50 years. Each of the major companies has developed its own in-house breeding pools, divergent from those of the others. They have based their breeding, in part, on different open-pollinated maize varieties and different key inbred progenitor families. Amounts and breadth of useful genetic diversity, therefore, have been increased overall.

But the seed companies have not been completely dependent on their own in-house gene pools. The universities and the USDA also have developed elite maize inbreds and/or germplasm lines and released them to the public. Public and private breeders alike have met at this common well of useful germplasm, introgressing its products into their divergent private pools, invigorating and broadening them.

Additionally, smaller companies without the means to maintain full-sized breeding operations have shared proprietary germplasm in several ways, usually though contract arrangements with foundation seed companies. This method also has increased genetic diversity in maize hybrids deployed on the farm, throughout the U.S.

It seems likely that if large-scale use of utility patents should restrict unauthorized use for breeding of released patented cultivars of self-pollinated crops such as wheat and soybeans, strategies like those used in U.S. maize breeding might be developed for such self-pollinated crops, as follows:

1. Public opinion would encourage continued release of open-licensed products of public breeding, either as finished cultivars or as elite breeding germplasm selected for specific attributes,
2. Numerous privately developed cultivars still would be protected solely with plant variety protection as the only method applicable to them. They would be freely available, upon release, for use as breeding germplasm,

3. Because some future cultivars also would merit and need the added protection of utility patents, companies additionally would create their own distinctive proprietary germplasm pools, thereby creating several divergent, new germplasm pools, in place of a single relatively narrow panmictic pool (per class or region) for each crop,
4. Smaller companies would cooperate in various ways to create additional proprietary pools useful for them,
5. From time to time, cross-licensing arrangements – structured, legal exchange of patented germplasm stocks – would be devised among owners of the various germplasm pools, allowing infusions of divergent, elite germplasm to reinvigorate the isolated germplasm pools. (Some evolutionists point to genetic isolation of populations, followed by limited hybridization, as likely to initiate far-reaching kinds of evolutionary change. Maybe this process is also good for progress in plant breeding!), and
6. Other licensing arrangements would be arranged to allow non-patent holders legally to use patented materials for breeding new lines.

The twin needs for germplasm diversity and germplasm exchange thus would force constructive accommodations to be made in the new game (new to plant breeders) of utility patents for intellectual property protection, a game which allows privately funded plant breeding research to be compensated through sale of proprietary products and processes but which also allows–in fact, is intended to encourage–continued progress in research and development.

Another concern about intellectual property protection for plants, plant parts, and plant processes is the effect such protection would have if it were to become generally available in third-world countries.

To date, little or no protection of any sort is available in developing countries. For this reason, commercial breeding and seed sales in these countries generally is restricted to the self-protecting hybrid crops such as maize, sorghum, and sunflower. (The inbred parental lines of hybrid crops physically can be protected, giving trade secret protection to hybrids made from those proprietary parents.)

Some individuals and organizations believe that plant variety protection and/or utility patent protection for cultivars in developing countries would be harmful to agriculture in those countries since it would attract monopolistically inclined international seed firms interested only in profits [12, 13]. Others say, on the contrary, that to attract private capital to breeding and sale of seeds in developing countries actually would increase the amounts of useful genetic diversity deployed on the farm in those countries, to the advantage of their agriculture and thus to their total economy [14].

To further complicate the matter, some of those who dislike intellectual property protection for modern cultivars nevertheless espouse some sort of ownership and control of indigenous landraces by those developing countries having such materials, with the goal of charging something akin to a royalty to those who use them for breeding [15]. Workable methods of control and of assessing charges have not yet been devised; indeed, to do so rationally and equitably seems to be a very difficult task. For example, how does one settle on the ultimate originator country of Asian or African landraces of maize that certainly came originally from multiple locations in the Caribbean or the U.S., and before that from what is now Mexico?

Regardless of the eventual outcome of these conflicting ideas about germplasm ownership in the developing countries, it does seem that if the developing countries are to join the industrial nations as equals in use of genetic diversity, they will need to do so within the general framework of current concepts and laws regarding intellectual property protection.

The challenge for plant scientists who are interested, as they must be, in the economics of agriculture is to learn how to operate with maximal effectiveness as scientists in a world economy increasingly interrelated and therefore increasingly governed by laws and regulations relating to property matters, a world increasingly concerned with matters of justice and social well being, and increasingly concerned, as well, with the ecological health of the planet. Proper balance perhaps can be reached in these matters if they are kept in perspective, that is, if the principle of constructively striving towards common final goals is followed and if, above all, one realizes that the struggle to accommodate is never-ending, because change begets change: time goes only forward.

Genetic diversity and farmer preference

We have referred to the trend toward the use of a small number of cultivars, each with broadly based genetic protection against pest damage and environmental stress. This trend seems to run counter to the often expressed belief that genetic uniformity brings on danger of genetic vulnerability – danger of epidemics or environmentally induced crop failure. The 1970 southern corn leaf blight epidemic is cited repeatedly as an example of the danger of genetic uniformity [16]. Most U.S. maize hybrids in 1970 had one cytoplasm–T–that proved to have genetic susceptibility to a previously unidentified race–race T–of southern corn leaf blight. The logical inference is that to grow heterogeneous cultivars, or to grow many genetically different cultivars of any given crop, is the best or even the only way to ensure safety from disastrous yield loss resulting from epidemics or weather stress.

Yet U.S. farmers, and those of industrial countries in general, persistently plant most of their acreage to only a few of the scores or hundreds of cultivars of wheat, soybeans, maize, and sorghum that are available. Nor have they demanded more highly heterogeneous cultivars. In fact, individual U.S. maize hybrids are more uniform today than they were in 1970. In 1970, most maize hybrids were double-cross (four parents); today they are single-cross (two parents).

Some critics have said that U.S. farmers are forced by monopolistic commercial seed firms to accept small numbers of highly uniform cultivars. But until recently, wheat and soybean cultivars were bred only by state and federal institutions. Yet farmers, even in those earlier times, concentrated on the few cultivars they believed to be most productive for them. The number of soybean and wheat cultivars now in use on the farm actually has increased over the past 15 years as a result of increased activity in commercial soybean and wheat breeding following passage of the 1970 Plant Variety Protection Act [17].

And the swing from double-cross to single-cross maize hybrids was initiated by the desires of farmers and led by entrepreneuring small seed companies. The large companies introduced single-cross hybrids to meet competition from energetic small firms, which had discovered that farmers were willing to pay more for seed of higher yielding single-cross hybrids. Commercial single cross hybrids yielded about 10% more than the double-cross hybrids they replaced.

The empirical facts (as compared to the theories) about genetic diversity and protection from disaster are that:

a) The genetics of a cultivar, not its degree of uniformity, is the major determinant of its ability to withstand pest attack or environmental stress,
b) Uniform genotypes (either self-pollinated or first-generation hybrid cultivars) can contain a broad array of genetic protective mechanisms and thus internally can be genetically diverse (buffered),
c) Plant breeders continually can add internal genetic diversity (additional protective systems) to successive new cultivars, and
d) Rapid turnover of cultivars (replacement by new cultivars of different genotype) helps prevent build-up of pests able to attack a specific, widely grown, uniform cultivar. (Even the best protected cultivars cannot last forever). 'Genetic diversity in time' thus is afforded to today's farmers at reasonably efficient levels.

One should note that cultivars can be replaced rapidly with improved ones only when breeders have sufficient numbers of superior new cultivars and numerous elite pools of breeding materials under development, that is, when breeders have sufficient 'genetic diversity in reserve'. Thus, genetic diversity in time requires genetic diversity in reserve, to properly support modern plant breeding.

Farmers who have chosen individually those cultivars that best satisfied their needs have affirmed collectively these statements of empirical fact. It is true that farmers have based their annual decisions on short-term experiences, but for 50 or 60 years, in the industrial nations at any rate, their collective decisions seem to have proven wise since few really broad-scale epidemics have occurred.

Even in 1970, the year of the southern corn leaf blight epidemic, U.S. maize yields were reduced by not more than 15% from trend. Some researchers think that poor maize-growing weather in 1970 actually was responsible for half of the loss; thus southern corn leaf blight race T may have reduced yield by only about 7%. In 1971, the problem was largely corrected by widespread substitution of hybrids with normal (non-T) cytoplasm. By 1972, all U.S. maize had normal (or at least non-T) cytoplasm [18].

Genetic diversity in time, combined with genetic diversity within genomes (or within zygotes), seems to be working reasonably well in the industrial countries, most of which lie in temperate zones. But solutions for avoiding catastrophic yield loss in temperate zones in industrial countries will not necessarily work in the tropical and subtropical environments in which most developing countries are located. In the lower latitudes, a combination of continuous plantings, no wintertime break in pest reproduction, and limited access to chemical pest controls increases the possibility of pest epidemics. These factors also increase the possibility that any very widely grown cultivar, even if fortified with genes for multiple pest resistance, soon will be overcome by a new disease or insect biotype for which it has no genetic defense.

The Green Revolution nevertheless has turned third-world crop agriculture firmly toward the industrial nations' practice of growing a very small number of cultivars on most of the production acres [10, 19]. For example, in 1984 Indonesia planted 66% of its rice land to two cultivars. And in 1984 Bangladesh planted 70% of its wheat lands to one cultivar.

If turnover of widely used cultivars were significant and rapid in the third-world countries, one might be satisfied that 'genetic diversity in time' could contain the potential problem. But it appears that the international agricultural research centers (the original sources of early high yield cultivars) cannot by themselves supply the needs of all countries. The developing countries traditionally have only rarely turned out significant numbers of elite replacement cultivars, annually. More investment in breeding and faster rates of production of superior new cultivars needed in the developing countries to maintain a steady and rapid flow of elite new cultivars of those crops most important to the developing countries [20]. Otherwise there may not be enough 'genetic diversity in time' to control or mitigate problems from cultivar-specific disease and/or insect pests in those regions.

Concluding remarks

The greatest need in the use of plant germplasm today is for its preservation: for development and maintenance of worldwide standards of germplasm storage, management, and use, including introgression of exotic germplasm into existing cultivars. Current disputes about location and ownership of plant germplasm collections tend to distract attention from the very poor state of most of those collections. Many of them are losing viability, are uncataloged in any useful sense, and largely are unavailable to qualified workers [21]. These problems usually arise not for political reasons but rather because of a simple lack of funds.

Knowledge of the practical utility of germplasm in the collections–identification of useful traits in individual cultivars–also is nonexistent in most instances, in part because of a lack of funding but also because of a lack of agreement or knowledge among breeders about what really should be known and recorded.

Finally, and most importantly, we plant breeders have done virtually nothing to develop efficient and effective ways of introgressing large blocks of exotic germplasm into elite adapted breeding pools. We have a world of work in store: work to fund and work to do.

References

1. Duvick, D. N. 1984. Genetic contributions to yield gains of U.S. hybrid maize, 1930 to 1980. Pp. 15–47, in W. R. Fehr (ed.), Genetic Contributions to Yield Gains of Five Major Crop Plants. Crop Science Society of America. Spec. Publ. 7. Madison, Wisconsin 101 pp.
2. Duvick, D. N. 1984. Genetic diversity in major farm crops on the farm and in reserve. Econ. Bot. 38: 161–178.
3. Darwin, C. 1979. Chap 1, pp. 50, Chap 11, pp. 171, in The Illustrated Origin of Species. Hill and Wang, New York. 240 pp.
4. Anderson, E. and W. L. Brown. 1952. Origin and significance of corn belt maize. Pp. 124–148 in J. W. Gowen (ed.), Heterosis. Iowa State Univ. Press, Ames, Iowa 552 pp.
5. Stephens, J. O. and R. F. Holland. 1954. Cytoplasmic male sterility for hybrid sorghum seed production. Agron. J. 46: 20–23.
6. Virmani, S. S. and I. B. Edwards. 1983. Current status and future prospects for breeding hybrid rice and wheat. Adv. Agron. 36: 145–214.
7. Specht, J. E. and J. W. Williams. 1984. Contribution of genetic technology to soybean productivity – retrospect and prospect. pp 15–47, in W. R. Fehr (ed.), Genetic Contribu-

tions to Yield Gains of Five Major Crop Plants. Crop Science Society of America Spec. Publ. 7. Madison, Wisconsin. 101 pp.
8. Sackston, W. E. 1981. The sunflower crop and disease: Progress, problems and prospects. Pl. Dis. 65: 643–648 pp.
9. Quinby, J. R. 1973. The genetic control of flowering and growth in sorghum. Adv. Agron. 25: 125–162.
10. Dalrymple, D. G. 1986. Development and Spread of High- Yielding Wheat Varieties in Developing Countries. Bur. Sci. Technol., Agency For Internat. Devel. Washington, DC. 99 pp.
11. Duvick, D. N. 1984. Progress in conventional plant breeding. pp. 17–31 in J. F. Gustafson (ed.), Gene Manipulation in Plant Improvement. Plenum Publishing Press, New York, New York 668 pp.
12. Mooney, P. R. 1983. The law of the seed. Dev. Dialogue 1983 (1–2): 1–172.
13. Hobbelink, H. 1987. New Hopes or False Promise? Biotechnology and Third World Agriculture. Internat. Coalition for Development Action, Brussels, Belgium. 72 pp.
14. Duvick, D. N. 1986. The nobility of seed research and its critics. The World and I, 7: 275–281.
15. Mooney, P. R., Ibid.
16. National Academy of Sciences. 1972. Genetic Vulnerability of Major Farm Crops. Committee on Genetic Vulnerability of Major Farm Crops, Agricultural Board, Natl. Res. Council, Natl. Acad. Sci., Washington, DC. 307 pp.
17. Butler, L. J. and B. W. Marion 1985. The impacts of patent protection on the U.S. seed industry and public plant breeding. North Central Reg. Res. Publ. 304. Univ. of Wisconsin, Madison. 128 pp.
18. Duvick, D. N. 1978. Current and future use of cytoplasmic male sterility for hybrid seed production. pp. 265–277 in D. B. Walden (ed.), Maize Breeding and Genetics. John Wiley & Sons, Inc. New York, New York 794 pp.
19. Dalrymple, D. N. 1986. Development and Spread of High-Yielding Rice Varieties in Developing Countries. Bur. Sci. Technol. Agency for Internat. Devel. Washington, DC. 117 pp.
20. Dalrymple, D. N. 1985. The development and adoption of high-yielding varieties of wheat and rice in developing countries. Am. J. Agric. Econ. 67: 1067–1073.
21. Plucknett, D. L., N. J. H. Smith, J. T. Williams, and N. M. Amishetty. 1987. Gene Banks and the World's Food. Princeton Univ. Press, Princeton, New Jersey 247 pp.

Summary of symposium

JOHN A. PINO
Board on Agriculture National Research Council, National Academy of Sciences, 2101 Constitution Avenue, N.W., Washington, D C 20418, U.S.A.

Biotic diversity is the base upon which modern agriculture is built. It encompasses both the diversity among species and that within species. It is this diversity from around the world that has given us a rich and variable agriculture. For the breeder, genetic variation allows the continuous improvement and development of crops and livestock. A large portion of the dramatic gains in world crop production over the past two decades has been ascribed to germplasm-based technologies.

The variability in extant life forms has accumulated over thousands of years of evolutionary development. That evolutionary process (nature's experimentation) has provided the mechanism by which living things may survive the physical environments presented to them or in which they may deliberately or inadvertently find themselves. In that process, some variant types cannot compete or survive and are lost, and others slowly emerge into types specifically adapted to new physical and biological environments. This trial-and-error, hit-or-miss process went on undisturbed until the beginning of agriculture only 10,000 years ago. It was then that humans sped up and redirected this process to produce forms suitable to fulfilling their needs. The essential process was the same, but with two important changes: people selected certain variants (useful or preferred) and gave them preferential treatment (i.e. protection, water, nutrients, etc.); and with the advent of the science of genetics and plant and animal breeding in more recent times, it became possible for related individuals to 'meet each other' through controlled breeding. This process further accelerated the accumulation of 'good' genes – those with utilitarian value. The process of controlled breeding and selection eventually led to the widespread use of high yielding, broadly adapted cultivars and animal breeds and the displacement of many of the earlier indigenous landraces. The potential for large-scale loss of genetic diversity in the process became a very real concern. But as Garrison Wilkes pointed out in this symposium, however, that plant-breeding technology itself is creating new sources of diversity.

Diversity must also be viewed in the broad context of natural systems. Most plants, animals, and other forms of life are more important to each other and to the niche they fill in nature than they are to man directly. Some co-evolved in rather dependent associations. We must not assume that if the organism is not of immediate interest to man, it has no purpose. In this context, Benjamin Matthews reminded us of the significance that species loss is the loss of genes peculiar to that species.

Our concern today with collecting and preserving the total variability extant in a species goes beyond the immediate needs of the breeders, however. It is based on the fact that we may be losing the range of diversity that has existed for that species. In part, this is the result of appropriation of land areas to support growing human populations. Even agricultural land is being lost to roads, construction, cities, erosion, salinization, desertification, etc. We are leaving less and less space for nature to experiment. Nature itself, in fact, exacts a selective toll on living organisms as a result of fire, drought, storms, plagues and so on. As eloquently

L. Knutson and A.K. Stoner (eds.). Biotic diversity and germplasm preservation, global imperatives.
 ISBN 0-7923-0178-1

described by Lawrence Busch, cultures in all of their social, economic, and political aspects are having a profound effect not only on how natural systems are viewed but on how they are used and abused. Thus the imperative to maintain natural systems that, to the best of our ability, will give some assurance for the conservation of biotic diversity and for the process by which new variation evolves.

Early collection and storage of seed provided the raw material to plant breeders for crop improvement. These materials were primarily of interest to the individual plant breeder and were often stored with little concern for their long-term preservation. Many of these collections were lost after the individuals who built them retired or died. Though we may be uncertain about the value of much of the material we preserve in seed banks today, we are also uncertain about what may be lost if we do not conserve. We do know that the frequency of individual alleles at any particular locus probably will vary with the ecogeographic location and the stresses to which organisms must respond. It is quite probable, therefore, that adequate sampling in each ecogeographic setting will capture the full range of allelomorphs possible. Under natural or artificial conditions, gene frequencies will change, and as we learned, genes in different companion allelic configurations may behave differently. Though this point was not elaborated fully, it was pointed out that species evolve not in isolation but in interactions with other species, microbes, pests, and pathogens. To be sure, these relationships will be different for different systems. As Trevor Williams stated, we need to give more thought to this issue.

The manipulation and use of germplasm has entered a new and dramatic era. Somatic hybridization and gene transfer techniques along with gene isolation, cloning, and recombination techniques promise to open up crop improvement work never before possible. In livestock, techniques such as embryo splitting, freezing, *in vitro* oocyte maturation, parthenogenesis, and cloning suggest a developing revolution in animal reproduction and breeding. Females of most domestic animal species have either single or a small multiple ovulation at each estrous cycle. It has been shown experimentally in sheep that the large number of oocytes present in the ovary may be matured and fertilized *in vitro*, thus offering the possibility of obtaining literally thousands of offspring from elite animals. In addition, the cloning of embryos suggests the possibility of generating a continuous line of identical individuals.

We have heard several presentations on the issues of sampling techniques, sample size, and numbers of accessions actually needed in a collection. Unfortunately, plant germplasm collections are rapidly growing beyond a size that can be easily managed; thus the task of quantifying the diversity captured by them has become daunting.

In the absence of readily available, low-cost technology to identify similarities or differences in accessions, collectors and curators have adopted the practice of keeping everything. Duplicate samples, multiple generations of the same material, and overlapping collections have swelled the shelves of seed banks. Paradoxically, as these collections have grown, so has the realization that unless carefully managed, their very size may hamper their use by the scientists for whom they are most often intended – the breeders.

The task of managing large collections is becoming more difficult and costly. Testing for viability, regrowing, maintaining data, and evaluating and characterizing each accession require greater resources than are available. The concept of a core collection (with a reserve) seems attractive and theoretically sound. It is the direction in which major collections should move.

In his elegant address, Gene Namkoong provided us with some of the theoretical concepts that should assist us in developing better sampling and conservation techniques. Since genes behave predictably (as Amy Rossman noted), except when they do not, we should be able to be more precise about 'how much of what' we need to keep in germplasm collections.

I suspect that as our knowledge of how to identify and measure variation improves, our procedures for what and how much to keep in cold storage, in living collections, or in some other condition will change. As pointed out by Peter Day, molecular biology should help us identify differences, shifts in gene composition in storage, and other aspects of germplasm management.

We have also heard that the cost and effort required for plant germplasm conservation is justified mainly by the effective use made of genes found in germplasm collections. A weak link in that process is the shortage of people interested in and capable of pre-breeding and evaluating materials in the genebanks to facilitate their use. The cost and complexity of operating an effective gene bank and plant-breeding program is one of the most compelling reasons for nations to work toward an open, integrated, global system rather than closed national systems.

Considerable progress has been made in developing countries in building plant-breeding capabilities. However, until recently they have lacked the ability to apply significant resources for collecting, processing, storing, and using germplasm, particularly wild types and even landraces. Notable exceptions to this are India, China, Brazil, Mexico, and Ethiopia. On the other hand, for many years the developed nations' plant research institutions have intensively collected specimens from around the world and have made extensive use of this germplasm in crop improvement research. With the help of international organizations, today's genetic wealth is becoming more widely distributed, much of it in a form readily useable by national programs. More than 40 plant genebanks have been erected worldwide, with long-term storage facilities distributed almost equally among developed and developing countries. In addition, five of the International Agricultural Research Centers maintain genebanks each for the crop(s) in which they specialize. Still, the global task is far from complete. Wonderful examples of the value of collections can be cited, especially those about the repatriation of parts of or entire collections to particular countries from which for different reasons they had been lost.

With regard to plant germplasm conservation in developing nations, our conferees have pointed out:

1. The need for continued exploration and collection not only of food, fiber, and fuel plants but for medicinal, ornamental, and other commercial species,
2. The need for improved management, documentation, and systematic evaluation of collections, and for improved communication among curators, plant breeders, and researchers,
3. The need for further assistance to build germplasm conservation systems that would facilitate cooperation and exchange of germplasm and research results,
4. The need for dialog and resolution of issues regarding patents, ownership, and global accessibility of germplasm, and
5. The need to develop concepts and strategies for *in situ* conservation, particularly in centers of diversity. Although programs for *ex situ* conservation are relatively well developed, it is apparent that *in situ* conservation will require a parallel effort. The role of the general society will be more important in that effort than in genebank work.

Except for Brazil, India, and several species-specific programs, no organized national program exists to sample, evaluate, cryopreserve, or maintain and use diverse sources of animal germplasm (including fishes). Unlike plant seed, from which a new plant may be derived, with mammals suitable living recipients for embroyos or sperm are required to produce a new individual.

The reason for conserving a particular breed or type within a species is to preserve traits unique to that particular group. It may be possible to preserve the isolated genes for the desired trait as DNA fragments. Technology is now developing to transfer such conserved genes into other individuals of the same or different species. Genomic libraries one day may have an important place in germplasm conservation programs.

Incipient programs in conservation of animal genetic resources need to be nurtured. Strategies must be defined and refined. Keith Gregory and John Acree pointed out the necessity of developing procedures to facilitate international movement of animal germplasm. Introduction of exotic germplasm has played an important role in crop improvement in most countries and it can do the same for livestock improvement. Quarantine precautions in plant and animal movement must be respected, while at the same time the germplasm must be maintained and access to it even promoted. The technology for safely moving animal germplasm without compromising the risk of disease transfer would seem to be at hand now.

Critical issues identified by the conferees

1. The development of the powerful tools of molecular biology is outpacing the professional and institutional capability to apply those technologies to germplasm evaluation, characterization, preservation, and use. Specialists in plant and animal breeding, physiology, pathology, etc., are dwindling in number. Linkage specialists are needed who can bridge two or more rapidly developing technological-scientific disciplines.
2. We need to improve methods for identifying valuable genes, for separating them from undesireable genes, for inserting and regulating them, and for reincorporating and stabilizing them.
3. The fascinating but underdeveloped field of microbial germplasm requires expansion and coordination of conservation strategies among various fields. Similarly we should expand work on conservation of invertebrate germplasm. Both areas require additional definition and research effort.
4. Concern and evidence are sufficient regarding the loss of diversity in animal species to suggest that urgent attention be given to global and national efforts to conserve species and breed types. Problems of cost, disease transmission, adequate technology, and attitudes have prevented organized efforts to document, evaluate, and preserve much of the existing animal genetic diversity.
5. Most of our attention regarding germplasm preservation has been oriented to those plants and animal species considered to be important for agriculture. Natural habitats may harbor additional, potentially useful species and others of value as keystone species for the conservation of those natural habitats. Strategies for *in situ* conservation and preservation of biological diversity in natural habitats need greater definition and support.
6. The active participation and leadership of the U.S. in international dialog and activities regarding genetic resources is vital not only to U. S. interests but to developing nations as well. In that connection, the importance to communicate U.S. interest and concern in all aspects of conservation of genetic resources should not be minimized. This meeting and others similar to it very well might be construed as making such a statement.

   Many elements of national and global systems need clarification and certainly more strategic, long-term thinking. Garrison Wilkes pointed out, for example, the absence of specific strategies to deal with major crop failures (vulnerability). The *in situ* conservation issue is another example, and access to germplasm yet another.
7. Research related to germplasm conservation needs to be expanded and strengthened in such areas as:
   - nondestructive tests for viability;
   - simple, cost-effective methods for preservation (are we moving toward greater, more costly sophistication rather than toward less costly, more effective methods?);
   - techniques to monitor and reduce excessive duplication; and maintenance of diversity in collections.

8. Training – although the conference did not dwell specifically on the subject of training needs, by implication the need is obvious and great for highly skilled researchers, curators, managers, collectors, and breeders.
9. Another area about which very little was said is that of the grassroots heirloom livestock and seed savers: the American Minor Breeds Conservancy, Seed Savers Exchange, Native Seeds/SEARCH, and others. I feel that official programs must somehow interface with these efforts as well as those in arboreta and botanic gardens.
10. Data and communications systems need to be improved in this fast-moving activity.

Finally, may I say that on behalf of the Board on Agriculture, NRC/NAS, we are pleased to have had the privilege of participating in this symposium. Peter Day, Chairman of the Committee on Managing Global Genetic Resources, the staff, and members of our study team who participated in the conference are delighted with the interest and enthusiasm that the participants have shown.

I hope that the Department of Agriculture will continue to give a high level of attention to the various dimensions of genetic resource conservation and use at both the national and international level. Other nations must join in the global task of conserving biotic diversity, and everyone should share in the fruits of this effort. We must not shy away from engaging in dialog and cooperation in germplasm conservation, no matter how complex or difficult the issues. Science and technology hold great promise for giving us a better understanding of the nature of genetic resources, which will improve our management and use of them. It will be up to the policymakers to provide the appropriate policy environment to facilitate and encourage the conservation of biotic diversity.

INDEX